DATA HISTORIES

EDITED BY

*Elena Aronova, Christine von Oertzen,
and David Sepkoski*

O S I R I S | 32

A Research Journal Devoted to the
History of Science and Its Cultural Influences

Osiris

Acknowledgments

This volume is the culmination of a working group at the Max Planck Institute for the History of Science titled "Historicizing Big Data," which convened between 2013 and 2015. First and foremost, our gratitude goes to the participants in two workshops—the first in October 2013, and the second in October 2014—most of whom have contributed essays to this collection. Many of the group's external members spent several months at the Max Planck Institute as visiting scholars while working on their contributions and, in addition to attending the formal workshops, participated in numerous reading groups and impromptu discussions on themes relating to the history of data. The institutional and intellectual support of Department II at the Max Planck Institute—and particularly its director, Lorraine Daston—allowed this publication to evolve through extended interaction among the group members, and we thank all the contributors to this volume for making this project a truly collaborative experience.

We have benefited from conversations about data in a variety of conferences and institutional settings, with interlocutors who are too numerous to list here. However, our particular gratitude goes to Soraya de Chadarevian, Lorraine Daston, Cathy Gere, Matt Jones, Sabina Leonelli, Ted Porter, Dan Rosenberg, and the late John Pickstone for their generous and stimulating interventions and comments. We would also like to express our deep thanks to Andrea Rusnock and the *Osiris* Editorial Board for their support for and engagement with this project. The entire volume benefited enormously from the detailed and constructive comments of two anonymous reviewers, who were exemplary referees.

Introduction:
Historicizing Big Data

by Elena Aronova,* Christine von Oertzen,§ and David Sepkoski#

ABSTRACT

The history of data brings together topics and themes from a variety of perspectives in history of science: histories of the material culture of information and of computing, the history of politics on individual and global scales, gender and women's history, as well as the histories of many individual disciplines, to name just a few of the areas covered by essays in this volume. But the history of data is more than just the sum of its parts. It provides an emerging new rubric for considering the impact of changes in cultures of information in the sciences in the *longue durée*, and an opportunity for historians to rethink important questions that cross many of our traditional disciplinary categories.

We live in a world of data. Or, more precisely, we live in a world where we have become used to understanding our lives, the economic fortunes of our societies, the information we surround ourselves with, and the objects and phenomena of science, as bits stored electronically on digital computers. Our era is not just an era of data, we are told, but of "Big Data," a phenomenon that was first named in the computer industry of the 1990s and now serves as a label for everything that is bad or good (depending on one's perspective) about twenty-first-century technological society.[1] From Google analytics to NSA surveillance, from biodiversity databases and genomics archives to the interactions of subatomic particles, we now often assume that anything worth knowing can be counted, quantified, digitized, and reduced to binary electronic signals residing on huge servers or floating somewhere "in the cloud." Big Data refers to the sheer scope of modern information technology—a world measured in terabytes and petabytes, or even yottabytes (a trillion terabytes)—as well as to the ubiquity of data in every aspect of modern existence.[2]

* Department of History, University of California, Santa Barbara, Santa Barbara, CA 93106-9410; earonova@history.ucsb.edu.
§ Max Planck Institute for the History of Science, Boltzmannstrasse 22, 14195 Berlin, Germany; coertzen@mpiwg-berlin.mpg.de.
# Max Planck Institute for the History of Science, Boltzmannstrasse 22, 14195 Berlin, Germany; dsepkoski@mpiwg-berlin.mpg.de.

[1] On the introduction of the term "Big Data" to general currency, see Rebecca Lemov, "Anthropology's Most Documented Man, Ca. 1947: A Prefiguration of Big Data from the Big Social Science Era," in this volume. We refer to "Big Data" in capital letters simply as a reference to current usage in the literature, which does not signify any endorsement of the exclusive significance of present-day data practices.
[2] The emphasis on the size of data derives its impulse from estimates of data flows in social media. As of 2012, Facebook collected more than 500 terabytes of data on a daily basis, Google processed

Big Data also signals, to some observers, a profound change in the very nature of science: it has been heralded as a "fourth paradigm" or even "the end of science," a model of investigation in which algorithms do much of the work of interpreting the world around us.[3] Practitioners in business and industry praise Big Data as a "new oil" or "new asset class" affording unprecedented opportunities for production, marketing, and social engineering.[4] Many of these enthusiastic accounts of the Big Data "revolution" bear more than a whiff of technological determinism. It sometimes seems as if Big Data is the inevitable product of the introduction of electronic computers, or that the developments in infrastructure—such as the giant server farms built to sustain and optimize the data flow—from the late twentieth century onward have led to a qualitative transformation in the nature of scientific practice and epistemology.

Sometimes called the "digital age," our era is perhaps more than anything defined by a reliance on a particular technology: the electronic digital computer. Computers are ubiquitous—they are in our pockets, on our wrists, in our cars, in our kitchens and bathrooms, at the supermarket, and on the subway. Computers large and small comprise nodes in a vast network of information services we access on an almost continual basis, from checking the weather forecast to predicting the performance of the global stock market. Smaller devices, faster processors, bigger volumes of data: we are so constantly bombarded by the forward-looking rhetoric celebrating the multitude of gadgets surrounding us that we can forget that these technologies have histories, and that those histories stretch back well before the advent of electronic computing. "Data," a term that originally referred to the "givens" in a geometric proof, does not necessarily reside in electronic computers. The same applies to "digitization," the process by which continuous phenomena (whether the light emitted by a distant star or a Beethoven sonata) are translated into discrete (or numerical) format. Even "computing" was performed in a variety of technological contexts, both pre- and postelectronic.[5]

The notion of Big Data calls to mind another alleged transformation in twentieth-century science that has been much discussed by historians of science and technology in the past several decades: the advent of "Big Science."[6] A term first coined in the

---

dozens of petabytes, and Twitter produced nearly 12 terabytes. See Hamid Ekbia, Michael Mattioli, Inna Kouper, G. Arave, Ali Ghazinejad, Timothy Bowman, Venkata Ratandeep Suri, Andrew Tsou, Scott Weingart, and Cassidy R. Sugimoto, "Big Data, Bigger Dilemmas: A Critical Review," *JASIST* 66 (2015): 1523–1746.

[3] Tony Hey, Stewart Tansley, and Kristine Tolle, eds., *The Fourth Paradigm: Data-Intensive Scientific Discovery* (Redmond, Wash., 2009); Chris Anderson, "The End of Theory: The Data Deluge Makes the Scientific Method Obsolete," *Wired*, 23 June 2008.

[4] World Economic Forum, "Personal Data: The Emergence of a New Asset Class" (2011), https://www.weforum.org/reports/personal-data-emergence-new-asset-class (accessed 17 January 2017); P. Rotella, "Is Data the New Oil?," *Forbes*, 2 April 2012, http://www.forbes.com/sites/perryrotella/2012/04/02/is-data-the-new-oil/ (accessed 17 January 2017).

[5] Often, computing tasks were performed by women. See Pamela Mack, "Strategies and Compromises: Women in Astronomy at Harvard College Observatory, 1870–1920," *J. Hist. Astron.* 21 (1990): 65–75; Jennifer Light, "When Computers Were Women," *Tech. & Cult.* 40 (1999): 455–83; David Alan Grier, *When Computers Were Human* (Princeton, N.J., 2005); Denise Kiernan, *The Girls of Atomic City: The Untold Story of the Women Who Helped Win World War II* (New York, 2013).

[6] See, e.g., James H. Capshew and Karen Rader, "Big Science: Price to the Present," *Osiris* 7 (1992): 3–25; Peter Galison and Bruce Hevly, eds., *Big Science: The Growth of Large-Scale Research* (Palo Alto, Calif., 1992); Catherine Westfall, "Rethinking Big Science: Modest, Mezzo, Grand Science and the Development of the Bevalac, 1971–1993," *Isis* 94 (2003): 30–56; Daniel Kevles, "Big Science and Big Politics in the United States: Reflections on the Death of the SSC and the Life of the Human Genome Project," *Hist. Stud. Phys. Biol. Sci.* 27 (1997): 269–97.

1960s by physicist Alvin Weinberg, Big Science represented a shift toward enormous scientific undertakings that were incredibly costly, involved hundreds or even thousands of investigators, adopted a corporate-style management structure, and tended to monopolize support and attention from public and private sources. There are parallels—and indeed direct overlaps—between Big Science and Big Data. Many projects that involve Big Data—the Human Genome Project, CERN, the Very Large Telescope array—unquestionably fit the definition of Big Science. Big Data has also become a buzzword whose deployment can loosen purse strings from funding agencies, attract journalists, and confer importance and prestige on scientific projects. But at a time when the smartphones most of us carry in our pockets are as powerful as 1990s-era supercomputers, it makes sense to question whether Big Data is "big" in the same way that Big Science is.

Historian Jon Agar has suggested distinguishing between Big Science as a mode of organization of scientific research that had pride of place in the nineteenth century, and the "labeling of 'Big Science' as something of concern . . . a product of the . . . long 1960s."[7] Likewise, it might be useful to draw a distinction between Big Data as a temporal cultural phenomenon of the twenty-first century and Big Data as a data-driven mode of doing science with deeper roots in the past. As a label, Big Science has specific connotations—including a quantitative threshold for size, costliness, and scope—that distinguish it from the science practiced (in most cases) before the Second World War.[8] In contrast, Big Data is something of a moving target. When the ENIAC (Electronic Numerical Integrator and Computer) was introduced in 1946, it boasted a processing capability some three orders of magnitude more powerful than its electromechanical predecessors. For the scientists and engineers engaged in nuclear weapons research, this increased processing capability enabled Big Data indeed. However, the 30-ton ENIAC was less powerful than a 1970s-era programmable calculator (in fact, a 1995 project at the University of Pennsylvania managed to re-create the original ENIAC on a 7.44- by 5.29-mm chip), and current smartphones and tablets have up to 10^9 times as much computing power.[9] Although computing technology has tended toward exponential growth in recent decades ("Moore's law" states that the number of transistors on a microchip will tend to double every two years), it is worth remembering that even exponential growth is a continuous function. As a phenomenon, then, Big Data seems to lack the decisive qualitative distinction present in the transition to Big Science. Unlike Big Science, which was typified, as a historical phenomenon, in the Manhattan Project and the National Laboratories, what constitutes Big Data is contested by practitioners and observers alike.[10]

---

[7] Jon Agar, *Science in the Twentieth Century and Beyond* (Cambridge, 2012), 330. Many prominent Big Science projects of the nineteenth century were philological or historical in nature, e.g., Theodor Mommsen's *Corpus Inscriptionum Latinarum*, a collection of all known ancient Latin inscriptions; see Lorraine Daston, "The Immortal Archive: Nineteenth-Century Science Imagines the Future," in *Science in the Archives: Pasts, Presents, Futures*, ed. Lorraine Daston (Chicago, 2017), 53–84. For botany as Big Science of the nineteenth century, see Londa Schiebinger and Claudia Swan, eds., *Colonial Botany: Science, Commerce, and Politics* (Philadelphia, 2004).

[8] Alvin M. Weinberg, "Impact of Large-Scale Science on the United States," *Science* 134 (1961): 161–4; see also Weinberg, *Reflections on Big Science* (Oxford, 1967).

[9] See http://www.seas.upenn.edu/~jan/eniacproj.html (accessed 17 January 2017).

[10] On the emergence of the term "Big Data," see Lemov, "Anthropology's Most Documented Man" (cit. n. 1).

What, then, is the history of Big Data? Is it a history of ever more powerful technologies, quantitative increase of data, and qualitative changes associated with these processes, as the proponents of Big Data suggest? It is undeniable that computing technologies have altered the practice of science in some profound ways: examples such as the Human Genome Project, the Large Hadron Collider, and the Global Biodiversity Information Facility, among many others, testify to the intellectual, political, and economic significance of Big Data. At the same time, however, practitioners and scholars have begun to critically reflect on the more breathless and hyperbolic proposals regarding the revolutionary and transformative character of today's data-driven science. Considering the legal, ethical, and political implications of today's information technologies and data practices for our societies, they ask how the growth of data is impacting the emergence of new elites, the growth of wealth and capital, as well as second-order harms, such as profiling, government surveillance, and discrimination. If data is a new "asset class," as some Big Data enthusiasts claim, what is the source of the new value?[11] The sheer volume and diversity of the expanding literature featuring Big Data in its titles attest to the fact that data in itself has become an object of study, manifesting an important and growing conceptual space in the natural and social sciences alike.

All these questions have important historical dimensions. Historians, however, have been a rare minority in the growing literature on Big Data. Humanities scholars are usually skeptical about claims framed in explicitly presentist terms. Historians of science have been especially reluctant to take part in such discussions: for a relatively young field that established its identity as part of history rather than the sciences, the projection of categories of the present into the distant past involved not only methodological concerns about producing bad history (what historian Herbert Butterfield famously dubbed "Whig historiography") but also political caution regarding disciplinary identity.[12] However, historians of science have an important stake in the conversation on and around data, big or small.

HISTORY OF DATA AND THE HISTORY OF SCIENCE

Data has long been a key category in the history and philosophy of science. A fundamental epistemological category, data has been, and is, a central notion of empiricist epistemology. At the same time, data figured prominently in struggles to free the history of science as a field from a particular vision of science as a steady accumulation of gathered empirical data—the "givens" in the original Latin sense of the word.[13] This vision of science narrated as a story of an upward, linear, and univocal progress was part and parcel of the Enlightenment project, which integrated a particular model of positivist epistemology with the notion of progress. No less than three generations of historians of science have been introduced to the profession through Thomas Kuhn's *Structure*

[11] For a useful review of the literature addressing these questions, see Ekbia et al., "Big Data" (cit. n. 2).
[12] See Peter Dear, "What Is the History of Science the History *Of*? Early Modern Roots of the Ideology of Modern Science," *Isis* 96 (2005): 390–406. On the divergence between the history of science and science studies in defining their respective subject matter and their "methods of explanation," see Lorraine Daston, "Science Studies and the History of Science," *Crit. Inq.* 35 (2009): 798–813. For the discussion of the changing attitudes toward "Whiggism" among historians of science and the need to consider these attitudes as historically situated, see Nicholas Jardine, "Whigs and Stories: Herbert Butterfield and the Historiography of Science," *Hist. Sci.* 41 (2003): 125–40.
[13] On the history of the notion of "data," see Daniel Rosenberg, "Data before the Fact," in *"Raw Data" Is an Oxymoron*, ed. Lisa Gitelman (Cambridge, Mass., 2013), 15–40.

of Scientific Revolutions (1962), which shuttered this "received view" of a cumulative progression of knowledge. Few, however, would recall now what philosopher Gerald Doppelt and others described as *the* central implication of Kuhn's *Structure*: "data loss" caused by change of paradigms.[14] According to John Zammito, the demonstration of "even the short-term failures of cumulation in data" was the main problem for philosophers who, for this reason, rejected Kuhn's proposal as a lapse into irrationalism, while at the same time turning their attention away from traditional philosophical appraisals of scientific theories and focusing instead on actual scientific processes.[15]

Historians and other students of science in the wake of Kuhn have confronted similar scenarios. When post-Kuhnian students of science reinvented the history of science as the history of scientific *practices* rather than scientific *ideas*, the observation of actual practices went hand in hand with discussions of how experience and observations have been shaped and transformed into scientific data, and what the stakes of these processes were. Laboratory studies pioneered by Bruno Latour, Karin Knorr-Cetina, Michael Lynch, and Sharon Traweek, among others, have demonstrated that data are not only gradually "construed" through routine negotiation and tinkering until stabilized; data-mining techniques, quantitative algorithms, and qualitative visual selections are all part of making data meaningful. Latour has argued that scientific abstractions and propositions result from ever more rapid merging and condensing of one set of "inscriptions" about empirical data into another, simpler one, made possible by ever more sophisticated techniques of quantification, condensing, and visualization. It is this merging and condensing of complex piles of inscriptions into compact, movable, and manipulable "immutable mobiles" that make scientific data from an array of empirical observations.[16]

Other scholars saw these types of transformations as transitions from "traces," "signs," and "indices" to "data" themselves.[17] Hans-Jörg Rheinberger, for instance, pointed out that "traces only start to make sense and acquire meaning as data, if one could relate and condense them into an epistemic thing, a suspected *Sachverhalt*, a possible 'fact' in the everyday language of the sciences . . . that in turn could be used to create data patterns: curves, maps, schemes, and the like—potentially the whole plethora of forms of visualization the sciences have been so inventive to generate."[18] From this perspective, a meaningful visualization of the data *is* what constitutes knowledge. As Latour argued, reducing the complexity and uncertainty of observations, visualizations—diagrams, technical drawings, lists, and graphs—can have an impact beyond the realm of language-based arguments. Consequently, for Latour the history of science *is* the history of

[14] Gerald Doppelt, "Kuhn's Epistemological Relativism: An Interpretation and Defense," *Inquiry* 21 (1978): 33–86; see discussion in John Zammito, *A Nice Derangement of Epistemes*: *Post-Positivism in the Study of Science from Quine to Latour* (Chicago, 2004).

[15] Zammito, *A Nice Derangement* (cit. n. 14), 71

[16] Bruno Latour, *Science in Action: How to Follow Scientists and Engineers through Society* (Cambridge, Mass., 1987). Several authors in this volume are taking Latour and other classic work in science studies as their starting point; see in particular Elena Aronova, "Geophysical Datascapes of the Cold War: Politics and Practices of the World Data Centers in the 1950s and 1960s"; Etienne S. Benson, "A Centrifuge of Calculation: Managing Data and Enthusiasm in Early Twentieth-Century Bird Banding"; and Dan Bouk, "The History and Political Economy of Personal Data over the Last Two Centuries in Three Acts."

[17] See Hans-Jörg Rheinberger, "Infra-Experimentality: From Traces to Data, from Data to Patterning Facts," *Hist. Sci.* 49 (2011): 337–48; and the discussion in Bruno J. Strasser and Paul N. Edwards, "Big Data Is the Answer . . . but What Is the Question?," in this volume.

[18] Rheinberger, "Infra-Experimentality" (cit. n. 17).

innovations in visualizations: "every possible innovation that offers any of these advantages will be selected by eager scientists and engineers: new photographs, new dyes to color more cell cultures, new reactive paper, a more sensitive physiograph, a new indexing system for librarians, a new notation for algebraic function, a new heating system to keep specimens longer. History of science is the history of these innovations."[19]

Kuhn's *Structure* did not have much to say about technology—indeed, Kuhn explicitly bracketed the entire question of the role of technology, as well as techniques of making and moving data, in understanding science and its change. Post-Kuhnian historians of science who turned to study the material cultures of science have found that strategies and technologies developed to deal with information and data played a vital role in making knowledge itself, demonstrating that the interconnection between data manipulation and ordering knowledge can be seen throughout the history of the natural sciences.[20] New technologies allowed not only new kinds of data analysis but also ever larger data production. Over the past decades, historians of science have explored how previous societies coped with their own problems of "information overload," whether that meant a superabundance of manuscripts and printed works in the medieval and early modern periods, the inexhaustible supply of observations of natural history during the age of European expansion, or the bureaucratic accumulation of an "avalanche of numbers" in the nineteenth century.[21] Against this backdrop, today's Big Data can be seen as a chapter in a longer history (or, rather, histories) of observation, quantification, statistical methods, models, and computing technologies.[22] What all these studies have shown is that the strategies and technologies developed to deal with information overload played a vital role in making knowledge itself.

Practices of observing, collecting, and sorting data are by nature collective endeavors, often involving long-distance cross-cultural interactions, sponsored by state governments, industries, or powerful stakeholders with their own interests. A long-distance collective empiricism enabled through distributed collection, classification, transportation, and negotiation of specimens, things, and images necessarily involves the arrangements of power and authority. New technologies that allowed new kinds of data analysis were thus coproduced with cultural and political orders. As Steven Shapin and Simon Schaffer put it succinctly in their classic *Leviathan and the Air-Pump*, "The problem of generating . . . knowledge is a problem in politics, and, conversely, . . . the problem of political order always involves solutions to the problem of knowledge."[23] Mak-

[19] Bruno Latour, "Visualisation and Cognition: Drawing Things Together," in *Knowledge and Society: Studies in the Sociology of Culture Past and Present*, vol. 6, ed. H. Kuklick (Greenwich, Conn., 1986), 1–40, on 20.

[20] Bruno J. Strasser, e.g., showed that the practices developed for the construction of the GenBank database were intertwined with the development of the classification system of protein sequences in the 1970s and 1980s. See Strasser, "The Experimenter's Museum: GenBank, Natural History, and the Moral Economies of Biomedicine," *Isis* 102 (2011): 60–96.

[21] See Ann Blair, *Too Much to Know: Managing Scholarly Information before the Modern Age* (New Haven, Conn., 2010); Staffan Müller-Wille and Isabelle Charmantier, "Natural History and Information Overload: The Case of Linnaeus," *Stud. Hist. Phil. Biol. Biomed. Sci.* 43 (2012): 4–15; Ian Hacking, *The Taming of Chance* (Cambridge, 1990); Lissa Roberts, ed., *Centres and Cycles of Accumulation in and around the Netherlands in the Early Modern Period* (Berlin, 2011).

[22] Lorraine Daston and Elizabeth Lunbeck, eds., *Histories of Scientific Observation* (Chicago, 2011); Theodore M. Porter, *Trust in Numbers: The Pursuit of Objectivity in Science and Public Life* (Princeton, N.J., 1995); Paul N. Edwards, *A Vast Machine: Computer Models, Climate Data, and the Politics of Global Warming* (Cambridge, Mass., 2010).

[23] Steven Shapin and Simon Schaffer, *Leviathan and the Air-Pump: Hobbes, Boyle and the Experimental Life* (Princeton, N.J., 1985), 21.

ing data, much like the conventions about how knowledge is made, is inherently political.[24]

The examination of the epistemologies, technologies, and politics behind the practices developed for disciplining, recording, and systematizing observational experiences, turning them into scientific evidence, has thus been embedded in post-Kuhnian history of science. Building on this rich historiography, this volume represents a new step toward a more comprehensive history of data in the natural, social, and human sciences. By focusing on data, this volume sharpens and develops further the insights that were often merely secondary themes, or were the *explanans* rather than the *explanandum*, in the historiography focused on questions of knowledge production and contextualization. A more explicit focus on data and questions motivated by current issues around Big Data helps us to revisit earlier historical studies in the light of these present concerns, and to connect the historiographies that had so far only little to say to each other. We hope to set out—in this introduction, in our individual contributions, and in the reflective essay at the end of the volume—some of the key questions that emerge when we historicize modern data culture, and to point to some preliminary conclusions about how and why data has come to matter so much for science and society at large.

THEMES AND QUESTIONS

In historicizing and problematizing our current fascination with Big Data over the *longue durée*, we are as committed to examining continuities in the practice and moral economy of data as we are to identifying points of rupture. We take as a starting point, for example, that a history of data depends on an understanding of the material culture of data—the tools and technologies used to collect, store, and analyze information—that makes data science possible. However, we do not share with many of the recent popular appraisals of Big Data the implicit view that this has been a progressive or teleological development. Rather, we see data as immanent to the practices and technologies that support it: not only are the epistemologies of data embodied in the tools that scientists use, but in a concrete sense data themselves cannot exist apart from that material culture. However, the precise relationship between technologies, practices, and epistemologies of data is complex.

Big Data is often, for example, associated with the era of digital electronic databases, but this association potentially overlooks important continuities with data practices stretching back to much earlier material cultures. While technologies have changed—from paper-based to mechanical to electronic devices—database practices have, as several contributions to the volume reveal, been more continuous than the technologies and tools scientists used to organize, analyze, and represent their data. What is apparent in the case of recent and dramatic technological innovation, though, is that electronic computers have accelerated and amplified features of data-driven science already present or latent in earlier data cultures. These features include the automation of data collection, storage, and analysis; the mechanization of hypothesis testing; an increasing division of labor between data collectors and consumers; and the technological blackboxing of many data practices. This volume aims to present a nuanced genealogy of

---

[24] This approach has been endorsed by scholars for some time; see, e.g., Michel Foucault, *The Order of Things: An Archaeology of the Human Sciences* (New York, 1970).

these features of modern data-driven science that recognizes underlying continuities (automation and division of labor are not unique to the computer era, for example), as well as genuine discontinuities (such as the black boxing of statistical analysis in many current scientific disciplines).

The very notion of size—of "bigness"—in the history of data is also contingent on factors that need to be historicized. Like the analogous phenomenon of Big Science, the term Big Data invokes the consequences of increasing economies of scale on many different levels. The term ostensibly refers to the enormous amount of information collected, stored, and processed in fields as varied as genomics, archaeology, climate science, astronomy, and geology. But it is also "big" in terms of the major investments of time, energy, personnel, and capital it requires to function and in the way it has reoriented scientific priorities: Big Data has contributed to a hierarchical organization of science in which large, collaborative projects involving impressive-sounding databases often have pride of place in funding decisions and media attention. In other words, not only have new technologies emerged that allow new kinds of analysis and production of data, but a new cultural and political landscape has shaped and defined the meaning, significance, and politics of data-driven science.

The making of Big Data has implied the creation of new communities—of scientists, technicians, and the public—committed to the project of collecting, indexing, organizing, and analyzing large amounts of data.[25] But while it has taken on new forms and contexts, the project of translating the world into data, as the contributions to this volume show, has been under way for centuries. More recent iterations of data culture may embody distinct political and technological idioms, but they are also built upon conventions, structures, epistemologies, and political relations of the past. We therefore gain insight from identifying continuities that run through distinct eras of data, as well as from comparing the historically contingent differences and relations that emerge from the constellation of case studies we have collected in this volume.

Four basic historical questions serve as our starting point. First, how have scientists understood what data are, and what are the relationships between data and the physical objects or phenomena that the data represent in this and earlier periods? Second, what technologies, techniques, or practices are emblematic of data-driven science, and how have they developed or changed over time? Third, what has been at stake—politically, epistemically, economically, and even morally—in the increasing orientation of the sciences toward data-driven approaches? And finally, to what extent (if at all) does the era of Big Data represent a genuinely new paradigm in modern science?

STRUCTURE OF THE VOLUME

This volume is organized around three major themes. The contributions in the opening section, "Personal Data," bear witness to a persisting, though often overlooked, feature of data history. While most scientific data may not be about individual people, the four essays in this section highlight that individuality, intimacy, and personal ownership not only figure prominently in current debates around social media but also formed a crucial part of earlier data practices in science, commerce, and bureaucra-

[25] See, e.g., Sabina Leonelli and Rachel A. Ankeny, "Repertoires: How to Transform a Project into a Research Community," *BioScience* 65 (2015): 701–8.

cies. Rebecca Lemov reconsiders the historical context of today's obsession with personal data, as she sets out to examine the sharing and amassing of uniquely intensive and private personal information in the history of modern anthropology. Hers is the case of a Hopi Indian who became the most-documented single individual in post–World War II social science, long before such practices were common. Lemov uses the lived biography and the pioneering data set compiled from it to scrutinize the differences and distinctness of Big Data in relation to the personal, psychological realm.

Joanna Radin reinserts the historical and political context of personal data into the allegedly neutral testing tools of today's Big Data machine learning. She reconstructs the circumstances that have led to the formation of a comprehensive long-term data set on rates of diabetes and obesity in the Native American Akimel O'odham (known in science as Pima) tribe living at the Gila River Indian Community Reservation. Originally collected in agreement with the data's donors in a strictly medical context, the database produced from this information has since been mobilized far beyond the reservation and the donors' reach to serve as an openly accessible tool for any kind of algorithmic manipulation. Taking the itinerary of the Pima data set as a case study, Radin argues for an approach to data ethics grounded in history and lived experience that helps us understand how data come into being.

Markus Friedrich examines Europe's early modern "genealogical craze," a time when ever more comprehensive reconstructions of family lineage were required and requested for the validation of claims to gentility, proofs often necessary to hold an office. Friedrich unveils how genealogical information was generated, circulated, vetted, and compiled: families were stripped down to individuals, who in turn were reduced to a few bare dates and names. To produce these basic data and to establish relationships among them required complex archival research, turning genealogy into a social and epistemological battleground where standards of evidence and the reliability of data were constantly challenged.

Dan Bouk completes this section by highlighting changes in the political economy of personal data over the last two centuries. Crucial to the operation of modern states and companies and deeply embedded in culture and practice, personal data, as Bouk claims, have come to exist in two types: as individual data sets ("data doubles") chained to individuals or as "data aggregates" generated from such individual inscriptions. Examining the relationship between these two kinds of personal data, Bouk reveals staggering shifts of power since the nineteenth century that gained enormous momentum during the 1970s. In Bouk's view, traditional aggregates of personal data have lost value, whereas data doubles of individuals rule our current era, often being out of the control of those they represent.

The contributions in the second section, "Epistemologies and Technologies of Data," focus on the methods, tools, and practices to collect and process data that have emerged since the late eighteenth century. In particular, these essays analyze the contexts and driving forces behind crucial changes in the history of data and ask what kinds of new insights are offered by the identification of such shifts. While the first three contributions zoom in on particular turning points in botany, population statistics, and bioinformatics, the second three span a wider temporal arc to showcase long-term continuities and ruptures in fields such as paleontology, linguistics, and librarianship. Staffan Müller-Wille reveals how the study of nature became a data-driven activity during the late eighteenth and early nineteenth centuries. Taking up the epochal "historical turn" in botany during that time, Müller-Wille offers a new explanation for why naturalists developed powerful temporal visions of the earth's flora and fauna. His close reading of the many posthumous updates, translations, and adaptations of Carl Linnaeus's taxonomic works reveals the practices of Linnaean nomenclature and classification as "information science," where names and taxa were reduced to labels and containers. As these infrastructures—brought into play to manage and enhance the circulation of data—themselves became a research subject and an object of experimentation and manipulation, they generated surprising and never-before-seen phenomena unsettling long-held intuitions of nature as stable or uniform.

Christine von Oertzen unearths a similarly unremarked transition in mid-nineteenth-century European population statistics. She considers the concepts, tools, and logistics of manual census compilation in Prussia from 1860 to 1914, showing that the use of the term "data" coincided with a fundamental reform in census taking. At the core of this transformation lay a new movable paper tool encompassing all relevant data of one person on a single page, which enabled statisticians to sort and combine data in new ways. This method yielded an unprecedented refinement of census statistics, produced by a carefully nurtured workforce, including many women toiling from home. Von Oertzen concludes that the concepts, techniques, and manual sorting methods introduced since the mid-nineteenth century revolutionized European census taking and statistical complexity long before punch cards and Hollerith machines came to the fore, a fact that also explains why the Prussian census bureau chose to abstain from mechanization for many years.

While Müller-Wille and von Oertzen explore the emergence of new paper-based infrastructures and novel, data-driven sensibilities and practices at the outset of what we might call the information age, Hallam Stevens fast-forwards us to very recent shifts he sees taking shape in skills associated with Big Data work in the biomedical sciences. Stevens uncovers complex interactions between human users and computers that rely on new ways of thinking and working. He demonstrates that computers and computational practices have generated specific approaches to solving problems: producing knowledge under the label of Big Data requires "thing knowledge" and a "feeling for algorithms," analogous to the intuition, tacit knowledge, and close attention required in working with organisms.

Juxtaposing Stevens's emphasis on distinct ruptures brought about by computers in current biomedical research, David Sepkoski reflects on the long-term continuities between recent technologies and practices of data and earlier archival or collecting practices. By reconstructing and contrasting the ways in which nineteenth- and late twentieth-century paleontologists collected, sorted, compiled, stored, and visualized their data on paper, Sepkoski traces a long and continuous genealogy of data-driven science. He shows how practices that originated in the nineteenth century migrated from collection to paper based, then to mechanical, and finally to computer-based information technologies and concludes that the history of databases in the natural sciences stretches back well before the advent of computers. This earlier context, he reasons, has contingently shaped the concepts and structures deployed in electronic tools.

Judith Kaplan unveils similar—yet unacknowledged—diachronic conjunctions in historical linguistics. She scrutinizes current claims that statistical phylogenetic methods as well as Big Data computing technologies and practices bring about a radical, revolutionizing data-driven change to this field. By tracing the methods of historical linguistics from the nineteenth century onward, Kaplan shows that the tools and the actual data of current large-scale research on historical linguistic evolution are by no means new. Wordlists of 100–200 terms defined as basic vocabulary go back to nineteenth-century comparative philology. Current large-scale computational efforts to unravel humanity's linguistic deep past continue to rely on such long-vetted lists, a fact that leads Kaplan to characterize such methods as "data drag" rather than recent innovations enabled by Big Data.

In the closing contribution of this section, Markus Krajewski argues for a still longer trajectory of data history with reference to librarianship. Krajewksi claims that ancient bibliometrics knowingly distinguished between data and metadata to control and navigate vast amounts of information: headwords and indexing made any book's content accessible, as did classification and ordering on shelves. Tracing continuities of information processing from eighteenth-century classical card catalogs to Online Public Access Catalogs (OPACs) in libraries, Krajewski identifies "smart" catalog software as an immanent and radically new feature of today's Big Data, capable of merging data (content) and metadata (structure) into one, infinitely browsable "absolute book."

The contributions in the third section, "Economies of Data," examine how changing data practices in specific disciplinary contexts, particular institutional cultures, and political frameworks impacted epistemologies as well as scientific and social values (and vice versa). Patrick McCray explores how changes in computer-assisted astronomy from the 1970s onward altered communal elements of scientific practice, such as collective empiricism, open access, and notions of intellectual property. Paying special attention to crucial moments of data and social friction, McCray analyzes practices and activities beyond the telescope itself to highlight how the moral economy in astronomy changed along with the knowledge infrastructure, creating new frameworks of data circulation and sharing.

Mirjam Brusius ventures into nineteenth-century archaeology to discuss the contested history of the excavation of the Ishtar Gate in Babylon and its subsequent reconstruction in Berlin's Pergamon Museum. Following the excavated rubble on its complex journey from Persia to Berlin, Brusius illustrates how difficult it can be to pin down the relationship between objects and data, especially if we take into account how epistemic values, political objectives, and power relations change from one context to another. She thus unravels the malleability of data: what the rubble found onsite represented to the excavators on the one hand and to museum curators on the other differed so dramatically that the reconstruction in Berlin barely resembled what the experts in the field considered authentic or truthful.

Etienne Benson examines the power relations between center and periphery with regard to early and mid-twentieth-century ornithology. Benson examines the efforts of amateur bird banders and professional ornithologists in North America in order to study the ways in which data-management practices balanced the fine line between disciplining collecting efforts and encouraging enthusiasms. With reference to today's buzz around citizen science projects, Benson uncovers a backhanded, "centripetal" dynamic of power: whereas early efforts to engage amateurs in large-scale collecting programs also encouraged individual and regional forms of scientific sociality, current data-centric visions of science privilege quantity and pattern matching over quality and interpretation—visions that narrow the role of amateurs to mere sensors for science and contribute to eroding participatory notions of citizenship.

Elena Aronova concludes this section by tracing economies of data sharing at the intersection of mid-twentieth-century Big Science and world politics. She explores how the charged political climate of the Cold War affected the ways in which data were produced, collected, analyzed, and used in geophysics. Focusing on the International Geophysical Year and the World Data Centers in the USSR and the United States, Aronova analyzes a regime of exchange and secrecy in which data became a form of currency, enabling an unprecedented global circulation of data on various aspects of the physical environment and shaping distinct technological approaches that differed on the two sides of the "Iron Curtain."

The volume closes with an epilogue by Paul Edwards and Bruno Strasser, who formulate eight provocations about what can be learned from the contributions in this volume regarding the changing meanings, characteristics, and practices of data to understand the current moment(um) of data-driven science. As they argue, the data histories presented here not only help us to see the multitude of practices and modes of knowledge production that are embedded in today's Big Data but also suggest new questions in need of attention from historians.

CONCLUSIONS

It would be difficult—if not impossible—to summarize the conclusions of such a diverse group of contributions as are represented in this volume, on a topic as voluminous as "data." However, we feel that several general lessons can be drawn from the essays presented here. We firmly believe that using "data" as a category for historical analysis not only opens up new conceptual space for considering a diverse array of practices, technologies, and cultures in *longue durée* context but also promotes new disciplinary conversations among scholars (e.g., historians of computing and technology, historians of natural and social sciences, sociologists, anthropologists, and philosophers) who have not always been in direct communication. We offer, then, our conclusions—as influenced by working with the essays in this volume—about three broad questions: how data should be defined, how its history can be periodized, and how the current culture of data in the sciences can (or should) relate to its past. Our conclusions are not meant to be definitive or prescriptive but rather should be seen as reflections or encouragements for further historical study of the role of data in the natural and human sciences over the past several centuries.

What Is Data?

What does this volume as a whole tell us about what data was and is? It might be worth noting that none of the authors offers a definition of the term. Nor do they always use the same vocabulary to identify the many different meanings of data, its attributes, and its behavior. Analogies and metaphors abound: Radin, Stevens, and Sepkoski appeal to Lisa Gitelman's evocative phrase of "cooking data" to analyze how physicians, bioinformatics specialists, and paleontologists aggregated their collected findings.[26] McCray, Bouk, and Aronova problematize the notion of "data flow," unpacking the challenges of moving information from one context to another. Müller-Wille and Krajewski similarly unpack and scrutinize the fluid metaphors "data flood" and "data deluge," pointing to the tension between scarcity and abundance of evidence inherent in most scholarly work. Benson contextualizes a case of "information over-

---

[26] Gitelman, "*Raw Data*" (cit. n. 13).

load" when he describes how data management structures proved inefficient to process the evidence provided by amateur bird banders, while von Oertzen uses the notion "data jam" to describe what happened when millions of records accumulated in the American census office in 1880. Lemov uses "data exhaust" to denote the current perception that the mundane details of our daily lives are worthy of constant monitoring and recording. Invoking an alternate meaning of the term, Radin refers to "digital exhaust" to characterize the silent accumulation of the digital traces produced by personal devices such as smartphones and computers, which manifest as a "virtual commons" and as a source that can be used to trace how people think and behave.

Moreover, some authors use identical notions to describe contrary procedures. To give just one example: whereas Friedrich refers to "basic data" as a hard-won product distilled by early modern genealogists from rich and messy family lives, Aronova employs this expression to characterize the assumption behind the archives of Cold War Big Science. In Aronova's case, "basic data" was seen as material waiting to be used and processed toward more refined "data products." Rather than imposing a common vocabulary and definitions, we as editors regard such divergences as highly productive. They underscore that the data histories presented in this volume refocus recent approaches in the history of science devoted to practices, technologies, and materiality on data itself—the most fundamental and apparently simplest entity of empirical research. And thus, instead of defining what data was or is, our authors analyze what data did or do, and what was or is done with them.

This discrepancy speaks to the fact that the authors in this volume emphatically refuse to see data as acting on their own accord, or as "givens," as the Latin translation of the term suggests. Rather, the essays highlight the complexity and diversity of processes involved in making data, spanning the natural and the human sciences to render visible the myriad choices that went into creating data in different ways and diverse contexts: whether they were extracted from individual blood samples donated by members of the native American Akimel O'odham tribe (Radin); derived from celestial photographs for astronomic calculations (McCray); culled from a Hopi Indian's descriptions of his dreams and visions to form a social science archive (Lemov); recovered from archeological rubble in drawings for reconstruction purposes far from the excavation site (Brusius); retrieved from aluminum bands in the search for patterns of bird migration (Benson); or selected from vast vocabularies and compressed into basic word lists for linguistic comparison (Kaplan). Data could also be "chained" to individuals (Bouk) or decontextualized from their original context (Radin). They were mustered to reconstruct narratives about the past—of human genealogy (Friedrich), languages (Kaplan), or life itself (Sepkoski)—and they were gathered and processed automatically without human intervention (Stevens).

What is (or are) data? On this question, we as editors are adamantly pluralistic.[27] Philosophers may find profit in debating the ontology of data, but as historians, we feel that it is more fruitful to adopt the principle that data is what its makers and users have considered it to be.[28] This point is underlined by the fact that the term "data"

---

[27] We are not sticklers about the use of the plural singular ("data is") construction in some contexts for talking about data, especially when discussing data as a category.

[28] Strasser and Edwards, in "Big Data" (cit. n. 17), argue in a similar vein, claiming that instead of asking what data was or is, historians should explore what counted as data, and how the meanings of data changed.

was, for nearly all of the cases considered in this collection, an actor's category—a term that changed its meanings across time and contexts in important ways. For example, Müller-Wille draws on Alexander von Humboldt's use of the term in 1790 to illuminate the beginnings of a new, innovative reflexivity toward cumulative scientific methods, often dismissed as a "naïve," Baconian approach. Sepkoski and von Oertzen spot similar conjunctions in nineteenth-century paleontology and population statistics, respectively, in the sense that seeing collections of fossils or gathered census information as "data" went hand in hand with developing methods, tools, and practices to classify, count, and correlate data in new ways. One important feature of this novel practice was to turn data into numerical values, a development that fed into the "avalanche of printed numbers" and nineteenth-century statistical thinking, phenomena that have been so aptly described by Porter, Hacking, and others.[29]

As Paul Edwards has helpfully clarified, data have "mass" and are subject to "friction."[30] The essays in this volume illustrate in multiple ways that data did not exist in some insubstantial Platonic realm but rather occupy physical space—in specimen drawers, in compendia and catalogs, in reams of census records or insurance forms, on microfiche cards, or on magnetic storage media. The contributions show that collecting, processing, recombining, and transferring data required energy and labor, which carry costs that are tangible, whether in direct economic terms or otherwise. Data could also serve as a kind of currency or medium of exchange, and a fairly elaborate economy of data transfer and sharing has emerged over the past several hundred years in many scientific disciplines. However, the very physicality of data made it volatile; some essays point to the challenges of data storage and preservation, shedding light on dead-end technologies, archives lost (Lemov), and roads abandoned (Aronova). Such narratives show that histories of data are open-ended and help us to see what was at stake at crossroads of change.

Data themselves are usually mute and require association, processing, reconfiguration, and representation to be made to speak. Expanding earlier insights of science studies scholars, a number of the essays in this volume show that the history of data is intimately intertwined with histories of visualization. In a range of disciplines, data were conceptualized in relational and spatial terms, a perception that led to new insights through visualization, whether in the form of statistical tables, genealogical pedigrees, paleontological spindle graphs, language trees, or computer-animated visuals for genomic comparisons. Several authors show that just as data themselves occupied tangible physical space, data representations occupied visual space—for example, the measured graphical space of a simple line graph, or the more complex and even multidimensional virtual space of computer models in genetics, physics, and other fields. Aiming at making hidden patterns visible, these products of data-driven analysis were by no means hypothesis free, as various examples in this volume document. Rather, they resulted from long and arduous processes, in which the data in question had to be made commensurable before they were recontextualized, recombined, and subjected to various sorting and classifying procedures. The resulting tables, graphs, maps, and images served both as an end in themselves and as new tools for the production of further knowledge.

[29] Theodore M. Porter, *The Rise of Statistical Thinking, 1820–1900* (Princeton, N.J., 1988); Hacking, *The Taming of Chance* (cit. n. 21); see also the works cited in nn. 22 and 23.
[30] Edwards, *A Vast Machine* (cit. n. 22).

What we see as a real strength of the collective contributions to the volume is that they illuminate data histories from multiple perspectives. Our contributors show us how data were collected, where they have been stored, who has handled them, how they have been processed and recombined, the practical economy of their exchange, and the moral economy of their use in scientific, political, and cultural contexts. Though not comprehensive or encyclopedic, this volume illuminates the entire life cycle of data through its historical case studies. We consider this to be the best way of illustrating the diverse forms and roles that data can occupy.

Continuities and Ruptures in the History of Data

Our second question relates to the periodization of data histories. Is the history of data essentially continuous, proceeding by incremental steps through succeeding techno-logical, scientific, social, and epistemological contexts, or is it instead marked by sudden rupture and discontinuity? To put it in the current language of Silicon Valley, has it been dependent on periodic technological or social "disruptions"? This has been one of the crucial issues in the workshops that preceded this volume, and it is a central concern in a number of the essays. It is fair to say that no collective consensus emerged on how to periodize the history of data, but there is general agreement among our con-tributors on a few key points.

In the first place, we see no reason to treat the choice of continuity versus rupture as a mutually exclusive one. Depending on a variety of factors—the discipline being ex-amined, the span of time considered, the social and political contexts—the history of data can appear either broadly continuous or punctuated by distinctive breaks. Some disciplines with long traditions—such as astronomy, geology, or linguistics—seem less prone to epistemic or practical disruption or reinvention by new technologies or ma-terial cultures, while others with shorter traditions (e.g., biomedicine) have been more dramatically shaped by particular technologies or institutional contexts. The appear-ance of continuity or discontinuity also sometimes relates to the level of resolution at which the history is being examined: as scientists who study prehistoric extinctions have learned, an event that might appear as an instantaneous and anomalous "spike" when viewed from the perspective of the immensity of geologic time can resolve to a more drawn out "smear," lasting millions or tens of millions of years, from the perspective of ecological time. The same, we feel, is true of our own more modest historical inves-tigations: in some of the *longue durée* case studies in this volume (e.g., Krajewski, Sepkoski, Kaplan), empirical, conceptual, and epistemological conventions established at an early point seem more stable against periodic disruptions by new technologies or practices when viewed from the perspective of centuries, but they might appear more volatile if decomposed to shorter units of time.

It is impossible, however, to ignore the impact that electronic computers (both an-alog and digital) have had in the history of data. While we suspect that the view of the computer as the paradigm-defining technological innovation of late modernity by many historians is partly attributable to some quite reasonable and forgivable present-ism, it is nonetheless the case that computers have transformed the storage, process-ing, and even interpretation of data in profound ways in many disciplines, as a number of the essays here show. What we urge, though, is to avoid making the introduction of computers a decisive Rubicon in a broader history of data—to avoid, in other words, thinking of data histories as being "B.C." (before computers) or "A.C." Computers take on enormous significance in disciplines that are roughly contemporaneous with their emergence: fields like molecular genetics or particle physics, as Hallam Stevens and Peter Galison have documented, had many of their essential concepts and practices (notions of causality, tools for analysis, and even models of phenomena) shaped by the computer technologies and applications of the 1940s and 1950s.[31] But for disciplines with longer histories—bibliometrics, taxonomy, paleontology, linguistics, population statistics, astronomy—the computer is just one of a number of innovations in technology or material culture over the past few centuries.

Furthermore—as essays by Sepkoski, Kaplan, Bouk, von Oertzen, and others in this volume show—in many disciplines, data practices involving computers were strongly conditioned and constrained by practices developed around earlier technologies, such as punched-card tabulators, printed tables, index cards, and even simple lists. We as editors view the history of the relationship between material culture and practice around data as an exemplary illustration of the phenomenon of historical contingency: there is no predetermined goal that historical development is destined to reach, but particular contingent decisions—the adoption of particular tools or techniques at one point in time—have often strongly constrained subsequent developments. We see strong evidence that particular collective decisions at one point in history—such as the adoption of numerical tables as the standard format for presenting quantitative data—had downstream consequences for later developments—such as the format and function of early computerized databases.[32]

We also broadly agree with Jon Agar's argument that the initial adoption of computers in disciplines with established traditions has tended to be followed by a period when older techniques were applied to the new technology.[33] That is to say, the arrival of computers did not immediately and automatically change the way scientists collected, analyzed, or interpreted data, even if genuinely new tools (simulation models, machine learning, automated algorithmic analysis) have eventually come to the fore in some disciplines. What we see in many of our case studies is that computers promoted other kinds of important changes around what Bruno Strasser has called the "moral economy" of data: new divisions of labor between investigators and technicians (or "data jockeys"), new political tensions and relationships (e.g., around data sharing and access), and new economies of scale for data collection and analysis. Additionally, our contributions yield a tangle of further deep and momentous changes with regard to today's Big Data, within and beyond science. In the pre-electronic era, huge amounts of data were collected on a global scale, with the explicit aim of creating data archives that could be endlessly mined, for future uses yet unknown, but data was also bound in space and time to physical archives and analog infrastructures (Aronova). In stark contrast, today's Big Data radically transcends the circumstances and locality of its production (Radin). What is more, data and metadata can be swapped at will, potentially erasing the distinction between content and structure (Krajewksi). Similarly, personal data chained to individuals empower consumers, but much more

[31] Hallam Stevens, *Life Out of Sequence: A Data-Driven History of Bioinformatics* (Chicago, 2013); Peter Galison, *Image and Logic: The Material Culture of Microphysics* (Chicago, 1997).
[32] Thomas Haigh, "How Data Got Its Base: Information Storage Software in the 1950s and 1960s," *IEEE Ann. Hist. Comput.* 31 (2009): 6–25. See also Martin Campbell-Kelly, Mary Croarken, Ramond Flood, and Eleanor Robson, eds., *The History of Mathematical Tables: From Sumer to Spreadsheets* (New York, 2003).
[33] Jon Agar, "What Difference Did Computers Make?," *Soc. Stud. Sci.* 36 (2006): 869–907.

so states and corporations (Bouk), and many social situations have come to resemble laboratories in which each individual who enters becomes a de facto experimental subject (Lemov).

Are We Presentists?

Finally, we want to say a few words about the relationship between our current moment of datafication—the Big Data era—and our studies of data's past. Is it acceptable to be inspired by the ubiquity of data in our own lives to investigate the history of data? Naomi Oreskes has made a compelling case for a kind of "motivational presentism," or allowing ourselves to be guided in our historical interests by the concerns of the present, that we believe applies well to our collective project.[34] One can be motivated as a historian to explore the genealogy of technologies, institutions, or social and cultural phenomena that matter to us today without resorting to the triumphalism or teleology associated with Whiggishness. Indeed, one way of consciously avoiding Whiggism is to avoid privileging the emblematic data technology of the twenty-first century—the computer—in our study of data in the past. The methodology of historicizing is genealogical, and the essence of genealogy is contingency. We do not attempt to show that the past inevitably led us to our current data moment but rather to highlight the many unpredictable, contingent decisions and events that produced the present that we actually have, rather than one of the innumerable ones that did not come to pass. This strategy serves also to "make the familiar strange" by showing how something as apparently self-evident as "data" has encompassed many surprising and unfamiliar contexts, associations, and practices. Time and again, these essays illustrate that changing technologies or practices often did not make the work of scientists easier or "better"—to say nothing of the societies they have participated in—but rather that they introduced new challenges, new social arrangements, and new kinds of "friction" to be confronted.

In the end, we as editors strongly believe that "data history" is an important category not because many of its individual components—histories of technology and quantification, bureaucratic history, history of measurement and assessment—have been overlooked, but rather because a focus around data brings many often independent conversations and perspectives together under a broader umbrella. This matters because tracing the genealogy of data-driven science reveals essential contexts and assumptions underlying our current material culture, politics, and epistemology of data. By presenting historical analyses of the materiality, practices, and political ramifications of data collection and analysis, we gain new insights in many distinct scientific disciplines and eras relevant to historians of science, while adding a much-needed comparative dimension and historical depth to the ongoing discussion of the revolutionary potential of data-driven modes of knowledge production. Our hope is that this volume will inspire scholarship that will further enrich our historical understanding of data and its consequences beyond the cases presented in this collection. If Big Data is the question, data histories hold the answers.

---

[34] Naomi Oreskes, "Why I Am a Presentist," *Sci. Context* 26 (2013): 595–609.

PERSONAL DATA

Anthropology's Most Documented Man, Ca. 1947:

A Prefiguration of Big Data from the Big Social Science Era

*by Rebecca Lemov**

ABSTRACT

"Big Data," a descriptive term of relatively recent origin, has as one of its key effects the radically increased harnessing of ever-more-personal information accrued in the course of pedestrian life. This essay takes a historical view of the amassing and sharing of personal data, examining the genealogy of the "personal" and psychological elements inherent in Big Data through the case of an American Indian man who (the reigning experts claimed) gained the status of the most documented single individual in the history of modern anthropology. Although raised a traditional Hopi Indian in Oraibi, Arizona, Don Talayesva (1890–1985) gave over his life materials to scientists at prominent universities and constituted in and of himself a "vast data set" long before such practices were common. This essay uses this pioneering data set (partially preserved in the Human Relations Area Files and its web-based full-text database, eHRAF) to examine the distinctiveness of Big Data in relation to the personal, psychological realm; finally, a comparison is made with twenty-first-century data-collection practices of quantifying the self.

Even buzzwords have a genealogy. Although the descriptive modifier "big" has been applied to "data" in scientific publications since the 1970s—sparingly at first and later more vigorously—the phrase "Big Data" saw its first deliberate, conceptual deployment as a computing term in 1997, when two NASA scientists, Michael Cox and David Ellsworth, laid out the challenges of visualizing large data sets and put the two words together as a phrase. There was no hype involved; "Big Data" arose in the context of the problem it constituted. "Visualization provides an interesting challenge for computer systems," they wrote. "Data sets are generally quite large, taxing the capacities of main memory, local disk, and even remote disk. We call this the problem of *big data*."[1] As soon as the phrase existed, arguably, it was accompanied by very

---

* Department of the History of Science, 371 Science Center, 1 Oxford Street, Harvard University, Cambridge, MA 02138; rlemov@fas.harvard.edu.

[1] Gali Halevi and Henk F. Moed, "The Evolution of Big Data as a Research and Scientific Topic: Overview of the Literature," *Res. Trends* 30 (September 2012). Through a search of the scientific literature, Halevi and Moed examine instances of the phrase "Big Data." Although the phrase was used during the 1970s–1990s, it was not clear at that time what "big data" or "Big Data" would come to

contemporary-sounding talk of headaches and taxed capacity, and a cultivated affect of being overwhelmed in the face of an overabundance of information. By 1998, a year later, John Massey, chief scientist at Silicon Graphics, a computer hardware and software firm specializing in 3-D representations, referred in a slide to Big Data as the "next wave of InfraStress."[2] Note that the neologism was still making headlines at this point more as a challenge than as an opportunity. Many commentators have observed that the phrase itself is significant more for its myth-promoting and obfuscating tendencies than its doubtful ability to demarcate a clearly defined field. Yet this very undefined, plastic nature of "Big Data" may be instrumental in its durability and appeal.

During the early years of the millennium, the term entered wider circulation, and by 2008, "big-data computing" became a rallying cry for "Creating Revolutionary Breakthroughs," to quote from the title of an influential talk by three computer scientists who helped spread the term.[3] After some twists and turns, an array of scientists, commercial data processors, and marketing professionals attached it, in a curiously circular manner, to the ever-larger conglomerations becoming possible. This reinforced a tendency toward tautology in definitions of big data, which often—still, today—seem to come down to, "Big data is . . . data . . . that is really big." When the term in 2013 finally entered the *Oxford English Dictionary*, its definition was along these lines, both circular and problem oriented: "*big data*: Computing data of a very large size, typically to the extent that its manipulation and management present significant logistical challenges." Soon, the fine edge on which Big Data teetered between problem and possibility would tip toward the latter, especially in the popular press. Large amalgamations of data engendered vast potentialities, with the future itself hanging on their eventual outcome. Visions of uploaded, archived "selves" were rife in science fiction and nonfiction equally. Revolutionary social science drafting off Big Data's sheer bulk and speed also unfurled its banner.[4]

As this last development suggests, another characteristic of Big Data's rise is missing from broad-beamed popular definitions yet common to its futuristic implications: Big Data's tendency, in many of its most prominent spheres, to harness ever-more-personal sources of data. This turn may be referred to loosely as "Big Social Data," but the moniker implies that it is merely a subset of the larger emporium; here, my aim is to show that the personal and social aspects of Big Data are far from peripheral

signify. The first use of the term found in the Association for Computing Machinery (ACM) database was in 1997 by Michael Cox and David Ellsworth (an article earlier that year by the same authors uses the term in a less deliberate manner). Cox and Ellsworth, "Application-Controlled Demand Paging for Out-of-Core Visualization," in *Proceedings of the 8th IEEE Visualization '97 Conference*, Phoenix, 18–24 October 1997, 235–44, on 235; emphasis in the original. See Gil Press, "A Very Short History of Big Data," *Forbes*, 9 May 2013, http://www.forbes.com/sites/gilpress/2013/05/09/a-very-short-history -of-big-data/ (accessed 13 September 2015). On the vagueness of definitions, see danah boyd and Kate Crawford, "Critical Questions for Big Data: Provocations for a Cultural, Technological, and Scholarly Phenomenon," *Inform. Comm. & Soc.* 15 (2012): 662–79, on 672.

[2] On this and other uses among 3-D graphics programmers, see Francis X. Diebold, "A Personal Perspective on the Origin(s) and Development of 'Big Data': The Phenomenon, the Term, and the Discipline," PIER Working Paper no. 13-003, 26 November 2012, http://papers.ssrn.com/sol3/papers.cfm ?abstract_id=2202843 (accessed 3 October 2016).

[3] Randal E. Bryant, Randy H. Cox, and Edward D. Lazowska, "Big-Data Computing: Creating Revolutionary Breakthroughs in Commerce, Science, and Society," 22 December 2008, http://cra.org/ccc/wp -content/uploads/sites/2/2015/05/Big_Data.pdf (accessed 21 September 2014).

[4] Gary King, "Ensuring the Data Rich Future of the Social Sciences," *Science* 331 (2011): 719–21; Evelyn Ruppert, John Law, and Mike Savage, "Reassembling Social Science Methods: The Challenge of Digital Devices," *Theory Cult. Soc.* 30 (2013): 22–46.

but lie at the heart of its productivity, mutability, and power (a fact often elided in discussions; but consider Dan Bouk's discussion of "three acts" in the history of personal data, as well as Markus Friedrich's take on genealogical data).[5] Another way to address this neglected aspect of Big Data is to ask: What is Big Data's relationship to the technology and science of the self? Even if this aspect is not always heralded with equal fanfare as that of its sheer bigness, whether measured in petabytes or larger bytes, it is of the essence of "Big Data"—especially when framed as an unparalleled opportunity—that it accumulates through the living of daily life and through the sharing or leaving of intimate details, traces, and clues to subjects' locations, tastes, behaviors, "likes," and encounters. To the extent one participates in communicative common life, each person generates data exhaust, like a trail of evidence waiting to be picked up. These things (when in fact picked up) are intensely revealing. "What we know about you" is, now, everything, or at least a lot of things.

At the same time, the entity of the "self" seems curiously to elude these trapping devices, even as new claims about the social scientific possibilities opened by Big Data multiply. Though the Quantified Self movement offers as its motto "Self Knowledge through Numbers," the result may appear radically thin (especially when compared with modern definitions of self once defined by depth and an essential or existential inscrutability). Anthropologist Natasha Schüll describes the emergence of a postontological, data-driven self she calls the "data self" as a way of characterizing ongoing experiments in self-construction through intensive data tracking that take place in communities of self-trackers, archive builders, and quantified selfers. Likewise, legal scholar Frank Pasquale identifies the "algorithmic self" as one with the potential to be engineered obligingly to embrace social and commercial forces. Cultural critic Rob Horning observes that all of this data harnessing, for personal as well as scientific users, results in a kind of how-to guide for framing and creating individual personality. Of watching the Facebook "year in review" videos, he comments, "These aren't intrusive re-scriptings of our experience but instructional videos into how to be a coherent person for algorithms—which, since these algorithms increasingly dictate what others see of you, is more or less how you 'really' are in your social networks." Meanwhile social media data increasingly populate "Big Data" stores, and these stores feed emerging generations of research on personal, social, cultural, and political phenomena—rebirthing the behavioral sciences.[6]

Increasingly, supplies of personal data are swelled by self-initiated data.[7] Gary Wolf, cofounder of the Quantified Self movement, calls this "self-collected data." At a recent symposium organized by Wolf and Ernesto Ramirez, self-collected data were defined mainly in terms of examples: "step and activity data from popular devices like the

[5] Lev Manovich, "Trending: The Promises and the Challenges of Big Social Data," http://manovich.net/content/04-projects/067-trending-the-promises-and-the-challenges-of-big-social-data/64-article-2011.pdf (accessed 9 November 2015). See also Dan Bouk, "The History and Political Economy of Personal Data over the Last Two Centuries in Three Acts," and Markus Friedrich, "Genealogy as Archive-Driven Research Enterprise in Early Modern Europe," both in this volume.

[6] Natasha Schüll, "Data for Life: Wearable Technology and the Design of Self-Care," in "Big Data," special issue, *BioSocieties* 11 (2016): 317–33; Frank Pasquale, "The Algorithmic Self," *Hedgehog Rev.* 17 (2015), http://www.iasc-culture.org/THR/THR_article_2015_Spring_Pasquale.php (accessed 1 May 2015); Rob Horning, "Permanent Recorder," *New Inquiry*, 5 March 2015, http://thenewinquiry.com/blogs/marginal-utility/permanent-recorder/ (accessed 1 May 2015).

[7] Jeremy Vetter, "Introduction: Lay Participation in the History of Scientific Observation," *Sci. Context* 24 (2011): 127–41.

Fitbit, Jawbone UP, and the Nike+ system; glucometer and CGM (continuous glucose monitor) data used by diabetics to manage their own care; sleep data from the many sleep tracking apps on mobile phones; lifelogs; food diaries; and the many other types of data originating from the dozens of other self-tracking tools now becoming popular."[8] This is, from another perspective, data exhaust—a constant outflow that reflects activity in the world—and accompanying it is the urgency of its harnessing. People are portrayed as leaching data all the time, and the awareness of this fact stimulates further efforts to harness outflow: "There's an abundant amount of data that we're losing . . . every moment," as Chris Dancy, sometimes referred to as the world's most quantified man, put it in his account of the "inner net," his self-created data accretion system that tracks the ripple effects reflected in data he generates.

For present purposes, perhaps the best way to examine the cross-cutting lineage of self-study and social science's study-of-others is to examine key individuals in whom the two imperatives—to seek out intimate data and to undertake self-initiated data—combined. First, we can discern a tradition of those who offer themselves as objects of intensive investigation. A prime example is Dancy (mentioned above), a computer IT worker who, initially in his spare time and more recently as his full-time occupation, pursued "self-quantifying" by layering more and more devices, platforms, gadgets, and services on his person and embedded in his surroundings, achieving a count that ultimately numbered (depending on how they are counted) between 300 and 700 different tracking technologies. As a result, epithets such as "The Versace of Silicon," "cyborg," and "Most Quantified Man" have been applied to him.[9] Others in the "extreme life hacking" field vie with Dancy in the realm of intensive self-documentation. For precursors, one might also look back a century to Buckminster Fuller, who, in the guise of self-described "Guinea Pig B," collected each scrap and sliver of data about his activities within the ever-expanding confines of his "Chronofile." More recently, anthropologist Josh Berson self-tracks, memoir-writes, and explores the theoretical dimensions of these twinned activities. Like Fuller, Berson wants to ask questions about what constitutes the well-lived life and intends his reflections to be reminders of the somatic dimension of data, for "when we focus on data—how much of it we're producing, what it says about us, who has access to it—we lose track of the bodily intimacy of instrumentation."[10] These avatars of tracking, to a greater or lesser degree, undertake the project with epistemological questions and transfer such questions to the data they gather; many were trained in social science or have imbibed the social scientific approach—isolating particular areas for intensive monitoring and experimentation.

Considering the ubiquity-mixed-with-curious-inexactitude of the term "Big Data," and its often lost or temporarily obscured orientation in subjective data—the data of

---

[8] Such data involve special problems. "The special issues of self-collected data arise when the data is gathered in the first place by individuals for their own purposes." Problems can result when the data are then later not available to the individual who generated them (having been absorbed by corporations or health institutions) or when they are not updated properly. See Gary Wolf and Ernesto Ramirez, *Quantified Self: Public Health Symposium*, http://quantifiedself.com/symposium/Symposium -2014/QSPublicHealth2014_Report.pdf (accessed 6 March 2017).

[9] For more on Dancy and his precursors in life hacking, life logging, and life blogging, along with their implications, see Rebecca Lemov, "Archives-of-Self: The Vicissitudes of Time and Self in a Technologically Determinist Future," in *Science in the Archives: Pasts, Presents, Futures*, ed. Lorraine Daston (Chicago, 2017).

[10] Josh Berson, *Computable Bodies* (London, 2015).

the self, so to speak—it is of interest, I argue, to examine a particular scientific case whose status as (equally radical) object and subject of study makes him potentially unique.

What follows is an account of the life of a man named Don Talayesva, who may qualify as the most intensively studied human subject in the history of predigital social sciences. At the same time, Talayesva's status as "object" was stronger and more continuously maintained than the kinds of self-trackers just mentioned. Thus, his story—in its push and pull between object and subject status—reveals neglected aspects of emerging data-driven social science. He lived the life of a subject who was also an object in a more radically observable and strikingly intimate way than most, as we will see. He operated at the very cusp of the self-initiated data domain, often going beyond the requests of anthropologists and sociologists in his zeal to engage the enterprise of documentation and (one might say) render his life as an immense and unprecedentedly comprehensive data set.

In what follows, I will trace the ways in which his data set traveled during and after his natural lifetime, bringing to mind Sabina Leonelli's concept of "data journeys" in another register. (The migration of data, too, recalls Joanna Radin's history of indigenous data.)[11] His story also highlights certain key differences between "big social science" of the mid-twentieth century and Big Data social science now emerging. Via Talayesva's efforts, the domain of subjective entrepreneurship began to extend beyond the scientific expert to the scientific subject himself, an extension that has only continued with the dawning of "crowd science" and "citizen science," each of which bolsters the claims—both epistemological and evidential—of Big Data.[12] (On proto-citizen science in bird-watching circles, see Etienne Benson's essay.)[13] Aside from the differences highlighted, his life also begins to undermine claims about the sheer newness—the breathless "never before in history" claims—of data-driven inquiries and their deterministic soldering to digital technologies. Nonetheless, his lifelong contribution may also help highlight elements of true novelty often obscured in full-fledged promotions as well as dragon-slaying critiques of "Big Data."

A LIFE OF DATA PROSPECTING

Through the work of many scholars, Don C. Talayesva, a Hopi Indian man who was born at the end of the nineteenth century and died in the late twentieth (1890–1985), achieved the status of the most intensively documented anthropological subject in the

---

[11] Sabina Leonelli, "Why the Current Insistence on Open Access to Scientific Data? Big Data, Knowledge Production, and the Political Economy of Contemporary Biology," *Bull. Sci. Tech. Soc.* 33 (2013): 6–11. See also Joanna Radin, "'Digital Natives': How Medical and Indigenous Histories Matter for Big Data," in this volume.

[12] On the links between data-driven social science and phenomena such as citizen science, crowdsourcing, and crowd science, see Jeffrey Young, "The Rise of Crowd Science," in "Big Data's Mass Appeal: A Special Report," *Chronicle of Higher Education*, 2 October 2015. On the blossoming of large-scale citizen science (especially in ecological and environmental sciences) see, e.g., Jonathan Silvertown, "A New Dawn for Citizen Science," *Trends Ecol. Evol.* 24 (2015): 467–71. The International Neural Network Society (INNS) in its inaugural INNS–Big Data Conference featured a workshop, "Crowd Behavior and Big Data," in 2015, but this topic is quite different from the shift of agency I am pinpointing here: the galvanizing of the crowd itself, or members of the crowd, to collect and self-identify new sources of data.

[13] Etienne Benson, "A Centrifuge of Calculation: Managing Data and Enthusiasm in Early Twentieth-Century Bird Banding," in this volume.

world.[14] This feat of subjecthood entailed 350 hours of formal interviews with Yale anthropologist Leo Simmons alone between 1938 and 1940, during which Talayesva used his life experiences as a Hopi to fill the taxonomic pigeonholes for "Hopi" within an encyclopedic "bank of knowledge," the Human Relations Area Files (HRAF) hosted at Yale (discussed further below). This wealth of documentation also included 8,000 diary pages Talayesva wrote at the request of Simmons; 341 dreams written in the wire-bound notebooks dream researcher Dorothy Eggan sent to Talayesva (many of these dreams were followed by exploratory Q and A sessions); a set of wider-ranging interviews conducted by ethnographer Mischa Titiev;[15] a full Rorschach protocol and other projective tests Talayesva completed with psychologist Bert Kaplan; collaboration with Simmons on a groundbreaking life-history account; and, as a result of this account and its translation, which was enthusiastically hailed abroad, a thriving correspondence with French surrealist André Breton. Over the course of the mid-twentieth century, Talayesva granted additional interviews to anthropologists Leslie White, Fred Eggan (Dorothy's husband), Edward Kennard, Volney Jones, and Harold Courlander. There was, as well, the unfinished "Voth project" for which a team hired Talayesva to document an encyclopedic collection of Hopi sacred secrets; and, finally, a series of late interviews extending into the 1970s. When his memory, once considered by some researchers "almost a miracle," began to fail noticeably, it was "even so . . . probably better than [that of] most men of his age."[16]

Talayesva's usual rate of pay was 7 cents per page of diary writing and 35 cents per hour of interviewing, adding some expense for the Rorschach test, all of which made him a relatively wealthy man by Hopi standards. Whether his dream collector, Eggan, paid Talayesva by the dream or by the page is unclear.[17]

---

[14] Dorothy Eggan to Bert Kaplan, 15 July 1959, Box 3, Folder 25, Dorothy Eggan Papers, Special Collections Research Center, University of Chicago Library (hereafter cited as "Eggan Papers"). Talayesva's data comprised "certainly the most complete record of any preliterate available." This claim is contestable, depending on how one defines the degree and kind of documentation being done and what constitutes "completeness" in certain cases, but the fact that the claim was made is, in itself, significant. Other contenders might include the participants in Progressive Era observational networks (e.g., baby studies, forgotten studies recently documented by Christine von Oertzen—in which babies were observed, at first casually, and then more intensively. As the movement's initiator, Milicent Shinn, recalled: "quite unexpectedly I found myself in possession of a large mass of data" (quoted in von Oertzen, "Science in the Cradle: Milicent Shinn and Her Home-Based Network of Baby Observers, 1890–1910, *Centaurus* 55 (2013): 175–95, on 177, 178). Another contender is a Polish immigrant who wrote a 300-plus-page account for the experts studying him; see William I. Thomas and Florian Znaniecki, *The Polish Peasant in Europe and America*, vol. 4, *Life-Record of an Immigrant* (Boston, 1919). On the history of the life history techniques, see Lemov, "Archives-of-Self" (cit. n. 9). Other American Indian autobiographies had been published that were longer in terms of sheer page count, as historian David Brumble (personal communication with Rebecca Lemov, 7 April 2014) points out: "I'm sure Eggan is right about D[on] T[alayesva]—but if we're only counting what's been published, Dyk & Dyk's work with the Navajo Left Handed puts DT altogether in the shade: LEFT HANDED = 578 pp; SON OF OLD MAN HAT = 378 pp." (See H. David Brumble, *An Annotated Bibliography of American Indian and Eskimo Autobiographies* [Lincoln, Neb., 1981].) Participants in other studies during the first half of the twentieth century were also intensively documented, but Talayesva is unique, I argue, in the participatory dimension and range of his self-collected data.

[15] See Mischa Titiev, *The Hopi Indians of Old Oraibi* (Ann Arbor, Mich., 1972), in which Talayesva serves as Titiev's main informant and translator, though he is called "Ned." It also includes Titiev's fieldwork diaries from 1933–34.

[16] Eggan to Kaplan (cit. n. 14).

[17] However, clues may be found in Dorothy Eggan's correspondence with Talayesva, which has yet to be examined (Eggan Papers).

Whether or not he remains the most documented native person in history, he was the font of an "enormous body of data," according to David Aberle, the author of a psychosocial reexamination of the Talayesva corpus.[18] Likewise, Talayesva's autobiography was "a storehouse of substantive data," according to the eminent anthropologist Clyde Kluckhohn.[19] Talayesva himself became a kind of data pipeline, and therefore, the fact that Simmons used only one-fifth of Talayesva's materials diminished the quality of the published "document," according to Kluckhohn, although it was certainly useful to other students of culture and personality. (Unfortunately, subsequent scholars who searched for the remaining four-fifths of Talayesva's Yale data have been unable to find it in the archives, which touches on a theme addressed here: the paradoxical ephemerality of data intended to solidify the ephemeral evidence of phenomena such as dreaming.)[20] According to a more recent commentator, Talayesva had a "virtual monopoly on supplying white scholars with information," and therefore one must watch for idiosyncrasies in his character and not automatically attribute them to Hopi culture.[21] Nonetheless, in many circles he has come to stand for the archetypal Hopi human being, the essence of "Hopi-ness," an accomplishment that both relies on and tends to efface the fact that he was, unlike most Hopi, so willing and even eager to work with social scientists collecting data on himself. Of the unparalleled number of dreams Eggan collected in her groundbreaking career, over half were from Talayesva.[22]

This was not an expected outcome considering his life's start in a Hopi one-room dirt-floor dwelling in the village of Oraibi (Orayvi)[23] on Third Mesa in 1890. Although there were omens that the boy called Chuka was destined for some special fate, "most documented anthropological subject" was likely not the expected one. Nor was it expected that he would become the first Hopi to have linoleum flooring in his home, which was seen as a sign of both wealth and assimilation with white culture. Nonetheless, he had a strong sense that he had been marked as "twice lucky." First, he believed that he had been conceived as a twinned serpent in his mother's womb (subsequently, as he recounted, dissolved into one child through the acts of a medicine man), and second, he recalled his mother telling him that he "just miss[ed] being born a girl; I was an antelope child with special powers."[24] This sense of being lucky may have contributed to his willingness, once he reached middle age, to participate in a myriad of scholarly enterprises. Also, he enjoyed writing.

[18] David Aberle, *The Psychosocial Analysis of a Hopi Life History* (Berkeley and Los Angeles, 1951), 21.

[19] Clyde Kluckhohn, review of *Sun Chief, Amer. Anthropol.* 45 (1944): 267–70, on 267. Still, aside from its provision of data, "Even as ethnography, its importance is not negligible," Kluckhohn wrote. He also found evidence in the book of latent homosexuality.

[20] Brumble recently recalled (personal communication with Rebecca Lemov, 4 April 2014), "Back when I was working on all this, I tried tracking it down—through the Anthro and Sociology Depts and through the Library—but no one knew anything about it, alas." See also H. David Brumble III, "Social Scientists and American Indian Autobiographers: *Sun Chief* and Gregorio's 'Life Story,'" *J. Amer. Stud.* 20 (1986): 273–289, 279n10.

[21] Frederick Dockstader, quoted in Margaret Jacobs, *Engendered Encounters: Feminism and Pueblo Cultures, 1879–1934* (Lincoln, Neb., 1999), 142n115.

[22] On Eggan's dream collecting methods more broadly, see Rebecca Lemov, *Database of Dreams: The Lost Quest to Catalog Humanity* (New Haven, Conn., 2015), chap. 8.

[23] Note that "Oraibi" was the common spelling of the important Hopi village during most of Talayesva's life (although U.S. government documents sometimes used "Arabi"); today it is normally spelled "Orayvi." In what follows I will stick to the historically common spelling for the events described unless otherwise specified.

[24] Don Talayesva, *Sun Chief: The Autobiography of a Hopi Indian* (New Haven, Conn., 2013), 33.

Early portents seemed borne out by Talayesva's somewhat unusual (for a Hopi) behavior as a young child. He was reluctant to hunt, which led other boys to tease him and
treat him with condescension. He hated to see animals hurt or male livestock gelded,
which was a very un-Hopi-like attitude, according to Dorothy Eggan.[25] Talayesva's
fear of seeing animals castrated echoed the trauma he recalled of being symbolically
emasculated when he was five or six at the hands of a "grandfather," an older clan member. One of the things Talayesva seemed to respect about white culture was its aversion
to cruelty to animals. He mentioned hearing that white authorities once put a man in jail
for abusing a donkey during a Hopi clowning ceremony, and in white culture it was
taboo to eat the testicles of a sheep or goat (in fact, he thought abusing a donkey was
inhumane, but he thoroughly approved of eating the testicles of sheep and goats). At
six, when typical Hopi boys were already hunting in packs and shunning their mothers,
he was still nursing, until his peers shamed him out of it. Finally, he was different because he was a poor runner. His feet had an unusual, toe-out stance that caused him
to look funny and hindered his ability to run, a great tragedy because long-distance running abilities were valued by Hopi.[26]

Talayesva's childhood coincided not only with major bureaucratic incursions by the
U.S. Indian Office into Hopi agricultural practices, ceremonial life, and educational
systems, but a devastating Hopi-on-Hopi struggle, the Oraibi Split, the consequences
of which are felt even today (fig. 1). On one side were the "Friendlies" and on the other
the "Hostiles." The terminology came originally from Indian Office administrators,
but, when adopted in the late 1890s by Hopi themselves, it reinforced the brewing dis-
dain each faction had toward the other, hardening attitudes and ending with physical
confrontation in the form of a fateful "push"-of-war on 8 September 1906. This event,
the culmination of decades of conflict resulting in a showdown that would fuel even more
conflict, took place after thirteen years of drought and fifty years of below-average rain-
fall, environmental conditions that intensified stressful relationships and exacerbated
preexisting sociopolitical disputes. Divided by four lines in the sand, the two leaders
battled, backed by their respective factions, and the Friendlies "won." As a result of
this violent conflict, 102 families were forced out of Oraibi. They moved down the road
to Hotevilla, their displacement an outward manifestation of the internal split that vi-
olently rent Hopi society, in some cases dividing family members from each other;
in other cases, the split appeared as troubled images in Hopi dreams (among the dreams
recorded, stored, and memorialized by anthropologists).[27] Oraibi eventually broke into
five factions. Some renegade leaders were banished to Alcatraz and to the Sherman

---

[25] At times, though, Talayesva did willingly participate in or initiate games in which boys tortured
animals. For example, at one point he decided he wanted a puppy. He found a dog nursing five puppies, shot her with arrows, and took the puppies. On the other hand, he reported feeling bad for a "poor
rooster" being tormented. Note Eggan's point that "all or most Hopi are rough with animals and exhibit tendencies which seem sadistic to us." Dorothy Eggan, "Notes and Comments on Don," Confi-
dential Document (40 pp.), undated (ca. 1941, as indicated by an accompanying note requesting con-
fidentiality), 30–4 approximately, Box 2, Folder 10, Eggan Papers.
[26] These abilities continue to be cultivated among Hopi; for a historical view of Hopi running, see
Matthew Gilbert, "Hopi Footraces and American Marathons, 1912–1930," *Amer. Quart.* 62 (2010):
77–101.
[27] For more on how the trauma of the split appeared for decades in Hopi dreams, see Lemov,
Database of Dreams (cit. n. 22), chap. 4. The terms "Friendlies" and "Hostiles" came from U.S. Indian Agency personnel as early as 1891 but were adopted by Hopi secondarily. The terms are found in
a telegram from Superintendent Collins to Honorary Commander of Indian Affairs, 22 June 1891,

Figure 1. German art historian Aby Warburg visited Oraibi in 1896, during a time of fierce conflict between the Friendlies and Hostiles that resulted in the Oraibi Split, although he did not mention the political situation in his contemporary diary or in his famous lecture of 1923. Warburg's guide in Hopi was the much-disliked Mennonite Rev. H. R. Voth, Don Talayesva's nemesis.

boarding school, which also functioned as a de facto prison.[28] "Hostile" families refused to send their eighty-two children to government schools thereafter. Eventually, however, almost all Hopi children went to school, supplying to the government's schools their "quota of children," in the words of the U.S. Secretary of War to Congress.[29]

---

regarding the conflict at "Arabi," in which "the rebellious faction" or "hostile" villagers were described as threatening the "friendly Arabi's" and their children, asking for U.S. troops to intervene. Reprinted in Peter Whiteley, *The Orayvi Split*, 2 vols. (New York, 2008), 2:851. See also Whiteley, *Deliberate Acts: Changing Hopi Culture through the Oraibi Split* (Tucson, 1988).

[28] On the Oraibi Split, see Matthew Saskiestewa Gilbert, *Education beyond the Mesas: Hopi Students at Sherman Institute, 1902–1929* (Lincoln, Neb., 2010), 51–70.

[29] Report of the Secretary of War, Fort Wingate, N. Mex., House Exec Docs, 1st Sess., 52nd Cong., Vol. 1, 1891 (258–60), reprinted in Whiteley, *The Orayvi Split* (cit. n. 27), 854–5.

Talayesva, who was born into a mixed Friendly-Hostile household whose members ultimately aligned themselves with Friendlies by deciding to accept white gifts and education, found this initially forced training made certain tools (such as literacy and the inculcation of a more "agreeable" personality) available in dealing with white authorities and teachers, while it discouraged others (such as direct rebellion). During his first years attending school, just before the Oraibi Split, he became, according to the somewhat partial accounts of Dorothy Eggan, an "apple polisher," spiritually torn and weak kneed, and this experience planted the seeds of a deep-seated lifelong desire, at least according to Eggan, to be white. Yet despite an apparently accommodating attitude, Don emphasized his primal fear and loathing for whites, emotions stretching back as long as he could remember.

> I grew up believing that Whites are wicked, deceitful people. It seemed that most of them were soldiers, government agents, or missionaries, and that quite a few were Two-Hearts. The old people said that the Whites were tough, possessed dangerous weapons, and were better protected than we were from evil spirits and poison arrows. They were known to be big liars too. They sent Negro soldiers against us with cannons, tricked our war chiefs to surrender without fighting, and then broke their promises. Like Navahos, they were proud and domineering—and needed to be reminded daily to tell the truth.[30]

As a very young child he acted as a lookout, "taught to give warning when a white person started up the mesa." To be a "Friendly" in Talayesva's family lineage was not, in fact, to be so very friendly to whites. Rather, it meant to agree to "coöperate a little" but not to compromise fundamentally one's Hopiness.[31] When his grandfather and the old chief Lolumai agreed to accept white gifts (tools, supplies), to send Oraibi children to Keams Canyon boarding school, and thus to embark on the Friendly path, they were clear about the limits of friendliness: Talayesva recalls that they "warned that we would never give up our ceremonies or forsake our gods."[32] Other Hopi could not forgive the surrendering of the village's children to the (essentially) forced assimilation program of U.S. government schools, which included a day school at the foot of the mesa, a boarding school forty miles away, and an industrial training school in Riverside, California. Authorities deputized "Negro troops" to take Hopi children away in wagons and deliver them to schools (Talayesva's recollection of these armed troops in his autobiography, quoted above, suggests the deep impression their presence in the village made.) Children arriving at a school would have their hair cut, their clothes burned, and their name changed. This happened twice to Talayesva's sister, who was sequentially christened Nellie and then Gladys, because the school principal at her second school forgot (or was never told) her initial white name.

Around 1902, Talayesva's parents took Don for the first time to school, in his case the agency boarding school at Keams Canyon at the foot of the mesa, in part because they were poor and could not afford clothes for him (the school provided clothing as well as food), and partly so he could learn to read (also, he mentions in his autobiography that he did not take to herding). His parents signed him over at a ceremony by making their marks of X, and government agents "gave my mother fifteen yards of

---

[30] Talayesva, *Sun Chief* (cit. n. 24), 88.
[31] Ibid.
[32] Ibid., 89.

dress cloth and presented an axe, a claw hammer, and a small brass lamp to my father. Then they asked him to choose between a shovel and a grubbing hoe. He took the hoe." They were also given some food—bread, meat, and coffee—and rode away. At the time Talayesva was eleven and worried that he would never see them again.[33]

With this began a life oscillating between Hopi and mainstream U.S. mores: Talayesva's assimilation in school followed by his rejection of it, his embrace of white standards and of the Christian God followed by a syncretic rediscovery of Hopi ways, his absorption of boarding-school fashions followed by a partial return to the food, clothing, and rituals (such as the salt trail pilgrimage)[34] with which he was raised. After a health crisis, Talayesva married and, having relinquished white culture, tried to return to Hopi farming. He and his wife had five children, all of whom died. This was considered an insurmountable black mark in Hopi culture, perhaps the price exacted for Talayesva's sins in flouting Hopi culture. The couple subsequently adopted a boy on whom Talayesva was known to dote.

Around 1929, Talayesva made his first white friend, George D. Sachs, who "was free with his money," and, in 1932, a likable young anthropologist named Fred Eggan came to the mesa. Talayesva began to see a possible route to prosperity through work with tourists and anthropologists, picking out those whose company he enjoyed. "Some who called themselves 'anthropologists' asked me to tell them stories and paid me very well." He worried that fraternizing and being involved in scientific work with whites would foment gossip within the Hopi tribes (it did). However, in addition to providing him with convivial company, the work was enjoyable: "I found it much easier to talk in the shade for cash than to cultivate corn and herd sheep in the hot sun," he remarked pragmatically in *Sun Chief*.[35]

In 1933, Titiev traveled from Chicago to Oraibi to conduct in-depth interviews with Talayesva. In addition, Titiev helped him order a complete cowboy outfit for his adoptive son, Norman, from the Sears, Roebuck catalog. No longer used only as a means to procure hygiene devices or as wind guards to protect nursling plants, the catalog was now a vehicle with which to conjure consumer goods. When Leo Simmons arrived from Yale in July 1938 to begin the life history project with him, Talayesva dropped his subsistence work entirely and "went about" with Simmons. More anthropologists, artists, botanists, and schoolteachers followed, and Talayesva would help them find lodging in the village, sometimes with his sister. Eventually, as a result, Talayesva would later say he had "friends in all the universities," but few friends in Hopi.[36] Whereas

[33] There has been an increasing amount of scholarship on the intentional destructiveness of American Indian boarding schools to American Indian cultures; see, e.g., David Wallace Adams, *Education for Extinction: American Indians and the Boarding School Experience, 1875–1928* (Lawrence, Kans., 1995); Brenda J. Child, *Boarding School Seasons: American Indian Families 1900–1940* (Lincoln, Neb., 2001); K. Tsianina Lomawaima and Brenda J. Child, eds., *Away from Home: American Indian Boarding School Experiences, 1879–2000* (Phoenix, 2000); and Gilbert, *Education* (cit. n. 28). A recent *New York Times Magazine* article profiling young Sioux activists who began the actions to block the Dakota Access Pipeline features an interview with a young Lakota Sioux woman, Jasilyn Charger; she described growing up with the legacy of trauma carried by parents and grandparents who had been sent to such boarding schools: "No one realizes what the repercussions of colonization have been, the repercussions of forced removal." Saul Elbein, "The Youth That Launched a Movement at Standing Rock," *New York Times Magazine*, 31 January 2017, https://www.nytimes.com/2017/01/31/magazine/the-youth-group-that-launched-a-movement-at-standing-rock.html?_r=0 (accessed 15 March 2017).
[34] See Mischa Titiev, "A Hopi Salt Expedition," *Amer. Anthropol.* 39 (1937): 244–58.
[35] Talayesva, *Sun Chief* (cit. n. 24), 310.
[36] Eggan, "Notes" (cit. n. 25), 19.

Hostiles handed down to their children a toughness (though not always bitterness) and a "satisfactory adjustment in their half-white economy," as Dorothy Eggan observed, "Don in aligning himself almost wholly with whites lost the mental comfort of shared Hopi regard."[37] He was an unadjusted, *unheimlich* man—at least in Eggan's portrayal.

In his dreams (which he sold to Eggan, mailing batches of them in manila envelopes from Oraibi to Chicago, where she and her husband normally lived), Talayesva was often pursued by the police. He dreamed of a white protector in a big yellow car. When Dorothy noted one of his dreams, in which "people in Europe had heard of him," he may or may not have suspected it would come true (fig. 2). Yet within twenty years, his autobiography was to spread through French- and German-language editions, graced by his visage and adorned—as it was not in the United States—by his name as primary author.[38] The heady combination of "specialness" in both its senses—based on not only childhood predictions of a special fate but also the childhood mark of being unusual, meaning that he was both more and less than a representative Hopi—likely made him feel at ease with visiting anthropologists, through whom (on one level at least) his sense of importance was reinforced and his fears were assuaged. As he wrote to Simmons on 4 November 1941,

> The picture that Mr. Grossman has gotten is very nice. Im proud to say that my picture will be placed in the Yales mans book and soon it will be all over the world. . . . I have gotten a lot of letters from the white people saying that they wanted to get a book that was published. I think the books are passing out from the Yale University and scatter all over the U.S. to most of the people whom I never met. I think the way it sounds to me is pretty soon Ill be a great man. I dont really mean that I like to be but the white people will call me great man. When they put my address they write Chief Don C. Talayesva you remember I dont like to be Chief or Mr. Talayesva. Don is good enough for me.[39]

For Dorothy Eggan, who planned to make his dreams and story the heart of her life's anthropological work, Talayesva was an imperfect opportunity. On the one hand, his ability and willingness to supply sumptuous amounts of data made him a rich subject of study. "I had used up almost three dozen pencils on it," Talayesva recalled of his diary for Simmons, "but he wanted to know more," and, working with Eggan on his dream notebooks at the same time, he was indeed capable of more.[40]

On the other hand, to put it frankly, he was odd. Eggan's private notes make clear that although she pitied him, she did not really respect him as she did other "real" Hopi. Along these lines, historian David Brumble compares Talayesva's life story with that of a Navaho man, "Gregorio the Hand Trembler," another in the "as told to" genre of American Indian autobiography, this one assembled by Harvard anthropologist Clyde Kluckhohn. Brumble argues that Talayesva's autobiography reveals a distinctly West-

---

[37] Ibid., 20.

[38] Talayesva's autobiography was published as *Soleil Hopi* in the French Terre Humaine series (Paris, 1959) and featured a preface by Claude Lévi-Strauss. During a 1959 television interview about Talayesva, Lévi-Strauss appears framed as a "technocrat" amid a swirling, star-studded background. The German edition was *Die Sonne der Hopi: Eine autobiographie* (Munich, 1985). In both the French and German editions, Talayesva's name appeared on the cover as author of his story.

[39] Talayesva, *Sun Chief* (cit. n. 24), 471.

[40] Ibid., 386. Note that Don did not refer to his dreams as data—rather, he called his productions "dream stories"; see Eggan, "Cultural Factors in Dreaming," Box 1, Folder 9, p. 10, Eggan Papers.

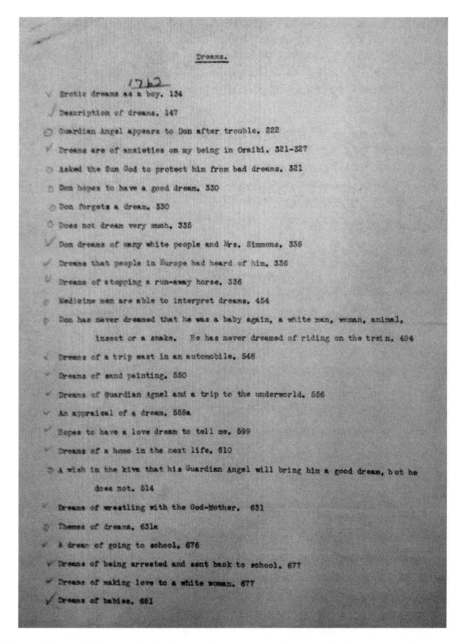

Figure 2. List of Talayesva's significant dreams written by Dorothy Eggan, including one in which he becomes famous in Europe, and "dreams of many white people" (Eggan Papers).

ern European notion of "self" and a style of self-revelation that is not seen in Gregorio's less structured, less reflective account. His autobiography, structured in three phases—Hopi boyhood, white-style school years, and return to Hopi—traces a dramatic arc through which critics have noted a strong sense of *Bildung* and self-development often notably absent in the life stories of preliterate people. Working with a roster of white academics such as Titiev and Jones further "enlarged his sense of himself."[41] In contrast to the life history of Gregorio the Hand Trembler, in which "we struggle against the strangeness of Gregorio's narrative, wondering at the silences, wondering if what is left unsaid is left unthought," Talayesva seems (according to this critic) prepared by life events to be further transformed by the experience of being intensively studied: he marks turning points and crises and resolutions in his life. It is structured, as the critic Robert Hine would observe in the foreword to the 1963 edition, like Rousseau's *Confessions* in that it evokes a universal human response—or, questions of universality aside, *Sun Chief*'s story is, as one might put it today, "relatable."[42] On the other hand, perhaps it is better to say that these techniques of social scientific study and life narration, as relentlessly worked through with Simmons, culled, and reimposed on the page, themselves structured Talayesva's sense of himself and further transformed him. Roy Pascal's dictum that the making of the autobiography re-creates the autobiographer suggests such an iterative process.[43] As Brumble notes, "Simmons' questions and his interest furthered the transformation that had begun for Talayesva in the white man's schools. Simmons helped Talayesva to discover or to create a self whose limits were not set by the tribe." In this way, Brumble concludes, "The social scientists' very questions could be powerful instruments of acculturation."[44] This process also raises the question of a collaboration that is not only literary but also existential.

THE TALAYESVA DATA'S JOURNEY TO WORLDWIDE DATABASE

Much as Don Talayesva contributed to a unique data cache—that "storehouse of substantive data" Kluckhohn spoke about[45]—the cache itself had a continuing afterlife. It did not simply rest in anthropological files or trunks of field notes. Instead, social scientists assiduously contributed his data to an ongoing and circulating database that would ultimately take digital form. By examining the sequelae of Don's data, as well as the senses in which it did and did not assimilate into repositories that can be called "Big Data," this essay will end with some reflections on the ontology and epistemology of data-driven social science and the role of personal data in the "driving."

While still in the field, anthropologist Simmons sent back installments of the collections secured from Talayesva. Once the life history was published, the clearinghouse filing system of the Human Relations Area Files (HRAF) absorbed it. (In subsuming his data under the Yale system, and filing it fresh from the field, Simmons was not alone, it should be noted: other prominent anthropologists who earmarked their data while still in the field, for future filing in the Yale database, include David Schneider on the island of Yap in the late 1940s and John W. M. Whiting during his fieldwork with Kwoma people in the early 1950s.) In order to understand this key step in the "data

[41] H. David Brumble III, *American Indian Autobiography* (Lincoln, Neb., 2008), 116. See also Brumble, "Social Scientists" (cit. n. 20).
[42] Robert Hine, foreword to *Sun Chief: The Autobiography of a Hopi Indian*, ed. Leo Simmons, 2nd ed. (New Haven, Conn., 1963).
[43] Roy Pascal, *Design and Truth in Autobiography* (New York, 2016), 182–3.
[44] Brumble, *American Indian Autobiography* (cit. n. 41), 108.
[45] Kluckhohn, review of *Sun Chief* (cit. n. 19), 267.

journey"[46] taken by Talayesva's information, a brief sketch of this data-gathering edifice is in order.

The Human Relations Area Files, later known affectionately as "the Files," or "Yale's Bank of Knowledge," began in 1928 when then-fledgling anthropologist George Peter Murdock imagined an organ that would comprise comprehensive bibliographical details for all works dealing with any known culture around the world. This he reimagined (by 1931) as a repository for the total sum of cultures, their traits and constituent parts broken down into categories. It went, thus, from bibliographic to encyclopedic enterprise in a few short years and would soon undergo even more extensive transformations. However, in these early years, Murdock saw his data-gathering enterprise—then known as the Cross-Cultural Survey—as a personal, lifelong, virtuosic project to be undertaken by himself and himself only, as he was wont to spend his nights haunting the library, and his days engaged in the other tasks that fell to scholars, hardly sleeping at all. Initially he targeted "2000 primitive tribes" as a way potentially to test "various theories of social evolution by statistical techniques."[47] Soon, all "cultures known to humankind" were said to be the project's remit; or (granting totality to be a daunting if worthy goal) a representative 10 percent sample of such. By the start of World War II, files were stacking up.

How were the data organized? Murdock assembled a team of six lower-ranking associates to map the full taxonomy of culture, resulting in the *Outline of Cultural Materials* (*OCM*; first edition published in 1938, modified in 1942, reprinted in 1945, and with subsequent editions in 1987, 2000, and 2004). The assembled group of graduate students and junior professors in anthropology and sociology undertook the task of cataloging cultural variety in all its forms, carving up the component bits that made up culture and—after seeking advice from a range of experts in each domain—publishing the taxonomic tree as their *Outline*. Beginning from scratch, they identified several "large blocks" of culture, such as Kinship, Magic, Politics, the Reproductive Cycle, and Material Culture and Technology, and then proceeded to break them down as logically as possible. Two-digit numbers from 10 to 88 marked each major heading, and a third or fourth digit marked each subsequent subdivision, thus communicating with that numerical coding a confidence in the impartiality and neutrality of the divisions provided. (Nonetheless, the *OCM* has undergone constant revision and fine-tuning over succeeding decades.) Three-digit *OCM* numbers 471 through 478, for example, served as subcategories of 47, "Reproduction," and corresponded, respectively, to "Theory of Reproduction," "Menstruation," "Conception," "Pregnancy," "Abortion, Infanticide, and Illegitimacy," "Childbirth," "Unusual Births," and "Postnatal Care." Each subtopic included component elements and could be broken into further parts: 4741, 4742, 4743, and so on. Filers retained "certain other marking conventions" such as asterisks, zeros, and cross-referencing marks used in the margins to regulate the reader's passage

[46] On the concept of "data journeys," and the idea that scientific data are not simply given or raw but undergo travels and tribulations ("the ways in which data are actually disseminated and used to generate knowledge"), see Sabina Leonelli, "What Difference Does Quantity Make? On the Epistemology of Big Data in Biology," *Big Data & Soc.* 1 (April 2014): 1–11, on 2.

[47] Letter regarding Murdock's project at the Institute of Human Relations, Mark May to George Peter Murdock, 20 January 1931 (May is describing Murdock's proposed project), YRG 37-V, IHR, series II, box 11, file 11-95, Yale University Archives, New Haven, Conn.

through the broken-down textual passages (in the basic unit of the paragraph) that were
then excerpted, typed, and filed on matching index cards.[48]

Early on, a processing issue beset Murdock's team; the question was whether—or
the extent to which—the original texts from which they worked should be allowed to
survive the top-down filing process intact, or whether the compiled cultural data
should be reduced to 0s and 1s. After lively debate, they decided to retain the original
textual context culled from full reports and other sources, excerpted and cross-
referenced: "So it was decided that whatever was done—even though the information
might be characterized and classified into pigeonholes, as it were—would be done in
the original text. . . . In other words, this would be an organization and ordering of original information and not simply a[n] . . . abstracting or coding system."[49] Any relevant
sentences or fragments of sentences would be marked by a processor, transcribed from
the source, typed on a 5- by 8-inch page, and filed according to the appropriate geographical area and *OCM* two-, three-, or four-digit heading. Meanwhile, a copy of
the excerpted text could be consulted in its entirety, save for certain parts deemed irrelevant, under category 116, "Texts." In a seeming paradox, it appears that this project
conceived to marshal "controlled data" amenable to statistical manipulation and leading to universal behavioral laws—laws that were to explain the entirety of what is human—still relied upon the texts *as they were written*. Part of the reason may have been
that the texts in question were often the work of highly peculiar individuals, so it would
be better to preserve their eccentricities. Part of it may have been Murdock's feeling
that the files would be all the more self-sufficient if the scholar had no need to look
up the referenced work. By preserving the "original information," the files would be
more authentic. At any rate, within each category, the original filers decided that
"the standard unit of analysis is the paragraph."[50] From these humble yet somehow in-
transigent "paragraphs, sections and chapters of . . . reports," the files would be built.[51]

Soon, and sometimes simultaneous with their gathering, the Talayesva data would
be tagged with the three- and four-digit codes[52] of the HRAF's *OCM*, the whole data
set marked with the Hopi area numeric NT09. As the HRAF complex underwent "migrations from format to format" over the years, moving from physical files to microfiche to the digital eHRAF network—resulting finally in a digitized "database of information on cultures"—the Talayesva data did too.[53] Now a researcher no longer needed
to read the autobiography to sample the materials Talayesva provided, for example,
about his sexual dalliances, his views on women, his acculturative attitudes, his encounters with deities—which now appeared duly coded, variously, among life history

---

[48] For an overview of this process and the HRAF, see Rebecca Lemov, "Filing the Total Human:
Anthropology Archives from 1928 to 1963," in *Social Knowledge in the Making*, ed. Charles Camic,
Neil Gross, and Michèle Lamont (Chicago, 2012), 119–50.

[49] Clellan S. Ford, "The Development of the *Outline of Cultural Materials*," *Behav. Sci. Notes* 3
(1971): 173–85.

[50] *HRAF Research Guide* (New Haven, Conn., 1965), 27.

[51] Clellan S. Ford, "HRAF: 1949–1969: A Twenty Year Report," *Behav. Sci. Notes* 5 (1970): 1–64,
on 5.

[52] Note that these numeric markers are called codes by HRAF personnel, but library scientists stip-
ulate that they are indices or "notations" and not, technically, codes, for a code entails the assigning of
a value to a variable based on a scale or score of frequency (i.e., always/sometimes/never). On this
point, see Sandra Roe, "A Brief History of an Ethnographic Database: The HRAF Collection of Ethnography," *Behav. & Soc. Sci. Libr.* 25 (2007): 47–77, on 51.

[53] Ibid., quotations on 52 and 48.

materials (159); social personality (156); drives and emotions (152); organized cere-monial (796); kin relationships (602); culture summary (105); infancy and childhood (850); family (590); sodalities (575); diet (262); sex (830); and behavior toward non-relatives (609). One can also cut a swathe across, for example, category 152, and dis-cover other sources of data at hand on drives and emotions either in Hopi or other groups around the world: for example, there is information about Hopi children's graves from the controversial Mennonite missionary-ethnographer Heinrich "Henry" Voth's miscel-laneous papers, which also found its way to files within the digital eHRAF. In the ex-cerpt provided, Voth described five instances of parents mourning a dying child, which led to the conclusion "that death, or even approaching death, strikes such terror to the Hopi heart, that he shuns and flees the sickbed and death-chamber as much as possible. For this reason he does not like to speak or hear others speak about the dead, however much he may have loved them and he prefers to say, 'they are gone' or 'they have gone to sleep' to saying, 'they have died.'"[54]

In light of the fact that Reverend Voth's notes came to populate a database represent-ing the totality of Hopi knowledge, it is worth noting that Hopi people particularly dis-liked Voth's behavior around graves. In the years Voth spent at Hopi, he alienated many, for example, by barging uninvited into sacred rituals and preaching loudly in the Hopi language; Talayesva's autobiography condemned "this wicked man [who] would force his way into the kiva [an underground room used for religious ceremo-nies] and write down everything that he saw. He wore shoes with solid heels, and when the Hopi tried to put him out of the kiva he would kick them."[55] His habit of selling Hopi ritual artifacts to the Bureau of American Ethnology, and building a reconstructed Hopi altar for the Field Museum in Chicago, did not further endear him. In particular, Talayesva, who himself lost five children, took vehement umbrage at Voth's intrusive data-gathering activities in Hopi and prided himself on personally running the reverend off the mesa one day after a showdown (by the time this happened, in the late 1920s, Voth was an old man). Moreover, the two had a long history: as a young child, Tala-yesva felt pressured to sell his bow and arrow to Voth for a few pennies when his family needed food. For this reason, it is a painful irony that he and Voth went on to be linked both in the minds of Hopi—condemned as betrayers of secrets—and in the files of the Yale project. To rest in a digitized eternity with the hated Voth's data was another, per-haps ironic fate of Talayesva's, and in what remains I will discuss its significance for "histories of data" and of Big Data specifically.

BIG DATA FUELING BEHAVIORAL SCIENCE: THREE PRELIMINARY CONCLUSIONS

Talayesva's story can be framed within growing self-documentation and self-quantification movements that feed the onrush of "Big Data." According to many prognosticators, Big Data is nothing else than the future of social science; and in laying out this future vision, such promoters often cite the ever-more-personal sources of data (everything from Amazon purchases of personal items to cell-phone geolocation data to online di-aries or public tweets, gathered in massive, depersonalized sloughs) to be tapped, man-

[54] See eHRAF database, http://ehrafworldcultures.yale.edu/ehrafe (accessed 19 July 2017).
[55] Talayesva, *Sun Chief* (cit. n. 24), 252. For more views of Voth, see Whiteley, *Deliberate Acts* (cit. n. 27), 82–4; and Fred Eggan, "H. R. Voth, Ethnologist," in Barton Wright, *Hopi Material Culture: Artifacts Gathered by H. R. Voth in the Fred Harvey Collection* (Flagstaff, Ariz., 1979), 1–7.

aged, and worked with.[56] Changing standards in the appropriateness of "sharing" intimate details of one's life, from family milestones to daily dog walks, contributes to an ever-growing barrage of data.

Thus, in becoming if not aspiring to be the most documented human being at the heyday of mid-twentieth-century social science, Talayesva lived through and by means of an intensity of documentation that makes him something of a pioneer. His case resonates with current instances of individuals who claim to be the "most quantified" person in the world. How does his experience, including the technologies of self that accompany and shape it, reflect both on the state of the sciences in a data-driven era and on the ontology of the self—what is private, what is shared, what is archived, what is forgotten, and what is, in a strange way, held in abeyance for future use—as it changes under these conditions?

Briefly, there are three main areas Talayesva's data serve to illuminate. First, they prefigure privacy concerns over sensitive materials, even as they highlight striking differences in data-gathering conditions and the legal assumptions surrounding them.[57] While he was fairly generously remunerated from his point of view, Talayesva paid a dear social price for his revelations to social scientists—including that paid for his imagined, not real, revelations—as it was the appearance of traitorous collaboration rather than the reality that damned him within many Hopi circles. In his ethnographic prime, he was even hated, as Dorothy Eggan observed: packs of boys hounded him at the post office, and many Hopi shunned him. In part, he served as a scapegoat for Hopi grief over the loss of their ceremonial secrets to untiring cadres of missionaries and anthropologists over the course of several decades: Talayesva stood accused of selling Hopi corpses and giving away Soyal (winter solstice) ceremony details, and his "proof" of his own innocence—pointing out that these secrets had already appeared in the zealously observational Voth's writings decades before, when Talayesva was still a boy—did not avail him. Still, he remains controversial today. It is likely he did not suspect the fate of his data, available across a worldwide network where, digitized and easily tapped into from any portal in the world, his own personal stories, from dalliances to divine interactions, could be accessed. If he had known, indeed, he would perhaps have been pleased.

Yet even as the ethics of prospecting for Native American samples—whether of life history materials or genetic materials—are increasingly contested, related concerns among Big Data researchers are also growing today.[58] Despite stronger institutional review board (IRB) protections, subjects who have unwittingly contributed data harnessed in Big Data studies may be unaware of their participation and of the "data journeys" their own preferences and personal contributions may make. In contrast to the singular figure of Talayesva, it is impracticable to query each such subject. As boyd

[56] Kevin Haggerty and Richard Ericson, "The Surveillant Assemblage," *Brit. J. Sociol.* 51 (2000): 605–22.

[57] For a point of contrast, see Radin, "'Digital Natives'" (cit. n. 11).

[58] See Sophie McCall, *First Person Plural: Aboriginal Storytelling and the Ethics of Collaborative Authorship* (Vancouver, 2011); Kim TallBear, *Native American DNA: Tribal Belonging and the False Promise of Genetic Science* (Minneapolis, 2013). See also Radin, "'Digital Natives'" (cit. n. 11), in which she discusses data mining of Akimel O'odham. On the history of IRBs, see Laura Stark, *Behind Closed Doors: IRBs and the Making of Ethical Research* (Chicago, 2012). For a history of attempts to reclaim cultural materials, see Chip Colwell, *Plundered Skulls and Stolen Spirits: Inside the Fight to Reclaim Native America's Culture* (Chicago, 2017).

and Crawford observed in 2012, "It may be unreasonable to ask researchers to obtain consent from every person who posts a tweet, but it is problematic for researchers to justify their actions as ethical simply because the data are available."[59] Debate is under way over solutions—librarians and social scientists struggle to articulate "best practices"[60]—as these challenges, already confronted by (if occasionally invisible to) earlier researchers, take new forms. This is due less to the unwieldy size of Big Data data sets than to the modes and technologies facilitating their gathering, via which the production of data occurs as a by-product of the social, political, and commercial interactions that drive it. The singular figure is eclipsed. The new subject of data-driven social science is a "data self."[61]

Second, the personal part of personal data: Talayesva's data highlight the quest to access ever-more-personal data, which certainly existed before the shift to digital mining operations. Yet, with that shift, personal information, and even the definition of the personal realm, can be seen to transform. Examining Talayesva's career affords a clearer view of the new ways in which personal data are used today as a scientific and technical object. The drive to data mine was there, but the fate of data was different. Most clearly, Talayesva's data ended up mixing in a large data set (eHRAF) with the contributions of his archenemy and with other ethnographers of Hopi. The HRAF system invited comparisons and extractions. Yet Talayesva's mass of data became neither completely abstracted nor completely alienated from the identifying markers—its metadata. (A subset of the HRAF, indeed, did go on to populate an all-purpose "Standard Cross-Cultural Sample," although Hopi were not one of the 186 groups chosen.) A certain order was maintained: a researcher can still tell which was Voth's and which was Talayesva's contribution. Despite its messiness and nonsystematic nature, and the fact that, in the second half of his life, Talayesva used it to create and maintain relationships—with visiting anthropologists, with schoolteachers, with journalists on road trips to New Haven, and more generally in populating his world with those he called "friends"—nonetheless the Talayesva corpus was contained.

A useful contrast may be seen with one of the most famous actors associated with the Quantified Self movement. As mentioned earlier, the moniker "Most Quantified Man" has often been applied to Chris Dancy. However, it is notable that since 2016 emphasis has shifted to the epithet "Most Connected Man." That is, the uniqueness of Dancy's contribution is not simply the degree of quantification to which he subjects his life but its effect in tethering him to myriad devices and other humans in new ways. In addition, it is arguably his inclination to reflect on connectedness that makes Dancy unique among those rendering their selves as data. One can watch him embrace the role in an existential way in an episode of the recent, brooding Showtime documentary series *Dark Net* (fig. 3).[62] The show's introduction showcases the fact that "every day on the Internet we create 2.5 quintillion bytes of data," after which Dancy describes on-screen how his data-collecting project emerged out of spiraling

[59] See boyd and Crawford, "Critical Questions" (cit. n. 1), 672. Cf. Lemov, *Database of Dreams* (cit. n. 22) for further discussion of historical debates over ethics of data storage and publication.

[60] Steven Ovadia, "The Role of Big Data in the Social Sciences," *Behav. & Soc. Sci. Libr.* 32 (2013): 130–34, on 132.

[61] Rob Horning, "Notes on the 'Data Self,'" *New Inquiry*, 2 February 2012, http://thenewinquiry .com/blogs/marginal-utility/dumb-bullshit/ (accessed 16 March 2017).

[62] "Upgrade," *Dark Net*, season 1, episode 2, aired 28 January 2016 (Showtime).

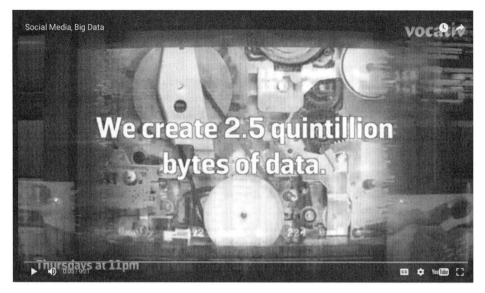

Figure 3. *Screenshot from promotion for* Dark Net *documentary series (2016), featuring Chris Dancy; see http://www.sho.com/dark-net (accessed 19 July 2017).*

personal anxiety and, for a time, led only to further anxiety. "The idea to bring more data into my life really came from massive amounts of anxiety. Whether it was food or alcohol or recreational drugs or cigarettes, I could feel this heaviness of getting older and being unhealthy. I needed to find a way to harvest my behavior out of systems." Aided by a practice of meditating on connectedness—watching the ripples that ran through his ongoing data amalgamations projected on screens—Dancy now uses his incoming data as a way of changing his own daily behavior, losing over 100 pounds as a result and generally becoming healthier, even as he embraces his role as a kind of sentinel of intensive data gathering on the self: "I realized data makes us lonely. We . . . need each other . . . badly."

Third, Big Data is usually poor data. One of the ongoing debates among critics and adherents of Big Data is the question of its significance. Moving beyond the rhetoric of clickbait promotions, its importance does not lie in sheer size or even the passing of a particular threshold (such as the petabyte). Nor is it, per se, velocity or other qualities (the "4 V's") attributed to Big Data.[63] Rather, for some, the advantage of Big Data is its poverty: it is unstructured, instantaneous, constantly generated, and decidedly uncurated. As Plantin et al. argue, Big Data is distinctive for "escaping the con-

---

[63] The point that Big Data is not simply about being "big" is made by many. See, e.g., Bruno Strasser, "Data-Driven Sciences: From Wonder Cabinets to Electronic Databases," *Stud. Hist. Phil. Biol. Biomed. Sci.* 43 (2012): 85–7; Sabina Leonelli, "Global Data for Local Science: Assessing the Scale of Data Infrastructures in Biological and Biomedical Research," *BioSocieties* 8 (2013): 449–65; J. C. Plantin, C. Lagoze, P. Edwards, and C. Sandvig, "Big Data Is Not about Size: When Data Transform Scholarship," in *Les données à l'heure du numérique: Ouvrir, partager, expérimenter*, ed. C. Mabi, J. C. Plantin, and L. Monnoyer-Smith (Paris, forthcoming). The "four V's" (volume, velocity, variety, veracity) that are said to define Big Data, at least among some in computer visualization and related fields, are visualized through an IBM analytic: http://www.ibmbigdatahub.com/infographic/four-vs-big-data (accessed 3 October 2016).

trol zone"—defined as a traditional realm secured to produce trust in research materials, buttressed by a cordon or guarantor of reliability (such as a library). When the control zone breaks down via Big Data, which circulates in new and unheralded ways, disruption to the chain of scholarly publication and the mode of judging reliability occurs. In some cases, this is because the data are relatively unclean, and indeed many claim data-driven science can make a positive virtue of "messiness." In other cases (such as climate science or epidemiology), ruptures of control take place at the point of analysis, storage, or publication.[64] Here, the Talayesva data set serves as a distinctly contrasting case. It retains rich metadata, its provenance is never unclear, and it concerns an unusual individual who then comes to stand for the echt Hopi—the last marking a retrospectively odd mental gymnastics characteristic of mid-century liberal social science. In the early twenty-first century, however, there is no such maneuver, because in the case of digitized Big Data the sources are very often elided and metadata may be slim or nonexistent. There are many Dons, and they are much more connected. Their most personal materials (wedding pictures and gas purchases) are being figured and recombined in new ways. Today, "personal data assemblages are never stable or contained. They represent a 'snap-shot' of a particular moment in time and a particular rationale of data practice," observes sociologist Deborah Lupton.[65] In this sense, Big Data reveals differences that are not just quantitative but qualitative, and this is especially true in the realm of the "data self."

The example of the densely data-fied Talayesva provides a way of discerning evolution from revolution in the field of Big Data, and looking for both gaps and continuities with the present. Certainly the urge to mine the subjective, to penetrate ever more persistently into the remotest or—perhaps even rarer, the most mundane—parts of human subjectivity, has a long genealogy. Talayesva's data are a harbinger of this (his dreams are often ho-hum, and the details of his life capture its everyday quality), but also remarkable as a different sort of activity, one with a relatively well-structured, relatively well-controlled fate. Ultimately what has been enabled by our own data practices feeding Big Data today is not the personal per se but the dislodging and ramifying of the personal.

Recently, digital humanities scholar Lev Manovich observed that past trade-offs concerning data will no longer be necessary in the new dawn of the social sciences: "The rise of social media along with the progress in computational tools that can process massive amounts of data makes possible a fundamentally new approach for the study of human beings and society. We no longer have to choose between data size and data depth."[66] He gives the example of Nielsen ratings in the past: they generated large amounts of data but little on the day-to-day and minute-to-minute experiences of television watchers. What I suggest in telling Talayesva's story and contextualizing

---

[64] Plantin et al., "Big Data" (cit. n. 63). On "messiness" as a characteristic advantage of Big Data, see Viktor Mayer-Schoenberger and Kenneth Cukier, *Big Data: A Revolution That Will Transform How We Live, Work, and Think* (London, 2013), 32–49.

[65] As Lupton further observes, "Shifting forms of selfhood are configured via these digital data assemblages, depending on the context in and purpose for which they are assembled." See Lupton, "Managing and Materialising Data as Part of Self-Tracking," *This Sociological Life* (blog), 22 March 2015, https://simplysociology.wordpress.com/2015/03/22/managing-and-materialising-data-as-part-of-self-tracking/ (accessed 3 October 2016). See also Lupton, *The Quantified Self: A Sociology of Self-Tracking* (Cambridge, 2016).

[66] Manovich, "Trending" (cit. n. 5).

it, however, is that earlier data-collecting efforts did not so easily fall into the either-or of size and depth, and that we can observe the impulse to gather the texture, warp, and woof of everydayness—on a large scale—in existing digital technologies, social media, and projects driven by Big Data.

Don Talayesva was a man who stood on the brink of an older century (the nineteenth) as it met a newer one (the twenty-first). The problems, epistemological and practical alike, that arose in the process of intensive documentation of a single human subject may bear on similar problems that arise in present-day data-driven projects, for today, many social situations resemble laboratories in which myriad and multifarious experimental projects unfold, and in which each individual who enters becomes a de facto experimental subject.

"Digital Natives":
How Medical and Indigenous
Histories Matter for Big Data

by Joanna Radin*

ABSTRACT

This case considers the politics of reuse in the realm of "Big Data." It focuses on the history of a particular collection of data, extracted and digitized from patient records made in the course of a longitudinal epidemiological study involving Indigenous members of the Gila River Indian Community Reservation in the American Southwest. The creation and circulation of the Pima Indian Diabetes Dataset (PIDD) demonstrates the value of medical and Indigenous histories to the study of Big Data. By adapting the concept of the "digital native" itself for reuse, I argue that the history of the PIDD reveals how data becomes alienated from persons even as it reproduces complex social realities of the circumstances of its origin. In doing so, this history highlights otherwise obscured matters of ethics and politics that are relevant to communities who identify as Indigenous as well as those who do not.

WHAT'S IN A NAME?

Several years ago I found myself in conversation with a mathematician. She was an expert in a field of problem solving called machine learning. As she explained it to me, applications of her work served to do things like optimize Google search rank orders, to make sure that people found what they were looking for or, perhaps, what they did not even know they were looking for. At the time of our conversation, she was using her expertise to help the electricity provider Con Edison to predict fires sparking in the underground power grid in New York City. Such fires, in addition to disrupting service, could lead to dangerous explosions of manhole covers.[1] To address the chal-

* Section of the History of Medicine, Sterling Hall of Medicine, Yale University, 333 Cedar Street, L132, New Haven, CT 06520; joanna.radin@yale.edu.

I would like to thank Cynthia Rudin and David Aha for conversations about their machine learning work and Jennifer Brown, Laurel Waycott, Laura Stark, and participants in the "Big Data and Invisible Labor" symposia at the Max Planck Institute for the History of Science.

[1] Cynthia Rudin, "21st Century Data Miners Meet 19th Century Electrical Cables," *Computer* 44 (2011): 103–5; Rudin, "Machine Learning for the New York City Power Grid," *IEEE Transactions on Pattern Analysis and Machine Intelligence* 34 (2012): 328–45. These circular, cast-iron caps cover holes constructed to provide repair workers access to the underground power grid when maintenance is required. For some data on the frequency and severity of fires beginning in manholes, see https://users .cs.duke.edu/~cynthia/docs/RudinEtAl2011ComputerMagazine.pdf (accessed 26 September 2014).

lenge of trying to anticipate and prevent such critical disruptions to the electricity supply, she was testing and optimizing the predictive algorithms that were the coin of her
professional realm. These tests required access to sufficiently complex and validated
data sets to feed to the algorithm. I asked her whether she collected the data to test these
algorithms herself. "No," she answered. Then where did the data come from? She re-
cited the names of several preexisting, freely available data sets. One struck a chord of
recognition: Pima.

Having studied the history of efforts to enroll Indigenous peoples in biomedical
knowledge projects, I knew that "Pima" could refer to members of an Indigenous community who live in the southwestern region of the United States.[2] The community has
come to experience extremely high rates of diabetes and obesity, which has made its
members the focus of extensive research by epidemiologists, geneticists, and medical
anthropologists.[3] In 1990, those living at the Gila River Indian Community Reservation, outside of Phoenix, Arizona, were defined as having "the highest recorded prevalence and incidence of non-insulin-dependent diabetes of any geographically-defined
population."[4] Members of this community, known to science as "Pima," refer to themselves as Akimel O'odham, which has been translated as River People.[5]

The political boundaries of the Gila River Indian Community, which was created in
the mid-nineteenth century, made it possible for public health officials to conceptualize the reservation, one of the oldest in the United States, as a laboratory for observing the epidemiology of diabetes. Since the 1960s, before the rise of genomics, medical information collected from Akimel O'odham bodies was regarded as a valuable
resource for improving and even defining general knowledge about the disease. In the
process, this medical information was translated into digital form that would enable it
to be reused for knowledge projects unrelated to diabetes or even biomedicine.

The short answer, then, to the question of whether or not there was a connection
between the "Pima" I had encountered in my studies and the "Pima" this mathematician was using in hers was "yes." A longer answer is provided in this essay, where I
describe historical circumstances that have led data about Akimel O'odham people to

---

[2] I describe the history of efforts to collect and freeze blood from members of Indigenous communities around the globe in Joanna Radin, *Life on Ice: A History of New Uses for Cold Blood* (Chicago,
2017).

[3] Mariana Leal Ferreira and Gretchen Chelsey Lang, eds., *Indigenous Peoples and Diabetes: Com-
munity Empowerment and Wellness* (Durham, N.C., 2006); Michael Montoya, *Making the Mexican
Diabetic: Race, Science and the Genetics of Inequality* (Berkeley and Los Angeles, 2011); Carolyn
Smith-Morris, *Diabetes among the Pima: Stories of Survival* (Tuscon, Ariz., 2006). Others may have
read Malcolm Gladwell's article about the struggle of the Pima with diabetes: Gladwell, "The Pima
Paradox," *New Yorker*, 2 February 1998.

[4] William C. Knowler, David J. Pettit, Mohammed F. Saad, and Peter H. Bennett, "Diabetes Mellitus
in the Pima Indians: Incidence, Risk Factors and Pathogenesis," *Diabetes/Metabol. Rev.* 6 (1990): 1–27.
For more recent statistics, see Centers for Disease Control and Prevention, "Diabetes Prevalence among
American Indians and Alaska Natives and the Overall Population—United States, 1994–2002," *Mor-
bid. Mortal. Week. Rep.* 52 (2003): 702–4.

[5] In the early 1600s, the first Spanish settlers questioned members of the community about their identity. Akimel O'odham, unfamiliar with the language of their inquisitors, are said to have responded with
the phrase "pi-nyi-matchi," translated as "I don't know." The Spanish colonizers assigned them the
name Pima, which is how they heard the response. Carl Waldman, *Encyclopedia of Native American
Tribes* (New York, 2014), 4. When I use the word "Pima," it is to point to how the community was described historically, by those who studied it, as well as certain social scientists who have since written
about its members. Tracking the label "Pima" matters for me as a historian for reasons that I will make
clear.

be produced and mobilized beyond the reservation where they live and where they have long participated in medical research. I argue that the broader enterprise of data collection, use, and reuse has reproduced certain patterns of settler colonialism. This insight is not merely consequential for those who identify as Indigenous but also provides inspiration for intervening in the creation and management of new digital technologies of representation and knowledge that perpetuate—often needlessly—exploitative ideas about property, innovation, and self-determination.[6]

With the title "Digital Nativity," I am referencing a phrase coined by education consultant Mark Prensky, who in 2001 observed of the most recent information age that "today's students are no longer the people our education system was designed to teach."[7] I engage in my own practice of reuse by resignifying the term to extend questions of sovereignty and justice that have focused on land and land rights to the digital territories of our current information age, which is so often inappropriately cast as a virtual Wild West or an uninhabited frontier. The practice of reproducing frontier narratives short-circuits the potential to unsettle ideas about how innovation works, including the potential to learn from Indigenous peoples who have been at the vanguard in encouraging innovative and decolonial approaches to research.[8] Today's research subjects are no longer the objects Euro-American knowledge systems designed to make invisible.

At the core of this history are questions about the origins, ownership, and reuse of the personal and bodily data that fuels information economies. The story of how Indigenous participants in the National Institutes of Health's longitudinal research on diabetes at Gila River became understood as donors of data used to study diabetes and later, how that data was used to refine algorithms that had nothing to do with diabetes or even to do with bodies, is exemplary of the history of Big Data writ large. What makes data "big" is not so much its size—though that is relevant too—but its ability to radically transcend the circumstances and locality of its production.[9] Computers and algorithms make that possible, but understanding the politics of Big Data also requires attention to the creation and processing of the data itself, including the recognition that it often

---

[6] In other words, I am arguing for the relevance of indigenous studies theory to the critical study of Big Data. E.g., Kim TallBear, "Narratives of Race and Indigeneity in the Genographic Project," *J. Law. Med. & Ethics* 35 (2007): 412–24; Audra Simpson, "The Ruse of Consent and the Anatomy of 'Refusal': Cases from Indigenous North America and Australia," *Postcolon. Stud.* published online 6 June 2017, http://dx.doi.org/10.1080/13688790.2017.1334283 (accessed 20 June 2017); Linda Tuhiwai Smith, *Decolonizing Methodologies: Research and Indigenous Peoples* (1999; rep., London, 2012); Jodi A. Byrd, *The Transit of Empire: Indigenous Critiques of Colonialism* (Minneapolis, 2011).

[7] Marc Prensky, "Digital Natives, Digital Immigrants: Part 1," *On the Horizon* 9 (2001): 1–6.

[8] For instance, what would it mean to understand these spatially distributed territories as "data country," with reference to "Indian country," broadly defined as any of the self-governing Indigenous communities in the United States? See Vine Deloria Jr. and Clifford M. Lytle, "Indian Country," in *American Indians, American Justice* (Austin, Tex., 1983), 58–79. A recent example of an Indigenous community's efforts to produce protocols that are animated by their values is described by Linda Nordling, "San People of Africa Draft Code of Ethics for Researchers," *Science*, 17 March 2017, http://www.sciencemag.org/news/2017/03/san-people-africa-draft-code-ethics-researchers (accessed 21 March 2017). Notably, the San refuse to grant broad consent for other researchers to reuse data for purposes not specified in their original agreements.

[9] This is a feature that emerges from but does not contradict the tripartite definition of Big Data— a function of technology, analysis, and mythology—provided by danah boyd and Kate Crawford, "Critical Questions for Big Data: Provocations for a Cultural, Technological, and Scholarly Phenomenon," *Inform. Comm. & Soc.* 15 (2012): 662–79, on 663.

comes from living, breathing people.[10] Not unlike recent conceptualizations of diabetes as a problem of food justice, wherein metabolic problems arise from decades of alienation from traditional food ways and land use, the history of a particular data set known as the Pima Indian Diabetes Dataset (often referred to as PIDD) makes political and economic subjectivity visible in an algorithmic age sustained by a steady diet of repurposed data. In doing so, it provides an approach—grounded in Indigenous practices of refusal as well as self-governance—for resisting or differently engaging with research in an age of Big Data.

INVISIBLE LABOR

A medical and Indigenous history of Big Data is, by necessity, one of invisible labor, in which the freedom or autonomy of participation of persons from whom data is generated is too often taken for granted or even sometimes celebrated as a form of "citizen science."[11] In the realm of computing, scholars have had to develop new methods to even identify let alone understand the vast and hidden human labor force that has become responsible for all kinds of tasks that enable our information infrastructures.[12] They are trying to access a realm that Ivan Illich called "shadow work": that which is functionally necessary to maintain institutions but is either not compensated or undercompensated.[13]

Sociologists Susan Leigh Star and Anselm Strauss demonstrated the utility of Illich's ideas in the realm of the hospital, where nurses—a traditionally feminized form of labor—struggled to "change work previously embedded under a general rubric of 'care' and usually taken-for-granted to work that is legitimate, individuated and traceable across settings."[14] The point of their research agenda was to make clear the often unseen—and uncompensated—social energy that was nonetheless necessary to sustain institutions.

More recently, these ideas have been applied to the role of individuals who engage with health care as patients and participants in biomedical research. Bioeconomic theorists Melinda Cooper and Cathy Waldby consider these individuals to be performing "clinical labor."[15] Cooper and Waldby begin with the premise that forms of "*in vivo* labor (either through the production of experimental data or the transfer of tissues) are

---

[10] See Malte Ziewitz, "Governing Algorithms: Myth, Mess, and Methods," *Sci. Tech. Hum. Val.* 41 (2016): 3–16, and other papers in that issue.

[11] The ambiguous valences and forms of recognition in recent citizen science efforts are described by Etienne Benson, "A Centrifuge of Calculation: Managing Data and Enthusiasm in Early Twentieth-Century Bird Banding," in this volume.

[12] Kavita Philip, Lilly Irani, and Paul Dourish, "Postcolonial Computing: A Tactical Survey," *Sci. Tech. Hum. Val.* 37 (2010), 3–29; Lilly Irani and M. Six Silberman, "Turkopticon: Interrupting Worker Invisibility in Amazon Mechanical Turk," in *CHI '13: Proceedings of the SIGCHI Conference on Human Factors in Computing Systems* (Paris, 2013), 611–20.

[13] Illich wrote, in 1981, "While for wage labor you apply and qualify, to shadow work you are born or are diagnosed for." See Ivan Illich, *Shadow Work* (Boston, 1981), 100. My interest in Illich's concept of shadow work is directly inspired by Laura Stark's history of studies involving "normal" human subjects at the NIH; Stark, *The Normals: A People's History of the Human Experiment* (Chicago, forthcoming).

[14] Susan Leigh Star and Anselm Strauss, "Layers of Silence, Arenas of Voice: The Ecology of Visible and Invisible Work," *Comput. Support. Cooperat. Work* 8 (1999): 9–30, on 15.

[15] Melinda Cooper and Catherine Waldby, *Clinical Labor: Tissue Donors and Research Subjects in the Global Bioeconomy* (Durham, N.C., 2014).

increasingly central" to late capitalist economies. They further argue that professional bioethical practices have played an ironic role in placing research on human subjects "under an exceptional regime of labor" that allows them to have been understood as exempt from the standard protections of twentieth-century labor law. The resulting practice of viewing participants in biomedical research as "volunteers"—part of the bioethical insistence that biological labor should not be waged—has, in their view, only served to facilitate inadequate forms of compensation. Irrespective of the forms of compensation participants receive—such as health care itself—they argue that "services should be considered as labor when the activity is intrinsic to the process of valorization of a particular bioeconomic sector, and when the therapeutic benefits to the participants and their communities are . . . incidental."[16]

Whether it is through participation in longitudinal diabetes research such as that which created the PIDD or more recent direct-to-consumer genomic services that offer information about genetic risk and ancestry to individuals in exchange for access to their genome, many people—Indigenous and otherwise—also participate in unintentional shadow work when they use Google, Facebook, and other ostensibly "free" services. It is shadow work because this freely given "digital exhaust"—the mundane evidence of how people live and breathe on the web—becomes part of what allows these platforms to be valued in the billions of dollars on the stock market.[17] "It is on . . . 'invisibilities' that the collective delusions and collusions of the modern economy run," historian of science Rebecca Lemov has argued, "particularly as that economy merges with the virtual realm."[18] Tracing the itineraries of Big Data derived from medical information about Indigenous bodies—and, crucially, not the knowledge of Indigenous persons—reveals commonalities and settler colonial pathologies that exist across genres of invisible labor, in biomedical research as well as machine learning. Those who make the investments in repositories of data, including their creation and maintenance, are not always those who help conceptualize their design or benefit from their use.[19]

If, as information historian Geoff Bowker has argued, "Data is always already cooked," the circumstances of the initial preparation of Pima diabetes data have made it a newly valuable ingredient to be used for different kinds of meals, even as Akimel O'odham persons have been effaced in the process. By mixing medical and Indigenous history with the history of Big Data, in which metaphors of consumption abound, I reckon with the following questions: What kind of embodied or clinical labor is involved in generating the ingredients? What kind of effort is recognized or is legible as having been required to cook them? Are these ingredients ones that may ultimately lead to serious health problems? Is there a point at which data has been processed so much that they become inedible or unhealthful? And, perhaps most important of all, who gets to be at the table when these meals are served?

[16] Ibid., 4–6.

[17] Jaron Lanier, *Who Owns the Future* (New York, 2014); Christian Rudder, *Dataclysm: Who We Are When We Think No One's Looking* (New York, 2014); Chris Anderson, *The Long Tail: Why the Future of Business Is Selling Less of More* (New York, 2006). See also Dan Bouk, "The History and Political Economy of Personal Data over the Last Two Centuries in Three Acts," in this volume.

[18] Rebecca Lemov, "On Not Being There: The Data Driven Body at Work and at Play," *Hedgehog Rev.* 17 (2015), 44–55. See also Lemov, "Anthropology's Most Documented Man, Ca. 1947: A Prefiguration of Big Data from the Big Social Science Era," in this volume.

[19] Christine L. Borgman, *Big Data, Little Data, No Data* (Cambridge, Mass., 2015), 229. See also Benson, "Centrifuge" (cit. n. 11).

Through this historical case study of data about diabetes and a discussion of its limitations, I am arguing for an alternative approach to "data ethics"—different from one that might be modeled on that of the normative bioethics critiqued by scholars like Waldby and Cooper. I ask instead for critical engagement with the metabolism of Big Data grounded in awareness of settler colonial history and lived experience that can unsettle approaches to research and forms of compensation that challenge the capitalist and colonialist logics that have animated biomedicine and Big Data alike. Before it is possible to determine best practices for the appropriate use, reuse, and maintenance of data, it is imperative to understand how data comes into being.

THE GILA RIVER COMMUNITY RESERVATION: AN UNNATURAL LABORATORY

Akimel O'odham refer to their ancestors as HuHuKam, known for their engineering prowess in supporting an agricultural lifestyle. For centuries they built and maintained miles of irrigation infrastructure (canals that were 10 feet deep and as much as 30 feet wide) drawing on the Gila River.[20] The annexation of the Arizona Territory to the United States following the Mexican-American War in the mid-nineteenth century coincided with the California gold rush, which brought thousands of prospectors through the region, greatly disrupting local life. The Pima Gila River Indian Community was established during this time, in 1859. Encompassing 372,000 acres along the Gila River, it was the first reservation in what is now Arizona (which did not achieve statehood until 1912). The Pima Gila River Indian Community was also one of four federally recognized tribes at the time (today there are 566).[21] The 1870s and 1880s marked a downturn in the agricultural prosperity of the tribe when the construction of upstream diversion structures and dams by non-Native farmers cut Pima off from the water of the Gila River with devastating consequences.[22]

Anthropologist Frank Russell came to Gila River in 1901 as part of a study authorized by the Bureau of American Ethnology (BAE). The reservation had already been in place for several decades and offered Russell the convenience of an apparently bounded field of ethnographic study. However, as museum curator Joshua Roffler has argued, Russell's book, *The Pima*, published posthumously in 1908 (Russell died of tuberculosis in 1903 at the age of 35), "elided the reality that the lives of the Gila River Pima were deeply intertwined with the non-Indian residents of central Arizona."[23] Even though missionaries and BAE officials urged assimilation (to the extent of granting a horse wagon to each man who cut his hair and built an adobe house), modern references were cropped from the already deliberately posed photos in *The Pima*. It was a work of salvage ethnography that relied upon making certain aspects of lived experience invisible.[24]

[20] David H. DeJong, *Stealing the Gila: The Pima Agricultural Economy and Water Deprivation, 1848–1921* (Tucson, Ariz., 2009).
[21] The consequential differences between tribes that are recognized by the federal government and those that are not are described in M. E. Miller, *Forgotten Tribes: Unrecognized Indians and the Federal Acknowledgement Process* (Lincoln, Neb., 2004).
[22] Ibid.; Cary Walter Meister, *Historical Demography of the Pima and Maricopa Indians of Arizona, 1846–1974* (New York, 1989); David H. DeJong, *Forced to Abandon Our Fields: The 1914 Clay Southworth Gila River Pima Interviews* (Salt Lake City, 2011).
[23] Joshua Roffler, "Frank Russell at Gila River: Constructing an Ethnographic Description," *Kiva* 71 (2006): 373–95.
[24] Jacob Gruber, "Ethnographic Salvage and the Shaping of Anthropology," *Amer. Anthropol.* 72 (1970): 1289–99.

While many community members participated as research assistants to Russell, to the point of effectively conducting ethnography on themselves, they did not have control over the final cultural description that Russell produced. He was aware of this, writing to his boss in 1902, "These people are starving so that I am getting specimens and information that would not otherwise be obtainable."[25] These specimens included farming implements rendered obsolete by the drying of the Gila River.

Around this time, Aleš Hrdlička, considered to be one of the founding leaders of American physical anthropology, reported only one case of diabetes in the population.[26] But over the next forty years, those who lived on the reservation experienced mass famine and starvation, to which the American government responded with canned and processed food assistance. Between 1908 and 1955, the Bureau of Indian Affairs ran a health program, and, when a forty-two-bed hospital opened in 1953, the Indian Health Service (initially the Division of Indian Health, a branch of the U.S. Public Health Service) became responsible for providing comprehensive health care for the eligible residents of the Gila River Reservation.[27]

Tribal members have since argued that the loss of their ability to productively farm the land, combined with the change of diet introduced by government assistance, spurred the rise of obesity and Type 2 diabetes.[28] Such explanations have intermingled with genetic theories, such as James Neel's controversial 1962 "thrifty genotype" hypothesis, which posited that such communities developed genetic adaptations to help them survive during periods of famine.[29] In fact, Neel visited the region as a consultant for the NIH at the beginning of their longitudinal study in 1965. He claimed that "certain problems relevant to Indian health can be much more easily approached on the reservation than in, e.g., Phoenix or Tucson," including questions of genetics.[30] Critics have more recently suggested that biomedical researchers' emphasis on genetically determinist explanations have led to a sense of "fatalism" among some members of the community, who report feeling that diabetes is an unavoidable component of their identity.[31]

What has come to be defined by biomedical scientists as "cooperative" research between community members at Gila River and the NIH began formally in 1963. This

[25] Quoted in Roffler, "Frank Russell" (cit. n. 23), 388.

[26] Aleš Hrdlička, *Physiological and Medical Observations among the Indians of the Southwestern United States and Northern Mexico*, Bureau of American Ethnology Bulletin 24 (Washington, D.C., 1908).

[27] Stephen J. Kunitz, "The History and Politics of US Health Care Policy for American Indians and Alaskan Natives," *Amer. J. Public Health* 86 (1996): 1464–73; Abraham B. Bergman, David C. Grossman, Angela M. Erdrich, John G. Todd, and Ralph Forquera, "A Political History of the Indian Health Service," *Millbank Memorial Fund Quart.* 77 (1999): 571–604.

[28] Smith-Morris, *Diabetes* (cit. n. 3); Melanie Rock, "Classifying Diabetes; or, Commensurating Bodies of Unequal Experience," *Public Cult.* 17 (2005): 467–86; Emoke J. E. Szathmary, "Non-Insulin Dependent Diabetes Mellitus among Aboriginal North Americans," *Annu. Rev. Anthropol.* 23 (1994): 457–82.

[29] Puneet Chawla Sahota, "Genetic Histories: Native Americans' Accounts of Being at Risk for Diabetes," *Soc. Stud. Sci.* 42 (2012): 821–42; Margery Fee, "Racializing Narratives: Obesity, Diabetes, and the 'Aboriginal' Thrifty Genotype," *Soc. Sci. & Med.* 62 (2006): 2988–97. Indeed, Neel visited Gila River to see how his hypothesis might apply to the community.

[30] Neel, memo to National Cancer Institute, 8 November 1965, MS Coll. 96, Box 92, Folder "Pima—Tohono O'odham—1965–1968," James V. Neel Papers, American Philosophical Society, Philadelphia.

[31] Diane Weiner, "Interpreting Ideas about Diabetes, Genetics, and Inheritance," in *Medicine Ways: Disease, Health, and Survival among Native Americans*, ed. Clifford E. Trafzer and Diane Weiner (Walnut Creek, Calif., 2001), 108–33.

was when the National Institute of Arthritis, Diabetes and Digestive and Kidney Diseases (now known as the National Institute of Diabetes and Digestive and Kidney Diseases [NIDDK]) made a survey of rheumatoid arthritis among the groups they referred to as "Pima" in Arizona and "Blackfeet" in Montana. Researchers were surprised to find an extremely high rate of diabetes. In 1965, the Epidemiology and Field Studies Branch (EFSB) of the Institute—in partnership with the Indian Health Service—sent a team to begin an observational study of the Pima community at Gila River. The research was supposed to last for ten years. It continued for more than forty.[32]

Beginning in 1965, every resident of the Pima study area (which refers to the Gila River Indian Community) of at least five years of age was asked to participate in a "comprehensive longitudinal study of diabetes," for which they were examined approximately every two years. In 1984, the EFSB became the current Phoenix Epidemiology and Clinical Research Branch (PECRB), which oversaw the study. By the 1990s, the PECRB had come to oversee prevention programs, incorporated in response to criticism that the observational nature of the epidemiological study had not yielded any findings that directly benefited its participants. At the turn of the twenty-first century, despite great advances in knowledge about diabetes more generally, there were no discernible decreases in obesity or diabetes rates among community members.

Writing in the early 1990s, two leaders of the project, Clifton Bogardus and Stephen Lillioja, argued that "the Pima Indian model of this disease [diabetes] affords . . . major advantages," not least of all because "the population is genetically homogenous compared to Caucasian populations, and therefore the causes of NIDDM [non-insulin-dependent diabetes mellitus] are less heterogeneous, simplifying linkage studies."[33] In other words, the legal and political boundaries of the reservation functioned to enclose the Pima, making them appear as a natural and perhaps captive laboratory in which to study diabetes.[34]

A government pamphlet from 1996, "The Pima Indians: Pathfinders for Health," quoted Bogardus, who reported that "NIDDK scientists . . . have studied well over 90 percent of the people on the reservation at least once."[35] This same document cast Akimel O'odham as magnanimous donors of biomedical knowledge:

[32] National Institute of Diabetes and Digestive and Kidney Diseases, "Prospective Studies of the Natural History of Diabetes Mellitus and Its Complications in the Gila River Indian Community," ClinicalTrials.gov, National Institutes of Health, https://clinicaltrials.gov/ct2/show/NCT00339482 (accessed 21 March 2017).

[33] Clifton Bogardus and Stephen Lillioja, "Pima Indians as a Model to Study the Genetics of NIDDM," *J. Cell. Biochem.* 48 (1992): 337–43.

[34] Historian Matthew Klingle is undertaking an environmental history of diabetes, including work at Gila River. This research, like my own, is indebted to a large body of literature on efforts to transform indigenous communities into living laboratories. See, e.g., Helen Tilley, *Africa as a Living Laboratory: Empire, Development, and the Problem of Scientific Knowledge, 1870–1950* (Chicago, 2011); Warwick Anderson, "The Colonial Medicine of Settler States: Comparing Histories of Indigenous Health," *Health Hist.* 9 (2007): 144–54; Roffler, "Frank Russell" (cit. n. 23); David S. Jones, "The Health Care Experiments at Many Farms: The Navajo, Tuberculosis, and the Limits of Modern Medicine, 1952–1962," *Bull. Hist. Med.* 76 (2002): 749–90; Christian W. McMillen, *Discovering Tuberculosis: A Global History 1900 to the Present* (New Haven, Conn., 2015). On the racialized history of diabetes more broadly, see Arleen Tuchman, "Diabetes and Race: A Historical Perspective," *Amer. J. Public Health* 10 (2011): 24–33; Tuchman, "Diabetes and 'Defective' Genes in the Twentieth Century United States," *J. Hist. Med. Allied Sci.* 70 (2015): 1–33.

[35] National Institute of Diabetes and Digestive and Kidney Diseases, *The Pima Indians: Pathfinders for Health* (Bethesda, Md., 1996), 7.

> Once trusted scouts for the US Cavalry, the Pima Indians are pathfinders for health. . . . The Pima Indians are giving a great gift to the world by continuing to volunteer for research studies. Their generosity contributes to better health for all people, and we are all in their debt. . . . The Pima Indians' help is so important . . . because of the uniqueness of the community. There are few like it in the world.[36]

As William Knowler, an NIH researcher at the reservation since 1975 who is also recognized as one of the world's 250 most highly cited researchers in clinical medicine, biology, and biochemistry, testified before Congress, "This study has contributed much to the world's current understanding of the causes and consequences of Type 2 diabetes and its complications, for which we are indebted to this community."[37] The year that "Pathfinders for Health" was published was the first one in which the NIH funded a large-scale prevention program involving Akimel O'odham participants, raising concern that the publication may have been produced to help assuage local feelings of exploitation, that they had been guinea pigs for the benefit of others.[38]

In the 1990s, the issue of compensation was addressed by ensuring that those who participated in the study continued to receive medical care and also $50, free transportation, and a meal each time they received diagnostic testing.[39] Before considering the significance of this form of compensation, I want to pause this epidemiological history to shift attention to a quite different area of expertise that was emerging in parallel, that of machine learning.

MACHINE LEARNING

Machine learning is a subdiscipline of artificial intelligence that practitioners date back to the late 1950s. It focuses on algorithms capable of learning and/or adapting their parameters based on a set of observed data without having been programmed to do so. An algorithm is, in its simplest form, a set of instructions or a code. It is a form of "software" that organizes data to generate meaningful information. It is what allows computers to do computational work.[40] As media theorist Tarleton Gillespie has explained,

> Algorithms are inert, meaningless machines until paired with databases upon which to function. For us, algorithms and databases are conceptually conjoined . . . but before results can be algorithmically provided, information must be collected and readied for the algorithm and sometimes excluded or demoted.[41]

[36] Ibid., 5.

[37] Knowler testimony before Committee on Indian Affairs, U.S. Senate, 8 February 2007, https://olpa.od.nih.gov/hearings/110/session1/testimonies/diabetesAI.asp (accessed 21 March 2017).

[38] Rachel Winer, "Diabetes, the National Institutes of Health, and the Arizona Pima Indians: A Study of Ethics and Experimentation in American Medicine, 1965 to the Present" (senior thesis, Yale Univ., 2006). In her thesis, Winer also analyzed a three-part exposé on the diabetes research undertaken at Gila River; Graciela Sevilla, "A People in Peril: Pimas on the Front Lines of an Epidemic," *Arizona Republic*, 31 October–2 November 1999.

[39] As observed by Winer, "Diabetes" (cit. n. 38).

[40] David Berlinski, *The Advent of the Algorithm: The 300-Year Journey from an Idea to the Computer* (New York, 2000).

[41] Tarleton Gillespie, "The Relevance of Algorithms," in *Media Technologies*, ed. Tarleton Gillespie, Pablo J. Boczkowski, and Kirsten Foote (Cambridge, Mass., 2013), 167–94, on 169. A more cynical view is offered in Cathy O'Neill, *Weapons of Math Destruction: How Big Data Increases Inequality and Threatens Democracy* (New York, 2016).

Machine learning theorists are concerned with issues such as computational complexity, computability, and generalization. Algorithms are the coin of their realm; data is used to refine them. The field, in ways that are described by Hallam Stevens in this volume, is a marriage of applied math and computer science.[42]

Machine learning and the related field of statistical pattern recognition have been the subject of increasing interest to the biomedical community because they offer the promise of improving the sensitivity and specificity of detection and diagnosis of disease, while at the same time purportedly increasing the objectivity of the decision-making process. According to an early editorial published in the *Journal of Machine Learning*, "unlike psychology, machine learning is fortunate in that it can experimentally study the relative effects of 'nature' versus 'nurture.'"[43] The author, Pat Langley, based at the University of California, Irvine (UCI), believed that by looking at unprecedentedly large amounts of supposedly raw data, machine learning could avoid human biases.[44] He looked forward to the day when standardized databases would facilitate such studies. And it was at UCI that an important resource was established to help achieve this goal.

The UCI Machine Learning Repository is, in the words of its stewards, "a collection of databases, domain theories, and data generators that are used by the machine learning community for the empirical analysis of machine learning algorithms."[45] It is an archive of data sets, created in 1987 by David Aha and several other graduate students at UCI, including Jeff Schlimmer and Doug Fisher. According to Aha, it began after he had heard several calls for such a repository at machine learning conferences, but nothing suitable materialized. "I became convinced it would not exist without some reasonable number of datasets," he recalled. "I requested all those I had read about in publications at that time and from other sources, and I believe I announced it only after it had 25 (a number I only vaguely recall; the actual number may have differed) mostly well-known datasets. . . . It was pre-web, available by ftp [File Transfer Protocol]. I was gratified by the community's interest, and Pat Langley (a mentor) [and the author of the editorial referenced above] was particularly encouraging and no doubt suggested others to contact me proactively, which many folks did."[46]

Among the major issues initially raised by the repository were those of attribution and ownership. Aha recalled that he "realized that UCI needed credit for this effort, and I broadly requested and publicized a reference format to cite the repository. . . . I reviewed many papers over the years where I requested this to be inserted, or contacted folks after they had published their paper to remind them to include its citation in their future papers, when appropriate."[47] Yet, as information theorist Christine Borgman has recently argued, such citation practices "largely presume that objects

---

[42] Paul Sajda, "Machine Learning for Detection and Diagnosis of Disease," *Annu. Rev. Biomed. Eng.* 8 (2006): 537–65. See also Hallam Stevens, "A Feeling for the Algorithm: Working Knowledge and Big Data in Biology," in this volume.

[43] Pat Langley, "Machine Learning as an Experimental Science," *J. Machine Learning* 3 (1988): 5–8, on 6.

[44] A form of "mechanical objectivity." See Lorraine Daston and Peter Louis Galison, *Objectivity* (New York, 2007).

[45] UCI Machine Learning Repository, "About," http://archive.ics.uci.edu/ml/about.html (accessed 21 March 2017).

[46] David Aha, UCI Repository History, as told to Padhraic Murphy, 21 June 2009, transcript of e-mail correspondence from the personal files of David Aha (used with his permission).

[47] Ibid.

are fixed, stable, and complete units. None of these conditions can be assumed with data."[48] The radical instability and alienability of this kind of data, as we will see, would become central to its power and also its critique.

Issues of accuracy in citing data sets were also of concern, especially for data sets that dealt with medical issues. "Some of the databases were donated only on the condition that their use was cited accurately," Aha explained, "as was especially the case for medical datasets I obtained. . . . Most of the others had no such condition."[49] One such medical resource is the "Heart Disease Data Set" from the Cleveland Clinic, which became part of the repository in 1988. In this case, Aha noted that those researchers who donated the data were primarily concerned with attribution—making certain that their work and funders were credited properly—not about compensating patients, whose data had been made anonymous and was therefore regarded as protected.

By the first decade of the twenty-first century, the UCI Machine Learning Repository had been cited over 1,000 times, making it on its own one of the top 100 most cited "papers" in all of computer science.[50] It is worth noting that Aha did not list it as a publication on his CV, making his own connection to the repository invisible, at least in a formal capacity. He did, however, acknowledge that the labor of encouraging people to donate their data sets—itself a kind of shadow work—actually contributed to raising his profile in the field of machine learning.[51]

Still known as the UCI Machine Learning Repository, it is now maintained by the University of Massachusetts, Amherst, with financial support from the National Science Foundation. The home page for the repository thanks "the donors and creators of the databases and data generators," which refers to data scientists, not research subjects.[52] Aha's sense is that, in recent years, the field has shifted in ways that have allowed other kinds of repositories to become more important, but the "discretized" data in the UCI repository make it appropriate for certain, specific research and also a resource for those new to the field of machine learning.[53]

One of the oldest data sets on file is the "Pima Indians Diabetes Data Set," which was donated in 1990. The data set is often referred to by its initials, PIDD, and "has become a standard for testing data mining algorithms to see their accuracy in predicting diabetic status from the 8 variables given."[54] These variables, or "attribute information," were extracted from paper patient records, not any kind of DNA-based data, and included (1) number of pregnancies, (2) plasma glucose concentration after two hours in an oral glucose tolerance test, (3) diastolic blood pressure, (4) triceps skin fold thickness, (5) two-hour serum insulin level, (6) body mass index, (7) diabetes pedigree function, and (8) age.

In a pregenomic era, the diabetes pedigree function (DPF) provided a synthesis of the diabetes history of relatives (including parents, grandparents, full and half siblings,

[48] Borgman, *Big Data* (cit. n. 19), 242.
[49] Aha, UCI Repository History (cit. n. 46).
[50] Google Scholar reports a much higher number, closer to 6,000, but many of those are redundancies.
[51] David Aha, telephone interview by Joanna Radin, 30 June 2015.
[52] UCI Machine Learning Repository, "About" (cit. n. 45).
[53] See, e.g., machinelearnringmaster.com/case-study-predicting-the-onset-of-diabetes-within-five -years-part-1-of-3/, 29 March 2014 (accessed 15 July 2015).
[54] Joseph L. Breault, "Data Mining Diabetic Databases: Are Rough Sets a Useful Addition?," in *Proceedings of the 33rd Symposium on the Interface*, ed. Edward J. Wegman, D. T. Gantz, and J. J. Miller (Fairfax Station, Va., 2002), 597–606.

full and half aunts and uncles, and first cousins) and the genetic relationship of those relatives to the subject. It has been described in academic publications as providing "a measure of the expected genetic influence of affected and unaffected relatives on the subject's eventual diabetes risk."[55] Effectively, the DPF was a technology that functioned to devalue Akimel O'ohdam understandings of kinship—which include familial relationships across the community as well as with nonhuman animals and landscapes—folding biogenetic information into the data set, enabling researchers to extrapolate—in the absence of DNA—patterns of heredity.[56] It also folded in assumptions about life course and gender in that "all patients [768 in total] were females at least 21 years old of Pima Indian heritage."

Experts in the field of machine learning pay little attention to these particular attributes but have observed that "diabetes is a particularly opportune disease for data mining technology. . . . First, because the mountain of data is there. Second, diabetes is a common disease that costs a great deal of money, and so has attracted managers and payers in the never ending quest for saving money and cost efficiency. Third, diabetes is a disease that can produce terrible complications . . . so physicians and regulators would like to know how to improve outcomes as much as possible. Data mining might prove an ideal match in these circumstances."[57]

Let us take a closer look at the PIDD as it appears on the UCI Machine Learning Repository (fig. 1). The donor of the data is listed as Vincent Sigillito, who was part of the Applied Physics Laboratory at Johns Hopkins. The Applied Physics Laboratory (APL) is a not-for-profit engineering research and development center, founded in 1942 to assist the military with ballistics detonation. According to the APL's website, its mission is to "provide solutions to national security and scientific challenges with systems engineering and integration, research and development, and analysis."[58] The APL is located not far from the Johns Hopkins School of Public Health, which was crucial for facilitating interactions between researchers in epidemiology and machine learning. Aha, I learned, did a postdoc at Johns Hopkins University in the early 1990s, and Sigillito was his sponsor. This is how the PIDD wound up in the UCI Machine Learning Repository. The data set was constructed from a larger database by the NIDDK. As of March 2017, it had been viewed nearly 260,000 times.[59]

From the information available on the UCI Machine Learning database it is possible to discern that data in the PIDD was based on research first reported by epidemiologists in 1981.[60] The 1981 paper, published in the *American Journal of Epidemiology*, indicated that diabetes incidence was computed for 3,137 subjects, each of whom had been examined at least twice. Estimates of age- and sex-adjusted incidence and prevalence rates with 95 percent confidence intervals were computed comparing Pima

[55] Jack W. Smith, J. E. Everhart, W. C. Dickson, W. C. Knowler, and R. S. Johannes, "Using the ADAP Learning Algorithm to Forecast the Onset of Diabetes Mellitus," *Johns Hopkins APL Tech. Digest* 10 (1988): 262–6.

[56] Deborah House, "'Know Who You Are and Where You Come From': Ties of Kin, Clan, and Homeland in Southwestern Indian Identity," *Rev. Anthropol.* 33 (2004): 371–91.

[57] Breault, "Data Mining" (cit. n. 54).

[58] Johns Hopkins Applied Physics Laboratory, "About APL," http://www.jhuapl.edu/aboutapl/default .asp (accessed 21 March 2017).

[59] This is up from about 90,000 times, when I began this research in 2015. The exponential uptick in views may track with the explosion of machine learning applications.

[60] William C. Knowler, David J. Pettitt, Peter J. Savage, and Peter H. Bennett, "Diabetes Incidence in Pima Indians: Contributions of Obesity and Parental Diabetes," *Amer. J. Epidemiol.* 113 (1981): 144–56.

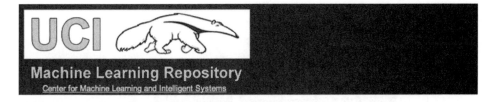

Pima Indians Diabetes Data Set
Download: Data Folder, Data Set Description

Abstract: From National Institute of Diabetes and Digestive and Kidney Diseases; Includes cost data (donated by Peter Turney)

Data Set Characteristics:	Multivariate	Number of Instances:	768	Area:	Life
Attribute Characteristics:	Integer, Real	Number of Attributes:	8	Date Donated	1990-05-09
Associated Tasks:	Classification	Missing Values?	Yes	Number of Web Hits:	96886

Source:

Original Owners:

National Institute of Diabetes and Digestive and Kidney Diseases

Donor of database:

Vincent Sigillito (vgs '@' aplcen.apl.jhu.edu)
Research Center, RMI Group Leader
Applied Physics Laboratory
The Johns Hopkins University
Johns Hopkins Road
Laurel, MD 20707
(301) 953-6231

Figure 1. Screenshot of Pima Indians Diabetes Data Set on UCI Machine Learning Repository, "About" (cit. n. 45).

data to the 1970 U.S. Census of the white population, including armed forces. Incidence rates were also stratified by body mass index. The authors concluded that their findings were "consistent with Neel's hypothesis that diabetes results from the introduction of a steady food supply to people who have evolved a 'thrifty genotype.'"[61]

In 1988, a group of researchers at the NIH and Johns Hopkins, with connections to the Applied Physics Laboratory, sought to further extract and digitize data from this resource and apply it to the machine-learning context.[62] The goal of that paper, published in a very different forum—the *Johns Hopkins APL Technical Digest*—was to test the ability of an early neural network model algorithm called ADAP (proclaimed to be "loosely" modeled after computation in the brain), to forecast the onset of diabetes in the high-risk population of Pima Indians. This required transforming the original, manually recorded data set into one that could be entered into a computational machine (fig. 2).[63]

---

[61] Ibid., 145. For an analysis of the remarkable endurance of the "thrifty genotype hypothesis," see Isabel Beshar, "A Tale of a Mutating Theory: The Evolution of the Thrifty Genotype Hypothesis from 1962–2007" (senior thesis, Yale Univ., 2014).

[62] Smith et al., "Using the ADAP Learning Algorithm" (cit. n. 55).

[63] The digitization step was essentially punch card technology. See Martin Campbell-Kelly and William Aspray, *Computer: A History of the Information Machine*, 1st ed. (New York, 1996). The early history of efforts to use computers to model living systems has been told in Timothy Lenoir, "Shaping

FIXED MATRIX & VARIABLE ARRAY

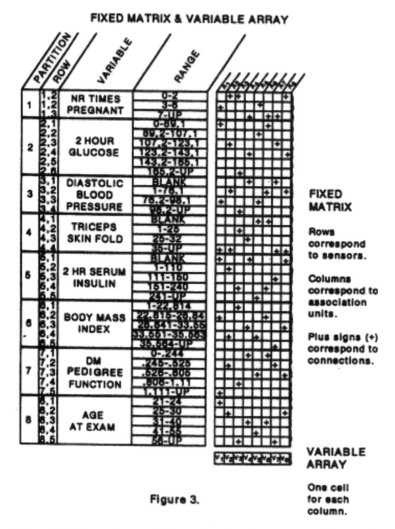

Figure 3.

Figure 2. Example of data reformatted for machine-based computation from Smith et al., "Using the ADAP Learning Algorithm" (cit. n. 55).

The authors explained that neural network models worked by using a "training" data set to discover patterns in data. Once the algorithm had been trained using 576 cases, it was used to forecast whether another 192 test cases would develop diabetes within five years. In the materials and methods section of the paper, the authors presented a four-part justification for privileging the study population. First and foremost, they argued that the fact that the Pima population could be "recognized as such due to its location on the reservation"—its ability to function as a natural laboratory— was crucial to the validity of the data set. Related was the matter of the longitudinal

Biomedicine as an Information Science," in *Proceedings of the 1998 Conference on the History and Heritage of Science Informaton Systems*, ed. Mary Ellen Bowden, Trudi Bellardo Hahn, and Robert V. Williams (Medford, N.J., 1999); Joseph November, *Biomedical Computing: Digitizing Life in the United States* (Baltimore, 2012).

aspects of the research, that Pima had been the subject of "continuous study since 1965 by NIHDDK." The third justification focused on the WHO's standardization of criteria for diagnosing diabetes (based on a Technical Report of a Study Group, TRS 727, 1985; a prior report was published in 1980, TRS 646).[64] The final source of legitimation was the prior ubiquity of the data set itself, which was already known to the investigators and served as a "well-validated" resource.[65]

REUSING AND REFUSING

Through its availability on the UCI Machine Learning Repository, the PIDD became a standard resource for testing algorithms of all kinds. The data set, by virtue of the fact that it was and continues to be freely available, has also been used to refine algorithms intended for "knowledge discoveries" that have nothing do with diabetes, including the prediction of manhole fires.[66] To David Aha, what made PIDD such a valuable data set in the world of machine learning beyond the realm of diabetes research was that it was data on a topic that seemed important, but, perhaps more relevant to its long life in the machine learning community, it was not overly large: "people could work with it. It was all tables and this made it easier to be used with the algorithms that were being created at the time." Furthermore, the attributes were straightforward, which made it "amenable from a number of perspectives." Even though there were some missing values, this became part of its appeal or a "feature" of the data set.[67] It was not too big or too small. Even what it did not include came to be exemplary, appearing to replicate the contingency and complexity of real-world situations.

Research on the prediction of manhole fires—which used algorithms developed with the use of PIDD—featured prominently in Viktor Mayer-Schonberg and Kevin Cukier's popular 2013 exposition, *Big Data*. In their book, they invoke the reuse potential of Big Data as one of its defining and most valuable features. Among the broader lessons they draw: (1) Big Data is being put to new uses to solve difficult real-world problems; (2) to achieve this, however, we need to change how we make knowledge; (3) we have to use all the data, as much as can possibly be collected, not just a small portion; (4) we need to accept messiness rather than treat exactitude as a central priority; and (5) we must put our trust in correlations without fully knowing the causal basis for predictions.[68]

The history that I have provided thus far highlights other equally consequential features of Big Data to which Mayer-Schoenberg and Cukier were not attuned. Chief among them is that today's machine learning scientists have been disincentivized from

---

[64] On the WHO's technical reports in shaping research agendas, see Joanna Radin, "Latent Life: Concepts and Practices of Human Tissue Preservation in the International Biological Program," *Soc. Stud. Sci.* 43 (2013): 483–508. The standards disseminated by such documents play an important role in the construction of disease categories. See Geoffrey C. Bowker and Susan Leigh Star, *Sorting Things Out: Classification and Its Consequences*, Inside Technology (Cambridge, Mass., 1999).

[65] Smith et al., "Using the ADAP Learning Algorithm" (cit. n. 55).

[66] Krzysztof J. Cios and Witold Pedrycz, *Data Mining Methods for Knowledge Discovery* (Boston, 1998); Jiawei Han and Micheline Kamber, *Data Mining: Concepts and Techniques* (San Francisco, 2001). On the subject of "open access," see Sabina Leonelli, "Why the Current Insistence on Open Access to Scientific Data? Big Data, Knowledge Production, and the Political Economy of Contemporary Biology," *Bull. Sci. Tech. Soc.* 33 (2013): 6–11.

[67] Aha, interview (cit. n. 51).

[68] Viktor Mayer-Schonberger and Kenneth Cukier, *Big Data: A Revolution That Will Transform How We Live, Work and Think* (New York, 2013), 70.

considering the meaning or origins of the data they feed to their algorithms. This may be a function of "algorithmic objectivity," upheld by members of the data mining and machine learning communities as "fundamental to the maintenance of these tools as legitimate brokers of relevant knowledge."[69] The fact that this data is also considered to be naturally occurring—a neutral product of the contingent circumstances of its acquisition—is seen as one of its additional advantageous qualities, even if the data itself is—as in the case of the PIDD—a product of settler colonialism, economic struggle, and biosocial suffering.[70] While it may be tempting to consider the PIDD—the data set itself—as a kind of model organism for machine learning, doing so closes off access to the shadow work of people, Akimel O'odham, who continue to live and die on the reservation and to circulate as disembodied data, stored on the servers of universities and corporations.[71] It also elides the work Akimel O'odham have undertaken to redefine the norms of research encounters with their community that are in keeping with values not grounded in capitalist or settler colonial logics.[72]

As Pima data moved off the reservation, it became available for new and unexpected uses in basic informatics research. As a result, Akimel O'odham lost "direct control over intervention in and treatment of serious diseases affecting their populations," as Richard Narcia, Lieutenant Governor of the Gila River Community, testified before Congress.[73] How do members of the Akimel O'odham community at Gila River, in particular, feel about their loss of control over data derived from studies of the serious diseases affecting them? I have not been able to speak with a representative of the community. The community has recently put into place strict procedures restricting researcher access. Even with a reference from a trusted colleague, the head of the Gila River Indian Community's Committee on Health and Social Standing, which controls IRB permission, did not respond to my inquiries. I also experienced this lack of response from non-Indigenous health workers who have long-term relationships with the community.

A decade ago, Rachel Winer, then an undergraduate at Yale, was able to conduct interviews with individuals from both of these groups for her Yale senior thesis project.[74] In 2015, they chose not to engage with my requests for interviews; it was their right, in accordance with a Health Care Research Ordinance enacted by the Gila River Indian Community in 2009.[75] The ordinance states that "the Community Council has found that Medical and Health Care Research has been conducted in ways that do not respect the human dignity of human subjects and that do not recognize the legitimate interest of the Community in the integrity and preservation of its culture."[76]

---

[69] Gillespie, "Relevance" (cit. n. 41).

[70] On the politics of biosocial suffering, see Duana Fullwiley, *The Encultured Gene: Sickle Cell Health Politics and Biological Difference in West Africa* (Princeton, N.J., 2011).

[71] Nathan Ensmenger, "Is Chess the Drosophila of Artificial Intelligence? A Social History of an Algorithm," *Soc. Stud. Sci.* 42 (2011): 5–30.

[72] See, e.g., Kim TallBear, "Beyond the Life/Not Life Binary: A Feminist-Indigenous Reading of Cryopreservation, Interspecies Thinking and the New Materialisms," in *Cryopolitics: Freezing Life in a Melting World*, ed. Joanna Radin and Emma Kowal (Cambridge, 2017), 179–202.

[73] Richard Narcia, Testimony of the Gila River Indian Community before the Senate Committee on Indian Affairs, Washington, D.C., 8 March 2000. Cited in Winer, "Diabetes" (cit. n. 38).

[74] Winer, "Diabetes" (cit. n. 38).

[75] Gila River Indian Community, "Ordinance GR-05-09," http://nptao.arizona.edu/sites/nptao/files/gila_river_indian_ord_gr-05-09_title_17_chapter_9_0.pdf (accessed 21 March 2017).

[76] Ibid. Section 9.107 includes a discussion of the information to be provided to the review board by prospective researchers, including "Who shall own the data from Medical and Health Care Research?"

In addition to specific circumstances leading to the ordinance, in part associated with reuse of tissues of members of other Indigenous communities without consent, research relationships are not stable over time and involve shifting priorities, allegiances, and desires. One facet of this is a phenomenon known as "research fatigue"—a feeling of exhaustion about being subjects of inquiry.[77] This position can grow out of a desire to avoid what has been referred to as "voyeurism," where researchers seek to learn lessons from the study of communities but fail to improve conditions or consider the points of view of community members.[78]

This is a perspective that I respect. However, I am also mindful of the risks of erasure of the role of Indigenous peoples from studies of existing and emerging technoscientific infrastructures.[79] Part of what it means to write the history of Big Data is to be attentive not only to the voices that have been silenced but also to those who have chosen not to speak, or to speak at this particular moment in time. This silence might be understood in terms of "refusal," a position that has been identified and explored by anthropologist Audra Simpson in the context of her research on Mohawk citizenship and sovereignty.[80] Simpson recognized the political dimensions of moments when her desired subjects resisted her inquiries. "Rather than stops, or impediments to knowing," she realized, "those limits may be expansive in what they do not tell us. . . . The refusals speak volumes because they tell us when to stop."[81]

Simpson's decision to embrace her ethnographic subjects' right not to give her the information she initially believed she needed to obtain appears to run counter to the maxim, upheld by digital entrepreneurs like Mark Zuckerberg, that "information wants to be free." The other, less frequently cited coda to that maxim is that information "also wants to be expensive," meaning that it can be hugely valuable to the recipient, though not necessarily to the donor.[82]

Simpson is aware of the apparent perversity of refusal in a Euro-American culture that values openness. However, she notes that the political theory of John Locke, which

and "What control will the individual medical and Health Care Research participants have over the use of their own data? What control will the Community or Medical and Health Care Research participants have over the current and future use of the data, and how will the control be exercised?" This does not include oversight of social science research. As documented in Naomi Tom, "Protecting Tribal Nations through Community Controlled Research: An Analysis of Established Research Protocols within Arizona Tribes" (MS thesis, Arizona State Univ., May 2015), https://repository.asu.edu/attachments/150598/content/Tom_asu_0010N_14771.pdf (accessed 1 October 2016).

[77] Tom Clark, "'We're Over-Researched Here!' Exploring Accounts of Research Fatigue within Qualitative Research Engagements," *Sociology* 42 (2008): 953–70. On uneven partnerships in global health research that can contribute to research fatigue, see Johanna Crane, "Unequal 'Partners': AIDS, Academia and the Rise of Global Health," *Behemoth* 3 (2010): 78–97.

[78] Lauri Gilchrist, "Aboriginal Communities and Social Science Research: Voyeurism in Transition," *Native Soc. Work J.* 1 (1997): 69–85, https://zone.biblio.laurentian.ca/bitstream/10219/472/1/NSWJ-V1-art6-p69-85.pdf (accessed 1 October 2016).

[79] The latent racism that has animated our information infrastructures can be interrupted by recognizing the contributions of Indigenous people. See, e.g., Lisa Nakamura, "Indigenous Circuits: Navajo Women and the Racialization of Early Electronic Manufacture," *Amer. Quart.* 66 (2014): 919–41.

[80] Audra Simpson, "On Ethnographic Refusal: Indigeneity, 'Voice,' and Colonial Citizenship," *Junctures* 9 (2007): 67–80; Simpson, *Mohawk Interruptus: Political Life across the Borders of Settler States* (Durham, N.C., 2014). The concept of refusal was first articulated within the context of anthropology by Sherry B. Ortner, "Resistance and the Problem of Ethnographic Refusal," *Comp. Stud. Soc. Hist.* 37 (1995): 173–93.

[81] Simpson, "On Ethnographic Refusal" (cit. n. 80), 78.

[82] Stewart Brand stated this in 1984 at the first Hackers' Conference. It was printed in the May 1985 *Whole Earth Review*, and again in 1987 in his book, *The Media Lab: Inventing the Future at MIT* (New York, 1987).

promotes shared standards of justice and truth in an intellectual commons, was derived from the violent enclosure of land and alienation of Indigenous peoples from their modes of governance.[83] It is a philosophy of property that has made it such that Indigenous communities' own laws of dominion over their bodies, including knowledge about those bodies, are not legible in settler colonial regimes of power.[84]

In Euro-American history, reaching all the way back to Montaigne and Rousseau, the Indigenous subject has often been situated as someone who is naturalized—taken as being closer to nature and outside of history, even as their existence forms the basis for modern political thought.[85] Naturalness is also a component of what makes Big Data valued by those who consume it; it is supposedly harvested from people going about their daily lives, in the "real world," allowing scientists to learn how people behave and think *in vivo virtual*. The activity of daily life releases exhaust or hidden data into a virtual territory idealized as a commons, yet it is those who have defined the digital commons who are setting the terms upon which it can be valued. Data exhaust is made of the digital traces that are silently, or invisibly, accumulated and given off by the devices people use every day, like smartphones and computers. Discourses of "naturalness" preempt conversation about appropriate forms of compensation and governance.[86] Perhaps more importantly, they often foreclose the question of whether or not it is ever possible to cease being productive if one is making data that is not even visible to oneself.

Today, Akimel O'odham, in part as a result of their long-term involvement in biomedical research, are deeply skeptical about continuing to participate in additional forms of research, including that which is undertaken by historians.[87] Their skepticism or refusal, and the difficult legacies of other Indigenous communities' involvement with biomedicine, anticipate concerns that are beginning to be articulated about the datafication of all kinds of lives.[88] These issues are implied by the idea of "digital natives," a means of suggesting that Indigenous people themselves have been compelled to generate alternative ways of conceiving of what it means to be a citizen in a digital age, including tactics for resisting its embrace.[89]

---

[83] Simpson, "On Ethnographic Refusal" (cit. n. 80), 74.

[84] Simpson invokes Locke's assertion that "amongst those who are *counted the civilized part of mankind* [emphasis added by Simpson], who have made and multiplied positive laws to determine property . . . is by the labour that removes it out of that common state nature left it in made his property who takes pains about it"; quoted in ibid., 70. "Thus," Simpson concludes, "property could be defined only as that which was mixed with labour and belonged to those who perceived it, in contradistinction to the living histories of Indigenous peoples in those places"; ibid.

[85] Johannes Fabian, *Time and the Other: How Anthropology Makes Its Object* (New York, 1983); Eric Wolf, *Europe and the People without History* (Berkeley and Los Angeles, 1982); Marc Rifkin, *Beyond Settler Time: Temporal Sovereignty and Indigenous Self-Determination* (Durham, N.C., 2017).

[86] See, e.g., William Cronon's arguments about European settlers in colonial America. Cronon, *Changes in the Land: Indians, Colonists, and the Ecology of New England* (New York, 1983).

[87] Anthropologist and STS scholar Puneet Sahota has done important research on Native Americans' perceptions of being at risk for diabetes. Because of concerns about stigmatization, she has anonymized not only the individuals she interviewed but also the tribal community to which they belong. Sahota, "Genetic Histories" (cit. n. 29), 821–42. An alternative approach taken by Kim TallBear "studies up," focusing on the ideas and values of scientists who work with Indigenous peoples. TallBear, *Native American DNA: Tribal Belonging and the False Promise of Genomic Science* (Minneapolis, 2013).

[88] See, e.g., Borgman, *Big Data* (cit. n. 19).

[89] The idea of the "biodefector," one who resists and, in doing so, may even reconfigure expectations of the research encounter, has been advanced in a thoughtful discussion by Ruha Benjamin, "Informed Refusal: Toward a Justice-Based Ethics," *Sci. Tech. Hum. Val.* 41 (2016): 967–90.

I emphasize this because it is not only indigenous studies scholars who have critiqued and sought to redefine the research encounter and the broader political economic regime by which it is bolstered. The implications of the history of the PIDD are intensified and compounded when we consider that even though the UCI Machine Learning Repository was extremely important in the early years of the field of machine learning, it was also sharply criticized within the community. By 1995, Aha recalled, "the problems 'caused' by the repository had become popularly espoused. For example, at . . . [the International Conference on Machine Learning] Lorenza Saitta had [in an invited workshop that Aha co-organized] passionately decried how it allowed researchers to publish dull papers that proposed small variations of existing supervised learning algorithms and reported their small-but-significant incremental performance improvements in comparison studies. But even before this concern became broadly recognized throughout the community (which sadly, implies that the Repository was successful), I recall others could see this coming."[90]

Aha was referring to a paper that would become a widely cited takedown of the "popular practice of exploiting ready-to-use data sets," published by Italian machine learning scientists Lorenza Saitta and Filippo Neri.[91] They interpreted the stakes of relying on what they called "off the shelf data" as impeding a two-way or iterative process of technology transfer. Drawing on their own experiences in developing and applying machine learning systems in fields that included industrial troubleshooting, molecular biology, medicine, industrial robotics, speech recognition, cognitive psychology, and knowledge discovery in large databases, Saitta and Neri invoked experiences "interacting with both domain experts and end-users who displayed different attitudes and different degrees of trust and understanding of the potential of the methodologies we were proposing."[92]

Saitta and Neri explained that "designers of ML [machine learning] systems usually envision a scenario including themselves, data, and possibly, a marginally useful expert, but almost never the user. On the contrary, the user should be a fundamental component of this scenario . . . the user must not only be present, but also actively participate in the development of a 'real-world' application."[93] The conclusions they reached were not radically different from those reached by indigenous studies scholars or historians of science and technology, who make arguments for, respectively, self-determination and the fact that users matter.[94] The case of reuse further complicates these insights, reminding us that what constitutes a "user," let alone a "self," is not stable across time or place, even digital ones.

From the perspective of a historian such as myself, this creates a conundrum: How to tell a history of Big Data in a way that highlights the central role members of Indigenous communities and other generators of data cast as "digital natives" have played

[90] Aha, UCI Repository History (cit. n. 46).

[91] Lorenza Saitta and Filippo Neri, "Learning in the 'Real World,'" *Machine Learning* 30 (1998): 133–63, on 133.

[92] Ibid.

[93] Ibid., 136.

[94] See e.g., Nelly Oudshoorn and T. J. Pinch, *How Users Matter: The Co-Construction of Users and Technologies*, Inside Technology (Cambridge, Mass., 2003); Smith, *Decolonizing Methodologies* (cit. n. 6); Debra Harry, "Acts of Self-Determination and Self-Defense: Indigenous Peoples' Responses to Biocolonialism," in *Rights and Liberties in the Biotech Age: Why We Need a Genetic Bill of Rights*, ed. Sheldon Krimsky and Peter Shorett (Lanham, Md., 2005), 87–97; James Clifford, *Returns: Becoming Indigenous in the Twenty-First Century* (Cambridge, Mass., 2013).

in a way that does not reproduce the very harms that have led to this realm of inquiry? Rather than viewing the purpose of this essay as delivering a statement on what residents of the Gila River Community believe is at stake in the reuse of data based on their bodies, or viewing the nonresponse to my inquiries as a lack of interest, I hope this essay will be read by Akimel O'odham and by machine learning experts alike as an invitation to engage this history in their efforts to collectively redefine what it means to engage or even to refuse research.[95] Such work begins with my acceptance that for some Indigenous people, "research," as Linda Tuhiwai Smith has argued in her important writing on decolonizing methods, has become "a dirty word."[96] The conclusions of this essay can only ever be understood as the partial perspective of a non-Indigenous historian of medicine.[97] I have called for a reconceptualizion of the questions seen as relevant to the study of Big Data, rather than an attempt to seek solutions to problems that do not admit the concerns of all affected groups, especially the people from whom data is made.

CONCLUSION

What kinds of activities are seen as relevant to the regulation of our biomedical and, increasingly, our information infrastructures? Which ones are readily made invisible and with what consequences?[98] How is the project of historicizing Big Data augmented when participation in biomedical research is recognized as a form of labor, or when the imperative to cast new generations of citizens as "digital natives" is considered in terms of Indigenous experience?[99] In particular, I am urging that greater attention be given to alterlives: what happens to bodies that have become alienated from their personhood as well as forms of personhood that have become alienated from bodies.[100] Receiving health care, looking for love, or even ordering takeout food over the Internet have all become activities that can and are being turned into data to ask questions of which many are largely unaware. As authors in this volume have emphasized, Rebecca Lemov and Dan Bouk especially, data are people too.

Information studies scholar Christine Borgman has rearticulated concerns raised in the 1990s with what she calls "enchantment" with the possibilities of the reuse of old data.[101] She calls for a broader conversation about what the reuse of data is intended to accomplish and for greater participation by those who produce the data. Borgman recently coauthored a "Joint Declaration of Data Citation Principles," which is focused primarily on ensuring that data are cited in ways that support the "need to give

[95] See, e.g., Sally M. Reid and Raymond Reid, "Practicing Participatory Research in American Indian Communities," *Amer. J. Clin. Nutr.* 69 (1999): 755S–9S.
[96] Smith, *Decolonizing Methodologies* (cit. n. 6).
[97] Donna Jeanne Haraway, "Situated Knowledges: The Science Question in Feminism and the Privilege of Partial Perspective," *Feminist Stud.* 14 (1988): 575–99.
[98] This is an opportunity to build on the long-standing interests of historians of science in invisible labor. See, e.g., Steven Shapin, "The Invisible Technician," *Amer. Scient.* 77 (1989): 554–63; Naomi Oreskes, "Objectivity or Heroism? On the Invisibility of Women in Science," *Osiris* 11 (1996): 87–113.
[99] For recent high-profile critiques of our existing models of the value derived from data, see Lanier, *Who Owns the Future* (cit. n. 17); Mayer-Schonberger and Cukier, *Big Data* (cit. n. 68).
[100] The concept of alterlives is Michelle Murphy's. See http://www.toxicsymposium.org/conversations-1/2016/3/1/alterlife-in-the-ongoing-aftermath-exposure-entanglement-survivance (accessed 20 September 2016).
[101] Borgman, *Big Data* (cit. n. 19), 222. See also boyd and Crawford, "Critical Questions" (cit. n. 9).

scholarly credit to contributors and the importance of data as evidence."[102] In keeping with long-standing concerns about the importance of credit in expert communities, these principles are intended to ensure that those who produce, maintain, and curate the data and those who use the same data to refine algorithms or to recognize patterns are seen as equal contributors to the knowledge production process. In practice, this means relying on citations to make the "invisible technicians" as visible and as credible as the scientists.[103] Until these principles are adopted, Borgman has stated, and perhaps even then, data provenance will continue to be a "cascading problem" as data sets continue to be reused and recombined. The overarching challenge, as she sees it, is to "understand the many roles associated with data and to reach consensus within communities for which of these roles deserve credit and the best way to assign it."[104]

This is an admirable goal, but it may not extend far enough, not least of all because it is ambiguous about the kinds of credit and compensation that stand to be valued by those whose bodies or behaviors generate data captured by digital infrastructures. Similarly, it may not be able to capture the ways in which the data that supports fields like machine learning can be used to cultivate expertise that serves an array of possible futures. David Aha, who as a graduate student built the freely available UCI Machine Learning Database to support the cultivation of algorithmic expertise, now works for the United States military designing what are known as "adaptive systems." These, he explained in more colloquial terms, include "drones." Drones are vehicles that are piloted without a person on board. They are essentially flying robots, created by humans to do the work of maintaining but also severing connections that have been made onerous or unpalatable. Drones, it is often assumed, are servants of human intention.[105] Yet in the ways that machine learning's drones reorganize the relationships between humans and their machines through artificial intelligence, the drone provides yet another important opportunity to reflect on the deeper histories of alienation that have guided the distribution of labor in data-driven societies.[106]

In the face of research that points to the unexpected and potentially undesirable consequences of data reuse, some scholars have called for a domain of "data ethics," modeled on "bioethics." These scholars, as well as Borgman, are quite right to insist on the importance of wrestling with matters of social and political consequence in the realm of "Big Data." However, rather than reproducing existing models of bioethics, here is an opportunity to rethink and revise institutionalized strategies for guiding the research enterprise. It is a chance to first evaluate the history of bioethics itself—what has worked and what has not.

A good example of a thorny problem for bioethics has been the doctrine of informed consent. Time and again, it has proven inadequate to the task of adequately involving research subjects and potential research subjects in decision making. To cite only a few recognized limitations, the Human Genome Diversity Project of the 1990s struggled to address concerns about indigenous participation in genetic research by devel-

[102] Christine L. Borgman, "An Introduction to the Joint Principles for Data Citation," *Bull. Amer. Soc. Inform. Sci. Tech.* 41 (2015): 43–5.
[103] Shapin, "The Invisible Technician" (cit. n. 98).
[104] Borgman, *Big Data* (cit. n. 19), 242.
[105] Markus Krajewski, "Master and Servant in Technoscience," *Interdiscipl. Sci. Rev.* 37 (2012): 287–98.
[106] Neal Curtis, "The Explication of the Social: Algorithms, Drones and (Counter-) Terror," *J. Sociol.* 52 (2016): 522–36.

oping a model ethical protocol that advocated "group consent." This protocol was meant to recognize that the bioethical principle of autonomy did not enjoy the same status in all societies. Since then, concerns about reuse of human-derived research materials such as blood and DNA have led scholars to consider the ways that it becomes impossible to be informed about future applications.[107] Similar concerns are now being addressed in the realm of Big Data, where the phrase "click here to consent forever" has been invoked to express dissatisfaction with existing models of informed consent.[108]

The implications of the history of the PIDD case have obvious relevance for Indigenous communities, who have been at the vanguard of innovating strategies concerning the unintended uses of bodily extracts, including refusal to provide them in the first place.[109] It is also relevant to the millions of citizens who are increasingly entreated to have their most intimate desires transformed into data. For whom and what are they pathfinders?[110] These are timely opportunities for historians of medicine and Indigenous studies scholars to join forces with those invested in producing kinds of data science that better serves those whose embodied lives make it possible. This approach requires starting with empirically thick descriptions of the conditions of possibility that have given rise to technoscientific infrastructure, rather than with philosophical principles that are grounded in the logic of settler colonialism. The real innovation would be to let go of the historical fantasy of the future frontier and create systems that are accountable to those who live and labor in the present.

[107] For an excellent recent review of the issues, see Klaus Hoeyer and Linda F. Hogle, "Informed Consent: The Politics of Intent and Practice in Medical Research Ethics," *Annu. Rev. Anthropol.* 43 (2014): 347–62. For studies that highlight alternative approaches, largely grounded in Indigenous experience, see Benjamin, "Informed Refusal" (cit. n. 89); Kristin Solum Steinsbekk and Berge Solbgerg, "Biobanks—When Is Re-consent Necessary?" *Public Health Ethics* 4 (2011): 236–50; Joan Cunningham and Terry Dunbar, "Consent for Long-Term Storage of Blood Samples by Indigenous Australian Research Participants: The DRUID Study Experience," *Epidemiol. Perspect. Innovations* 4 (2007), https://epi-perspectives.biomedcentral.com/articles/10.1186/1742-5573-4-7 (accessed 21 March 2017). For research specifically concerned with indigenous participants, see Laura Arbour and Doris Cook, "DNA on Loan: Issues to Consider When Carrying out Genetic Research with Aboriginal Families and Communities," *Community Genet.* 9 (2006): 153–60; Michelle Mello and Leslie E. Wolf, "The Havasupai Indian Tribe Case—Lessons for Research Involving Stored Biological Samples," *New Engl. J. Med.* 363 (2010): 204–7; Constance MacIntosh, "Indigenous Self-Determination and Research on Human Genetic Material: A Consideration of the Relevance of Debate on Patents and Informed Consent, and the Political Demands on Researchers," *Health Law J.* 13 (2005): 213–51; Jenny Reardon, *Race to the Finish: Identity and Governance in an Age of Genomics* (Princeton, N.J., 2005); Annie O. Wu, "Surpassing the Material: The Human Rights Implications of Informed Consent in Bioprospecting Cells Derived from Indigenous People Groups," *Washington Univ. Law Quart.* 78 (2000): 979–1003.
[108] Bart Custers, "Click Here to Consent Forever: Expiry Dates for Informed Consent," *Big Data* 3 (2016): 1–6.
[109] Jenny Reardon and Kim TallBear, "'Your DNA Is Our History': Genomics, Anthropology, and the Construction of Whiteness as Property," *Curr. Anthropol.* 53 (2012): 233–45; Emma Kowal, Joanna Radin, and Jenny Reardon, "Indigenous Body Parts, Mutating Temporalities, and the Half-Lives of Postcolonial Technoscience," *Soc. Stud. Sci.* 43 (2013); Rebecca Tsosie, "Cultural Challenges to Biotechnology: Native American Genetic Resources and the Concept of Cultural Harm," *J. Law Med. Ethics* 35 (2007): 396–411; Debra Harry, "Indigenous Peoples and Gene Disputes," *Chicago Kent Law Rev.* 84 (2009): 147–96; Harry, "Acts of Self-Determination" (cit. n. 94); Emma Kowal, "Orphan DNA: Indigenous Samples, Ethical Biovalue and Postcolonial Science," *Soc. Stud. Sci.* 43 (2013).
[110] For example, recent controversies surrounding Facebook's manipulation of its users' advertising experience, conducted in the name of scientific research. Reed Albergotti, "Furor Erupts over Facebook's Experiment on Users," *Wall Street Journal*, 30 June 2014, http://www.wsj.com/articles/furor-erupts-over-facebook-experiment-on-users-1404085840 (accessed 21 March 2017).

Genealogy as Archive-Driven Research Enterprise in Early Modern Europe

ABSTRACT

This essay approaches the history of data by looking at research practices of early modern genealogists. Genealogy was (and is) a data-intensive enterprise. It requires the accumulation, management, organization, and display of vast amounts of personal data reaching back many generations. While recent research has studied in detail the social and political functions of premodern genealogy, and while the staging of genealogical information for purposes of representation has been investigated carefully, next to nothing is known about the work invested in actually researching family lines and dynastic relations. Yet the research strategies and practices, the constraints, and the possibilities of genealogical study had a crucial impact on the shape and understanding of genealogical information. By looking particularly carefully at archives as sites of genealogical data production, this essay attempts to shed light on this understudied question and contributes to the premodern history of data.

In 1710, François de Salignac de La Mothe-Fénelon (1651–1715), the French bishop and author, wrote several letters to Pierre Clairambault (1651–1741), the royal genealogist. They all dealt with details of Fénelon's family history and the difficulties of acquiring precise information about some of his more distant forebears:

> I advise you that the filiation before Menfroy de Salignac isn't very exact, as you will see from a little explanation that I attach. The filiation can be proven by solid documents until Aimery, seigneur de Salagnac, who lived in 1260 and who had then at least one son and a married daughter. Thus, he could have been born around 1200. Before that date, there are a number of people bearing the name de Salagnac and who go back until the year 997, but so far no proof of filiation has been found.[1]

He then continued that in the Abbey of Saint-Martial in Limoges "an old legend" was told that supposedly provided further details about those distant ancestors. This was, however, to Fénelon's mind, nothing but "a ridiculous story." Instead of relying on hearsay, he suggested, "one could find more specific proofs in several cartularies of abbeys or chapters in the vicinity of Salagnac and in other archives where we could have searches executed."

* Department of History, University of Hamburg, Von-Melle-Park 6, 20146 Hamburg, Germany; markus.friedrich@uni-hamburg.de
[1] Jean Orcibal, ed., *Correspondance de Fénelon* (Geneva, 1992), 14:235–7, on 236 (4 May 1710).

Fénelon's exchange with Clairambault is a fairly typical example of what J. H. Plumb, over a generation ago, called early modern Europe's "genealogical craze."[2] This craze had different manifestations and eventually included several forms of genealogical interests. Princely, noble, and, increasingly, bourgeois families wanted and needed to know about their ancestors.[3] People also felt the need to be informed in greater detail about other families. Politicians and diplomats had to know about dynastic relationships and relations among the sovereign houses in order to assess and predict current and future political alliances and conflicts. European families on a wide social spectrum thus became preoccupied, even obsessed with the genealogies of other families as well as their own. In short, an ever increasing amount of family data and genealogical information was required and requested, produced and processed.

Historians have long been aware of this burgeoning interest in all things genealogical.[4] They have debated the reasons for this growing concern with lineage and have found explanations in social as well as, more recently, cultural history. Scholars have also studied the symbolic meaning of genealogy, the forms of graphical presentation, and the function of genealogies in early modern struggles for power and prestige among noble factions. What scholars have been reluctant to investigate, however, is perhaps the most basic question to be addressed: How did people actually find and create all this genealogical data? Specifically, how did European families learn about their past, and how was genealogical information generated, circulated, vetted, discussed, and compiled? In other words, while the final products of genealogical research—polished genealogies in print or manuscript form—are well studied from multiple perspectives, the process of genealogical knowledge production has received only scant attention.[5] Fénelon's letters suggest, however, that this was evidently a major concern for people. It is the principal aim of this essay to provide a first sketch of early modern genealogy as a research enterprise invested in producing, evaluating, organizing, and exchanging vast quantities of family-related data. The sheer effort involved in the production of satisfying and convincing genealogies, the necessary size of "investments of time, energy, personnel, and capital," as well as the stamina required by the researchers frequently made genealogical investigation a "big" enterprise, if judged by early modern standards.[6]

---

[2] J. H. Plumb, *The Death of the Past* (Harmondsworth, 1973), 27–9. On the early modern English "pedigree craze," see Daniel Woolf, *The Social Circulation of the Past* (Oxford, 2003), 99–120.

[3] Nonnoble groups, such as the clerical, administrative, and university elites of the Holy Roman Empire and France, also started to produce genealogies. A good example can be found in Forschungsbibliothek Gotha (hereafter cited as "FB Gotha"), Chart B 44, containing the "Genealogia Gerhardiana et Cognatorum," which covers a wide-ranging network of well-known families of theologians.

[4] Valerie Piétri and German Butaud, *Les enjeux de la généalogie, XIIe–XVIIIe siècle* (Paris, 2003); Kilian Heck, *Genealogie als Monument und Argument: Der Beitrag dynastischer Wappen zur politischen Raumbildung der Neuzeit* (Munich, 2002); Kilian Heck and Bernhard Jahn, eds., *Genealogie als Denkform in Mittelalter und früher Neuzeit* (Tübingen, 2000); Beate Kellner, *Ursprung und Kontinuität: Studien zum genealogischen Wissen im Mittelalter* (Munich, 2004); Olivier Rouchon, *L'opération généalogique: Cultures et pratiques européennes entre XVe et XVIIIe siècle* (Rennes, 2014); François Weil, *Family Trees: A History of Genealogy in America* (Cambridge, Mass., 2013); Rouchon, "L'enquête généalogique et ses usages dans la Toscane des Médicis: Un exemple pisan de 1558," *Ann. Hist. Écon. Soc.* 54 (1999): 705–37.

[5] Olivier Poncet, "Cercles savants et pratique généalogique en France (fin XVIe siècle–milieu du XVIIe siècle)," in Rouchon, *L'opération généalogique* (cit. n. 4), 101–39.

[6] Elena Aronova, Christine von Oertzen, and David Sepkoski, "Introduction: Historicizing Big Data," in this volume.

A detailed study of early modern genealogical research practices contributes significantly and in several ways to a history of data, both big and small. It highlights how the production of data, information, and knowledge relies on specific media and creates its own unique "little tools of knowledge."[7] Practically speaking, genealogical investigation in early modern Europe morphed into the systematic exploration of libraries, archives, and other repositories of handwritten documents that were exploited by relying on the well-developed paper technologies of erudite research, such as notes, excerpts, commonplace books, and marginalia.[8] This meant that genealogists worked on and with thousands of pages filled with excerpts of archival documents containing data about a family's members. As this essay examines the daily routines of working with these papers, it will become clear that these impressive paper-based technologies of knowledge also had important unintended consequences and frequently even self-defeating implications. In showing that the same paper technologies that enabled much early modern genealogical research were ultimately responsible for some of its difficulties and inherent limitations, this essay follows a growing scholarly skepticism about the efficiency of early modern paper technologies.[9] Moreover, as will be seen in detail below, genealogical data gathering and knowledge making was hardly a solitary process. Quite to the contrary, practitioners were embedded in a pan-European republic of genealogists that transcended, sometimes in fascinating and surprising ways, borders of language, faith, and political allegiance.

Finally, a study of genealogical research practices sheds light on the complex interplay between two distinct yet necessarily connected epistemological operations, both of which relied on different yet related paper technologies. Genealogical research, in order to be effective and efficient, was conducted by fragmentizing families into individuals that were later reassembled into larger patterns. Each family member and his/her familial relationships first became a separate, distinct "research problem." This fragmentizing of families, in turn, was frequently accompanied by a stripping down of individual lives to a few key details. For each individual, the genealogist required only a handful of basic facts, such as complete name and date of birth, name of father and mother, dates of marriages, and names of spouses. Everything else, including broader historical narratives about the lives and deeds of individual persons, was desirable but dispensable.[10] While there was a broader concept of genealogy that tended to use family trees as scaffolding for more extensive narratives, this was not considered essential for successful genealogies.

---

[7] Peter Becker and William Clark, eds., *Little Tools of Knowledge: Historical Essays on Academic and Bureaucratic Practices* (Ann Arbor, Mich., 2001).

[8] Ann Blair, *Too Much to Know: Managing Scholarly Information before the Modern Age* (New Haven, Conn., 2010); Fabian Krämer, "Ein papiernes Archiv für alles jemals Geschriebene," *NTM* 21 (2013): 11–36; Fabian Krämer and Helmut Zedelmaier, "Instruments of Invention in Renaissance Europe: The Cases of Conrad Gesner and Ulisse Aldrovandi," *Intel. Hist. Rev.* 24 (2014): 1–21; Richard Yeo, *Notebooks, English Virtuosi, and Early Modern Science* (Chicago, 2014). None of these works discusses genealogical research. For early modern archives and their contribution to the culture of knowledge, see Markus Friedrich, *Die Geburt des Archivs: Eine Wissensgeschichte* (Munich, 2013); an English translation is forthcoming.

[9] For a largely optimistic assessment, see Jacob Soll, *The Information Master: Jean-Baptiste Colbert's Secret State Intelligence System* (Ann Arbor, Mich., 2010). For a more nuanced assessment, see Arndt Brendecke, *Imperium und Empirie: Funktionen des Wissens in der spanischen Kolonialherrschaft* (Cologne, 2009); an English translation is forthcoming.

[10] For a clear example, see Johannes Wallmann, ed., *Philipp Jakob Spener: Briefe aus der Frankfurter Zeit* (Tübingen, 1992), 1:264.

Genealogy was ultimately not about individual people but about the connections among them. The most important type of genealogical data was relational, for it arranged isolated individuals into families: X married Y, A was the daughter of B. What puzzled genealogists most in their everyday work was establishing the correct "conjunctions" (connections between two individuals) or "junctures" (proven relationships between two people). Ideally, genealogists ended up with vast quantities of data about individuals and their connections. Data, as many other contributions to this volume show, had limited impact if it remained isolated and disconnected.

* * *

Acquiring trustworthy genealogical data could mean different things and rely on different strategies depending on social and cultural circumstances. At least three types of genealogical research can be distinguished: legal or juridical genealogy, erudite or historical genealogy, and encyclopedic or reference genealogy. The first category refers to genealogies produced for the well-known "proofs of nobility"—if a noble family wanted to place a son or daughter into a noble convent or at the French court, their birth from purely noble ancestors had to be documented for several generations according to the highest evidentiary standards.[11] Moreover, in France Louis XIV established the *recherches de noblesse*, a mostly state-centered procedure requiring every family of the second estate to prove its claims to nobility.[12] "Erudite genealogy" refers to projects of genealogical investigation that were interested in the broadest possible reconstruction of a family or dynasty not for the imminent purpose of supporting legal claims or fulfilling administrative requirements but rather to enhance a family's reputation or show its intimate connection with a particular historical region. These genealogies, of course, were not politically innocent, for they were produced to bolster reputations and support social aspirations. Indeed, most of the fields of inquiry that later became known as auxiliary sciences, including numismatics and diplomatics as well as genealogy and heraldry, were deeply grounded in political contexts.[13] The final category of "reference genealogy" covers projects of genealogical research that were not so much focused on reconstructing single families but instead attempted to pull together vast encyclopedias of all family histories of Europe or even the entire world. Here, research was less concerned with documenting single genealogical facts through original research than with securing entire stemmata from reliable sources. This type of research relied mostly on secondhand practices of accumulation, evaluation, and assemblage of extant materials.

---

[11] Elizabeth Harding and Michael Hecht, eds., *Die Ahnenprobe in der Vormoderne* (Münster, 2011); Benoît Defauconpret, *Les preuves de noblesse au XVIIIe siècle: La réaction aristocratique: avec un recueil de tous les ordres, honneurs, fonctions, écoles, chapitres, réservés à la noblesse* (Paris, 1999); Robert Descimon, "Élites parisiennes entre XVe et XVIIe siècle: Du bon usage du Cabinet des Titres," *Bibl. École Chartes* 155 (2010): 607–44; Robert Descimon and Elie Haddad, eds., *Épreuves de noblesse: Les expériences nobiliaires de la robe parisienne (XVIe–XVIIIe siècle)* (Paris, 2010). None of these works deals with the practical questions of how those noblemen wishing to prove their nobility actually went about finding the relevant information.

[12] Valérie Piétri, "Bonne renommée ou actes authentiques: la noblesse doit faire ses preuves (Provence, XVIIe–XVIIIe siècles)," *Genèses* 74 (2009): 5–24; Piétri, "Vraie et fausse noblesse: l'identité nobiliaire provençale à l'épreuve des reformations (1656–1718)," *Cah. Méditerranée* 66 (2003): 79–91.

[13] See, e.g., Blandine Barret-Kriegel, *Les historiens et la monarchie*, 4 vols. (Paris, 1988).

Even though these various types of genealogies did involve different types of evidence and thus implied distinct research strategies, there nevertheless existed significant overlap. One crucial point seems to be beyond doubt: the "fable genealogies" with their mythical origins of European royal families that were so prominent in the later Middle Ages and the early sixteenth century were becoming less and less credible, even though they can be found up until the eighteenth century.[14] New standards of proof emerged for family data to be considered trustworthy. If they were to be taken seriously, genealogies increasingly had to be supported by solid written evidence, in particular on legal titles [*titres*].[15] As Père Joseph de Blainville (d. 1752), genealogist of the family La Tour-Auvergne, declared around 1740 in a typical statement, "The world by now refuses to be fooled by fictions that have recently been invented by mercenary historians. Instead, we want old proofs and we distrust every genealogist who does not have them."[16] "Old" in this context meant documents that were coeval with the fact that they purported to prove.

In practice, this meant that genealogical data had to be gathered in and from archives and libraries by exploiting manuscript sources. As Pierre Palliot (b. 1608), one of the most prolific genealogical researchers and writers of his time, boasted in 1664:

> I claim to advance nothing that is not true and I will not dream up relations based on *Memoires* that are often handed to me, but usually contain only words as evidence. [Instead] I will continue visiting your archives and will make excerpts from your record to inform posterity about your forebears and make future generations know where they come from, including marriages and the most important activities. To do all this, I invite you to allow me access to your legal titles once I arrive at your house.[17]

In early modern Europe, genealogy became an archive-based research activity. This process had several important consequences for the history of archives, the history of genealogy, and the history of historical studies in general. Thus, genealogy is a striking example of how socially embedded projects of data gathering relied on, and in turn changed the structure, function, and meaning of, key infrastructures of knowledge such as archives and libraries.

First, genealogical research helped create what might be called "archival consciousness" among the higher echelons of society. For a long time, most of the nobility had shown little interest in proper record keeping. The dire consequences of widespread neglect became obvious when the nobility was confronted with the new requirements for genealogical proof. Genealogy depended on the proactive preservation of records, and the availability of documents was not a natural given, but a cultural and social achievement. Historians, archivists, and genealogists used strong moral language to drive home this point. Disregard for record keeping was, according to one German author, a sign of the nobility's moral laxness.[18] Another thought that documentary ne-

---

[14] Roberto Bizzocchi, *Genealogie incredibili: Scritti di storia nell'europa moderna* (Bologna, 2009).

[15] Bibliothèque nationale de France, NAF 69, fol. b r-v: "dresser Vostre Genealogie sur la verité des Tiltres."

[16] Archives nationales (hereafter cited as "AN"), R2 71, no fol.

[17] Bibliothèque de l'Arsenal, MS 5046, 169–72: "Dessein ou Idée historique et genealogique du Duché de Bourgogne, projeté par Pierre Palliot, Parisien, Historiographe du Roy, & Genealogiste dud. Duché, MDCLXIV."

[18] Johann Georg Estor, *Practische anleitung zur Ahnenprobe* (Marburg, 1750), 23.

glect amounted to "ingratitude" toward previous generations.[19] Similar accusations were widespread, and they in turn forced the nobility to adopt exculpatory rhetorical gestures to explain the lack of sources and data. Excuses often took the form of rather stereotypical stories about unfortunate accidents. Robin de Bellair de Fressinaux wrote of one such accident in 1786: when a boat capsized, his father drowned—and scores of family papers were lost. He also blamed the sixteenth-century French Wars of Religion as a period of archival turmoil—a point frequently made by others trying to explain and excuse their documentary difficulties.[20] Whether or not this was correct in every single case, the Religious Wars clearly seem to have provided a powerful and largely credible explanation for cultural shortcomings that might, at least occasionally, have been used to mask other, more mundane reasons for archival chaos such as simple neglect and overt abuse.

And yet, partly inspired by the rise of archive-based genealogy, a significant number of noblemen became at least moderately interested in their records and the technologies of their exploitation. Fénelon's letters quoted above again provide a good example. He was, after all, able to produce a significant number of documents about his family; he managed to locate potentially significant holdings even in distant archives; and he had at least a relatively detailed grasp of his family history. He was personally involved in producing, discussing, and evaluating a wide range of genealogical data concerning his family. Quite a few other noblemen, too, used similar documents to do their own genealogical research and consulted experts about specific points.[21]

Still, even with considerable personal involvement, Fénelon and his noble peers were often helplessly overwhelmed by the requirements of archive-based research. Extracting genealogical data from the archives was a prolonged, problematic, and perplexing task for several reasons. First, dealing with old manuscript sources required a specific form of "archival literacy" not generally available: "the document that I sent to you has given me and some others a lot of pain while reading it entirely," as one nobleman admitted in a typical comment.[22] Genealogical evidence, moreover, was usually spread over several repositories, as Fénelon's letters once again make clear: "Further research has to be conducted in London, Pau, at the Chambre des Comptes in Paris, in the Abbeys of Userches, S. Amand, Terrasson, S. Martial de Limoges, and in the archive of Turenne," as he personally advised his correspondent, Clairambault.[23] Much travel was required in order to do the research necessary to prove a single specific genealogical detail.

And even if genealogical researchers braved infrastructural hardships and overcame practical challenges, results were not necessarily forthcoming. It was easy to fall into despair. Robin de Bellair de Fressinaux eventually gave up. He was a colonel in the army, trying to prove his exclusive noble status as *Duc et pair*. In 1786 he vented his exasperation with archive-based genealogical research and declared that he was just "tired after thirty years of researching papers that would enable my family to

---

[19] Reiner Reineccius in FB Gotha, Chart A 680, fol. 5r.

[20] AN 1 O 282, no fol.

[21] See, e.g., the letters of the Culant family in AN M 609, no fol. These letters reveal that detailed knowledge about ancient family connections existed and that well-established knowledge about the available documentary evidence also existed.

[22] AN M 609, no fol. (1750).

[23] Orcibal, *Correspondence* (cit. n. 1), 14:462–3 (16 November 1711).

go back to 1230."[24] He decided to forgo further investigation and asked instead for the royal grace of a noble title. Johann Georg Estor (1699–1773), a German expert on these procedures, summed this up conclusively when he wrote, "the proofs of nobility are riddled with innumerable difficulties." There were so many problems, in fact, that he called the procedure "almost diabolical."[25]

And even more disturbing, the archival mode of research could be a double-edged sword. Data gathered from the archive was potentially counterproductive. Genealogical research, thus, had to be well hedged, as François de Alouete, author of the well-known *Traité des Nobles et des Vertus dont ils sont formés* from 1577, observed:

> Old genealogies are usually the most cherished. Still, he who wishes to describe old genealogies needs to be circumspect and take care that he does not try to discover nor research too far beyond the first origin and source of a dynasty for it could have a dishonest, infamous or otherwise improper origin from which shame instead of honor will flow upon the family, contrary to one's intentions.[26]

This advice was directed especially at those involved in legal or judicial research—it was prudent to stop inquiries once the required number of noble ancestors had been documented. Otherwise, unsuitable family members might be discovered. Not knowing and not having to know were sometimes a privilege.

* * *

If contemporaries insisted that early modern genealogy distinguished itself from earlier forms by its growing reliance on written forms of evidence and proof and if all participants, too, agreed that this would make genealogical reconstruction more "trustworthy," everyone also knew full well that a new set of problems presented itself: Which documents to trust and whose interpretation of the evidence to follow? It is well known that huge interpretative battles were fought over supposedly well-documented genealogical reconstructions, for no genealogy was free from political and social implications. There were ample opportunities for the "creative" reading and researching of documents. Considerable pressure was accordingly put on genealogical practitioners, for it was common knowledge that "the art of genealogists" could give lowly born children "a social status that their birth would refuse them."[27]

In extreme cases, this could involve outright forgery. The case of the family of La Tour d'Auvergne is a well-known example, for it caused a great stir around 1700.[28] The La Tour family came up with a genealogy that claimed to be even older than that

---

[24] AN 1 O 282, no fol.

[25] Estor, *Practische anleitung* (cit. n. 18), 23.

[26] Quoted in Elie Haddad, "The Question of the Imprescriptibility of Nobility in Early Modern France," in *Contested Spaces of Nobility in Early Modern Europe*, ed. Matthew P. Romaniello and Charles Lipp (Farnham, 2011), 147–66, on 153.

[27] Pierre Jacques Brillion, *Dictionnaire des arrêts, ou jurisprudence universelle des Parlemens de France, et autres tribunaux* (Paris, 1727), vol. 5, s.v. "généalogie"; Defauconpret, *Les preuves* (cit. n. 11).

[28] Martin Wrede, "Zwischen Mythen, Genealogie und der Krone," *Z. Hist. Forsch.* 32 (2005): 17–43; Olivier Poncet, "Des Chartes pour un Royaume: Les prétentions de la famille de la Trémoille sur le royaume de Naples au XVIIe siècle," *Annu. Bull. Soc. Hist. France* (2007): 145–72; Patrizia Gillet, *Étienne Baluze et l'histoire du Limousin: Desseins et pratiques d'un érudit du XVIIe siècle* (Geneva, 2008).

of the reigning house of Bourbon—a fact that the Sun King Louis XIV did not appreciate. Ultimately, the king ordered all copies of the printed *Histoire généalogique de la Maison d'Auvergne* to be destroyed. The La Tour case was fairly simple insofar as the family had relied on newly found documents that ultimately turned out to be forgeries. These documents had raised suspicion since their first appearance in 1695. The forger, one Jean-Pierre de Bar, was sent to prison in 1700 after his home was searched.[29] Years later, after still defending the spurious genealogy, the family's main research specialist and author of the offending book, Étienne Baluze (1630–1718), a prominent scholar and former librarian of Jean-Baptiste Colbert, was condemned to live out his days in exile in the provinces. This was no isolated case. Suspicions of forgery were rampant in early modern genealogy. Simply pointing to handwritten documents and claiming their archival origin in itself were not enough to authenticate contested genealogical data. Researchers remarked critically, for instance, that the archives of interested parties were less trustworthy than "enemy" archives, the reason being that convenient forgeries were much easier to place in one's own repository than in those of competing families.[30]

Much more frequently, however, skepticism against individual pieces of evidence was based on an emerging typology of sources. Neither all archives nor all historical records were considered equal regarding their force of proof in genealogical matters. This was particularly strict in juridical genealogy. Institutions that required proofs of nobility for three, four, or five generations usually wanted each and every one of these noble forebears attested by impeccable legal documentation, including acts of notaries, notarized contracts, baptismal and marriage records, and death certificates.[31] Not every kind of archival record would convince military orders or cathedral chapters of an applicant's noble origins. It was clear, however, that in many cases such documents were not readily available—the older a family was, the less picky one could usually be regarding genealogical documentation. In fact, genealogists and state officials alike openly acknowledged the incommensurability between nobility's age and provability. Alexandre de Belleguise even stated that the inability to provide proof of ancient nobility was a form of social distinction:

> Time, which usually wears out everything, has an entirely different effect on nobility. The oldest nobility is the most prestigious; while knowledge of things is generally considered to be an advantage, this is not true in this case, for the most illustrious nobility is that whose origins are entirely unknown and which, thus, can properly be compared to the river Nile whose sources the curious have been seeking in vain for centuries. This was "fortunate ignorance."[32]

---

[29] De Bar seems to have inserted fakes into the otherwise authentic *Cartulaire de Brioude* that had been in the possession of Jean de Bouchet, De Bar's erstwhile mentor; see Arthur de Boislisle, "Appendice VIII: Le Cardinal de Bouillon, Baluze et le procès des faussaires," in *Saint-Simon: Mémoires: Nouvelle édition collationnée sur le manuscrit autographe augmentée des additions de Saint-Simon au journal de Dangeau* (Paris, 1899), 14:533–50.

[30] AN M 378, no fol. (fasc. marked "7"), a criticism of Du Bouchet's proof that supposedly showed the connection of the Chevillon and Bléneau branches of the Courtenay family: "mais ces actes estans tires du tresor de Chevillon on n'y adiouste point de foy."

[31] Defauconpret, *Les preuves* (cit. n. 11), 20 and throughout. For an easily accessible example, see Estor, *Practische anleitung* (cit. n. 18), 157–60.

[32] Alexandre de Belleguise, *Traité de la Noblesse suivant les préjugez rendus par les Commissaires deputez pour la verification des titres de Noblesse en Provence* (Paris, 1669), 51.

In Germany, Johann Georg Estor concurred.[33] What could be done? Simply upholding the strictest evidentiary standards was, from Estor's point of view, dangerous because it ultimately threatened to undermine the archival base of historical knowledge. Estor feared the threat of radical historical skepticism, which became more and more palpable around 1700. If genealogists excluded most of the available archival record from their professional practice, this would surely support skeptical assumptions about the intrinsic impossibility of historical research.[34] Thus, the standards of genealogical proof had to be calibrated. Often, the strictest form of proof was required only for a clearly specified number of years or generations. In the seventeenth century, uncontested documentation was usually required for a hundred years or so, but later, in the last decades of the ancien régime, the year 1400 came to be accepted as a crucial divide, separating different regimes of genealogical proof.[35] Before that date, a kind of genealogical "deep past" allowed for a more relaxed standard of provability:

> Up to 1400, for each family relation at least three originals have to be presented. . . . For times above [i.e., before] 1400, we follow the legal maxim *In old things narrative statements provide proof* [in antiquis enunciativa probant]. If a contemporary person tells of a [genealogical] fact and no other contemporary contradicts and if no diplomatic doubts regarding dates, witnesses and facts are raised, then the maxim is applicable and even copies are acceptable if the writer could reasonably have witnessed the fact he is writing about or if he could have heard it from eye-witnesses, that is, if it occurred within about one hundred years. These copies and narrative statements [*enunciativa*], however, are considered proof only before 1400.[36]

Legal experts involved in judicial genealogy thus basically constructed taxonomies of document types and discussed whether and under which conditions which records might be considered acceptable proof. In Germany, during the eighteenth century, a veritable genre of specialized legal treatises appeared discussing such matters. Justus Christoph Dithmar (1678–1737), for instance, anticipated the just quoted French position in 1737 when he accepted that "scriptores coaevis" were just as convincing as "public archives and untainted legal documents."[37] Estor, writing in 1744, also thought that at least "historians established by public authorities, based upon the faithfulness of archives" should be considered acceptable as proofs for claims to nobility.[38] And Ernst Friedrich Knorre in 1751 produced an entire dissertation of several dozen pages discussing in full detail different types of documents and their legal value for genealogical reconstruction. Just like the other authors, Knorre favored a rather lenient

---

[33] Johan Georg Estor, *De probatione nobilitatis avitae, et veteri et hodierna, ab illis potissimum, qui dignitatem ecclesiasticam appetunt, expedienda* (Marburg, 1744), 56.

[34] Estor, *Practische anleitung* (cit. n. 18), 16, 23: "wo man nicht allen historischen glauben übern haufen werfen wollte." On (radical) skepticism in historiographical context, see Markus Völkel, *"Pyrrhonismus historicus"* und *"fides historica"*: *Die Entwicklung der deutschen historischen Methodologie unter dem Gesichtspunkt der historischen Skepsis* (Frankfurt am Main, 1987), 313.

[35] De Belleguise, *Traité de la Noblesse* (cit. n. 32), 52; Defauconpret, *Les preuves* (cit. n. 11), 135–6.

[36] AN 1 O 281 ("Plan d'un reglement [. . .] des Preuves de Noblesse," fol. 7r–8r).

[37] Justus Christoph Dithmar, *Churmärckische Adels Historie, oder Genealogische Beschreibung derer in der Chur und Marck Brandenburg blühenden ältesten und ansehnlichsten Adelichen Geschlechter* (Frankfurt an der Oder, 1728), 20, n. "l." This was later quoted by Estor, *De probatione nobilitatis* (cit. n. 33), 52–3.

[38] Estor, *De probatione nobilitatis* (cit. n. 33), 65. On Estor, see Johannes Burkardt, *Die historischen Hilfswissenschaften in Marburg* (Marburg an der Lahn, 1997), 50–5.

approach, claiming that the standards of proof should not be too high.[39] Ultimately, Knorre seemed to accept almost every type of written document as potential proof of nobility, even "letters of private men, especially if they are fairly old."[40]

Scholars who were not researching for judicial purposes but attempting a more "erudite" and historical reconstruction of a family's past had an even broader concept of what was acceptable evidence. Historically researching and juridically proving genealogical data were two closely related yet somewhat different activities. For erudite research into family histories, almost any kind of material and even intangible sources were welcome. Caspar Sagittarius (1643–94), a professor of historiography in late seventeenth-century Jena, penned the following general statement regarding his preferred set of sources:

> If a reliable dynastic history is to be produced, one can rely on 1. old coins, 2. old monuments and inscriptions, 3. old charters, papers and seals, 4. old chronicles and 5. other books and narratives which discuss the family either directly or indirectly, extensively or only in passing, 6. oral histories from trustworthy persons.[41]

This passage invites several observations. First, while current scholarship generally agrees that manuscripts and material evidence became more and more important in early modern genealogical research,[42] it is nevertheless also true that oral testimony still held some appeal, if only as a means of last resort. In France, for instance, proofs of nobility could involve "the depositions of five or six gentlemen"—that is, oral statements of well-regarded noblemen concerning the noble status of the person under review—well into the eighteenth century.[43] Hence, the general drive for better genealogical documentation, while mostly calling for written and material sources, never quite gave up on the importance of orally transmitted knowledge. Much recent literature has stressed the persistent and frequently also innovative use that the early modern knowledge economy continued to make of oral testimony.[44] As Sagittarius's statement indicates, the question was ultimately who would count as a "trustworthy" witness. Often enough, data provided by oral testimony were authenticated by nothing else but the lack of contradicting voices. For orally transmitted knowledge to be acceptable, it sufficed that "no man then in life, hath not heard any thing, nor known any proof to the

---

[39] Ernst Friedrich Knorre, *De probatione nobilitatis per instrumenta* (Halle, 1751), 18n50. Knorre relied on Estor, *De probatione nobilitatis* (cit. n. 33), 62–6 for these sections.

[40] Knorre, *De probatione* (cit. n. 39), 47–8.

[41] Universitäts- und Landesbibliothek Kassel (hereafter cited as "ULB Kassel"), 2° Ms. Hass. 20/3, fols. 471r–474v, on 471v. I acknowledge the help of my doctoral student Jacob Schilling in locating and procuring this valuable document. A somewhat similar enumeration appears in Adam Rechenberg and Ernst Friedrich Kindermann, *Disputatio de studii genealogici praestantia* (Leipzig, 1697), fol. A3v.

[42] For the broad background of changing notions of nobility in regard to these questions of proof, see Haddad, "The Question" (cit. n. 26).

[43] Archives departementales Loir-et-Cher F 202, no fol. (7 March 1723; Joubert to the sieur de Montarnal). See also Estor, *De probatione nobilitatis* (cit. n. 33), 55–62. The role of oral transmission of familial and genealogical knowledge has been stressed by Michel Nassiet, "La généalogie entre discours oral et écrit (XVe–XVIe siècles)," in *La généalogie entre science et passion*, ed. Tiphaine Barthélemy and Marie-Claude Pingaud (Paris, 1997), 207–19. See also Piétri, "Bonne renommée" (cit. n. 12), 12–3, on the "persistence of proof by testimony."

[44] See, e.g., Steven Shapin, *A Social History of Truth: Civility and Science in Seventeenth-Century England* (Chicago, 1999); Andy Wood, *The Memory of the People: Custom and Popular Senses of the Past in Early Modern England* (Cambridge, 2013); Adam Fox, *Oral and Literate Culture in England, 1500–1700* (Oxford, 2000), esp. 258–98.

contrary," as was stated in 1635.[45] There were in effect no criteria other than the one used by the French lawyers quoted above when they discussed the circumstances under which "weak" manuscript evidence could be employed. Establishing trust in written and oral sources, thus, could easily rely on similar criteria.

Second, the vast range of material sources to be included in the production of genealogical data is remarkable. Coins were regularly considered to be key evidence. Friedrich Hortleder (1579–1640), researching genealogical questions on behalf of the Weimar branch of the princely house of Wettin, relied heavily on archives and coins. His important internal memorandum from 1619, *Gruntvest Sächssischer Weimarischer Praecedentz*, seamlessly combined references to archival and numismatic sources.[46] Furthermore, genealogical data mining also focused on stone inscriptions and particularly on epitaphs. For the older times with little or no written evidence on parchment or paper available, these epigraphic sources shed new light on hitherto unanswerable questions. Finding, studying, and interpreting epitaphs thus became a genealogical research strategy that was as important as archival visits. The Jesuit Claude-François Ménestrier (1631–1705), probably the most important French expert on heraldry in the seventeenth century and also a well-regarded genealogist, visited the churches of Lyon, his hometown, searching for epigraphic documentation.[47] However, deciphering epitaphs and other inscriptions, much like reading and understanding archival sources, was a challenging task. War, natural forces, and other circumstances had often badly damaged the stones. And yet genealogists had to make do with whatever they found. A bad inscription was better than no reference at all. Sagittarius accordingly suggested that stone inscriptions had to "be copied in drawing and writing." He insisted that this task should be entrusted only to people who were "well-versed in German antiquities"—genealogical data mining clearly required some expert knowledge. In copying stones and inscriptions, he added, careful attention should be paid to all the details of the original object. In fact, the copyist should patiently include "all lines, points, and the smallest minutia with utmost exactness."[48]

Sagittarius was certainly no isolated case. Gabriel Bucelin (1599–1681), one of the better-known German historians and genealogists of the seventeenth century, praised burial monuments and epitaphs as rich sources of genealogical data, although he cautioned researchers that "error and misinformation is not infrequent [on such stones]."[49] Likewise, in France, Jean Du Bouchet (1476–1557) learned about families from chance visits to local churches.[50] Georg Helwich (1588–1632), working in Hesse in the seventeenth century, even created a vast collection of inscriptions that was later used to compile genealogical works. He systematically traveled the countryside transcribing inscriptions, noting dates, and copying coats of arms.[51]

---

[45] Charles Calthrope, 1635, quoted in Fox, *Oral and Literate Culture* (cit. n. 44), 262. For a practical application of that criterion, see ibid., 281.

[46] The manuscript is in Weimar, Thüringisches Hauptstaatsarchiv, Nachlass Hortleder 13, fols. 141r–156v.

[47] P. Allut, *Recherches sur la vie et les oeuvres du P. Claude-François Menestrier de la Compagnie de Jésus* (Lyon, 1856), 299–300.

[48] ULB Kassel 2° Ms. Hass. 20/3, fol. 472r.

[49] Quoted in Estor, *De probatione nobilitatis* (cit. n. 33), 68–9.

[50] AN M 609, no fol. (cit. n. 22).

[51] Rüdiger Fuchs, "Georg Helwich. Zur Arbeitsweise eines Inschriftensammlers im 17. Jahrhundert," in *Deutsche Inschriften: Fachtagung für mittelalterliche und neuzeitliche Epigraphik* (Mainz, 1987), 73–99. Helwich never managed to write the wide-ranging genealogical publication that his col-

Coats of arms frequently served as sources for genealogical reconstructions. Indeed, early modern heraldry and genealogy developed apace and in close connection as innovative and dynamic research areas.[52] Coats of arms, genealogists discovered, could help solve genealogical riddles because they documented marriages and exchanges of territories and provided information about family connections. Yet one had to be careful, too, for coats of arms were complex sources.[53] Carelessness could easily lead to heraldic anachronisms.[54] Estor was acutely aware of the practical problems. He argued for a measured appreciation of coats of arms. They were, indeed, valuable sources, but they had to be checked carefully against other evidence, especially archival documents.[55] In general, he believed, knowledge based on heraldic matters never left the realm of the "probable."[56]

Early modern genealogical research thus found itself increasingly entangled in the study of multiple types of source materials. Yet with such an abundance of means to investigate the past, questions of hierarchy were bound to arise. How should the genealogist evaluate all the different types of sources against each other? Was there a hierarchy of sources? These issues affected the daily work of genealogical researchers as well as their more theoretical debates. Among those who addressed the problem was Jacob Friedrich Reimmann (1668–1743), a minor figure in early modern German genealogy. He wrote a short treatise *De necessitate scepticismi in studio genealogico.*[57] Again, genealogy was embedded in the broader debate about historical skepticism and the possibility of evidence-based study of the past. Reimmann acknowledged the difficulties of genealogical research in general and the frequent problems with archival documents in particular. He accepted that errors were impossible to avoid when trying to extract genealogical information from parchment and paper manuscripts—a common concession. Yet, with all these shortcomings acknowledged, should archives and manuscript sources be entirely abandoned? According to Reimmann, at least some scholars seemed to hold such views. He was particularly offended by a statement by

---

lection was probably meant to support. For the most part, his vast collection of papers remained with his superior, the archbishop of Mainz, Georg Friedrich Greiffenclau à Volrats (who openly acknowledged his indebtedness to Helwich). Nevertheless, his manuscripts were used by several regional genealogists, including Johann Maximilian Humbracht, *Die höchste Zierde Teutsch-Landes, Und Vortrefflichkeit des Teutschen Adels Vorgestellt in der Reichs-Freyen Rheinischen Ritterschafft, Auch auß derselben entsprossenen und angränzenden Geschlechten, so auff hohen Stifftern auffgeschworen, oder vor 150. Jahren Löblicher Ritterschafft einverleibt gewesen, Stamm-Taffeln und Wapen* (Frankfurt am Main, 1707).

[52] For a broader assessment of coats of arms and names as identifiers of consanguinity and lineage, see Michel Nassiet, "Nom et blason: Un discours de la filiation et de l'alliance (XIVe–XVIIIe siècle)," *L'Homme* 34 (1994): 5–30. The modern standard work on heraldry is Michel Pastoureau, *Traité d'héraldique* (Paris, 1997). There is, however, little treatment of early modern heraldic science and research in this valuable book.

[53] For a rather skeptical view, see Estor, *De probatione nobilitatis* (cit. n. 33), 32–3.

[54] See the scathing review of Engelbert Flacchio's arguments by Père Blainville, AN 2 R 71, no fol.

[55] The heraldic description must be based upon "authentic charters"; see Estor, *Practische anleitung* (cit. n. 18), 39. For a case study in which a wide range of existing assumptions about specific coats of arms are discussed and invalidated by juxtaposing them with archival evidence, see Johann Georg Estor, *Probe einer verbesserten Heraldic an dem Hoch Fürstlich Hessisch und Hochgrävlich Hanauischen, sodann Chur Mayntzisch wie auch Hochfürstlich Brandenburg Anspachischen Wappen* (Giessen, 1728).

[56] Estor, *Probe einer verbesserten Heraldic* (cit. n. 55), n.p. (preface).

[57] Jacob Friedrich Reimmann, *Historia literariae exotericae & acroamaticae particula sive de libris genealogicis vulgatioribus & rarioribus commentatio: Accedit disquisitio historica de necessitate scepticismi in studio genealogico* (Leipzig, 1710), 208–50.

Johann Christoph Olearius (1668–1747) from Halle, Germany. Olearius was an expert on coins and had recently published a manual on numismatics. In this book, Olearius had highlighted the general usefulness of coins for all kinds of research, including genealogy: "Genealogies or dynastic summaries cannot be produced more fruitfully than from coins." Whereas "books" were full of errors, starting with those made during typesetting, coins were, according to Olearius, "infallibly true."[58]

Reimmann vehemently disagreed. Coins were just as frequently forged as archival documents, he argued. They were indeed valuable but hardly infallible. Even more outrageous, to his mind, were the claims of those scholars who wanted to dismiss documents and coins entirely and rely exclusively on stone inscriptions. Although Reimmann could not associate this extremist position with a particular individual, he confuted it anyway. Such a conviction was nonsense, he said, because inscriptions were so hard to decipher, especially when it came to names and dates. Ultimately, Reimmann found no single type of sources entirely without fault. The best genealogists could do was to combine all types of sources and be careful (or, as Reimmann said, "skeptical") when evaluating genealogical data. Cross-referencing different types of sources was of paramount importance.

* * *

Early modern genealogists were embedded scientists, engaged by interested parties and often working in highly charged political, social, and erudite contexts. And yet cooperation and exchange of information were common. Few if any genealogists worked in isolation. Information was shared widely, often in unexpected ways. In 1660 or 1661, for instance, two men met in Lyon on the basis of their common interest in genealogy and heraldry; they were not, at first glance, very likely to meet on friendly terms. Philipp Spener, then a young German scholar and theologian who was to become the founding father of German pietism, met Claude-François Ménestrier, a Jesuit priest and leading authority on genealogical and heraldic issues. Spener, according to his own account, was fascinated by the Jesuit's highly acclaimed writings.[59] This inspirational meeting seems to have had a lasting impression on Spener, for he mentioned his connection to Ménestrier several times in the following decades.[60] Ménestrier, on the other hand, might have found this meeting to be one step toward educating Germany in proper scientific heraldry and genealogy, for the Jesuit was very dissatisfied with the extant literature on these fields in the Holy Roman Empire.[61]

Perhaps less spectacular but no less remarkable than this international and interconfessional collaboration was the flow, exchange, and debate of genealogical information within more restricted networks. Ménestrier himself was well connected to a

---

[58] Johann Christoph Olearius, *Curiose Müntz-Wissenschaft Darinne Von dero Unfehlbahren Nutzbarkeit, allerhand merckwürdigen Müntz-Arten, so auch nöthigsten darzugehörigen Mitteln* (Jena, 1701), 35–7.

[59] Philipp Jacob Spener, *Insignia Serenissimae Familiae Saxonicae* (Frankfurt am Main, 1668), "To the Reader"; Johannes Wallmann, *Philipp Jakob Spener und die Anfänge des Pietismus* (Tübingen, 1986), 151–2.

[60] Wallmann, *Spener: Frankfurter Zeit* (cit. n. 10), 1:131, 654. See also Johannes Wallmann, ed., *Philipp Jakob Spener: Briefe aus der Dresdner Zeit* (Tübingen, 2009), 2:244–5.

[61] Allut, *Recherches* (cit. n. 47), 292. Spener considered his own activities at least partly as a transfer of French heraldic and genealogical knowledge into German lands; see Wallmann, *Spener: Frankfurter Zeit* (cit. n. 10), 1:202–4.

myriad of French genealogical and heraldic enthusiasts. Even with his rival, Claude Le Laboureur, he was trading data and books in addition to compliments and insults.[62] Ménestrier also relied on the many aficionados of genealogy and heraldry within the Society of Jesus through the order's vast communication network.[63] Through these associations, he could easily reach beyond France's borders whenever necessary.

Spener also communicated frequently with other members of the early modern *république des généalogistes*. In his case, even more than in Ménestrier's, genealogical correspondence served a very specific purpose. Spener used his prominent position as one of Germany's leading genealogical experts to compile vast encyclopedias of genealogical charts. With Spener and his correspondence, we touch upon the third type of genealogical research mentioned at the beginning of this essay: encyclopedic genealogy. Spener strove to present as much genealogical material as possible, and he offered it to the reader as an endless series of charts, printed one after another for hundreds of pages. These were reference works. If one needed to know the genealogy of, say, a French dynasty or a great Spanish noble house, handbooks like Spener's would most likely have offered a relevant family tree, if nothing else.[64]

Such compendia had a long tradition going back well into the sixteenth century. During the seventeenth century, the genre underwent significant expansion and change at the same time. The need for simple references to genealogical information became increasingly evident. Data about who was related to whom in which ways became a (felt) necessity for a growing number of people from different social backgrounds. With intermarriage among members of European royal families on the rise, questions and conflicts of succession became more complicated and widespread, especially with ever more powerful state and military apparatuses supporting such claims—this was the era of the wars of succession, after all. Genealogy was obviously of utmost importance in navigating this increasingly dense network of frequently intermarried European dynasties.[65] Another context in which quick access to a wide variety of genealogical data sets came to be considered mandatory was the emerging market for journals and newspapers, as Volker Bauer has recently shown.[66] Avid newspaper readers were expected to have genealogical reference works close at hand to identify persons mentioned in the articles. The evolving political and media environments of early modern Europe thus contributed to a growing demand for easy access to large quantities of genealogical data for reference purposes. The new genre of ency-

---

[62] See the letters edited by Allut, *Recherches* (cit. n. 47). More generally, Stéphane van Damme, *Le Temple de la Sagesse* (Paris, 2004); van Damme, "Les livres du P. Claude-François Ménestrier (1631–1705) et leur cheminement," *Rev. Hist. Mod. Contemp.* 42 (1995): 5–45.

[63] Philipp Palasi, *Jeux de cartes et jeux de l'oie héraldiques aux XVIIe et XVIIIe siècles* (Paris, 2000). There have been no studies of Jesuit genealogies. On Jesuit networks of communication, see Markus Friedrich, "Government and Information-Management in Early Modern Europe: The Case of the Society of Jesus (1540–1773)," *J. Early Mod. Hist.* 12 (2009): 1–25; Friedrich, *Der lange Arm Roms? Globale Verwaltung und Kommunikation im Jesuitenorden 1540–1773* (Frankfurt am Main, 2011).

[64] Volker Bauer, *Wurzel, Stamm, Krone: Fürstliche Genealogie in frühneuzeitlichen Druckwerken* (Wiesbaden, 2013); Bauer, "Strukturwandel der höfischen Öffentlichkeit: Zur Medialisierung des Hoflebens vom 16. bis zum 18. Jahrhundert," *Z. Hist. Forsch.* 38 (2011): 559–84.

[65] Johannes Limnaeus, *Ius publicum imperii Romano–Germanici*, vol. 2 (Strasbourg, 1657), at the beginning of sec. V/15 (no pages). Stephan Kekulé von Stradonitz, *Ausgewählte Aufsätze aus dem Bezirk des Staatsrechts und der Genealogie* (Berlin, 1905), 89–103.

[66] Volker Bauer, "Jetztherrschend, jetztregierend, jetztlebend: Genealogie und Zeitungswesen im Alten Reich des ausgehenden 17. Jahrhunderts," *Daphnis* 37 (2008): 271–300.

clopedic genealogy was created to satisfy this demand. These publications were of relatively small quarto format (though often very voluminous), fairly mundane in layout and quality, and comparatively cheap and quick to produce. Eventually, this format evolved into periodically updated calendars of European nobilities, with the *Gotha* being the most famous example.[67]

Spener probably acquired most of his information secondhand. Given the enormous number of noble families he included and the vast geographical reach of his work, it was not possible for him to personally research genealogical data in archives, churches, or libraries. His letters contain little evidence to suggest that he himself did much research of that kind. Rather, his letters show that he had an extremely well-functioning network of individuals supplying him with the required information. He openly solicited up-to-date information about coats of arms and family relations from well-placed correspondents at several princely courts.[68] For example, while waiting for new information about the counts of Ronow, he wrote that he was looking forward to what (unnamed) sources would "send to me."[69] With others, he debated genealogical and heraldic details.[70] A great deal of additional information, it seems, was culled from existing publications and carefully cross-examined. Spener's volumes were produced by systematically collecting and analyzing preexisting data on individual families.

Other genealogists of the encyclopedic type proceeded similarly. Secondhand acquisition of carefully evaluated stocks of data is explicitly indicated in Daniel Mithoff's (1595–1673) well-known genealogical table from 1636, showing a genealogy of the dukes of Saxony.[71] Mithoff's work became a standard reference for anyone interested in Saxon genealogy, including Spener and Gottfried Wilhelm Leibniz. In the left top corner we can read, "To the reader: In this table the two following signs are used * + The first indicates where proof has been taken from archival evidence, the second where it has been taken from historiographers." Indeed, we can find the two signs throughout this vast print, indicating where Mithoff had used archival evidence and where he had not.

Yet another example is Johann Hübner (1668–1731) who, when preparing the fourth edition of his *Genealogische Tabellen*—a work similar to Spener's—seems to have sent letters of inquiry asking unnamed correspondents for up-to-date information.[72] His compilation of genealogical stemmata was truly universal. Starting with biblical dynasties, he presented family trees for ancient Greece and Rome and also in-

---

[67] No good analysis of the Gotha calendar exists; meanwhile, see Thomas von Fritsch, *Die Gothaischen Taschenbücher, Hofkalender und Almanach* (Limburg, 1968).

[68] Johannes Wallmann, ed., *Philipp Jakob Spener: Briefe aus der Frankfurter Zeit* (Tübingen, 2010), 5:204. Spener was very meticulous regarding the quality of the data forwarded to him: "hoc vero scire necesse est: von was vor Kraut oder Baum die 3 Blätter und 2 rothe blümlein sind. Es scheinen Weidenblätter (folia salicis) zu seyn. Si vero exprimendum scutum, nomen foliorum designandum est"; Wallmann, ed., *Philipp Jakob Spener: Briefe aus der Dresdner Zeit* (Tübingen, 2003), 1:142.

[69] Wallman, *Spener: Frankfurter Zeit* (cit. n. 68), 5:84.

[70] Ibid., 330–1, concerning the coats of arms and genealogy of the Osterode family, which Spener thought were not connected to the counts of Ostrorog.

[71] Daniel Mithoff, *Geschlechts-Tafel: Nach welcher Fürbildung der Stamm-baum der durchleuchtigen unnd hochgebornen Fürsten und Herrn / Herrn Hertzogen zu Sachsen [. . .] besage inhaltes unverwerfflicher urkunden der Archiven / die im Original mit fleiß darzue durchgesehen / zu enderen und wider in richtigkeit zu bringen* (n.p., 1636), one single sheet.

[72] Johann Hübner, *Genealogische Tabellen, Nebst denen darzu gehörigen Genealogischen Fragen, zur Erläuterung Der Politischen Historie* (Leipzig, 1725), fol. a2r.

cluded tables for Japanese, Moghul, and Turkish royal houses. Hübner's case, however, is also a reminder of the shortcomings of this genealogical genre and its supporting research strategies, for his books, though well regarded, were full of errors. Working with borrowed data could be self-defeating; obviously, Hübner had not been able to carefully check all the information he received against archival sources. After the information had been checked, many facts had to be corrected; indeed, a new set of "truths" and "improvements" emerged.[73] Numerous people checked Hübner's volumes against additional references, archival or otherwise, and improved upon his data.[74] The scrutiny and analysis of genealogical data by readers and recyclers of available publications was a widespread practice.

The growing demand for genealogical compendia created economic opportunities. Thus, competition between researchers for the most marketable genealogical data created the potential for overlap and redundancy. Spener and Jakob Wilhelm Imhoff, another well-known genealogical expert, found themselves at one point in friendly competition.[75] The authors were personally acquainted and seemed to have a fairly cordial relationship. In 1686, Spener received advance notice of Imhoff's forthcoming collection of genealogical tables. He suspected that this work might duplicate his own research, which had been done many years earlier but had not yet been printed. "Both projects are rather similar and thus might impede each other and damage the publishers," he concluded.[76] In an unusually collaborative move, Spener then offered to forgo publication of his own work. Indeed, when Spener finally printed his own material, he left out the tables that Imhoff had already made available. Although Spener's offer might seem rather generous at first glance, he admitted that it was also pragmatic and calculating. With a discerning public scrutinizing genealogical works and economically savvy editors assessing publication plans, simply duplicating existing data was not an attractive option, even for highly regarded authors such as Spener. In any case, this episode shows that data competition had become an (unwelcome) reality in the field of genealogy once a large pool of data had been created and was accessible, through printed resources and the cadre of genealogists.

* * *

Genealogy was clearly part of the multifaceted early modern endeavor to manage increasing amounts of knowledge through ever more sophisticated paper technologies. The genealogical expert Johann Maximilian Humbracht (1654–1714) described these efforts with some agitation:

> The hardships I faced in this work will be understood only by persons who know about working with old documents. I had to separate thousands of folios and uncounted numbers of pages of paper, I had to stack together what belonged to each family, I had to reread everything three and four times, take notes and collate my excerpts, before I could finally draft a first version of the family trees.[77]

---

[73] On Hübner's work, see Estor, *Probe einer verbesserten Heraldic* (cit. n. 55), preface (no pages).

[74] The Gotha exemplar of the fourth edition (FB Gotha, Gen 2° 00017/04), for instance, is full of handwritten additions, including small slips of paper that were glued into the books.

[75] Wallmann, ed., *Spener: Dresdner Zeit* (cit. n. 68), 1:151–3. I am currently preparing a larger project on Imhoff's strategies for managing genealogical information.

[76] Ibid., 179; Kekulé von Stradonitz, *Ausgewählte Aufsätze* (cit. n. 65), 187.

[77] Humbracht, *Die höchste Zierde* (cit. n. 51), "To the Reader" (no pages).

Genealogical research, in practice, meant excerpting and taking notes, which in turn meant transforming physically unwieldy objects such as medieval charters or stone inscriptions into manageable items such as slips of paper or sheets in notebooks. While recent studies have familiarized us in detail with the strategies and practices employed by early modern scholars and compilers to manage and work with immense amounts of data, the functioning of genealogical information management has been infrequently studied.[78] On the one hand, genealogists faced many of the same difficulties as scholars and compilers in other disciplines, and several of their strategies for managing information resembled those of other scholars. On the other hand, genealogical knowledge production faced some peculiar challenges. The sheer amount of data they needed to compile and handle could be stunning. Spener's work, for instance, concerned hundreds of families and tens of thousands of individuals, often with similar names, who were related to one another through multidimensional connections. This certainly was "Big Data" from an early modern perspective.

Genealogists like Spener were fond of their "first little sheet[s], upon which it [i.e., the data] was excerpted" and never let them out of their hands. These excerpts thus accumulated, and their increasing number made transforming them into presentable knowledge a daunting task. Further complicating this task were the many findings unrelated to the core genealogical questions as well as genealogical and historical data "noise."[79] Most practitioners were struggling with the difficult task of keeping track of what they knew already. Genealogists employed different strategies to navigate their mountains of notes. Some were quite rudimentary. Spener, for instance, relied heavily on memory. He simply knew by heart that he had certain genealogies "among my papers," although likely only in general terms.[80] In addition to memory, basic external indicators like the size of the sheet in question ("a full folio page and a quarter") or a rough description of handwriting ("not in my own hand") further helped when searching for something specific.[81] Handling stacks of notes and avoiding complete information overload for him depended as much on the power of memory and simple visual clues as it did on advanced systems of indexing or note taking.

Among the most complex tasks for genealogists was the process of reassembling the myriad of excerpts into larger patterns—in other words, transforming data into knowledge by agglomerating human individuals into "families." Early modern genealogists usually relied on a graphic tool to create families from personal data. Genealogists used (and of course still use) the diagrammatic form of the "family tree" to represent familial connections and patterns. While it is well known that finalized "family trees" were an ideologically charged visual technology to present kinship ties, social ambition, and political power,[82] a second function of family trees is much less often appreciated: the process of actually drafting, and redrafting, family trees, often multiple times, was one of the most effective aids in the process of ordering and se-

---

[78] See n. 8 for some of the most important recent publications in this field.

[79] Ann Blair, "Reading Strategies for Coping with Information Overload, Ca. 1550–1700," *J. Hist. Ideas* 64 (2003): 11–28.

[80] All this comes from his letter to Jakob Wilhelm Imhoff in Wallmann, *Spener: Dresdner Zeit* (cit. n. 68), 1:179.

[81] See the letter to Imhoff in ibid., 263.

[82] On these aspects, see Christiane Klapisch-Zuber, "The Genesis of the Family Tree," *I Tatti Stud.* 4 (1991): 105–29; Klapisch-Zuber, *Stammbäume: Eine illustrierte Geschichte der Ahnenkunde* (Munich, 2004).

lecting, integrating and evaluating the paper-based results of genealogical research. Drawing stemmata meant establishing relations between individuals. It also meant focusing the vast amount of data and notes by highlighting only what served the most basic genealogical purpose, that is, the reconstruction of families. Sketching a family tree helped to distill the core of relevant knowledge from the background noise of genealogical and historical evidence produced by wide-ranging research activities. Drawing and sketching genealogical diagrams was, thus, an act of organizing, prioritizing, and selecting information, long before the stemmata could serve as eye-catching representations of dynastic aspirations. The physical act of sketching a stemma—a widely used practice also in other fields—created order from piles of excerpts.

The immense organizational role played by diagrammatic family trees in the constructive process of ordering data-bearing archival excerpts is most obvious if one looks beyond the finalized and often visually stunning stemmata that were printed in family histories, painted on walls, carved on epitaphs, woven into tapestry, or drawn in manuscript proofs of nobility. For a study of genealogical knowledge making, the unappealing, confusing, incomprehensible, preliminary manuscript sketches that preceded and supported these finished works are much more interesting. The working papers of many early modern genealogists contain more or less comprehensive sketches of entire families or particular "filiations" or "junctures" that sought to establish graphically how a handful of people were related. Such sketches were frequently embedded in genealogical notebooks or scribbled directly below archival excerpts or research notes in order to immediately link individual findings to larger patterns.

Beholding such drafts, one can imagine the genealogist sitting at his desk, quill in hand, browsing his notes—as Humbracht mentioned—and trying to make sense of the research data by starting to integrate them diagrammatically. Vacant placeholders for unknown individuals might signal where additional data were still needed. A newly found note sheds light on a conjunction already drawn. Maybe a daughter was actually a grandchild. An only son seemed to have had siblings. In many families, moreover, certain names were quite common, so it was easy to confuse individuals. All this resulted, almost inevitably as the genealogist tried to (quite literally) place each person onto the page, in an increasingly confusing sketch. Disorderly lines, necessitated by initially overlooked data, quickly started to disturb a clear layout. Drafting a family tree, thus, helped to transform personal data sets, those cherished results of archival research, into coherent networks of familial relationships (fig. 1). Family trees in the making display the process of ordering and organizing personal data into genealogical knowledge, and as such they were indispensable tools for building up the families to be presented in the first place.

* * *

As we have seen, the hunt for documents substantiating sets of personal data became a quintessential part of genealogical research. Originally, kinship had produced the archival record—the birth of a child resulted in certain documents, and social relationships were documented in written wills. From the genealogical researcher's perspective, however, the process went the other way around, from retrieved documents to claims about biological and social relations. No documentation, no data—and hence no reliable genealogy, once the age of "fable genealogies" had passed. Thus, biological relationships and social facts (birth, marriage, name) became identified with

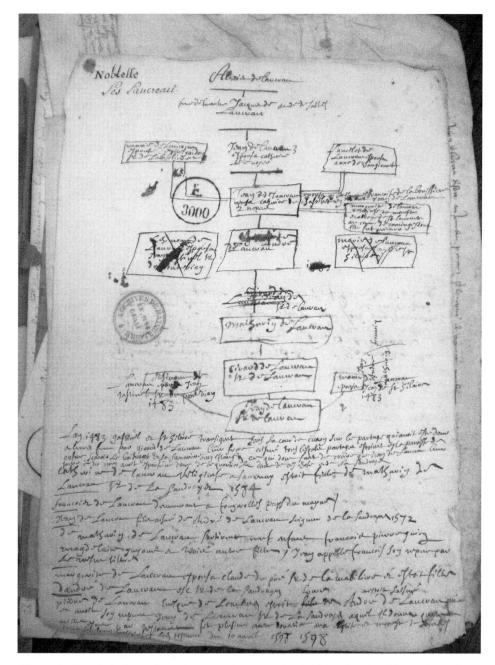

Figure 1. *Joseph Audouys from Angers orders genealogical data by sketching and altering a rough draft of a genealogical schema. Archives départementales de Maine et Loire, Angers, E 3000, n.p.*

the documents that evidenced them. The media bearing or "proofing" genealogical data became a focus of fascination. Early modern practitioners of genealogy thus acknowledged the enormous degree to which a world of paper had been superimposed upon family experiences, communally shared memories, binding emotions, and the lived reality of kinship relations. At least for the purposes of (archivally) documented genealogy, the family tended to be equated with its paper-based representations and traces. Trustworthy genealogical data resided in the archival documents to such a degree that the two tended to merge. Thus, genealogy was as much a systematically constructed network of documents as an imagined community of people.

The History and Political Economy of Personal Data over the Last Two Centuries in Three Acts

by Dan Bouk*

ABSTRACT

This essay uses the sociological concept of the "data double" to investigate the changing political economy of personal data in the United States in the nineteenth and twentieth centuries and up to today. It reviews secondary sources and examines select primary sources to argue for the ascendance of the double over the last two centuries and establishes a three-part periodization for personal data. In the first act, personal data worked mainly in aggregated form to help construct social imaginaries. From World War I until the 1970s, the power of personal data shifted toward the use of aggregates to fit individuals into the mass. Today, in the third act, individuals generate many data doubles that are commodified, capitalized, collected, celebrated, and often out of the control of those they represent.

Personal data is neither new nor newly powerful. The essays that precede this one have already proven as much. As Markus Friedrich shows, genealogists in early modern Europe worked feverishly to generate enough personal data to satisfy the demands of individuals and families who often needed that data to win or maintain political or economic status and position. In the modern era, personal data became even more crucial to the operation of states or companies, and to universities, too. Rebecca Lemov's essay invites us to recognize the embeddedness of personal data in our institutions and practices and to reconsider Big Data as a profoundly social phenomenon through a close investigation of the datafied life of Don Talayesva, perhaps the mid-twentieth century's most documented person. Joanna Radin's essay, in turn, further evokes an embeddedness so deep that even in standardized machine learning data one can discover the lingering shadow of colonial exploitation. The power of personal data (to predict, among other things, manhole fires) derives in no small part,

* Department of History, Colgate University, 13 Oak Drive, Hamilton, NY 13346; dbouk@colgate
.edu.

I thank the Max Planck Institute for the History of Science's "Historicizing Big Data" working group and "History of Data" reading group, especially Lorraine Daston, Christine von Oertzen, Elena Aronova, David Sepkoski, Joanna Radin, Judith Kaplan, Markus Friedrich, and Etienne Benson. Two anonymous reviewers, Sarah Igo, Dan Hirschman, Danielle Keats Citron, and especially Martha Poon also provided helpful comments.

her essay reminds us, from the now nearly invisible labor of those whose lives generated that data.[1]

But while personal data is neither new nor newly powerful, its political economy has been anything but static. In fact, we live today amidst a series of profound changes in the political economy of personal data, as the preceding essays have also made clear. This essay aims to help us understand those changes better by telling a story in broad strokes of how personal data has worked in the world, and especially how it has worked within changing modes of power and regimes of valuation. I have focused my narrative on the last two hundred years and mostly on the United States, where my primary expertise lies. But the framework developed in this essay will be useful in other contexts and the narrative sweep familiar in other places.

To begin, I must offer a few general words on personal data. All personal data exists as some sort of inscription. In making personal data, someone writes something down or enters values into a spreadsheet or sketches lines on a chart. Each such inscription represents a person or multiple people, or even a prior such inscription. As we delve into the history and political economy of personal data, it is convenient to think of personal data inscriptions in two broad types or classes. First, we can think about those inscriptions that describe only a single individual. I find the concept of the "data double," invented by the sociologists Kevin D. Haggerty and Richard V. Ericson, to be particularly compelling when thinking about such inscriptions.[2] Data doubles stand in for us in bureaucracies. They represent us on paper or in computer systems. They can be powerful, when, for instance, they convince a bank to provide us a mortgage or serve as evidence for a prosecution. Sociologists have offered other useful terms, too, like the concept of the "shadow body," employed by Ellen Balka and Susan Leigh Star to think through the way hospitals create records about us that linger.[3] Still others use terms like the "digital dossier," "data exhaust," or "data shadows," but in this essay I will use data double exclusively.[4]

The second type of personal data inscription derives from the aggregation and analysis of data doubles. Such inscriptions can include charts, tables, maps, or even algorithms, and so there is no single obvious term we can use to lump them all together. Let's call them data aggregates here.

Which have more power: data doubles or data aggregates? It's tempting to side with data doubles. They relate directly to an individual and so link a person to whoever possesses that double, whether a government, corporation, or some other bureaucracy. When a student takes a standardized test, for instance, we think of the test and the raw score it yields as being very powerful, since the raw score represents (in a

---

[1] Markus Friedrich, "Genealogy as Archive-Driven Research Enterprise in Early Modern Europe"; Rebecca Lemov, "Anthropology's Most Documented Man, Ca. 1947: A Prefiguration of Big Data from the Big Social Science Era"; Joanna Radin, "'Digital Natives': How Medical and Indigenous Histories Matter for Big Data," all in this volume.

[2] Kevin D. Haggerty and Richard V. Ericson, "The Surveillant Assemblage," *Brit. J. Sociol.* 51 (2000): 605–22. I use "inscription" in the spirit of Bruno Latour, *Science in Action: How to Follow Scientists and Engineers through Society* (Cambridge, Mass., 1987), chap. 6.

[3] Ellen Balka and Susan Leigh Star, "Mapping the Body across Diverse Information Systems: Shadow Bodies and How They Make Us Human," paper presented at the 4S Annual Meeting, Cleveland, Ohio, 2 November 2011, http://www.4sonline.org/files/print_program_abstracts_111007.pdf (accessed 8 February 2017).

[4] See, e.g., Daniel J. Solove, *The Digital Person: Technology and Privacy in the Information Age* (New York, 2004).

heavily limited and reduced fashion) the test taker. Yet later aggregations can be quite powerful as well. The average score for this student's school might well determine whether that school can remain open or how well it will be funded and thereby affect the school, its community, and the student's life through her school. To add a further complicating factor, the student's raw score may matter less than a percentile score derived by interpreting her performance in the context of all other students' performances—a new data double, in other words, may have to be constructed with reference to a data aggregate. There are no simple answers to questions of power, and so of economics and politics, relating to personal data.

Still, we can discern shifts over time in the distribution of power among data doubles and data aggregates and in the relationship of doubles to aggregates. First and foundationally, we can say that over the last two centuries, data doubles have become longer lived and more powerful than they previously were. With that trend in mind, we can further trace out three periods, three "acts" in the history of personal data, each characterized to a significant degree by the particular relationship of doubles to aggregates. The first act, which spanned the nineteenth century, saw the reign of aggregates derived from doubles and the invention of common means for ordering society. In the second act, which lasted until the 1970s, individuals were increasingly viewed through a statistical lens, which meant aggregates spawned a growing proportion of doubles used to fit people into their proper places in a mass society. Finally, in our own third act of personal data, more doubles met more (newer, nimbler) aggregates, inspiring some to dream of a dawning, fairer era of unbounded individualism. Yet such dreams required ignoring the persistent power of traditional aggregates in data-driven technologies—data, for instance, has not unmade race or racism. Nor has it unmade personal data's managerial drive or capitalist character. On the contrary, third-act dreams have fueled a more extensive extraction of value and a more intense (perhaps tragic) management of the people whom they were meant to free from their fetters.

ACT I: STRUCTURING THE SOCIAL

The story of personal data in the nineteenth century has been told before, but never as a story about personal data. First, it appeared as part of the history of statistics, occasioned to some extent by the desire of historians in the 1970s and 1980s (around the time the third act opened) to uncover histories that might shed light on the rise of mathematical statistics across the sciences and in independent academic programs beginning in the decades after World War II.[5] Writing at a moment when statistics were closely associated with the physical sciences, these scholars revealed the field's surprising origins: in law, gambling, statecraft, and the human sciences, as well as astronomy and mathematics. Personal data, they explained, came in a rush, an "ava-

---

[5] Lorraine Daston, *Classical Probability in the Enlightenment* (Princeton, N.J., 1988); Alain Desrosières, *The Politics of Large Numbers: A History of Statistical Reasoning* (Cambridge, Mass., 1998); Gerd Gigerenzer, Zeno Swijtink, Theodore Porter, Lorraine Daston, John Beatty, and Lorenz Krüger, *The Empire of Chance: How Probability Changed Science and Everyday Life* (Cambridge, 1989); Ian Hacking, *The Emergence of Probability: A Philosophical Study of Early Ideas about Probability, Induction, and Statistical Inference* (Cambridge, 1975); Hacking, *The Taming of Chance* (New York, 1990); Theodore M. Porter, *The Rise of Statistical Thinking 1820–1900* (Princeton, N.J., 1986); Stephen Stigler, *The History of Statistics: The Measurement of Uncertainty before 1900* (Cambridge, Mass., 1986). On independent statistics programs, see Alma Steingart, "Conditional Inequalities: American Pure and Applied Mathematics, 1940–1975" (PhD diss., MIT, 2013), 28–9.

lanche of printed numbers," as Ian Hacking has memorably put it, and provided much grist for the early statistical mill starting in 1820–40 Europe. The avalanche marked the ascent of new bureaucracies given life by fears of social disorder and dreams of controlling subject or troubled populations, including Native Americans in the United States or potential revolutionaries in France and Germany.[6]

In the United States, the census of 1980, which provoked controversy over "undercounting" of minority groups, drew attention from mainstream historians to once rather technical topics, like the history of census taking or the evolution of vital registration systems, that had previously been most useful only as guides to the sources employed by early social historians.[7] Leading works after 1980 turned their attention to questions of power: to the roots of censuses in subject colonies, to the inherently political nature of state-administered counts, and to the resistance to enumeration that plagued nineteenth-century counters.[8] The influence of Foucauldian ideas of biopolitics and governmentality, turning attention to the capacity of quantification to make and manage new kinds of abstract populations or subjects, intersected with postcolonial perspectives to highlight the subtle forms of influence and coercion flowing through even such apparently trivial objects as a questionnaire or so unintimidating a person as a Civil War veteran-cum-1890s census enumerator.[9]

Having decided to look, historians realized that some sort of paper trail had once followed in one way or another a wide variety of Americans via the manuscript census (after 1850, when the census starting recording individuals); church or government records of births, marriages, and deaths; mortgage records or cotton-picking quotas tied to chattel slaves; anthropological ledgers containing measures of dead Native Americans' crania; asylum files for the mentally ill; school records for children; credit reporters' catalogs of merchants' characters; small town farmers' account books tracking webs of debt; life insurers' files on farmers' family health histories; and federal pension files for Union soldiers.[10] And then there were, of course, personal records like the family bible, which tracked in private homes the sort of genealogical data that Markus Friedrich discusses in this volume. Most such records would seem quite porous by

[6] Ian Hacking, "Biopower and the Avalanche of Printed Numbers," *Hum. Soc.* 5 (1982): 279–95; Oz Frankel, *States of Inquiry: Social Investigations and Print Culture in Nineteenth-Century Britain and the United States* (Baltimore, 2006).

[7] On the census of 1980, see Margo J. Anderson, *The American Census: A Social History* (New Haven, Conn., 1988); William Alonso and Paul Starr, eds., *The Politics of Numbers* (New York, 1989). The most prominent earlier work is James Cassedy, *Demography in Early America: Beginnings of the Statistical Mind, 1600–1800* (Cambridge, Mass., 1969). On its use by social historians, see Michael Kammen, review of *Demography in Early America*, by James Cassedy, *J. Amer. Hist.* 57 (1970): 402–4.

[8] Anderson, *American Census* (cit. n. 7); Patricia Cline Cohen, *A Calculating People: The Spread of Numeracy in Early America* (Chicago, 1983).

[9] Geoffrey C. Bowker and Susan Leigh Star, *Sorting Things Out: Classification and Its Consequences* (Cambridge, Mass., 1999); Martha Hodes, "Fractions and Fictions in the United States Census of 1890," in *Haunted by Empire*, ed. Ann Laura Stoler (Durham, N.C., 2006), 240–70.

[10] Anderson, *American Census* (cit. n. 7), 36; Bonnie Martin, "Slavery's Invisible Engine: Mortgaging Human Property," *J. South. Hist.* 76 (2010): 817–66; Edward E. Baptist, *The Half Has Never Been Told: Slavery and the Making of American Capitalism* (New York, 2014), 131–43; Ann Fabian, *The Skull Collectors: Race, Science, and America's Unburied Dead* (Chicago, 2010); Theodore M. Porter, "Funny Numbers," *Cult. Unbound* 4 (2012): 585–98; Josh Lauer, "From Rumor to Written Record: Credit Reporting and the Invention of Financial Identity in Nineteenth-Century America," *Tech. & Cult.* 49 (2008): 301–24; Winifred Barr Rothenberg, *From Market-Places to a Market Economy* (Chicago, 1992); Sharon Ann Murphy, *Investing in Life: Insurance in Antebellum America* (Baltimore, 2010); Drew Gilpin Faust, *This Republic of Suffering: Death and the American Civil War* (New York, 2008).

modern standards, and most have been lost to us today. Still, they demonstrate that the creation and gathering of personal data are anything but recent phenomena. Indeed, the evidence of long-established bureaucratic drives to track populations confronts us as we wander the streets, as Anton Tanter and Reuben Rose-Redwood have shown in their studies of house numbering in the United States and Europe: house numbers arose first to aid in governing households, both in the collection of taxes and in the maintenance of military rolls.[11]

In more recent years, we have looked to the past with eyes made curious about older data-based technologies—like charts, maps, and even paper forms—by our encounters with computers, the Internet, data analysis software, and smartphones. Some scholars have looked to the past for inspiration, like the data visualization researcher Michael Friendly who has called attention to the era after the avalanche of printed numbers, from 1850 to 1910 (or so) as the "golden age of statistical graphics."[12] Others have excavated deep histories for our current fascinations with geographic information system (GIS) mapping, the Internet "form," or even the PDF.[13] Slowly, then, we have been sketching personal data's first act.

Let me trace its outlines. In the nineteenth century, states, corporations, and voluntary organizations (the NGOs of the past) began producing data doubles to meet their governing ambitions. Those doubles tended to be short lived, to travel only short distances, and to reflect or record the exercise of power, rather than serve as a source of power. Data aggregates derived from all those doubles mattered much more during the "golden age of statistical graphics."[14] Aggregations helped justify abstractions like the "nation," and they created new categories to describe and class people.[15]

An example will add some texture to my sketch. Examine figure 1, a map illustrating the prevalence of hernia in the American north in the mid-nineteenth century at the time of the Civil War. The map's story began in January 1864. That month the Provost Marshal General's Bureau, which had been charged by Congress during the Civil War with administering a nationwide draft and overseeing military volunteers, created a Medical Branch under the supervision of Jedediah Hyde Baxter. That branch in turn built a network of surgeons who administered medical exams for draftees and volunteers. In the process of each exam, the surgeon, if he followed Baxter's instructions, asked the by-then-naked draftee to "stand erect, place his heels together, and raise his hands vertically above his head, the backs together, in which position he was required to cough and make other expulsive movements, while the abdomen, the inguinal rings, and the scrotum were examined for hernia."[16] Examining surgeons in turn submitted detailed

---

[11] Anton Tantner, "Addressing the Houses: The Introduction of House Numbering in Europe," *Histoire & Mesure* 24 (2009): 7–30, on 17–8; Reuben Rose-Redwood and Anton Tantner, "Introduction: Governmentality, House Numbering, and the Spatial History of the Modern City," *Urban Hist.* 39 (2012): 607–13.

[12] Michael Friendly, "The Golden Age of Statistical Graphics," *Statist. Sci.* 23 (2008): 502–35.

[13] Susan Schulten, *Mapping the Nation: History and Cartography in Nineteenth-Century America* (Chicago, 2012); Ben Kafka, "Paperwork: The State of the Discipline," *Book Hist.* 12 (2009): 340–53; Lisa Gitelman, *Paper Knowledge: Toward a Media History of Documents* (Durham, N.C., 2014).

[14] Friendly, "Golden Age" (cit. n. 12).

[15] Hacking, "Biopower" (cit. n. 6).

[16] J. H. Baxter, *Statistics, Medical and Anthropological, of the Provost-Marshal-General's Bureau, Derived from Records of the Examination for Military Service in the Armies of the United States during the Late War of the Rebellion, of over a Million Recruits, Drafted Men, Substitutes, and Enrolled Men* (Washington, D.C., 1875), 1:iv.

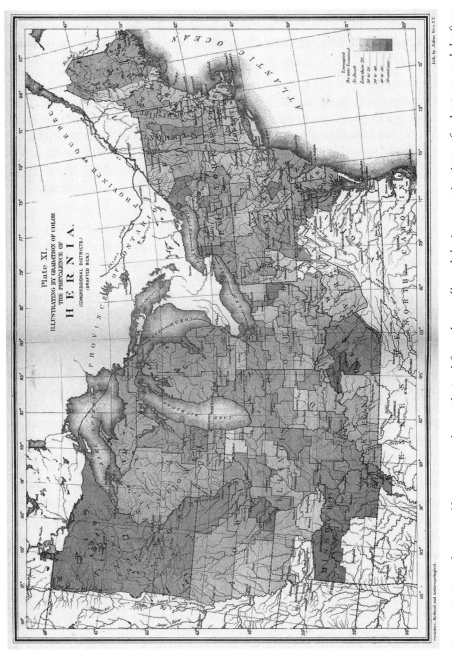

Figure 1. Baxter's map of hernia prevalence, derived from data collected during examinations of volunteers and draftees during the American Civil War. Baxter, Statistics (cit. n. 16), pl. xi.

records to Baxter after they accepted or rejected each candidate for military service. Baxter compiled these examination records and proceeded to make a series of statistical analyses.

Baxter's hernia map depended on blanks of paper filled out by surgeons for over half a million men. That meant half a million data doubles. Only one question on each paper really mattered to the individuals being examined and to the United States government's military apparatus: question 17 asked whether the examined individual was accepted or rejected and then, if the individual was rejected, queried the cause. The rest of the blanks were, in a sense, ancillary to the examination for war. They asked about age, nativity, occupation, "complexion," chest girth, "white or colored," and quite a few more factors, including eye color.[17] The blanks reflect medical ideas of the time and evidence Baxter's keen interest in using the war examinations to conduct statistical investigations about the American population. This appears to have been the surgeons' understanding as well, since in the letters they wrote Baxter to accompany their doubles, the few who commented about the blank forms either requested a better system of disease classification "for statistical purposes," or complained about the time lost to filling out the blanks.[18]

Baxter's data doubles themselves had little power or value. They might have had some disciplining effect in keeping examining surgeons honest in their judgments. Yet the fact that Baxter corresponded extensively with those surgeons, asking their advice about procedural matters, suggests he trusted his examiners.[19] The blanks mainly recorded a decision (reject or accept) that had already been made. They then went on to live secluded lives. Many blanks persist today in the National Archives, but they have been relegated to the Provost Marshal General's files, while many other individual service records were long ago integrated into individual files used to administer pensions or track Civil War service.[20] The short active life span and limited range of these data doubles appear typical of personal data's first act, just as it is typical that the data aggregates produced from those doubles proved powerful and traveled widely.

Baxter's blanks became much more significant when he brought them together in the 1870s for his statistical inquiries. Together, they enabled the creation of new ways of conceiving of the nation and of those within it. Baxter argued that his "statistical matter" related to "the people; the men engaged in every occupation; the professional man and the man of letters, the trader, the merchant, the clerk, the artisan, and the unskilled laborer; the rich man and poor man; the robust and the crippled; in short to the citizens of the United States, both native and foreign-born, and does, it is believed, illustrate the physical aptitude of the nation for military service."[21] Baxter crafted an image of a nation, one that conceived of the United States in terms of a statistical population, in the manner championed by European thinkers like Adolphe Quetelet,

---

[17] Ibid., v.

[18] Ibid., 167, 345.

[19] On quantifying as a "technology of distance" and as a means by which a bureaucracy disciplines its field agents, see Theodore M. Porter, *Trust in Numbers: The Pursuit of Objectivity in Science and Public Life* (Princeton, N.J., 1995), ix.

[20] For discussion of the compiled military service records and of the Provost Marshal General File, see http://www.archives.gov/research/military/civil-war/resources.html#discuss (accessed 8 February 2017); http://www.archives.gov/research/guide-fed-records/groups/110.html (accessed 8 February 2017). On Civil War personal data, efforts to recover information about veterans, and the compiled military service records generally, see Faust, *Republic of Suffering* (cit. n. 10), 255.

[21] Baxter, *Statistics* (cit. n. 16), vi.

presented as a map, echoing earlier Prussian examples, and set to depict a distinct set of social fears: here of a nation beset by disability and disease.[22] His lithographer, Julius Bien, printed a series of similar and if anything even more elaborate maps and charts around the same time for the census director Francis Amasa Walker, whose maps "aided state governance by transforming people into an abstract 'population' with particular attributes that could be managed and administered," according to historian Susan Schulten.[23]

Baxter's map structured social imaginings, blending medical knowledge with political boundaries to make the nation visible in a new way. In his reading of statistical work through a Foucauldian lens, Hacking explained statisticians' creative potential to define social entities, a potential made real in every new category generated by enumerators numbering a population.[24] In Baxter's map, this power manifested itself as a transformation of a political designation (the congressional district) into a medical object, with the possible transformation, in turn, of citizens' perceptions of the places in which they lived.

Data aggregates could also exert power through the governing powers of states and localities, by making populations "legible," to use James Scott's terminology.[25] Placed in the hands of the burgeoning public health apparatus, such perceptions could become a reason for state intervention in daily affairs.[26] Local elites apparently understood the potential power of data aggregates even at the time of the surgeons' examinations, and that understanding may have perverted Baxter's collected hernia data by making some districts appear more herniated than they were. During the draft, Baxter explained, local quotas for recruits were set by comparing ratios of men serving to eligible men in each district. Hernia, he wrote, was an easily identified disqualifier, and so encouraging extra scrutiny for hernias became a likely means by which "local authorities, or even interested private individuals" could lower a locality's quota.[27] So, as an aggregated mass, data doubles could exert power over the individuals they described, but that power came through group classifications and governing regulations.

Baxter's maps and the rest of his data aggregates proved much more durable and mobile than the Provost Marshal General's data doubles. Medical journals praised Baxter's work and particularly its charts for the vision of America and its citizens they laid bare.[28] The book and maps were distributed widely, as the *New York Medical Journal* explained, by senators and congressmen to interested constituents.[29] Such mobility was typical.

Friendly's graphical "golden age" qualified as such because bureaucracies with money had the will and resources to print nice lithographs, whether maps, charts, or diagrams, to accompany a new genre of social reform literature and statistical re-

[22] See ibid., lxxvii–lxxix; Schulten, *Mapping the Nation* (cit. n. 13), 164.

[23] Schulten, *Mapping the Nation* (cit. n. 13), 184.

[24] Hacking, "Biopower" (cit. n. 6).

[25] James C. Scott, *Seeing Like a State: How Certain Schemes to Improve the Human Condition Have Failed* (New Haven, Conn., 1999), 2.

[26] Charles E. Rosenberg, *The Cholera Years: The United States in 1832, 1849, and 1866* (Chicago, 1987).

[27] Baxter, *Statistics* (cit. n. 16), 88–9. This is a case of funny numbers. See Porter, "Funny Numbers" (cit. n. 10).

[28] See, e.g., *New York Med. J.* 24 (1876): 70–3, on 72. Schulten cites other reviews in *Mapping the Nation* (cit. n. 13), 182n54.

[29] *New York Med. J.* (cit. n. 28), 73.

porting.[30] The will to publish came out of a will to govern and control, to be sure. Hacking has pointed to the widespread belief that "one can improve—control—a deviant subpopulation by enumeration and classification."[31]

But it also came out of a desire to display. It is more than a coincidence that the golden age bears the same dates as the craze for Victorian exhibitions and world fairs set off by the so-called Crystal Palace exhibition in London in 1851.[32] When states, corporations, or reform organizations printed big, beautiful books or artful statistical graphics, they knew they could display their wares at the next exhibition, which was never very long in coming. The history of statistical display in the exhibition era remains to be written, and yet evidence from guidebooks, official descriptions, and printed catalogs suggests statistical artistry regularly on display across Europe and the Americas.[33] Large statistical presentations even showed up in parades.[34] States used statistical charts and books to showcase their modernity and their capacity to know their subjects, while corporations made similar moves to enhance their prestige. One prime case is the publication, paid for by seventeen American life insurers, of a volume on American vital statistics. The insurers printed it not only in New York, but also in London, Paris, and Madrid, apparently so they could brag at home of the reach of their influence, since they sold little insurance in Europe.[35]

Having been fashioned either to tend to populations or to enhance prestige, data aggregates usually traveled within exchange economies driven by a spirit of collective scientific inquiry. When, for instance, the American Geographical and Statistical Society in New York tried to acquire statistical materials (still a rare commodity in the mid-nineteenth-century United States), it could buy some volumes, but it relied even more on donations and exchanges.[36] The group's librarian reported accessions for a single month in 1859: "By donation, 200 volumes, 1,050 pamphlets, and 34 maps and prints; and by purchase 201 volumes and one globe."[37] Many donated materials came from governments (some foreign, some national, some more local, as in the case of materials submitted by the mayor of Lawrence, Mass., about city schools and finances),[38] others from regional statistical or scientific societies like the Academy of Sciences in Stockholm or "a German natural history society,"[39] and others from private corporations and private authors.[40] Data doubles did not usually move in these networks.

[30] Frankel calls this "print statism" in *States of Inquiry* (cit. n. 6), 8–13.

[31] Hacking, *Taming of Chance* (cit. n. 5), 3.

[32] See Robert W. Rydell, *All the World's a Fair: Visions of Empire at American International Expositions* (Chicago, 1984).

[33] For an example of a printed catalog, see Mutual Life Insurance Company of New York, *Introduction to the Exhibit of the Mutual Life Insurance Company of New York* (Paris, 1900); for references or images of statistical charts in displays, see, e.g., John D. Philbrick, *The Catalogue of the United States Collective Exhibition of Education* (London, 1878), 48; E. Monod, *L'Exposition Universelle de 1889* (Paris, 1890), 545.

[34] Willard C. Brinton, *Graphic Methods for Presenting Facts* (New York, 1914), 343.

[35] James Wynne, *Report on the Vital Statistics of the United States, Made to the Mutual Life Insurance Company of New York* (New York, 1857).

[36] Paul J. FitzPatrick, "Statistical Societies in the United States in the Nineteenth Century," *Amer. Stat.* 11 (1957): 13–21.

[37] "Proceedings," "Library Department," *J. Amer. Geogr. Stat. Soc.* 1, no. 10 (1859): 287.

[38] "Library Department," *J. Amer. Geogr. Stat. Soc.* 1, no. 7 (1859): 223.

[39] "Library Department," *J. Amer. Geogr. Stat. Soc.* 1, no. 9 (1859): 285.

[40] "Library Department" (cit. n. 38), 223.

They normally remained local in scope, seldom traveled in exchanges, and were, except in the case of credit reporting, not made into commodities themselves. We might attribute data doubles' quiet lives to the same sort of fear of centralized power that stimulated the American Revolution, to nineteenth-century individuals who refused to be recorded, numbered, or tracked.[41] Yet that cannot be the whole story, since those same revolutionaries were willing to allow themselves to be counted by the census in exchange for representation, and their near descendants proved open to bargaining away their personal data for valuable commodities like life insurance.[42] More likely, data doubles may not have moved much because of the expense of moving them: as the economists might say, information, storage, and transport costs remained high.[43] To support this conclusion, we need only look to Prussia, which regularly sold its counting cards "as spoilage" once aggregations were completed, as Christine von Oertzen notes.[44]

Two processes controlled the political economy of personal data in its first act. First, personal data provided the fodder necessary to structure emerging ideas of the "social" within and between nations. Maps and charts circulating in expositions and among bureaucrats provided tools for making sense of a nation's inhabitants in broad terms, using big categories like race, class, sex, state, and region and smaller categories related to health or occupation. They disseminated biopolitical tools and taught the means for making people into "populations," whether for the purpose of allocating them political power, providing them relief, or seeking reform. They also advertised the power of states, corporations, and reformers to intervene in the lives of the masses (or at least dream that they could), winning for their creators praise and prestige in an economy where both served as currency.

Second, personal data made possible new kinds of subjectivity. Cultural power flowed through data doubles, even if they were fleeting, local, and individually inconsequential. As bureaucracies came to fill their files with individual records, for instance, they participated in making the nineteenth century the "age of the first person singular," as Ralph Waldo Emerson dubbed it.[45] Data doubles in this period came to take individuals as their primary targets, rather than older units such as the household. The act of enumeration can, in and of itself, teach people to see themselves through the lens of new categories, as Martha Hodes has argued.[46] Individuals might therefore have come to see themselves, as individuals, through the options allowed on a mass-produced blank form. With the widespread publication of inscriptions based on aggregates, even more people—like the many who attended international expositions—

[41] On fears of tyranny and corruption, see Bernard Bailyn, *The Ideological Origins of the American Revolution* (Cambridge, Mass., 1967), 48–54; on earlier concerns about censuses, see Cohen, *Calculating People* (cit. n. 8), 34–40, 47–80, 231n50.

[42] "The Apportionment of Members among the States," *Federalist Papers*, no. 54 (1788); Dan Bouk, *How Our Days Became Numbered: Risk and the Rise of the Statistical Individual* (Chicago, 2015), xxiv.

[43] Naomi R. Lamoreaux, Daniel M. B. Raff, and Peter Temin, "Beyond Markets and Hierarchies: Toward a New Synthesis of American Business History," *Amer. Hist. Rev.* 108 (2003): 404–33, on 414.

[44] Christine von Oertzen, "Machineries of Data Power: Manual versus Mechanical Census Compilation in Nineteenth-Century Europe," in this volume.

[45] Louis P. Masur, "'Age of the First Person Singular': The Vocabulary of the Self in New England, 1780–1850," *J. Amer. Stud.* 25 (1991): 189–211, on 205–6.

[46] Hodes, "Fractions" (cit. n. 9), 256–8.

encountered social structures and categories that they could use to define themselves, as individuals.

ACT II: FITTING INTO THE MASS

In the first half of the twentieth century, a new adjective joined the abstraction "society," the term that was a focus in personal data's first act. Society became "mass," and mass pops up throughout our twentieth-century histories explaining a series of fundamental transitions in ordinary life. In the United States, investment capital flowed into gigantic trusts around the turn of the century, facilitating a "great merger movement" that integrated industries horizontally and created monopolies or oligopolies. Those companies in turn integrated vertically, controlling within a single firm the entire supply chain and distribution system.[47] They brought to maturity mass production, mass distribution, and mass marketing.[48] With radio, nickelodeons, and new techniques in advertising, mass society came into being. Which is not to say that mass conformity took hold, or that all people truly had the same experiences.[49] But for those selling the same product to the large mass, it became convenient to identify or even construct uniformities. And so our dominant stories of personal data in this era highlight the collection of data—by social scientists, pollsters, and advertisers especially—meant to constitute and describe the "mass."[50]

The urge to understand the mass went hand in hand with a belief that universals trumped particularities in social and human science research, and that archives might in fact be filled to a point approaching total coverage.[51] Existing historical accounts point to the early years of the Cold War as the apotheosis of the drive to know and measure everything, from natural phenomena like the aurora borealis, as Elena Aronova points out in this volume, to the most personal of data, like the dreams about which Lemov writes.[52] A fantasy of preservation flourished at midcentury, whether in popular literature, such as Isaac Asimov's *Foundation*, or in enthusiasms for time capsules and seed banks.[53]

The principal urge of personal data's second act was to fit individuals into the mass. Fitting could involve sorting and prediction—it could mean determining how a particular person related to and would in the future relate to the sort of social categories

[47] Naomi R. Lamoreaux, *The Great Merger Movement in American Business, 1895–1904* (Cambridge, 1988).
[48] Alfred D. Chandler Jr., *The Visible Hand: The Managerial Revolution in American Business* (Cambridge, Mass., 1977), pt. 3.
[49] Lizabeth Cohen, *Making a New Deal: Industrial Workers in Chicago, 1919–1939* (Cambridge, 1990), 120–47.
[50] Sarah E. Igo, *The Averaged American: Surveys, Citizens, and the Making of a Mass Public* (Cambridge, Mass., 2007); Olivier Zunz, *Why the American Century?* (Chicago, 1998).
[51] This was a good moment, in other words, for "sciences of the archive." See Lorraine Daston, "The Sciences of the Archive," *Osiris* 27 (2012): 156–87.
[52] See Elena Aronova, "Geophysical Datascapes of the Cold War: Politics and Practices of the World Data Centers in the 1950s and 1960s," in this volume; Lemov, "Anthropology's Most Documented Man" (cit. n. 1). On the widespread gathering of personal data in the early Cold War, see also Lemov, "Filing the Total Human: Anthropological Archives from 1928 to 1963," in *Social Knowledge in the Making*, ed. Charles Camic, Neil Gross, and Michelle Lamont (Chicago, 2011), 119–50; Joel Isaac, "Epistemic Design: Theory and Data in Harvard's Department of Social Relations," in *Cold War Social Science*, ed. Mark Solovey and Hamilton Cravens (New York, 2012), 79–95.
[53] Isaac Asimov, *Foundation* (New York, 1951); Jack Ralph Kloppenburg Jr., *First the Seed: The Political Economy of Plant Biotechnology, 1492–2000* (Cambridge, 1988), chap. 7.

created in the prior century. Fitting could also mean something more procrustean—it could mean managing and controlling individuals such that they acted more like the mass.[54] In many cases, the makers of personal data sought to fit their contemporaries to the mass mold in both ways at once.

For those building a mass society, it made sense to view individuals through statistical lenses. By making people into "statistical individuals," it became possible to sort them according to the futures the statistics predicted they would have.[55] World War I created fertile bureaucratic soils for advocates of sorting and prediction, most notably in the field of mental testing, which promised to produce representations of thinking selves to aid officers in selecting and placing soldiers.[56] After the war, mental testing inspired other sorts of statistical individuals. Educational Testing Service (ETS), for instance, succeeded in making the SAT examination, which scored test takers in normalized percentiles, into a necessity for getting into college. Buoyed by success, ETS's first president, Henry Chauncey, imagined designing a "Census of Abilities" capable through extensive testing of a "great sorting," as Nicholas Lemann has described it.[57] In the meantime, social scientists in colleges and universities used mental tests to predict future success in corporate life, which drew the financial support of men like W. T. Grant, who used his department store wealth to fund the most famous such study at Harvard beginning in the 1930s.[58]

Of course, one could control or manage men without making statistical individuals, as becomes clear when we consider the production of representations of working selves in the early twentieth century. The spread of "scientific management" and "time" and "motion" study, as propounded by F. Winslow Taylor and Frank Gilbreth, meant that data doubles came to fill company files, with papers describing the motions of workers in the machine shop or surgeons in their operating chambers, while standardized evaluation forms defined some workers as exemplary and others as in need of improvement.[59] In Henry Ford's extreme case, workers' private and working lives blended together in the records of the Ford Sociological Department, which was charged with keeping workers on the straight and Americanized narrow.[60] With the coming of automation, some individuals' motions even became the pattern for a later machine's operation.[61] The novelist Kurt Vonnegut captured concerns over the appropriation and reproduction of a data double—so as to put the model for that double

[54] Beniger casts the urges to control and manage, even in what this essay would call the third act of personal data, in terms of a long "control revolution" beginning with the Industrial Revolution. See James R. Beniger, *The Control Revolution: Technological and Economic Origins of the Information Society* (Cambridge, Mass., 1986).

[55] Bouk, *Our Days* (cit. n. 42).

[56] John Carson, *The Measure of Merit: Talents, Intelligence, and Inequality in the French and American Republics, 1750–1940* (Princeton, N.J., 2006), chap. 6.

[57] Nicholas Lemann, "The Great Sorting," *Atlantic Monthly* (September 1995): 84–100, on 88–90.

[58] Joshua Wolf Shenk, "What Makes Us Happy?," *Atlantic Monthly* (June 2009): 36–53; Heather Munro Prescott, "Using the Student Body: College and University Students as Research Subjects in the United States during the Twentieth Century," *J. Hist. Med.* 57 (2002): 3–38, on 19–21.

[59] Frank B. Gilbreth, *Primer of Scientific Management* (New York, 1912); W. E. Bloom, *Lincoln Factory Executive Service: Time, Motion, and Methods Study* (Cleveland, 1960); Caitjan Gainty, "'Going after the High-Brows': Frank Gilbreth and the Surgical Subject, 1912–1917," *Representations* 118 (2012): 1–27.

[60] Stephen Meyer III, *The Five Dollar Day: Labor Management and Social Control in the Ford Motor Company, 1908–1921* (Albany, N.Y., 1981), chap. 6.

[61] David F. Noble, *Forces of Production: A Social History of Industrial Automation* (New York, 1984).

out of work—in *Player Piano*: "Makes you feel kind of creepy, don't it, Doctor, watching them keys go up and down? You can almost see a ghost sitting there playing his heart out."[62] In such data doubles, prediction mattered less than control. Managers studied existing work methods to find the "one best way" and recorded efficient workers' actions so as to spur other laborers, or eventually machines, to work to the same standard.[63]

Overlapping efforts to predict, sort, and manage produced a profusion of data doubles in personal data's second act, each tied to a particular kind of self in the making. The Eugenics Record Office, for instance, set out to trace Americans' heredity, inscribing on paper a eugenic self for about 35,000 people.[64] Colleges measured new entrants as part of physical education curricula or took "posture photos" of students as a step toward literally straightening America's backbone, leaving behind records of embodied selves.[65] Consuming selves emerged in mail order companies' files, in the 7.5 million customer records created by Sears, Roebuck & Company around 1920.[66] By 1938, the tabulating cards kept by the Social Security Board recorded the pay records of 40 million workers, even as the Internal Revenue Service described the taxable selves of a growing proportion of the population, over one-third of Americans in 1945.[67] In some cases, people crafted identities from these bureaucratic selves, like those who displayed their social security numbers as tattoos on their bodies.[68]

Industrial consolidation beginning at the turn of the century followed by two world wars encouraged the growth of managerial apparatuses dependent on the constant flow of paper people. Those factors, as part of the larger process of making a mass society, made data doubles more valuable, while new technologies—first and foremost, the paper card, in conjunction with the card filing system, the tabulator, and the microcard—made data doubles more practical. If the nineteenth century was the era of the ledger, and our own times are dominated by computers and "smart" devices, the vast majority of the twentieth century belonged to the card file, to systems born of librarians' organizing ambitions coincident with the search for new efficiencies by banks and insurers.[69]

Life insurers were early adopters of card-filing technologies, having realized the capacity of cards to allow for wider sharing of data. Medical directors from the nation's largest insurers banded together in 1902 to form the Medical Information Bureau (M.I.B.), which came into being as a series of paper cards (a complete set for each company, supplemented and updated through Melvil Dewey's Library Bureau) con-

---

[62] Kurt Vonnegut, *Player Piano* (1952; repr., New York, 2006), 32.

[63] Robert Kanigel, *The One Best Way: Frederick Winslow Taylor and the Enigma of Efficiency* (New York, 1997).

[64] Garland E. Allen, "The Eugenics Record Office at Cold Spring Harbor, 1910–1940: An Essay in Institutional History," *Osiris* 2 (1986): 225–64, on 251.

[65] Ron Rosenbaum, "The Great Ivy League Nude Posture Photo Scandal," *New York Times Magazine*, 15 January 1995.

[66] On mail order as a model site, second only to life insurance, for the use of statistics and personal data in business planning, see J. George Frederick, *Business Research and Statistics* (New York, 1920), 70–3.

[67] Arthur J. Altmeyer, "Three Years' Progress toward Social Security," *Soc. Security Bull.* 1 (1938): 1–7, on 1; Daniel J. Boorstin, *The Americans: The Democratic Experience* (New York, 1974), 209.

[68] Bouk, *Our Days* (cit. n. 42), 230–6.

[69] Gerri Flanzraich, "The Library Bureau and Office Technology," *Libraries & Cult.* 28 (1993): 403–29; Markus Krajewski, *Paper Machines: About Cards and Catalogs, 1548–1929* (Cambridge, Mass., 2011).

taining data on any known impairments discovered during participating companies' routine medical examinations of applicants. The card exchange helped life insurers ferret out potential fraud or medical examiner error and served as a basis for decisions to deny applicants insurance or to issue a substandard policy charging higher premiums. The companies, recognizing that sharing sensitive medical data was not likely to be popular, forbade medical directors from talking about M.I.B. even with their own medical examiners, agents, or managers.[70] Because of the sensitivity of these data doubles, combined with their value to medical directors hoping to better select long-lived applicants, they were well guarded. One medical director aptly explained that his company kept its cards "behind iron doors" to restrict access to a few competent "medical men." Another described a veritable gauntlet of gates and watchful clerks guarding access to the M.I.B. card files.[71] Companies valued these data doubles highly enough to grant them their own safes.[72] Yet even as they shared cards within their private club, M.I.B. members resisted the commodification of their data doubles. When the American Life Convention, an organization composed mostly of smaller, more recently established firms, tried to create a similar exchange and asked to purchase access to some M.I.B. files, the M.I.B. refused the request and used its data power to lure the larger American Life Convention companies into the M.I.B. stables.[73]

Sharing data on cards also facilitated the mass production of statistical individuals. Consider the tabulating card pictured in figure 2. Life insurers used these cards in a 1912 study that collected data on individuals (identified by policy numbers) in order to determine which factors could best be used to predict longevity. The card's categories suggest all the possible factors, from occupation to the term of the life insurance policy purchased to the kind of "impairments" noted by examining doctors at the time of application. A team of life insurance medical directors and actuaries used the study's findings to improve a numerical rating system that could partially automate decisions about whom to grant a policy and that policy's terms. The tiny column "BLD," which stood for "build" and derived from a ratio of weight to height, held the data that served with age as the baseline for the rating systems. These cards, these data doubles, helped create a new data aggregate, as with Baxter's maps. But in this case, a bureaucracy already existed that could immediately take advantage of that new statistical finding. Many life insurance companies' applicants' data doubles would soon be interpreted using the results of the 1912 study. In viewing millions of doubles through the aggregate, insurers made statistical individuals from a significant proportion of the American population.

The second act's characteristic mobilization of data doubles to organize and control became evident in its changing modes of statistical visualization. An AT&T statistician named M. C. Rorty reflected on early efforts to reform statistical graphics and codified

[70] "Record of M. I. B. In the Proceedings of the Association of Life Insurance Medical Directors from May 1892- to December 1916," 7–8, in Folder "Medical Information Bureau 1916," Box RG/13-Subject Files M11, MetLife Archives. On the M.I.B., see Bouk, *Our Days* (cit. n. 42), 78–86.

[71] Dr. Dwight, speaking at 7 October 1914 executive meeting, in "Record of M. I. B. in the Proceedings of the Association of Life Insurance Medical Directors from May 1892- to December 1916" (cit. n. 70), 38.

[72] Frank Wells, speaking at 7 October 1914 executive meeting, in "Record of M. I. B. in the Proceedings of the Association of Life Insurance Medical Directors from May 1892- to December 1916" (cit. n. 70), 38.

[73] R. Carlyle Buley, *The American Life Convention 1906–1952: A Study in the History of Life Insurance* (New York, 1953), 1:440, 451–2, 466–7.

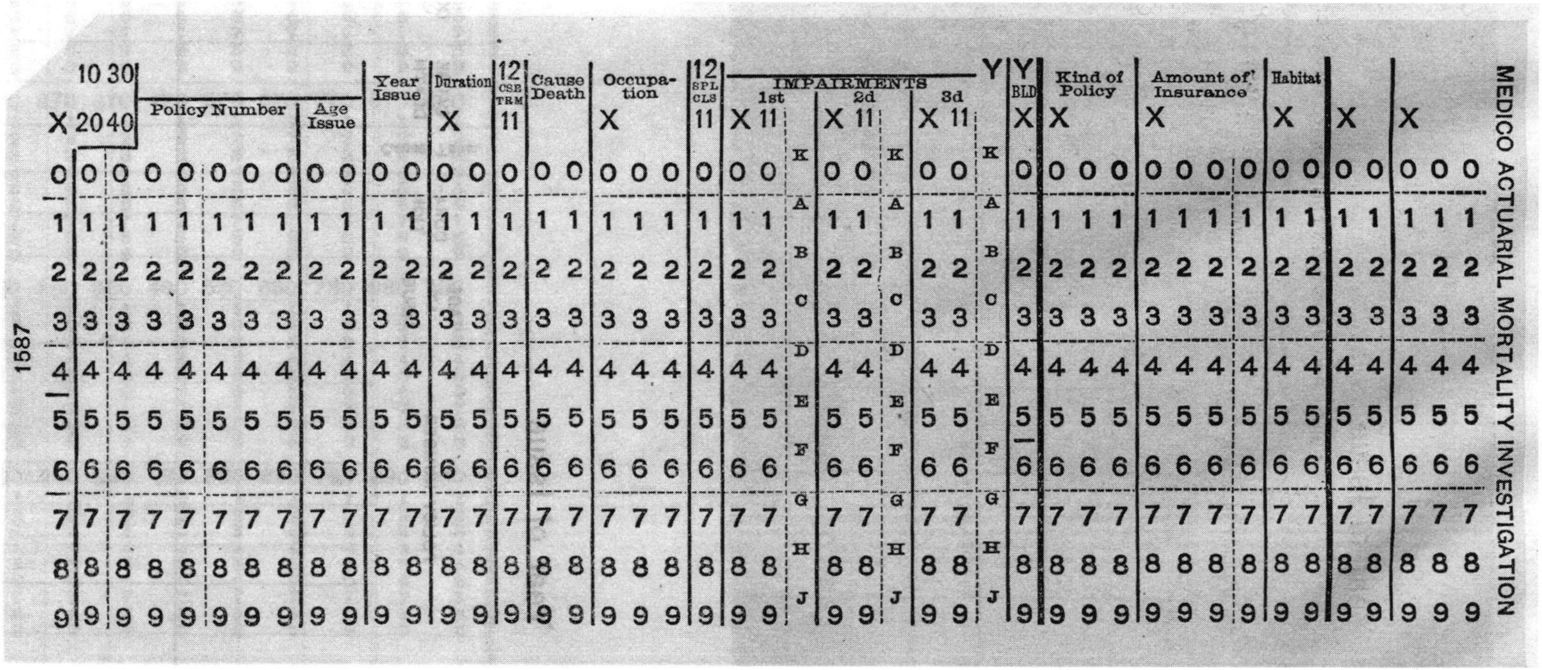

Figure 2. A replica of one of the cards used to record policyholder data for the Medico-Actuarial Mortality Investigation. Association of Life Insurance Medical Directors and Actuarial Society of America, Medico-Actuarial Mortality Investigation (New York, 1912), 1:136.

a new set of graphical norms in a set of articles in the journal *Industrial Management* in 1920. Energized by wartime experience in logistics, Rorty noted the new importance for the industrial executive of readily understandable data. He explained, "even the most thoroughly trained executives, backed up by the best systems of records and accounts, will be aided very greatly in carrying the load of administration if masses of figures are made easy of comprehension by reduction to graphical form."[74] Popular audiences still mattered, even if the world's fairs took a break after World War I, but graphics intended to reach wider audiences tended to be less baroque with fewer ornate maps and many more line graphs and tables (and the occasional cartoon), at least in part because rather than being printed on fine paper in leather bindings, many of the new style graphics were slated to be mass printed and distributed widely, as was the case with Metropolitan Life Insurance Company's *Statistical Bulletin*.[75]

Rorty was one of a new generation of administrative statisticians, about whom we still know too little, tasked with making possible "statistical control" of products and people. Textbooks written to teach this new breed of professional underline the value ascribed to data collection as an end in itself. "Millions of dollars a year are spent in the collection of data," opened one such text.[76] Another, in its distribution of topics, made clear that the collection of data would often have to be contracted and paid for. While some enterprises created data doubles as a matter of course, many more needed to hire surveyors, market researchers, or in-house statisticians to actively gather personal data, sometimes with a hefty charge attached.[77] Frederick L. Hoffman, a life insurance statistician, was like many of his contemporaries in thinking that what we might now call the "data exhaust" of his industry (life insurance) was not enough for effective administration. He built a library with over a hundred thousand resources, while taking regular rambles through Southern cemeteries typing out personal data (birth and death dates) from each gravestone.[78] Storing and maintaining collected personal data added further expense. As Rorty warned his zealous statistical brethren, "it will not do to forget the man who said that his card catalogues were working out beautifully, but he was afraid he would have to give up his business in order to keep them up to date."[79]

While World War I stimulated interest in the production and analysis of statistical individuals and spread the gospel of statistical control widely, World War II provides dramatic examples of how much ideas about the value and power of data doubles changed in the twentieth century. It is now widely understood that census data and registration data, whether gathered by the Nazis or by occupied governments, facilitated the Holocaust.[80] In such cases, military officials gained access to data doubles and used them to find and capture the individuals those doubles represented. In the United States, it does not appear that those responsible for interning Americans of Japanese ancestry had access to data doubles directly. However, historians have discovered that census officials cooperated, avidly, with the internment process, using

---

[74] M. C. Rorty, "Making Statistics Talk," *Indust. Manage.* 60 (1920): 394–8, on 394–5.
[75] To see both the older and newer, spare style, see images in Brinton, *Graphic Methods* (cit. n. 34).
[76] Ibid., 1.
[77] Frederick, *Business Research* (cit. n. 66), 96–7, 101.
[78] Bouk, *Our Days* (cit. n. 42), 116–22, 145.
[79] Rorty, "Making Statistics Talk" (cit. n. 74), 396.
[80] Götz Aly and Karl Heinz Roth, *The Nazi Census: Identification and Control in the Third Reich* (Philadelphia, 2004).

existing census rolls (and the data doubles therein) to generate new inscriptions telling officials how many Japanese Americans lived in a given census block, for instance.[81] Where the Prussian government had once used data doubles as waste paper after census activities, such bits of paper came in the mid-twentieth century to be seen as governing tools that could not only create social categories as before but could also play active roles in finding, disciplining, and destroying individuals. I argue that such a revaluation of the data double was, in fact, the norm in personal data's second act.

Changing modes of resistance further illumine the shifting political economy of personal data. Toward the end of the first act, to take one example, African Americans in the 1880s and 1890s successfully pushed state governments to outlaw the use of race in predicting life spans and pricing insurance.[82] Resistance came as a rebellion against a social category given power over a population's economic opportunities. In contrast, personal data's second act closed only a little after the so-called Free Speech Movement, led by students at the University of California, Berkeley, took as one of its targets the IBM punch card and its warning not to "fold, spindle, or mutilate."[83] The data double, rather than the social structure, became the object of protest.

ACT III: EXALTING AND EXPLOITING THE INDIVIDUAL

Personal data's third act has yet to fully play out. Still, we can draw tentative conclusions about its origins, trace its basic characteristics, and mull over its potentially revolutionary implications. The third act began in the 1970s amid economic crisis, technological change, and liberatory politics. At its center, we find the individual represented by increasingly persistent, interconnected data doubles or sometimes by (seemingly oxymoronic) personalized aggregations. We find the exaltation of new possibilities for individuals alongside growing concerns about the extraction, commoditization, and capitalization of personal data in a manner that amplifies inequalities and could imperil the idea of the liberal subject that data's first act served to establish.

What world wars were to the second act, economic crises were to personal data's third act. The global crises of the 1970s began a series of structural changes in the American (and global) economy that at once shifted capital toward services and finance and away from manufacturing,[84] intensified efforts to cut advertising costs and increase sales by targeting customers more carefully,[85] cut loose a generation of physicists and mathematicians set to work on government projects or at universities and sent them to work in corporate data analysis instead,[86] and limited funding from private and public sources for the production of new data while creating incentives for corporate raiders and cost-conscious boards of directors to take data left as a by-product

---

[81] William Seltzer and Margo Anderson, "The Dark Side of Numbers: The Role of Population Data Systems in Human Rights Abuses," *Soc. Res.* 68 (2001): 481–513, on 492, 498.

[82] Bouk, *Our Days* (cit. n. 42), 41–8.

[83] Steven Lubar, "'Do Not Fold, Spindle or Mutilate': A Cultural History of the Punch Card," *J. Amer. Cult.* 15 (1992): 43–55, on 45–6.

[84] Greta R. Krippner, "The Financialization of the American Economy," *Socio-Econ. Rev.* 3 (2005): 173–208.

[85] Joseph Turow, *Breaking up America: Advertisers and the New Media World* (Chicago, 1997), chap. 6.

[86] David Kaiser, "Booms, Busts, and the World of Ideas: Enrollment Pressures and the Challenge of Specialization," *Osiris* 27 (2012): 276–302, on 285, 298; Emanuel Derman, *Models Behaving Badly: Why Confusing Illusion with Reality Can Lead to Disaster, on Wall Street and in Life* (New York, 2011).

of other activities out of company magnetic tapes and hard disks and market it directly.[87] What had been seen as boring business segments—from banking to media buying to data analysis—came to be seen as among the most creative.[88] Customer loyalty programs, which offer benefits such as discounted prices at the grocery store or rewards for frequent flyers, soon looked like so many golden geese because they also produced extensive data doubles that advertisers, market researchers, and political groups were willing to pay for.[89] Media analysis and credit reporting companies, like A. C. Nielsen and TRW, expanded their business, becoming brokers of data doubles, some gleaned from government sources (the remnants of personal data's first act) and others from new market sources.[90]

Data brokers could reproduce, sell, and transfer data doubles more easily as digital, networked storage displaced card systems and ledgers. In the 1960s, IBM released the System/360 line of computers, which in turn became standard tools for large organizations.[91] Around that same time, work with military projects, building on corporate efforts to more effectively keep track of their workers and materials, led to the creation of the "database" concept and software for manipulating large volumes of data.[92] Networking new and more powerful computers soon allowed for real-time tracking of personal data, most notably with the creation of airline reservation systems in the mid-1960s.[93] Personal data might have found a centralized home in the federal government, but plans for a National Data Center drew controversy that killed the chance of such a project. Personal data would instead reside, in many pieces, in many different (public and private) databases.[94]

At the same time, power and potential resided more in data doubles. Demand for commoditized data doubles flowed from institutions bent on turning doubles into predictions. Highly engineered "risk" systems built on developments of the "actuarial age" that historian Caley Horan has identified in postwar America.[95] Risk systems promised to extend lives, improve justice, and expand credit. They generated personalized risk scores that located in the individual such disparate possibilities as heart

[87] Wendy Nelson Espeland and Paul M. Hirsch, "Ownership Changes, Accounting Practice and the Redefinition of the Corporation," *Accounting Org. Soc.* 15 (1990): 77–96; Philip Mirowski, *Science-Mart: Privatizing American Science* (Cambridge, Mass., 2011), 15–6; Doogab Yi, "Who Owns What? Private Ownership and Public Interest in Recombinant DNA Technology in the 1970s," *Isis* 102 (2011): 446–74.

[88] This shift is illustrated for banking, media buying, and data analysis, respectively, by Scott Patterson, *The Quants: How a New Breed of Math Whizzes Conquered Wall Street and Nearly Destroyed It* (New York, 2010); Joseph Turow, *The Daily You: How the New Advertising Industry Is Defining Your Identity and Your Worth* (New Haven, Conn. 2011); Michael Lewis, *Moneyball: The Art of Winning an Unfair Game* (New York, 2003).

[89] Turow, *Breaking up America* (cit. n. 85), 138–44.

[90] Paul Starr and Ross Corson, "Who Will Have the Numbers? The Rise of the Statistical Services Industry and the Politics of Public Data," in Alonso and Starr, *The Politics of Numbers* (cit. n. 7), 415–47, on 419.

[91] Martin Campbell-Kelly and William Aspray, *Computer: A History of the Information Machine* (New York, 1996), 144–7.

[92] Thomas Haigh, "How Data Got Its Base: Information Storage Software in the 1950s and 1960s," *IEEE Ann. Hist. Comput.* 31 (2009): 6–25; Haigh, "Charles W. Bachman: Database Software Pioneer," *IEEE Ann. Hist. Comput.* 33 (2011): 70–80.

[93] Campbell-Kelly and Aspray, *Computer* (cit. n. 91), 174–5.

[94] Simson Garfinkel, *Database Nation: The Death of Privacy in the 21st Century* (Cambridge, 2000), 16.

[95] Caley Horan, "Actuarial Age: Insurance and the Emergence of Neoliberalism in the Postwar United States" (PhD diss., Univ. of Minnesota, 2011).

disease (via the risk factor popularized in the 1960s and 1970s), criminal activity (with many states tying parole to actuarial scoring systems from the 1970s on), or credit default (with FICO scores developed in the late 1970s).[96] The resulting proliferation of statistical individuals turned attention away from systemic or social causes for disease, recidivism, or credit default. Individuals were supposed to have gained the power—and responsibility—to heal themselves. Many commentators invoke a political project of "neoliberalism" to explain this trend, while Daniel T. Rodgers has argued that the shift away from social explanations with an accompanying skepticism of larger social categories (like race or gender) in favor of a focus on individual choice-making agents with fluid identities and weak social ties characterized an "age of fracture" that transcended political movements.[97]

What had been merely a demand for data doubles came to look more like a new logic of extraction and accumulation with the growth of the Internet and of the tools we now associate with Big Data. New firms found in the Internet the capacity to make money by simultaneously commoditizing personal data while using such data internally in the style of old-school statistical control. Facebook exemplified that blending. It built a trove of customer data, harvested at little cost from its legions of users. Whether they acknowledged the deal or not, Facebook users traded their data for the right to craft an online persona, make connections, and see pictures of their "friends." Facebook may not have sold that data directly, but it did monetize that data. It sold, and sells, finely tuned profiles to advertisers so they can target ads according to the peculiarities of each data double (and presumably of the person that data double represents).

Shoshana Zuboff has argued that a new form of capitalism is in the making. Surveillance capitalism works by hoarding user data to produce "behavioral surplus" (here Facebook's data doubles), a form of capital gleaned from users that the capitalist can then use to produce "prediction products" for sale to third parties, like advertisers or insurers.[98] Facebook's annual reports do not list the value of its data doubles—its surveillance capital—directly, but we can infer that value by looking at the company's revenue per user, which in 2014 averaged $9.45 worldwide, and quite a bit more for just the United States and Canada.[99] When the Organisation for Economic Co-operation and Development (OECD) wanted to determine the dollar value of personal data in 2010, it presented a range of possible metrics, from the cost to companies of having personal data stolen (around $1.70 per person) to the cost of purchasing data from brokers (an average of $35 for a robust record) to the outlying value implied by the cost of identity theft insurance ($155).[100] Many marvel at how apparently valueless "digital

[96] William G. Rothstein, *Public Health and the Risk Factor* (Rochester, N.Y., 2008), 359–67; Robert A. Aronowitz, *Making Sense of Illness: Science, Society, and Disease* (New York, 1998), chap. 5; Bernard E. Harcourt, *Against Prediction: Profiling, Policing, and Punishing in an Actuarial Age* (Chicago, 2006), 40–1; Martha Poon, "Scorecards as Devices for Consumer Credit: The Case of Fair, Isaac & Company Incorporated," in "Market Devices," suppl. 2, *Sociol. Rev.* 55 (2007): 284–306, on 294–5.

[97] Daniel T. Rodgers, *Age of Fracture* (Cambridge, Mass., 2012).

[98] Shoshana Zuboff, "Big Other: Surveillance Capitalism and the Prospects of an Information Civilization," *J. Inform. Technol.* 30 (2015): 75–89; Zuboff, "The Secrets of Surveillance Capitalism," *Frankfurter Allgemeine Zeitung*, 5 March 2016.

[99] Facebook, *2014 Annual Report*, 29 January 2015, 37–8, http://investor.fb.com/secfiling.cfm ?filingID = 1326801-15-6&CIK = 1326801 (accessed 9 February 2015).

[100] OECD, "Exploring the Economics of Personal Data: A Survey of Methodologies for Measuring Monetary Value," *OECD Digital Economy Papers*, no. 220 (2013), 5, http://dx.doi.org/10.1787 /5k486qtxldmq-en (accessed 8 February 2017).

exhaust" can lead to such riches, while Zuboff argues that the label of "exhaust" is it-self a way of hiding the inherent value in data doubles and justifying its extraction in acts of "dispossession by surveillance."[101]

The prediction products that make money for Facebook or Google, while answer-ing our questions or filling our newsfeeds, can be understood as algorithmic aggre-gations that themselves sometimes resemble data doubles. The geographer Louise Amoore explains how state security apparatuses produce these aggregates, too: scan-ning millions of travelers for possible terrorists, they follow retail data miners in treat-ing much personal data as essentially disposable, focusing instead on the search for niche correlations. Linking together a series of correlations, new-style security creates a data aggregation capable, by definition, of encompassing only a very small popula-tion, or even just a handful of individuals.[102] The quest for individual improvement through self-quantification might be thought of as carrying this trend to its logical con-clusion.[103] Bernard Harcourt identifies a "doppelgänger logic" at the heart of Big Data, as algorithms search for "our perfect match, our twin, our look-alike, in order to de-termine the next book we want to buy . . . the perfect answer to our question."[104] By this logic, the best aggregation mirrors the data double so that the individual can be simultaneously exalted and exploited.

Enthusiasts in personal data's third act see a world of new possibilities dawning. Big data, exclaimed one influential book, is "poised to shake up everything from businesses and the sciences to healthcare, government, education, economics, the hu-manities, and every other aspect of society."[105] Big Data, some claimed, could empower individuals to find and understand love, uncover new self-knowledge, or unleash the democratic yearnings of oppressed masses.[106] The 2011 film *Moneyball* encapsulates a popular vision for that shake-up. Its lead character, the general manager of the Oak-land Athletics baseball team, Billy Beane, sets out a problem of inequality early in the film: "There are rich teams and there are poor teams." An ensuing montage poses the solution, as grainy, digitized images of players flash across a monitor, alongside ta-bles of data describing every aspect of their on-field performance. "This is a code that I've written for our year-to-year projections. This is building in all the intelligence that we have to project players," explains Beane's data-loving assistant. "It's about getting things down to one number, using the stats the way we read them, we'll find value players that nobody else can see. People are overlooked for a variety of biased reasons and perceived flaws: age, appearance, personality. Bill James and mathematics

[101] Zuboff, "Secrets" (cit. n. 98).

[102] Louise Amoore, "On the Emergence of a Security Risk Calculus for Our Times," *Theory Cult. Soc.* 25 (2011): 24–43, on 36. On concerns about similar methods in the retail sphere, see Ryan Calo, "Digital Market Manipulation," *George Washington Law Rev.* 82 (2014): 995–1051.

[103] Gary Wolf, "The Data-Driven Life," *New York Times Magazine*, 28 April 2010; Natasha Dow Schüll, "Data for Life: Wearable Technology and the Design of Self-Care," *Biosocieties* 11 (2016): 317–33.

[104] Bernard Harcourt, *Exposed: Desire and Disobedience in the Digital Age* (Cambridge, Mass., 2015).

[105] Viktor Mayer-Schonberger and Kenneth Cukier, *Big Data: A Revolution That Will Transform How We Live, Work, and Think* (New York, 2013), 11.

[106] Christian Rudder, *Dataclysm: Who We Are (When We Think No One's Looking)* (New York, 2014); on excitement (and subsequent disillusion) about Big Data's potential for supporting social movements, see Zeynep Tufekci, "Engineering the Public: Big Data, Surveillance and Computational Politics," *First Monday* 19 (2014), http://firstmonday.org/ojs/index.php/fm/article/view/4901/4097 (accessed 8 February 2017).

cuts straight through that."[107] Armed with data and with the algorithms to make sense of such data, Beane humiliates the team's manager, centralizes power in the executive, and gets offered a big payout in the end, while his lovable band of misfit players do get (lower-wage) opportunities that they deserve. It is a parable: personal data will save us. Inequality can be solved by data put in the hands of a technocratic, centralized elite who will look past older social structures and past old prejudices to bring new success, and who will be paid well for their efforts.

The third act of personal data has not in fact transcended old social biases, however, or solved all problems of inequality. The language and spirit of the civil rights movement did make data-intensive companies and their regulators skeptical of traditional categories and practices, while a movement to win equality for women set in motion new antidiscrimination laws, notably the 1974 Equal Credit Opportunity Act, which eventually extended the promise of nondiscrimination to a host of "protected categories," including gender, race, religion, age, and national origin.[108] New algorithms promised a formal indifference to those categories that would still allow for profitable discrimination (and so an increase in some forms of inequality), which we must understand as the essence of many Big Data projects. Auto insurers, for instance, have enthused over the possibility of "telematics": devices that precisely monitor individual driving behavior, to provide more perfectly individualized metrics (how long one drives, how fast, how often one stops short, etc.) to be used in setting premiums.[109] Critics maintain, however, that data mining may still hurt protected groups defined in personal data's first act and thus run afoul of the antidiscrimination laws that data mining was supposed to avoid.[110] Indeed, a recent investigation of an algorithm used in many states to inform decisions about bail or sentencing discovered racial bias.[111] Moreover, indifference to legacies of injustice could be a hindrance in and of itself to justice.[112]

Critics fear that Big Data could even displace the liberal self and social order that took its modern form in personal data's first act. Some worry about a corrosive "computational politics," that the drive toward personalizing media could destroy the public sphere or at least exaggerate political, social, and economic inequalities within it.[113] They worry that the "liberal ideal . . . no longer has traction in a world in which commerce cannot be distinguished from governing or policing or surveilling or just simply living privately."[114] They worry that Big Data "threatens the existential and

---

[107] *Moneyball*, directed by Bennett Miller (Los Angeles, 2011), DVD.

[108] Mary L. Heen, "Ending Jim Crow Life Insurance Rates," *Northwest. J. Law & Soc. Policy* 4 (2009): 360–99; Louis Hyman, *Debtor Nation: The History of America in Red Ink* (Princeton, N.J., 2011), chap. 6.

[109] Bill Franks, *Taming the Big Data Tidal Wave: Finding Opportunities in Huge Data Streams with Advanced Analytics* (Hoboken, N.J., 2012), 54–5.

[110] Solon Barocas and Andrew D. Selbst, "Big Data's Disparate Impact," *Calif. Law Rev.* 104 (2016): 671–732; Danielle Keats Citron and Frank Pasquale, "The Scored Society: Due Process for Automated Predictions," *Washington Law Rev.* 89 (2014): 1–33.

[111] Julia Angwin, Jeff Larson, Surya Mattu, and Lauren Kirchner, "Machine Bias: There's Software Used across the Country to Predict Future Criminals: And It's Biased against Blacks," *ProPublica*, 23 May 2016, https://www.propublica.org/article/machine-bias-risk-assessments-in-criminal-sentencing (accessed 8 February 2017).

[112] Jonathan Simon, "The Ideological Effects of Actuarial Practices," *Law Soc. Rev.* 22 (1988): 771–800.

[113] Tufekci, "Engineering the Public" (cit. n. 106); Turow, *The Daily You* (cit. n. 88).

[114] Harcourt, *Exposed* (cit. n. 104), 26.

political canon of the modern liberal order defined by principles of self-determination."[115] They wonder whether we are all on the verge of being reduced to servants of an engineered system fueled by our own data.[116] If personal data served in building the liberal self and our ideas of society, might it also help to dismantle them?

CONCLUSION

This essay has set out a tripartite scheme describing the changing political economy of personal data, a scheme that tracks in unsurprising ways with recent attempts to describe broad changes in America's broader political economy.[117] In broad strokes, it might well describe transitions in the political economy of personal data in many nations. Beginning in the early nineteenth century and lasting through the first decade of the twentieth, states and corporations bent on ordering the wide world they hoped to rule made aggregates and fashioned new social categories that characterized the first act of personal data. World war and unprecedented corporate expansion brought the second act of personal data into being by focusing more on the use of data doubles, often interpreted through the lens of older social aggregations, to manage and control individuals. In our own act—personal data's third, which began in the 1970s—data doubles became more powerful and more valuable, while the exaltation of the individual obscured both the lingering influence of older social categories and the capitalist fantasies fulfilled by flitting, fleeting "Big Data" aggregates, which are no less imposing for their evanescence.

[115] Zuboff, "Secrets" (cit. n. 98).
[116] Martha Poon, "Corporate Capitalism and the Growing Power of Big Data," *Sci. Technol. & Human Values* 41 (2016): 1088–1108.
[117] Lamoreaux, Raff, and Temin, "Beyond Markets" (cit. n. 43); Jonathan Levy, "Accounting for Profit and the History of Capital," *Crit. Hist. Stud.* 1 (2014): 171–214.

EPISTEMOLOGIES AND TECHNOLOGIES OF DATA

Names and Numbers:
"Data" in Classical Natural History,
1758–1859

by Staffan Müller-Wille*

ABSTRACT

The late eighteenth and early nineteenth centuries saw the transition from natural history to the history of nature. This essay analyzes institutional, social, and technological changes in natural history associated with this epochal change. Focusing on the many posthumous reeditions of Carl Linnaeus's *Systema Naturae* that began to appear throughout Europe and beyond from the 1760s onward, I will argue that Linnaean nomenclature and classification reorganized and enhanced flows of data— a term already used in natural history—among individual naturalists and institutions. Plant and animal species became units that could be "slotted" into collections and publications, reshuffled and exchanged, kept track of in lists and catalogs, and counted and distributed in new ways. On two fronts—biogeography and the search for the "natural system"—this brought to the fore new, intriguing relationships among organisms of diverse kinds. By letting nature speak through the "artificial" means and media of early systematics, I argue, new and powerful visions of an unruly nature emerged that became the object of early evolutionary theories. Natural history was an "information science" that processed growing quantities of data and held the same potential for surprising insights as today's data-intensive sciences.

> He gathered rocks, flowers, beetles of all kind for himself, and
> arranged them in series in manifold ways.
> —Novalis, *Die Lehrlinge zu Sais*, 1802[1]

FROM NATURAL HISTORY TO THE HISTORY OF NATURE

It has long been a trope in the historiography of the life sciences that classical natural history underwent a massive transition, if not a revolution, around 1800. Key con-

* Department of Philosophy, Sociology and Anthropology, University of Exeter, Byrne House, St Germans Road, Exeter, Devon, EX4 4PJ, United Kingdom; S.E.W.Mueller-Wille@exeter.ac.uk.

I would like to thank Lorraine Daston and the editors of this volume for the opportunity to develop and discuss ideas within the context of their project, "The Sciences of the Archive," at the Max Planck Institute for the History of Science, Berlin. Without their generosity, encouragement, and critical perseverance, this essay would never have come about. Very special thanks go to Katrin Böhme, for her keen eye for precious holdings of the Staatsbibliothek and for level-headed historiographical advice. I am also grateful to Joeri Witteveen, Polly Winsor, Sabina Leonelli and her "data science" group at the University of Exeter, as well as two anonymous reviewers for critical feedback on earlier drafts.

[1] Novalis (Friedrich von Hardenberg), *Novalis' Schriften* (Berlin, 1802), 2:162: "Er sammelte sich Steine, Blumen, Käfer aller Art, und legte sie auf mannichfache Weise sich in Reihen." Unless otherwise noted, all translations are my own.

cepts such as species, distribution, or adaptation changed from designating stable forms to denoting fluid processes extending over generations and across populations. In its ancient sense of a trustworthy account, *historia* had of course always had to do with tradition and hence with the passage of time. This is reflected in the methods early modern naturalists used, which were essentially the same as those used by humanists and antiquarians.[2] But only in the latter half of the eighteenth century was the subject matter of natural history—the diversity of species, their properties and uses, and their geographic, temporal, and ecological distribution—infused with a sense of historicity.

This transition has been captured succinctly in the catchphrase "from natural history to the history of nature."[3] Explanations as to why it happened remain scant, however. Michel Foucault deliberately abstained from causal explanations in order to highlight the transition as a "mutation in the space of nature of Western culture."[4] In a similar vein, an older Anglophone tradition has emphasized paradigmatic shifts in metaphysical outlook as the precondition for the historicization of nature.[5] An interesting early attempt to close the explanatory gap that such accounts left can be found in Wolf Lepenies's book, *End of Natural History* (1976). Lepenies likewise regards natural history as going through a "crisis" around 1800 but identifies it as a self-inflicted "growth crisis."[6] Pointing to the series of new editions and supplements that eighteenth-century naturalists produced of their works, he explains how each attempt to reduce observations to a timeless classification system precipitated further observations that were at odds with the system adopted.[7] Increasing "experiential pressure" thus ultimately exhausted the capacity of spatial classification systems and forced naturalists to open up a temporal dimension.[8]

Lepenies's causal association of "experiential pressure" with far-reaching paradigmatic changes is highly suggestive for any attempt to historicize the contemporary discourse of "Big Data." After all, this discourse is also rife with expectations—and fears— that "data-driven" science will be ushering in a new era in the history of knowledge.[9] And there is indeed evidence that late eighteenth- and early nineteenth-century natural history can be understood as data driven since knowledge it accumulated grew at exceptional rates. While the number of species described by European naturalists has been rising ever since the Renaissance, the growth curve is steepest for the period between 1760 and 1840, before it experiences a slackening from the late nineteenth century onward.[10] But there are also problems with Lepenies's explanation. As suggestive

[2] Gianna Pomata and Nancy G. Siraisi, eds., *Historia: Empiricism and Erudition in Early Modern Europe* (Cambridge, Mass., 2005).

[3] John Lyon and Phillip R. Sloan, *From Natural History to the History of Nature: Readings from Buffon and His Critics* (Notre Dame, Ind., 1981).

[4] Michel Foucault, *Les Mots et les choses: Une Archéologie des sciences humaines* (Paris, 1966), 150.

[5] Phillip R. Sloan, "Buffon, German Biology, and the Historical Interpretation of Biological Species," *Brit. J. Hist. Sci.* 12 (1979): 109–53.

[6] Wolf Lepenies, *End of Natural History* (Cambridge, 1980), 74.

[7] Ibid., 76.

[8] Ibid., 15.

[9] See Elena Aronova, Christine von Oertzen, and David Sepkoski, "Introduction: Historicizing Big Data"; and Judith Kaplan, "From Lexicostatistics to Lexomics: Basic Vocabulary and the Study of Language Prehistory," both in this volume.

[10] Given how often growth in species number is invoked to explain historical developments in natural history, actual data are surprisingly scarce. I am relying on Sara T. Scharf, "Identification Keys and the Natural Method: The Development of Text-Based Information Management Tools in Botany in the Long Eighteenth Century" (PhD thesis, Univ. of Toronto, 2006), 31–42, who analyzes data for plants, mushrooms, insects, fish, birds, and mammals.

as the association is, mere quantitative growth of knowledge does not provide a compelling reason to adopt a particular worldview, whether historicist or not.[11] More interesting problems arise when we confront Lepenies's account with the following statement by the young Alexander von Humboldt (1769–1859):

> Every plant is certainly not allocated to every rock as its domicile. Nature follows unknown laws here, which can only be investigated by means of botanists subjecting more data to induction [*Data zur Induction darreichen*].[12]

Humboldt's statement first of all shows that data talk is not a hallmark of modernity narrowly understood. The Latin past participle of *dare*, simply meaning "given," had long been in use in natural history to refer to any kind of information—a detailed description, a drawing, a preserved specimen, or just the name—that had been handed down about a particular subject.[13] Second, and more important, it is notable that Humboldt employed the language of "data" not to complain about its overabundance, as one would expect from Lepenies's account, but on the contrary, to complain about its scarcity. Such a call for "more" data in a world that otherwise bemoaned "too much" data is not at all exceptional and only seemingly paradoxical. Naturalists like Humboldt were both creators and users of data and thus were involved in an endless cycle of consuming data for the sake of producing them. The crucial problem of any data-driven science is therefore not just to come to terms with ever-growing bodies of data but also to make those data applicable to as many contexts of inquiry as possible.[14] The target of Humboldt's statement was a highly specialized subject—the distribution of plant species as a function of geological substrate, and hence their use as indicators in the search for mineral deposits—but it was hardly untrodden terrain; quite to the contrary, knowledge of correlations between particular plant varieties and particular types of rock had a very long and rich tradition in mining.[15] Hence, if there was a scarcity of data, it was a scarcity of data produced in a manner that could readily be consumed and processed. Finally, Humboldt's statement also suggests that producing general knowledge from data through induction is not simply a matter of individual psychology and experience but relies on the results of a collective endeavor of trained specialists, a group that Humboldt himself was aiming to become part of.[16] Naturalists were not passively exposed to a data deluge but collectively shaped the channels through which data

[11] Phillip R. Sloan, review of *Das Ende der Naturgeschichte*, by Wolf Lepenies, *Isis* 72 (1981): 123–4.
[12] Alexander von Humboldt, *Mineralogische Beobachtungen über einige Basalte am Rhein* (Braunschweig, 1790), 86.
[13] See, e.g., Carl Linnaeus, *Hortus Cliffortianus* (Amsterdam, 1737), "Bibliotheca botanica" (n.p.), who refers to Johannes Bauhin's *Historia Plantarum Universalis* (Yverdon, 1650–1) as containing "all that was given by [his] forebears" [*omnis data a praecessoribus*]. More specifically, Humboldt's language of "data" and "induction" reveals the influence of Immanuel Kant, who argued that empirical sciences are uncertain and incomplete because they rely on "data of intuition" [*datis der Anschauung*]; see Ursula Klein, "The Prussian Mining Official Alexander von Humboldt," *Ann. Sci.* 69 (2012): 27–68, on 54–5. For further discussion of the history of the word "data," see Aronova, von Oertzen, and Sepkoski, "Introduction" (cit. n. 9); and Markus Krajewski, "Tell Data from Meta: Tracing the Origins of Big Data, Bibliometrics, and the OPAC," in this volume.
[14] Sabina Leonelli, "Integrating Data to Acquire New Knowledge: Three Modes of Integration in Plant Science," *Stud. Hist. Phil. Biol. Biomed. Sci.* 44 (2013): 503–14.
[15] Ursula Klein, *Humboldts Preußen: Wissenschaft und Technik im Aufbruch* (Darmstadt, 2015), 95.
[16] Klein, "Alexander von Humboldt" (cit. n. 13), 29.

flowed, thus themselves defining the conditions under which they perceived data as abundant or scarce.[17]

Humboldt's early call for "more data" thus reminds us that solutions to epistemological problems of "data-driven science"—whether in its early modern or contemporary incarnations—are not only conceptual or theoretical, but also technological and infrastructural. Taking this conclusion on board, the next section is going to explore social and institutional changes that natural history underwent in its classical period from Linnaeus to Darwin. In particular, I want to highlight the integrative role that Linnaean nomenclature and taxonomy played in this period, which otherwise saw a diversification of agents, institutions, and cultures of natural history. The third and fourth sections will then focus on how Linnaean names and taxa were used as tools to organize exchange and retrieval of data. I will show that the adoption of these tools not only enhanced data circulation but also had peculiar epistemic effects, by turning species and other taxa into objects that were numbered and counted to reveal intriguing patterns in the geographic and taxonomic distribution of life forms. Only then, in a concluding section, will I return to the question whether one can claim that a causal connection exists between the data-driven nature of classical natural history and the discursive ruptures it underwent.

THE CHANGING LANDSCAPE OF CLASSICAL NATURAL HISTORY

Late eighteenth- and early nineteenth-century natural history experienced social and institutional changes that involved both diversifying and centralizing tendencies. On the one hand, its base of practitioners grew massively and came to include non-university-trained men and women as well, both within and outside of Europe, and across social classes. Amateur naturalists not only engaged in collecting specimens, maintained epistolary exchanges, and eventually published their observations; they also began to organize themselves from the bottom up in local and regional associations that often maintained their own periodical publications.[18] Rising levels of literacy and the spread of inexpensive print widened the potential audience for, and made it easier to contribute to, natural history.[19] At the same time, there was an increasing demand for experts trained in natural history to fill a growing number of professional positions, in state bureaucracies like mining boards; within the management of agricultural, industrial, and commercial enterprises; and, as we will see next, in large collections and museums. Needless to say, this held in particular for organizations and enterprises engaged in long-distance trade and colonial expansion. Participation in the global "information economy" of natural history, and the "logistical power" this bestowed upon its practitioners, thus provided a stepping stone for the middling classes to enter various occupations and careers of an administrative, brokering, or entrepreneurial nature.[20]

---

[17] For the parallel case of early modern genealogy, see Markus Friedrich, "Genealogy as Archive-Driven Research Enterprise in Early Modern Europe," in this volume.

[18] Ann B. Shteir, *Cultivating Women, Cultivating Science: Flora's Daughters and Botany in England, 1760 to 1860* (Baltimore, 1999); Roger L. Williams, *Botanophilia in Eighteenth-Century France: The Spirit of the Enlightenment* (Dordrecht, 2001); Bettina Dietz, "Making Natural History: Doing the Enlightenment," *Cent. Eur. Hist.* 43 (2010): 25–46.

[19] Denise Phillips, *Acolytes of Nature: Defining Natural Science in Germany, 1770–1850* (Chicago, 2012).

[20] Simon Schaffer, Lissa Roberts, Kapil Raj, and James Delbourgo, eds., *The Brokered World: Go-Betweens and Global Intelligence, 1770–1820* (Sagamore Beach, Mass., 2009); Ursula Klein and

While these developments led to a growing diversification of both objects and sources of natural history, a counterbalance existed in the rise of a new set of central nodes around which natural history exchange revolved. Until the mid-eighteenth century, exchange of specimens, letters, and publications was centered upon individuals who presided over large collections, such as Sir Hans Sloane (1660–1753), Georges Buffon (1707–88), and Carl Linnaeus (1707–78). By the early nineteenth century, central and permanent institutions had taken over this role—the Jardin des Plantes and Muséum d'histoire naturelle in Paris, Kew Gardens and the British Museum in London, or Berlin University with its gardens and collections in Prussia, to name just a few. Two important structural features distinguished these "new" collections from their early modern counterparts.[21] First, they represented collections of collections rather than collections *tout court*. Often starting out with the acquisition of a large, single collection—Sloan's collection in the case of the British museum, or Linnaeus's collection in the case of the Linnean Society (London)—these museums expanded by acquiring entire collections or commissioning naturalists to hunt for specimens on a global scale.[22] The most striking case of this is provided by the Muséum d'histoire naturelle in Paris after the French Revolution, which received a boost to its possessions from the confiscation of aristocratic collections, whose provenances and contents were carefully noted in a card catalog.[23]

Second, and concomitantly, museums were increasingly organized into specialized departments offering a hierarchy of positions for curators or "keepers" and various amanuenses who administered and enriched the collections. A new generation of professional naturalists emerged, often socialized through participation in long-distance natural history exploration, during which they collected for their patrons and then moved on to curatorial positions in metropolitan collections and libraries. Daniel Solander (1733–82), who accompanied Joseph Banks on Cook's first circumnavigation as one of the many traveling students or "apostles" of Linnaeus, is often cited as the first exemplar. Robert Brown (1773–1858)—who went with Flinders's expedition to Australia (1801–05), followed Solander as Banks's librarian, and finally, after the latter's death, became "Keeper of the Banksian Botanical Collection" at the British Museum in 1827—is another well-known example.[24]

Emma Spary, eds., *Materials and Expertise in Early Modern Europe* (Chicago, 2010). For an intriguing case study, see Minakshi Menon, "Medicine, Money, and the Making of the East India Company State: William Roxburgh in Madras, c. 1790," in *Histories of Medicine and Healing in the Indian Ocean World*, vol. 1, *The Medieval and Early Modern Period*, ed. Anna Winterbottom and Facil Tesfaye (London, 2015), 151–78. Menon borrows the concept of "logistical power" from Chandra Mukerji, "The Territorial State as a Figured World of Power: Strategics, Logistics, and Impersonal Rule," *Sociol. Theory* 28 (2010): 402–24.

[21] Dorinda Outram, "New Spaces in Natural History," in *Cultures of Natural History*, ed. Nicholas Jardine, Jim A. Secord, and Emma C. Spary (Cambridge, 1996), 249–65.

[22] On Sloane, whose collection already was a collection of collections, see James Delbourgo, "Collecting Hans Sloane," in *From Books to Bezoars: Sir Hans Sloane and His Collections* (London, 2012), 9–23; on Linnaeus, see Paul White, "The Purchase of Knowledge: James Edward Smith and the Linnaean Collections," *Endeavour* 23 (1999): 126–9. On traveling collectors, see Daniela Bleichmar and Peter C. Mancall, eds., *Collecting across Cultures: Material Exchanges in the Early Modern Atlantic World* (Philadelphia, 2011).

[23] Pierre-Yves Lacour, *La République naturaliste: Collections d'histoire naturelle et Révolution française, 1789–1804* (Paris, 2014).

[24] Edward Duyker, *Nature's Argonaut: Daniel Solander 1733–1782: Naturalist and Voyager with Cook and Banks* (Melbourne, 1998); David J. Mabberley, *Jupiter Botanicus: Robert Brown of the British Museum* (Braunschweig, 1985).

The knowledge networks that underwrote natural history were thus not just expanding and diversifying. At the same time, central institutions emerged that provided positions for "information brokers" who saw their task primarily as serving an imagined community of naturalists by mediating and organizing flows of data.[25] This double process of diversification and centralization turned natural history into an increasingly disparate field. Classical natural history never constituted a homogeneous and uniform knowledge regime, governed by a common paradigm or episteme. Peter F. Stevens coins the interesting phrase "continuity in practice" to highlight how naturalists discarded the idea of one timeless and universal system in which every conceivable species would find its place and began to join species one by one into open-ended series instead.[26] The urge to synthesize particulars, to be sure, persisted, but increasingly found expression in the development of highly specialized "tools of conjecture" deployed in narrowly defined subject areas.[27]

There is one element of unity to classical natural history, however, that has been recognized widely ever since the late eighteenth century. Within two decades of their introduction in *Philosophia Botanica* (1751), the two innovations that formed the cornerstones of Linnaeus's self-styled "reform" of natural history—the naming of plant and animal species by "trivial" names composed of genus name and specific epithet (as in *Homo sapiens*) and their ordering by variety, species, genus, order (or family), and class, the so-called Linnaean hierarchy of taxonomic ranks—had been universally adopted by naturalists, even by prominent opponents of Linnaeus like Buffon or Jean-Baptiste de Lamarck (1744–1829).[28] It is telling, however, that these innovations have habitually been characterized as being of pragmatic value only. According to the botanist Frans Stafleu, author of the most comprehensive history of the reception of Linnaean taxonomy, Linnaeus conceived of natural history "primarily as a device to register and to remember, to store and to retrieve."[29]

Such claims imply that stable, arbitrary names and a nested hierarchy of taxonomic units are of obvious practical value in communication, but neutral with respect to the knowledge they transport. And indeed, precisely this feature seems to have made both innovations so attractive to naturalists in the first place.[30] Yet it seems highly improbable, after all we know from work in the history and philosophy of science, that

[25] The dynamic continues. On citizen science in twentieth-century ornithology, see Etienne Benson, "A Centrifuge of Calculation: Managing Data and Enthusiasm in Early Twentieth-Century Bird Banding," in this volume; see also Geoffrey C. Bowker, "Biodiversity Datadiversity," *Soc. Stud. Sci.* 30 (2000): 643–83; Sabina Leonelli, "Classificatory Theory in Data-Intensive Science: The Case of Open Biomedical Ontologies," *Int. Stud. Phil. Sci.* 26 (2012): 47–65.
[26] Peter F. Stevens, *The Development of Systematics: Antoine-Laurent de Jussieu, Nature and the Natural System* (New York, 1994), 153.
[27] On "tools of conjecture," see Lorraine Daston, "The Empire of Observation, 1600–1800," in *Histories of Scientific Observation*, ed. Lorraine Daston and Elizabeth Lunbeck (Chicago 2011), 81–113, on 104–6.
[28] On the reception of Linnaeus in France, see Pascal Duris, *Linné et la France (1780–1850)* (Geneva, 1995).
[29] Frans A. Stafleu, *Linnaeus and the Linnaeans: The Spreading of Their Ideas in Systematic Botany, 1735–1789* (Utrecht, 1971), 33.
[30] William T. Stearn, "The Background of Linnaeus's Contributions to the Nomenclature and Methods of Systematic Biology," *Syst. Zool.* 8 (1959): 4–22; Lisbet Koerner, *Linnaeus: Nature and Nation* (Cambridge, Mass., 1999), chap. 2.

their adoption should have had no epistemic consequences at all.[31] In the following section, I will adopt a perspective that looks at binary names and the Linnaean hierarchy as tools to process information on paper. This will prepare the ground for my argument in the subsequent section that the way in which information brokers in classical natural history deployed these tools—both in order to collect and process data on plants and animals and in order to navigate the increasingly complex social landscape of natural history—did have epistemic consequences by turning species and other taxa into objects that could be counted and whose numbers mattered.

PAPER TOOLS AND PAPER EMPIRES

One of the most astonishing aspects of Linnaeus's taxonomic publications is the success they enjoyed in terms of print runs, especially if one considers that these were not books made for leisurely reading, but catalogs filled with names of genera and species, references to earlier literature, short morphological descriptions, and cryptic remarks about geographic and ecologic distribution. Linnaeus himself counted twelve editions of his *Systema Naturae* (which grew between 1735 and 1768 from an eleven-page folio volume to four octavo volumes of 2,441 pages in all), six editions of *Genera Plantarum* (1737–64), and two editions of *Species Plantarum* (1753, 1762).[32] But the success went far beyond Linnaeus as a person. From the late 1760s onward, but especially after his death in 1778, other naturalists began to publish editions, translations, and adaptations of these works, often adopting their main title and citing Linnaeus as author on the title page, or acknowledging their debt to his work in subtitles or prefaces. The most complete bibliography of Linnaeana lists about fifty posthumous items of this kind for *Systema Naturae* alone.[33]

The lasting success of Linnaeus's taxonomic works is often explained by claiming that they provided naturalists with the means to refer unambiguously to the various kinds of plants and animals, thus clearing the previous chaos of synonymy and conflicting classifications.[34] But what allows for unequivocal reference in modern taxonomy are not binary names as such but the type method, that is, the method of associating taxonomic names with fixed taxon elements, such as type specimens deposited in museums, and this method only began to emerge in the second half of the nineteenth century.[35]

---

[31] For the same reason, it is unlikely that contemporary digital technologies will result in nothing but a "scaling-up of pen-and-paper methods"; see Hallam Stevens, "A Feeling for the Algorithm: Working Knowledge and Big Data in Biology," in this volume.

[32] See Carl Linnaeus, *Systema Naturae*, 12th ed., 4 vols. (Stockholm, 1766–8). See Ratio editionis in vol. 1 (1766), n.p., for a list of "authorized" editions of *Systema Naturae*.

[33] B. H. Soulsby, *A Catalogue of the Works of Linnaeus (and Publications More Immediately Relating Thereto) Preserved in the Libraries of the British Museum (Bloomsbury) and the British Museum (Natural History) (South Kensington)*, 2nd ed. (London, 1933), nos. 64–169, 284–327, 480–529, 573–619. Linnaeus's works were also printed in North and South America, and one of the editions listed by Soulsby for the twelfth edition of *Systema Naturae* was printed in Jakarta in 1783 (nos. 104–5). A search of the online Linnaeus Link Union Catalogue (http://www.linnaeuslink.org/ [accessed 22 April 2017]), which builds on Soulsby's *Catalogue*, produces thirty-eight results for titles containing the words "systema" and "naturae" published between 1768 and 1859.

[34] For a succinct statement of this view, see the epilogue in Stafleu, *Linnaeus* (cit. n. 29), 337–9.

[35] Joeri Witteveen, "Suppressing Synonymy with a Homonym: The Emergence of the Nomenclatural Type Concept in Nineteenth-Century Natural History," *J. Hist. Biol.* 49 (2015): 135–89.

Linnaeus himself, when introducing binary names and the five-tiered hierarchy of taxonomic ranks, advertised an advantage that was quite different from disambiguation, namely, decontextualization. Traditional species names were composed of the genus name and a diagnostic phrase spelling out traits by which the named species differed from all other known species of the same genus. The function of such names was thus not only to designate a species, but also to distinguish it from already known species; without this context, legitimate names did not make much sense. The "trivial" or binary name, in contrast, just added a "single word . . . freely adopted from anywhere" to the genus name, usually in the form of an adjective. Hence, as Linnaeus emphasized, it was not only shorter and more easily reproduced, but above all more stable, since it did not have to be changed with the discovery of new species.[36] With his "systematic" arrangement by class, order, genus, species, and variety, Linnaeus created a series of multiple taxa nested within higher taxa. A class, for example, could contain ten orders, each of these orders another ten genera, and so on, in the same way that countries or armies form nested hierarchies of multiple administrative and military units. The meaning of the ranks constituting the Linnaean hierarchy was thus likewise not determined by any particular differences they exhibited, but by what they contained and came to contain.[37] Linnaean names were mere indexes or labels, whereas the Linnaean hierarchy simply provided a nested set of containers, or "boxes within boxes," defined extensionally only by the set of objects they contained.[38] In short, Linnaean nomenclature and taxonomy emphasized equivalence, not difference, a point to which I will return.

To gain a better understanding of how binary nomenclature and the hierarchy of ranks facilitated communication among naturalists, it is useful to look at the role they played in the creation of paper tools—devices made from paper and ink, whether in manuscript or print—that were employed in practices of extracting and processing written information like note taking, listing, cataloging, or tabulating.[39] Up to the early eighteenth century, the predominant methods scholars used for annotation had been marginalia and topically organized commonplace books, that is, media that tended to fix information in relation to a relevant (con)text.[40] The late seventeenth and eighteenth centuries witnessed a transition to more flexible paper tools, like loose files and card catalogs, and to more complex techniques of extracting, rearranging, and display-

---

[36] Carl Linnaeus, *Philosophia Botanica* (Stockholm, 1751), 98. Linnaeus indeed chose trivial names quite "arbitrarily," i.e., from a wide variety of sources, including vernacular languages; see Alexandra Cook, "Linnaeus and Chinese Plants: A Test of the Linguistic Imperialism Thesis," *Notes Rec. Roy. Soc. Lond.* 64 (2010): 121–38.

[37] Linnaeus, *Philosophia Botanica* (cit. n. 36), 202. For an English translation of the relevant aphorisms on trivial names and the five-tiered system of ranks, see *Linnaeus' Philosophia Botanica*, transl. Stephen Freer (Oxford, 2005), 99–100, 207–8.

[38] I am borrowing the language of labels and containers from Sabina Leonelli, "Packaging Small Facts for Re-use: Databases in Model Organism Biology," in *How Well Do Facts Travel? The Dissemination of Reliable Knowledge*, ed. Peter Howlett and Mary S. Morgan (Cambridge, 2010), 325–48. For an eighteenth-century case study that employs similar analytic categories, see Anke te Heesen, "Boxes in Nature," *Stud. Hist. Phil. Sci.* 33 (2000): 381–403.

[39] On the concept of paper tools, see Christine von Oertzen, "Machineries of Data Power: Manual versus Mechanical Census Compilation in Nineteenth-Century Europe," in this volume; Anke te Heesen, "The Notebook: A Paper-Technology," in *Making Things Public: Atmospheres of Democracy*, ed. Bruno Latour and Peter Weibel (Cambridge, Mass., 2005), 582–9; Volker Hess and Andrew Mendelsohn, "Paper Technology und Wissensgeschichte," *NTM* 21 (2013): 1–10.

[40] Ann M. Blair, *Too Much to Know: Managing Scholarly Information before the Modern Age* (New Haven, Conn., 2010); Richard Yeo, *Notebooks, English Virtuosi, and Early Modern Science* (Chicago, 2014).

ing information, like forms, tables, diagrams, and maps, often employed for highly idio-syncratic purposes.[41] Linnaeus participated in this transition by experimenting through-out his career with a variety of annotation and filing systems, different forms of lists and tables, and, toward the end of his life, paper slips that resemble index cards. In all of these media, Linnaean taxa carved out an allocated paper space, labeled with the name of a genus or species, and then used to collect pieces of information contained under that name. Because the name itself was a mere label, the resulting packages of data could be freely extracted from their context, and their contents inserted, or even redis-tributed, elsewhere, without losing their identity, as long as the label stuck.[42] As Lin-naeus put it in a remarkable metaphor in 1737, defining the role of generic names:

> The generic name has the same value on the market of botany, as the coin has in the com-monwealth, which is accepted at a certain price—without needing a metallurgical assay—and is received by others on a daily basis, as long as it has become known in the common-wealth.[43]

What this metaphor clearly expresses is that Linnaean names and ranks did not de-rive their value from any information they contained in themselves, but by providing others with the material means to access, accumulate, and exchange information.[44] Linnaeus's taxonomic works were designed to serve as templates for communal anno-tation, whether this took the form of creating a numbered list of the names of specimens or seeds sent to a correspondent, or whether an interleaved copy of one of these works was used to absorb new data gathered from the latest literature, from a letter received from a correspondent, or during field excursions. Linnaeus himself employed his publications for this purpose, thus being able to churn out one edition after another on the basis of data received from correspondents, and there is growing evidence that other naturalists quickly adopted the same kinds of strategies.[45] Drawing on an analogy from our digital age, one might claim that the formal structure of Linnaean nomencla-ture and taxonomy provided naturalists with the rows or "objects" of a crowd-sourced database; the columns, in turn, were constituted by "variables" such as morphological features, economic uses, and habitat or geographic origin of the species in question.[46]

---

[41] Anke te Heesen, "Accounting for the Natural World: Double-Entry Bookkeeping in the Field," in *Colonial Botany: Science, Commerce, and Politics in the Early Modern World*, ed. Londa Schiebinger and Claudia Swan (Philadelphia, 2005), 237–51; Volker Hess and J. Andrew Mendelsohn, "Case and Series: Medical Knowledge and Paper Technology, 1600–1900," *Hist. Sci.* 48 (2010): 287–314; Isa-belle Charmantier and Staffan Müller-Wille, "Worlds on Paper: An Introduction," *Early Sci. & Med.* 19 (2014): 379–97.

[42] Staffan Müller-Wille and Isabelle Charmantier, "Natural History and Information Overload: The Case of Linnaeus," *Stud. Hist. Phil. Biol. Biomed. Sci.* 43 (2012): 4–15; Müller-Wille and Charmantier, "Lists as Research Technologies," *Isis* 103 (2012): 743–52; Charmantier and Müller-Wille, "Carl Linnaeus's Botanical Paper Slips (1767–1773)," *Intellect. Hist. Rev.* 24 (2014): 215–38.

[43] Carl Linnaeus, *Critica Botanica* (Leiden, 1737), 204.

[44] The metaphor of data as currency is also found in twentieth-century sciences; see Elena Aronova, "Geophysical Datascapes of the Cold War: Politics and Practices of the World Data Centers in the 1950s and 1960s," in this volume.

[45] On the important role of specimen lists in correspondence, see Bettina Dietz, "Contribution and Co-production: The Collaborative Culture of Linnaean Botany," *Ann. Sci.* 69 (2012): 551–69. On Lin-naeus annotating interleaved copies of his own taxonomic works, see Staffan Müller-Wille and Sara Scharf, "Indexing Nature: Carl Linnaeus and His Fact Gathering Strategies," *Svenska Linnésällsk. Årssk.* 2011 (2012): 31–60.

[46] For an analysis of the limitations and potentials of applying this metaphor to predigital media, see David Sepkoski, "The Database before the Computer?," in this volume.

This explains one curious aspect of the many "editions" and "translations" of Linnaeus's taxonomic work, namely that, strictly speaking, they were not editions or translations at all. As Bettina Dietz has emphasized in a recent article, they rather continued his taxonomic project by incorporating new data.[47] Many of the editors of these works pointed this out explicitly. The Dutch physician and naturalist Martinus Houttuyn (1720–98), for example, stated in his *Natuurlyke Historie*—issued from his cousin's printshop between 1761 and 1785—that he had adopted Linnaeus's "system" [*Samenstel*] and "Latin bynames" [*Latynsche Bynaamen*], but only to add that he had also inserted information from publications by other naturalists such as Buffon in Paris, or Jacob Theodor Klein (1685–1759) in Danzig, whose works rivaled those of Linnaeus in scope and authority.[48] Philipp Ludwig Statius Müller (1725–76) made similar remarks in the preface to his German edition of *Systema Naturae*, stating that the reader should "not expect a translation," either of the twelfth edition of Linnaeus's *Systema Naturae* or of Houttuyn's *Natuurlyke Historie*. Instead, Müller's work also incorporated information gathered from other naturalists, above all from contributions to journals that scientific "academies" [*Sozietäten*] edited in Paris, Stockholm, St. Petersburg, and Vienna.[49] In the preface to a supplementary volume that appeared in 1776, Müller even asked his readers to report any new discoveries, whether made from reading, in collections, or in the field, directly to him by providing at least a short description and indication of the new species' taxonomic position.[50] When Müller died shortly after, his publisher Raspe commissioned Johann Friedrich Gmelin (1748–1804), professor of medicine at the University of Göttingen, and other naturalists to continue the endeavor, adding over the following decades more than seventeen volumes to Müller's expansion of Linnaeus's work, including a German translation of Gmelin's "thirteenth" edition of *Systema Naturae*.[51]

One can see from this short sketch that translations and editions of Linnaeus's *Systema Naturae* were products of intense and complex paper work. They often built on one another, rather than directly on Linnaeus's own publications, and they relied on a wide array of additional written sources—other general works in natural history, local floras and faunas, journal articles, and letters from correspondents—to integrate the latest discoveries. Müller coined a revealing expression for the unflagging compilatory activity that lay behind such works. In advertising his supplementary volume, he emphasized that "all *Addenda, Appendices* and *Mantissae* of the Knight von Linné have been properly slotted in [*gehörig eingeschaltet*]," and that the same had happened to new species reported by other naturalists.[52] *Einschalten* is a verb with over-

[47] Bettina Dietz, "Linnaeus' Restless System: Translation as Textual Engineering in Eighteenth-Century Botany," *Ann. Sci.* 73 (2016): 143–56.

[48] Martinus Houttuyn, "Voorreden," in *Natuurlyke Historie, of Uitvoerige Beschryving der Dieren, Planten, en Mineraalen, Volgens het Samenstel van den Heer Linnæus*, vol. 1, pt. 1 (Amsterdam, 1761), n.p.

[49] Philipp Ludwig Statius Müller, "Vorbericht," in *Des Ritters Carl von Linné vollständiges Natursystem . . . Erster Theil. Von den säugenden Thieren* (Nuremberg, 1773), n.p. For a more detailed account of Müller's sources and his manner of compilation, see Dietz, "Linnaeus' Restless System" (cit. n. 47), 148–9.

[50] Philipp Ludwig Statius Müller, "Vorbericht," in *Des Ritters Carl von Linné vollständiges Natursystem . . . Supplements- und Registerband* (Nuremberg, 1776), n.p.

[51] Soulsby, *Catalogue* (cit. n. 33), nos. 96–100, 577. The volumes on botany are analyzed by Dietz, "Linnaeus' Restless System" (cit. n. 47), 150–2.

[52] Müller, "Vorbericht" (cit. n. 50), n.p. I thank Sabina Leonelli for coming up with an ingenious English translation for *einschalten*.

tones of mechanical or bureaucratic labor and simply means to insert an object into a preexisting series of other objects.[53] It expresses vividly how easy it had become to compile data on plant and animal species after the Linnaean reform. This is not to say that the adoption of Linnaean nomenclature was a smooth and immediate process. It was through a long, protracted, and regionally diverse process, in which social and political relations were at stake, rather than through a mere technological fix, that the full potential of Linnaean nomenclature was realized.[54] A key element in this process was the fact that Linnaean names and taxa empowered naturalists who were situated in peripheral contexts or subaltern positions to build their own "paper empires" on the basis of purely derivative literary techniques like extraction, compilation, and rearrangement of names and accompanying descriptions. Even Buffon, an ardent lifelong opponent of the Linnaean reform, did not escape the maelstrom of information processing that was set free in this way. From 1801 to 1803, the poet René Castel, once deputy of Calvados for the Assemblée legislative, published a twenty-six-volume "new edition" of Buffon's *Histoire naturelle* "classified by orders, genera and species according to Linnaeus's system and with . . . Linnaean nomenclature."[55]

COUNTING SPECIES

Gmelin compared Linnaeus's work to an "admirably contrived edifice" constructed in such a manner "as to suffer . . . necessary additions, alterations, and improvements, without injuring its strength, permanency, or symmetry." Critics of Linnaean natural history, he argued, should consider "that such alterations, additions and improvements, as the *System of Nature* has hitherto required, have been made by the disciples of that great master"—disciples like himself, that is.[56] This did not keep Lamarck from heavily criticizing Gmelin for having composed his work without "preliminary research," simply "by attaching to the genera and species already determined by Linnaeus all that he found indicated as new in the works he consulted."[57] Similar attitudes shine through when Kant speaks of systems in natural history as mere "depositories" [*Registraturen*], or when Humboldt addressed contemporary naturalists as "sordid registrars" [*elende Registratoren*].[58] Such invectives became more and more common in the late eighteenth century. They reflect how the Linnaean way of doing natural history

---

[53] Johann Christoph Adelung, *Grammatisch-kritisches Wörterbuch der hochdeutschen Mundart*, 4 vols. (1793–1802), 1:1735. Adelung points out that the word was used primarily in the context of inserting "written sentences."

[54] Linnaean nomenclature shares this with other information technologies. On punch card technology, see von Oertzen, "Machineries" (cit. n. 39); on biogeographic maps, see Nils Güttler, *Das Kosmoskop: Karten und ihre Benutzer in der Pflanzengeographie des 19. Jahrhunderts* (Göttingen, 2014).

[55] René Richard Louis Castel, *Histoire naturelle de Buffon, classée par ordres, genres et espèces, d'après le systéme de Linné, avec les charactères génériques et la nomenclature Linnéenne*, 26 vols. (Paris, 1801–3).

[56] Johann Friedrich Gmelin, "Ratio hujus novae editionis," in *Caroli a Linné: Systema Naturae per Regna Tria Naturae. . . . Editio decima tertia, aucta, reformata* (Leipzig, 1788), n.p.

[57] Jean-Baptiste Lamarck, "Sur les ouvrages généraux en histoire naturelle, et particulièrement sur l'édition du Systema Naturae de Linneus [*sic*], que M. J. F. Gmelin vient de publiér," *Act. Soc. Hist. Nat. Paris* 1 (1792): 81–5, on 82. Indeed, Gmelin's thirteenth edition did build almost entirely on published information; see T. J. Spilman, "Gmelin's 13th Edition of Systema Naturae: A Case of Neglect," *Entomol. News* 78 (1967): 169–72.

[58] Quoted from Güttler, *Das Kosmoskop* (cit. n. 54), 57–8.

increasingly lost its former prestige and was relegated to the netherworld of mere manual labor. Therefore, naturalists began to foreground other concerns besides mere description and cataloging, notably questions relating to the "natural affinities" among
organisms and the "laws" governing their global and regional distribution.[59]

But it is worth taking the invectives seriously for a moment. A striking feature of
eighteenth-century taxonomic literature, which reflects its compilatory nature well,
is the increasing role that numbers began to play in it (see fig. 1). As in any proper
register, species and genera were numbered consecutively to create an additional layer
of indices that could be used to establish chains of references across field notes, correspondence, collections, annotations, and publications.[60] Numbering specimens in collections and gardens, or species entries in lists and catalogs, was a long-standing tradition in early modern natural history, to be sure.[61] But with Linnaean nomenclature and
taxonomy, such numbers acquired a new level of meaning that is best explained by
turning to a revealing example of their day-to-day use in late eighteenth-century natural
history.

In 1768, the German naturalist Johann Reinhold Forster (1729–98), then teaching at
the Dissenter's College in Warrington, was asked by Thomas Pennant (1726–98) to
assist him in producing a volume on insects for his *British Zoology*. In 1770, Forster
published a curious first product from his labors, entitled *A Catalogue of British Insects*
(see fig. 2). It consisted of a list of exactly 1,004 Linnaean names of insect species, neatly
lined up in two columns and numbered consecutively, both throughout and within each
genus. In addition, the list was structured by headings stating the genus name, again
numbered consecutively. The purpose of the catalog, as well as the meaning of the abbreviations set against many of the species entries, was succinctly explained by Forster
in the preface to his book:

> The author of this catalogue . . . presents his most respectful compliments to all ladies
> and gentlemen who collect insects, and begs them to favour him, if possible, with spec-
> imens of such insects, as they can spare, and which he is not possessed of: for this pur-
> pose he has made this catalogue, and put no mark to the insects in his possession; those
> which he has so plentifully as to be enabled to give some of them to other collectors,
> are marked with a (*d*); those which he has not, are marked either *Berk*. signifying
> Dr. Berkenhout's Outlines of the Natural History of Great Britain; or B. signifying a

[59] Janet Browne, *The Secular Ark: Studies in the History of Biogeography* (New Haven, Conn.,
1983); Philip F. Rehbock, *The Philosophical Naturalists: Themes in Early Nineteenth-Century British
Biology* (Madison, Wis., 1983); James L. Larson, *Interpreting Nature: The Science of Living Form
from Linnaeus to Kant* (Baltimore, 1994).

[60] Dietz, "Contribution" (cit. n. 45), 551–69; on Linnaeus's own use of genera and species numbers,
see Charlie Jarvis, "A Concise History of the Linnean Society's Linnaean Herbarium, with some Notes
on the Dating of the Specimens It Contains," in "The Linnaean Collections," ed. Brian Gardiner, special issue, *The Linnean* 7 (2007): 5–18; Charmantier and Müller-Wille, "Linnaeus's Botanical Paper
Slips" (cit. n. 42), 224–5. Naturalists collecting for the Jardin des plantes in Paris were commanded
to number herbarium specimens and seed sacks; see Marie-Noëlle Bourguet, "La collecte du monde:
voyage et histoire naturelle (fin XVIIème–début XIXème siècle)," in *Le muséum au premier siècle de
son histoire*, ed. Claude Blanckaert, Claudine Cohen, Pietro Corsi, and Jean-Louis Fischer (Paris,
1997), 163–96, on 174. Humboldt was particularly obsessed with numbering specimens during his excursions in the Amazon region; see H. Walter Lack, "Botanische Feldarbeit: Humboldt und Bonpland
im tropischen Amerika (1799–1804)," *Ann. Naturhist. Mus. Wien* 105B (2004): 493–514.

[61] Staffan Müller-Wille, "Reproducing Species," in *Secrets of Generation: Reproduction in the
Long Eighteenth Century*, ed. Raymond Stephanson and Darren N. Wagner (Toronto, 2015), 37–58.

Figure 1. *Frontispiece of the first volume of* Caroli Linnaei . . . Systema Naturae, *edited by Johann Joachim Lang (Halle, 1760). This edition was a pirated reprint of Linnaeus's tenth edition and may be the one Linnaeus himself referred to as the eleventh. The frontispiece shows a statue of Diana, taken from the frontispiece of Linnaeus's* Fauna suecica *(Leiden, 1746), and adds a human figure taking notes and pointing at the monkey in the top of the tree to the right. The heading refers to "names and numbers" [numeros et nomina] as essential elements of Linnaean natural history. The accusative is odd but may simply convey the idea that naturalists should work "toward names and numbers."*

manuscript catalogue of *British Insects* communicated to the author; or *B. B.* which signifies *Berkenhout*, together with the manuscript catalogue. *N. S.* is put to such insects as have not yet been described by Dr. *Linnaeus*, and are *new species* with new specific names.[62]

At a glance, then, Forster's catalog informed its readers of species he possessed in abundance, including species that were "new" to natural history, as well as species he knew existed and hoped to acquire through exchange to complement his own collection. The "d" probably stood for "duplicate;" that duplicates were expendable collection items as advertised by Forster is a notion that notably does not seem to have existed in pre-Linnaean natural history.[63] There is evidence that Forster had used the same communication strategy in earlier correspondence, with one similar manuscript list preserved in the Linnaean collections at London.[64]

---

[62] John [*sic*] Reinhold Forster, *A Catalogue of British Insects* (Warrington, 1770), 2; emphasis in the original. The "manuscript catalogue of *British Insects*" probably referred to the insect collection of Anna Blackburne (1726–98); see Arthur MacGregor, "Five Unpublished Manuscripts of Johann Reinhold Forster (1729–1798) in the Archives of the Linnean Society of London," *Arch. Nat. Hist.* 42 (2015): 314–30, on 320.

[63] Giuseppe Olmi, "From the Marvellous to the Commonplace: Notes on Natural History Museums (16th–18th Centuries)," in *Non-Verbal Communication in Science prior to 1900*, ed. Renato G. Mazzolini (Florence, 1993), 235–78, on 252–61; Claudia Swan, "From Blowfish to Flower Still Life Paintings," in *Merchants and Marvels: Commerce, Science, and Art in Early Modern Europe*, ed. Pamela Smith and Paula Findlen (London, 2001), 109–36, on 118; Delbourgo, "Collecting" (cit. n. 22), 14.

[64] MacGregor, "Five Unpublished Manuscripts" (cit. n. 62).

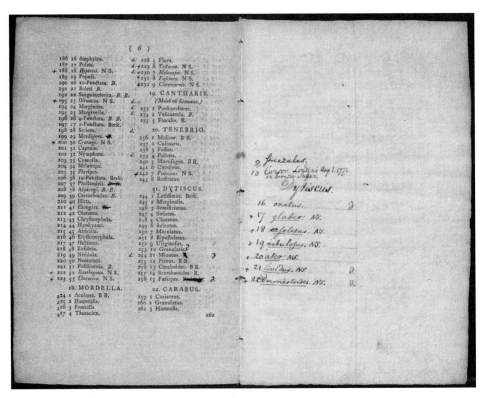

Figure 2. *Two pages from Forster,* A Catalogue of British Insects *(cit. n. 62) with annotations by its author. The printed text lists genera and species of insects, employing Linnaean trivial names. The notes document additional species that Forster came across after publication, many of them marked as new species ("NS."), and in one case reporting when and where a species was found: "10. [Tenebrio] Cursor. Londini Aug 1. 1771. in brown sugar." The latter remark refers to a beetle from Florida that established itself in Europe as a pest of stored foods. Forster, or his informant, may have come across this species in a shipment of sugar. Source: Staatsbibliothek zu Berlin—PK, Abteilung Historische Drucke, Signatur: Lt 12373: R. Courtesy Staatsbibliothek zu Berlin, Preußischer Kulturbesitz.*

The strategy was apparently successful; an interleaved and annotated copy of Forster's *Catalogue* has been preserved, in which he carefully noted species he had received or come across, either by deleting the abbreviations *Berk.*, *B.*, and *B. B.*, sometimes adding a "*d.*," or by noting additional species names on the interleaves, often followed by an "*N. S.*" or a "*d.*" A note on the flyleaf of this copy states "Aug. yᵉ 28. 1771. 42 more insects," and a calculation at the very end of the catalog registers "43 additional Insects" below the 1,004 already listed and draws up a new sum total of 1,047.[65] In the same year, 1771, Forster published a book presenting full species descriptions of one hundred "new" insect species. Again, an interleaved and annotated copy has survived from Forster's library, although in this copy the annotations do

---

[65] John [*sic*] Reinhold Forster, *A Catalogue of British Insects* (Warrington, 1770), Staatsbibliothek Berlin, Abteilung Historische Drucke, call no. Lt 12373R.

not record accessions to his insect collection but instead trace references to his descriptions of new species in entomological literature.[66]

Forster's *Catalogue*, with its extreme reduction of content to species names arranged according to the Linnaean hierarchy, illustrates the degree to which the discourse of classical natural history was dominated by naturalists' concern for their own position within the "market" of natural history. Linnaeus concisely, if slightly disparagingly, defined collectors in his *Philosophia Botanica* as those "who were primarily concerned with the number of species."[67] How many species of a particular genus were out there "on offer," whether in the hands of other collectors or out in the field? How many species had one already "acquired" in the form of specimens, and how many specimens could one "dispose of" as a kind of collector's capital to acquire specimens of other, preferably "new" or "rare" species? A whole new genre of taxonomic literature—consisting, like Forster's *Catalogue*, of nothing but taxonomic names, arranged in variously numbered and structured lists—emerged to answer these kinds of questions. Often openly advertising their poverty of content by incorporating terms such as "Index," "Nomenclator," or "Catalogue" in their title, these works, but especially their use, still await analysis by historians of science.[68] The fact that some of them were actually auction catalogs produced to support the sale of a collection clearly indicates that the genre catered to the desires of collectors.[69]

But there is more to Forster's *Catalogue* and its countless cognates. His list of insect genera and species shows striking structural similarities with what is certainly one of the most intriguing visual representations of the "order of nature" in late eighteenth-century natural history, the "genealogical-geographical table of plant affinities" ("Tabula Genealogico-Geographica Affinitatum Plantarum"), which Paul Dietrich Giseke (1741–96) produced on the basis of notes from private lectures that he and the entomologist Johann Christian Fabricius (1745–1808) had received from Linnaeus (see fig. 3). The table represents the plant kingdom in the form of fifty-eight circles of different sizes and slightly irregular shape, distributed over the sheet in an unruly manner, a little bit like an archipelago. The accompanying "explication" of the table does indeed speak of a "map," and of the circles as "provinces" or "islands," each of them standing for a particular "natural order" of plants, their "size" corresponding to the number of genera within each of these orders, and their mutual relative positions expressing relations of "affinity" [*affinitas*].[70] The orders differ strikingly in "size," that is, the num-

[66] Johann Reinhold Forster, *Novæ Species Insectorum: Centuria I* (London, 1771), Staatsbibliothek Berlin, Abteilung Historische Drucke, call no. Ls 3924.

[67] Linnaeus, *Philosophia Botanica* (cit. n. 36), 4.

[68] For an intriguing study of the catalogs produced of the avian collections in the early history of the British museum, see Jennifer M. Thomas, "The Documentation of the British Museum's Natural History Collections, 1760–1836," *Arch. Nat. Hist.* 39 (2012): 111–25; in the 1830s and 1840s, John Edward Gray (1800–1875) closed a political debate around the British Museum's authority simply by publishing a catalog of its natural history collections; see Gordon R. McOuat, "Cataloguing Power: Delineating 'Competent Naturalists' and the Meaning of Species in the British Museum," *Brit. J. Hist. Sci.* 34 (2001): 1–28.

[69] On auctions in natural history, see John Michael Chalmers-Hunt, ed., *Natural History Auctions 1700–1792: A Register of Sales in the British Isles* (London, 1976); Samuel J. M. M. Alberti, "Objects and the Museum," *Isis* 96 (2005): 559–71, on 564. On the history of economic, and even commercial, dimensions of natural history, see Ina Heumann and Nils Güttler, eds., *Sammlungsökonomien* (Berlin, 2016).

[70] Paul Dietrich Giseke, ed., *Caroli a Linne . . . Prælectiones in Ordines Naturales Plantarum* (Hamburg, 1792), 625. Giseke's "Tabula Genealogico-Geographica" became the model for many

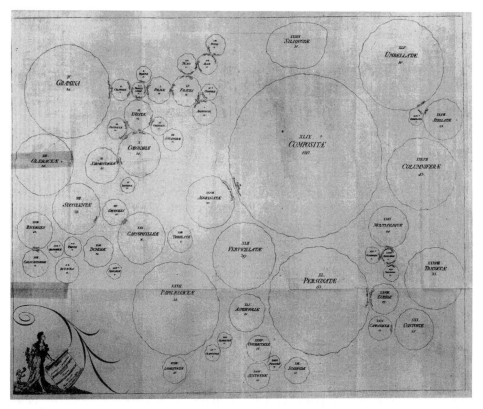

Figure 3. *"Tabula Genealogico-Geographica Affinitatum Plantarum," in Giseke*, Caroli a Linnaei Prælectiones *(cit. n. 70). The circles represent "natural orders" or plant families, and their size represents the number of genera they include. This number is also noted in the center of each circle, along with the family name and a roman numeral. The relative position of each circle indicates its taxonomic relationship with other families, sometimes highlighted by inscribing the names of closely related genera on the inside of two circles that face and almost touch each other.*

ber of genera they include (from eight to 120), just as the numbers of species per insect genus differ conspicuously in Forster's *Catalogue*. Both documents thus create an impression of a landscape of abundance and scarcity, of remoteness and propinquity, knowledge of which the old Linnaeus apparently imparted to his disciples with the unmistakable air of a sage privy to the mysteries of nature.[71] The objective of Linnaeus's speculations about a "natural" plant system may have been loftier than that of Forster's *Catalogue*, but his manuscript explorations of plant affinities took exactly the same form

map- or network-like representations of the "natural system" that were published in the first half of the nineteenth century; see Giulio Barsanti, *La Scala, la mappa, l'albero: Immagini e classificazioni della natura fra sei e ottocento* (Florence, 1992); Theodore W. Pietsch, *Trees of Life: A Visual History of Evolution* (Baltimore, 2013), 26–65. For a scrutiny of a particularly impressive "map" of bird affinities produced by Hugh Strickland in 1840, see Mary P. Winsor, "Considering Affinity: An Ethereal Conversation," *Endeavour* 39 (2015): 69–79 (issue 1), 116–26 (issue 2), 179–87 (issues 3–4).

[71] See especially the dialogue on plant affinities and their recognition that Giseke reports having had with Linnaeus in the summer of 1771 (*Prælectiones* [cit. n. 70], xv–xx). Linnaeus, Giseke claims, was constantly "chuckling" at the naïveté of his student's answers.

of numbered lists structured by headings and were certainly of equal strategic importance in his dealings with other plant collectors.[72]

There is a further way in which Forster's *Catalogue* connects with the higher aspirations of late eighteenth- and early nineteenth-century naturalists. The catalog he
produced was one of British insects and thus patently displayed a distribution of genera
and species that was peculiar to the British Isles. The shares that certain plant families
held in the overall number of genera and species of a certain climate or region played a
fundamental role in the attempts of Augustin de Candolle, Alexander von Humboldt,
and Robert Brown to establish "laws" that governed the geographic distribution of plants
in the second and third decades of the nineteenth century.[73] And again, the relationship
of these endeavors to the practice of numbering species and genera in taxonomic
works, especially local and regional floras and faunas, was not accidental. All three naturalists had themselves been involved in large-scale floral projects—de Candolle assisted
Lamarck in the third edition of his *Flore française* (5 vols., 1805), Browne prepared a
survey of the Australian flora (*Prodromus Florae Novae Hollandiae*, 1810), and Humboldt and his travel companion Aimé Bonpland issued seven volumes on South American plants (*Nova Genera et Species Plantarum*, 1815–25) as part of their landmark
travel account—and all three naturalists relied on floral catalogs for their calculations.
A palpable example of the kind of labor that was involved in this endeavor can be found
in a footnote that Humboldt added to his preface to the first volume of *Nova Genera et
Species Plantarum* when presenting a table comparing the absolute and relative number of species per "natural family" for France, Germany and Lapland:

> Since our floras are for the most part arranged according to the artificial system of Lin-
> naeus, [Karl Sigismund] Kunth, to whom I am much obliged for being in my service,
> transcribed the plants growing spontaneously under diverse [climatic] zones into natural
> orders; a labor which is truly cumbersome and protracted and if it had not been carried
> out in the most accurate manner, I could in no way have set out the arithmetic ratios of the
> geography of plants here.[74]

Karl Sigismund Kunth (1788–1850) also appears on the title page of *Nova Genera
et Species*, but in a subaltern position, as the one who "put [the volume] into order
from the handwritten paper slips of Aimé Bonland." Humboldt's remarks not only illustrate the longevity of Linnaeus's sexual system as a handy diagnostic tool but also
show how its limitations could be overcome by simple, if tedious, reallocation of species to their "natural families" or "orders."[75] One of the sources that Humboldt cites
on the German flora, Heinrich Adolf Schrader's (1767–1836) *Flora Germanica*, pro-

---

[72] Müller-Wille and Charmantier, "Lists" (cit. n. 42), 750–2.

[73] Browne, *The Secular Ark* (cit. n. 59); James L. Larson, "Not without a Plan: Geography and Natural
History in the Late Eighteenth Century," *J. Hist. Biol.* 19 (1986): 447–88; Güttler, *Das Kosmoskop*
(cit. n. 54), chaps. 1–3, esp. 181–5.

[74] Alexander von Humboldt, "De Instituto Operis et de Distributione Geographica Plantarum Secundum Coeli Temperiem et Altitudinem Montium Prolegomena," in *Nova Genera et Species Plantarum*, 7 vols. (Paris, 1815–25), 1:iii–lviii, n. 6 on xiii.

[75] While I focus in this essay on data-driven change in classical natural history, it is worth noting that
Linnaeus's sexual system provides an excellent example of the kind of "data drag" that natural history
was experiencing as well. The sexual system remained in use in natural history for almost a century,
although most naturalists, including Linnaeus, readily admitted that it was thoroughly "artificial." On
data drag, see Kaplan, "Lexicostatistics" (cit. n. 9).

vides a glimpse of how this task was sometimes made easier for Kunth. Schrader included a list that numbered species and genera in exactly the same way, as explained above for Forster's catalog.[76] Kunth could thus easily extract species numbers for each genus and only needed to add these numbers for each of the natural families. Humboldt planned to publish a stand-alone, second edition of their biogeographic treatise once Kunth had returned from Paris to Berlin to become professor of botany and vice director of the botanical garden, and throughout the rest of his life Kunth provided Humboldt with species numbers, partly drawn from what was to become his own magnum opus, a multivolume "Enumeration of all plants hitherto known arranged according to their natural families."[77] The second edition never materialized, but the surviving letters and manuscripts show that Kunth and Humboldt's speculations about relative and absolute species numbers involved the keen observation of how many species were known, above all, to naturalists at other important centers of botany, especially Paris.[78]

Kunth clearly exemplifies one of the "sordid registrars" that a younger Humboldt had despised, but on whose activities he, like other naturalists with higher aspirations, had to rely. "Registering" species with the help of Linnaean nomenclature and taxonomy was an activity that created the very condition for treating species and higher taxa as objects that could be meaningfully counted. As long as names and taxa had diagnostic functions, the number of species per genus was only a trivial consequence of the diagnostic criteria adopted. Once names and taxa were reduced to labels and containers in order to enhance the exchange of information—once the system they formed became a system of relations of equivalence, rather than difference—species numbers began to take on new, empirical meanings. The Linnaean reform, that is, was a pragmatic affair, serving the needs of an emerging landscape of central institutions and increasing levels of division of labor in natural history, but its widespread adoption also changed the ontological status of species from logical category to countable object.

CONCLUSION: DATA IN NATURAL HISTORY AND THE HISTORY OF NATURE

It is well known that the irregular patterns that emerged from late eighteenth- and early nineteenth-century attempts to document the geographic and taxonomic distribution of species formed the chief explanandum of Darwin's theory of evolution by natural selection.[79] Paleontology, with its observations on the stratigraphic distribution of species, followed a similar trajectory; as David Sepkoski has argued, it grew into a "substantially 'data-driven'" discipline in the early nineteenth century that contrib-

[76] Heinrich Adolf Schrader, *Flora Germanica* (Göttingen, 1806), 83–100.

[77] Karl Sigismund Kunth, *Enumeratio Plantarum Omnium Hucusque Cognitarum Secundum Familias Naturales Disposita*, 5 vols. (Stuttgart, 1833–50).

[78] See the various letters and manuscripts by Kunth preserved in Alexander von Humboldt's papers (Staatsbibliothek Berlin, Nachl. Alexander von Humboldt, gr. Kasten 6, 8, and 13). They have recently been made available online at http://humboldt.staatsbibliothek-berlin.de/werk/ (accessed 22 April 2017). On "counting" data in contexts of international competition, see Aronova, "Geophysical Datascapes" (cit. n. 44).

[79] Janet Browne, "Darwin's Botanical Arithmetic and the 'Principle of Divergence,' 1854–1858," *J. Hist. Biol.* 13 (1980): 53–89; R. Alan Richardson, "Biogeography and the Genesis of Darwin's Ideas on Transmutation," *J. Hist. Biol.* 14 (1981): 1–41; Wolfgang Lefèvre, *Die Entstehung der biologischen Evolutionstheorie* (Frankfurt am Main, 1984); Mary P. Winsor, "Darwin and Taxonomy," in *The Cambridge Encyclopedia of Darwin and Evolutionary Thought*, ed. Michael Ruse (Cambridge, 2013), 72–9.

uted equally to the formation of evolutionary theories.[80] Are we then to assume after all, in the spirit of Lepenies, that it was increasing "experiential pressure" from ever-heightened levels of accumulated and articulated data that sparked the historicization of nature?

In response to this question, it is worth pointing out two things. First, it was perfectly possible to remain "ahistorical" in face of the strikingly irregular patterns of species distribution; during the first half of the nineteenth century, most naturalists actually did so, and ideas of divine creation and directed evolution have survived the Darwinian revolution to this day. What does it mean to "historicize" nature anyway, if even Darwin and Wallace could not agree on some quite elementary points of their respective evolutionary theories? What meaning was assigned to the data that systematists, biogeographers, and paleontologists accumulated clearly depended on cultural factors other than the mere form that these data took once they were assembled to create new representations of the order of nature.[81] On the other hand, it is equally clear that the ways in which data on the distribution of species were presented with the help of Linnaean nomenclature and taxonomy held an enormous potential for generating surprises. Giseke's map, or even Forster's little *Catalogue of British Insects*, was a clear affront to the old idea that nature formed a continuous and unchanging scale of perfection.[82]

The second point I would like to make concerns the nature of "data" in natural history. Humboldt held on to his early views of induction, writing in 1808 that "the physics of the earth has its numerical elements, just like the world system, and one will only gradually reach knowledge of the true laws that determine the geographic and climatic distribution of plant forms through the collective labor of traveling naturalists." One such traveling naturalist, Friedrich Sellow (1789–1831), a protegé of Humboldt collecting specimens in Brazil from 1817 to 1831, has been described as having had an "obsession with data." The journals left from his travels show that these "data" consisted, among other things, of endless numbered lists of the names of species collected, as well as where and when they were collected.[83] Just as with Forster's *Catalogue*, almost nothing can be gleaned from these entries about the properties of the plants and animals encountered, their local environments, or their local uses. So, the data that were recorded in this way were not data that provided information about organisms, but rather what we would call "metadata" today, which in classical natural history consisted of a proper name, allocation of taxonomic position, and information on date and place of provenance.[84] Humboldt's early call for "more data" to unravel the unknown "laws" of nature from them essentially did not ask for much more than this. The infrastructure of "labels" and "containers" created by the Linnaean reform began to acquire a life of its own, producing phenomena that could not have been produced without it. This is true in particular for the taxonomic distribution of species, since stating the

[80] David Sepkoski, "Towards 'A Natural History of Data': Evolving Practices and Epistemologies of Data in Paleontology, 1800–2000," *J. Hist. Biol.* 46 (2013): 401–44.

[81] Wolfgang Lefèvre, "Das 'Ende der Naturgeschichte' neu verhandelt: Historisch-genealogische oder epigenetische Neukonzeption der Natur?," Max Planck Institute for the History of Science Preprint 476 (Berlin, 2016).

[82] Harriet Ritvo, *The Platypus and the Mermaid and Other Figments of the Classifying Imagination* (Cambridge, Mass., 1998).

[83] Hanns Zischler, Sabine Hackethal, and Carsten Eckert, eds., *Die Erkundung Brasiliens: Friedrich Sellows unvollendete Reise* (Berlin, 2013), quotations on 67 and 113.

[84] On the concept of metadata in library science, see Krajewski, "Tell Data from Meta" (cit. n. 13). There are striking parallels with early modern genealogical practices as well; see Friedrich, "Genealogy" (cit. n. 17).

number of species per genus, or the number of genera per natural family, remains totally within the ontology that this infrastructure created in the first place.

Classical natural history, and its post-Darwinian heir, the discipline of systematics, can thus indeed be considered an information science, that is, a science whose primary aim consists in the storage, organization, and mobilization of knowledge.[85] But if this is true, it can also be considered inherently "experimental," in the sense of building on art and artifice to produce new knowledge. Through the accumulation of specimens, containers, labels, and other inscriptions, naturalists bring together objects—on the page of a handwritten or printed text, in a drawing or diagram, within the drawer of a museum depot, or in the showcase of an exhibition gallery—that normally would never have coexisted. It is this peculiarity that endowed classical natural history, despite the occasionally dull appearance of its products, with its very own condition of creativity. The epochal shift from natural history to the history of nature was thus not produced with a kind of teleological necessity through the accumulation of data; rather, the instruments and infrastructures brought into play to manage and enhance flows of data—Linnaean names and taxa, above all—generated unforeseen and, indeed, never-before-seen phenomena that were difficult to reconcile with long-held intuitions.

[85] Ernst Mayr, "Systems of Ordering Data," *Biol. & Phil.* 10 (1995): 419–34; Quentin D. Wheeler, ed., *The New Taxonomy*, vol. 76 of Systematics Association Special Volume Series (Boca Raton, Fla., 2008).

Machineries of Data Power:
Manual versus Mechanical Census Compilation in Nineteenth-Century Europe

by Christine von Oertzen*

ABSTRACT

The advent both of punch cards and of the electric tabulating machine, which was invented in 1889, are typically described as key milestones in the development of modern data processing, bringing about a fundamental and inexorable transformation of information technology. This essay aims to decenter the American Hollerith revolution by assessing precisely how punch cards and machine processing transformed established manual techniques and practices of census compilation. By focusing on the Prussian census bureau and its long-standing reluctance to mechanize, this essay reveals an unremarked European revolution in data processing during the 1860s, when a new notion of "data," novel paper tools, and a carefully nurtured workforce, including many women working from home, yielded unprecedentedly refined census statistics. The essay argues that manual concepts, technologies, and practices of data power—rather than punch cards and Hollerith machines—heralded the modern information age.

INTRODUCTION

The history of data processing from Hollerith machines to electronic mainframe computers is a familiar tale of technology's forward march. The advent both of punch cards and of the electric tabulating machine, which was invented in 1889, are typically described as key milestones in the development of modern data processing, bringing about a significant transformation of information technology. The compilation of numerical data via mechanical means, as the account goes, recast the making of statistics in government, industry, and science, enabling the omnipresence of computation we increasingly have come to embrace as the starting point for virtually all inquiry, scientific or otherwise.[1]

* Max Planck Institute for the History of Science, Boltzmannstrasse 22, 14195 Berlin, Germany; coertzen@mpiwg-berlin.mpg.de.

I would like to thank all members of the "Historicizing Big Data" working group for their comments and suggestions; special thanks go to Dan Bouk, Daniel Rosenberg, Ted Porter, and two anonymous reviewers.

[1] Prominent examples of this kind of history are Martin Campbell-Kelly, *ICL: A Business and Technical History* (Oxford, 1989); Geoffrey Austrian, *Herman Hollerith: Forgotten Giant of Information Processing* (New York, 1982); William Aspray, ed., *Computing before Computers* (Ames, Iowa, 1990); Hartmut Petzold, *Rechnende Maschinen: Eine historische Untersuchung ihrer Herstellung und An-*

The teleological bias of such accounts aside, most tend to focus on the technology itself, placing machines and a handful of prominent inventors at center stage. In this essay, I will juxtapose the machine-centered approach to the triumphant success story of the American Hollerith system with an assessment of precisely how punch cards and machine processing transformed established techniques and practices used to convert large quantities of data into meaningful statistics and tables. I will do so by focusing on the technicalities and logistics of nineteenth-century census compilation. Until the out-break of global war in 1914, census bureaus were among the first agencies to place data on assembly lines, using machines to produce what became ever more ambitious and highly detailed statistical surveys. This was especially the case in the United States, where the federal census developed from a narrowly defined project, as mandated by the U.S. Constitution, to a far-flung effort to map and mold the frontier nation via the tools of enumeration. It was in the context of the ninth federal census of 1890 that the punch cards and counting machines of young engineer Herman Hollerith were in-troduced to prevent the collapse of the national census bureau, overwhelmed by the del-uge of enumeration lists containing gathered information.[2]

Historians have traced how the new technology of mechanical data processing spread across other fields and continents.[3] As applied to census taking, Hollerith's invention was introduced to European statisticians via hands-on demonstrations at the 1889 World's Fair in Paris; the inventor himself crisscrossed Europe trying to convince govern-ments and their statistical experts to follow America's lead and invest in his machines. However, Hollerith's innovation appealed much more readily to some European gov-ernments than others. Whereas his machinery was quickly embraced in Vienna, Saint Petersburg, and Oslo, census officials in London and particularly in Berlin regarded the new machines with suspicion. The apparent paradox that Prussia, the military and eco-nomic powerhouse that constituted nearly three-fifths of the German Empire, opposed mechanization for decades has long defied explanation. The reluctance of Prussia and other European states to implement the new system has been brushed aside with passing reference either to bureaucratic inertia with regard to innovation or, in more recent ac-counts, to the overall efficiency of government organization, as well as technical limi-tations, operational hurdles in marketing, and maintenance of the machines.[4]

In what follows, I offer a fresh perspective on the dynamics of innovation at the crossroads of manual and machine data processing. In line with Philip Scranton's ap-

---

wendung vom Kaiserreich bis zur Bundesrepublik (Düsseldorf, 1985). Although recent accounts offer more balanced narratives on the transition from manual to mechanical data processing, they credit Hol-lerith's machinery with a "momentous change," mentioning manual methods of data processing merely in passing; see Lars Heide, *Punched Card Systems and the Early Information Explosion, 1880–1945* (Baltimore, 2009).

[2] Margo J. Anderson, *The American Census: A Social History* (New Haven, Conn., 1988); Keith Reid-Green, "The History of Census Tabulation," *Sci. Amer.* 260 (1989): 98–103.

[3] Lars Heide, "Punched Cards for Professional European Offices: Revisiting the Dynamics of Infor-mation Technology Diffusion from the United States to Europe, 1889–1918," *Hist. Tech.* 24 (2008): 307–20. For the insurance industry, see JoAnne Yates, *Structuring the Information Age: Life Insur-ance and Technology in the Twentieth Century* (Baltimore, 2008); Dan Bouk, *How Our Days Became Numbered: Risk and the Rise of the Statistical Individual* (Chicago, 2014).

[4] Heide, "Punched Cards" (cit. n. 3), 308. For Great Britain's reluctance to introduce Hollerith ma-chinery, see Martin Campbell-Kelly, "Information Technology and Organizational Change in the Brit-ish Census," *Inform. Syst. Res.* 7 (1996): 22–36; Campbell-Kelly, "Data Processing and Technological Change: The Post Office Savings Bank, 1861–1930," *Tech. & Cult.* 39 (1998): 1–32.

peal to reject grand narratives of technological determinism by paying close attention to the contingencies and particulars of the local, I focus on the Prussian case to uncover how the census bureau in Berlin mastered the compilation of census data.[5] I show that the Prussian tools and methods proved perfectly up-to-date in technological terms, very similar to the ones integrated in Hollerith machinery. I claim that the Prussian rejection of punch card machinery, rather than being backward-looking, points to patterns of innovation similar to those described by Jon Agar with regard to the introduction of computers. Computerization, according to Agar, has only been attempted in settings where methods and material practices and technologies capable of the computational task at hand were already in place, and where well-tried older techniques were taken over by the new technology.[6] The same applies, as this essay demonstrates, to the transition from manual to mechanical data compilation.

With Scranton's and Agar's work in mind, I contend that in order to grasp Prussia's unwillingness to comply with mechanical innovation, we need to widen our perspective beyond machine technology. To that end, I consider the "machinery" of census compilation in a very broad sense, exploring both manual and mechanical data processing. Rather than focus on machines alone, I consider the technicalities, material practices, and politics of manual census compilation to scrutinize the processing of census data when tabulating machines arrived on the scene. Examining Prussia's machinery of data power from this angle reveals an unremarked European revolution in data processing during the 1860s, which in turn yielded a radical transformation of population statistics. At the heart of this fundamental reshaping of mid-nineteenth-century census taking lay a new concept of "data," as well as a novel methodology to turn information gathered in enumeration lists into complex tables with many variables. The technology that enabled this transformation of data was a new paper tool serving as a movable data carrier: a counting slip or counting card, very similar to the machine-readable American punch cards introduced a good twenty-five years later.

As an explicit contribution to the material culture of data, this essay unveils the concepts and workings of Prussia's machinery of data power as deeply embedded in their social and political contexts.[7] Inspired by Deleuze's reading of Kafka, I regard technology (in Kafka's case, the typewriter, but in the context of this essay, the paper tools used for manual compilation in census bureaus) as inextricably linked to infrastructures: labyrinths of corridors with clerks bustling from office to office, laborers hauling heavy boxes with paper materials, horse carriages, white-collar workers and their families toiling within private homes, administrators, orders, and signatures, a whole set of social relationships, rationalities, and desires that precede and embrace the tech-

---

[5] Scranton's approach shares similarities with Clifford Geertz's "thick descriptions"; see Philip Scranton, "Determinism and Indeterminacy in the History of Technology," in *Does Technology Drive History? The Dilemma of Technological Determinism*, ed. Merritt Roe Smith and Leo Marx (Cambridge, Mass., 1994), 143–68.

[6] Jon Agar, "What Difference Do Computers Make?" *Soc. Stud. Sci.* 36 (2006): 869–907. See also Hallam Stevens, "A Feeling for the Algorithm: Working Knowledge and Big Data in Biology," in this volume; he argues that Agar's views do not reflect the emergence of new forms of computing practice after 1970.

[7] My approach to technology is also indebted to Ken Alder's treatment of it: "our modern technological life is a political creation, and we can read in its artifacts the history of the struggles and negotiations which gave it birth"; see Alder, *Engineering the Revolution: Arms and the Enlightenment in France, 1763–1815* (Chicago, 2010), 15.

nical equipment.[8] The processing of gathered census information becomes palpable as a complex undertaking. In logistical terms, it involved many hands in a host of different locations, carefully planned choreographies of sorting, sustained labor management, skillful counting techniques, relentless tracing of errors, and strict control. Socially and politically, the workings of this expansive machinery proved prone to friction, because efficiency was achieved by clandestinely undermining core stipulations of the authoritarian Prussian welfare system with highly gendered implications. The rich source material that has survived in the Prussian Secret State Archive in Berlin demonstrates that, in order to ensure accuracy and timeliness while minimizing costs, the Prussian census bureau entrusted the bulk of compilation work to the wives and other female relatives of its male officials. Having these women do the state's paper work in their homes was cost-effective, but it meant that the census bureau was ignoring its formal mandate to hire first and foremost impoverished veterans to perform this task.

This essay consists of four parts. First, I sketch how Prussian statisticians responded to the arrival of the American invention in Europe and examine their assessment of Hollerith's tabulating machine. Second, I place the Prussians' views about the weaknesses in the punch card system in historical context by describing the conceptual and technological features of the 1860s census data revolution marked by the emergence of "data," the distinction between list and table, and the introduction of movable paper tools enabling new methods of counting and sorting. Third, I examine the period from 1871 onward, when the Prussian census, based on these novel methods, centralized all compilation work in Berlin. I pay close attention to the materials and practices in use and the men and women involved in the work by analyzing the planning, manpower, and skills necessary to organize and perform manual census compilation, an operation encompassing Berlin's vast metropolitan area. The Prussian government's choice against mechanization emerges as a well-thought-out strategy. Conceptual, political, and fiscal considerations shaped the desired statistical output based on manual technologies and practices, and led the state to doubt the benefits of the new machines. Finally, I turn to Austria-Hungary, the first European empire to introduce punch cards and mechanical tabulating machines in 1890, to show that fundamentally, the Prussian and Austrian—as well as the American—stories presented here were predicated on the same data-processing revolution, starting around the mid-nineteenth century, during the manual era of data compilation. Machine-readable punch cards accelerated the sorting and counting process considerably—and, in the long run, enabled statisticians to accomplish tasks that were impossible to perform manually. But they rested firmly on the concepts and paper tools developed for manual use.

THE PUNCH CARD CROSSES THE NORTH ATLANTIC

Herman Hollerith's mechanical tools were designed to address specific shortcomings of the American method of census taking: until 1903, the federal census bureau in Washington lacked a permanent organizational structure, and at the end of the nineteenth century, producing detailed tables with short-term, poorly trained staff had be-

[8] Gilles Deleuze and Félix Guattari, *Kafka: Toward a Minor Literature*, trans. Brian Massumi (Minneapolis, 1986). The term "machinery" is also used by Jon Agar to emphasize the close links between "order," "framework," "structure," and "machine" in states and governments; see Agar, *The Government Machine: A Revolutionary History of the Computer* (Cambridge, Mass., 2003), 3.

come a significant organizational challenge.[9] In 1880, the imbalance brought about by the combination of rapid population growth, statistical ambition, and the absence of permanent infrastructure culminated in a massive data jam.[10] In response to these challenges, Hollerith's tabulating machines sought to accelerate the counting and sorting of the collected data while at the same time reducing the number of skilled personnel needed to perform this work. The key element in his scheme was the punch card—a movable paper device designed to carry all relevant data of one individual.

Punch cards had been used to control weaving looms in early eighteenth-century France, but Hollerith insisted that the inspiration for his data carrier was distinctly American: the Pullman railroad ticket.[11] Just as the railroad ticket was used to store information about one specific railway journey at a set fare, Hollerith's punch card archived an anonymized binary record of an individual citizen. And just as conductors punched holes in different parts of the card to note price and class of service, Hollerith developed a mechanical card punch, enabling human operators to punch holes via either a hand gear or keyboard. For the 1890 U.S. Census, nearly sixty-three million cards of this sort were punched by hand. A second element of the Hollerith system was the tabulating machine, into which each card was individually entered. This mechanism made use of electrical connections to trigger a counter clock, recording information in numerical code (which in 1890 was written down and added by hand). By punching holes in specific locations (i.e., rows and columns), cards could in turn be separated by sorting machines, the third element of Hollerith's system. The cards were also fed by hand and subsequently—by means of electric relay connections—filed into boxes according to specific combinations.[12]

Even before the first mechanical numeration effort in Washington was successfully completed, Hollerith's innovations prompted enthusiasm, but also skepticism, among European statisticians.[13] The head of the Prussian statistical office at this time, Emil Blenck, rejected the adoption of electric tabulating and sorting machines. In Blenck's view, the American census had turned into societal surveys going well beyond the mandate in the U.S. Constitution, burdening the federal government with avoidable expenditures. Blenck's objections were also technical in nature. Hollerith's tabulators did not, in his opinion, lead to a lower error count owing to the fact that information transferred from enumeration lists on to the punch cards continued to be performed by hand. Instead, Blenck advocated the continued use of pen and paper for census compilation. This method allowed Prussian authorities to avoid the transmission of data from questionnaires to punch cards, an intermediary step Blenck considered the Achilles' heel of the Hollerith system.[14]

According to Blenck, the sole virtue of using punch-card-driven machines was the possibility to combine an infinite number of characteristics. Having conceded this point, Blenck set out to illustrate how superfluous this ostensible advantage proved in real-world statistical practice. States and governments, he argued, did not need

[9] Anderson, *American Census* (cit. n. 2), 52; Campbell-Kelly, *ICL* (cit. n. 1), 8–13.
[10] Reid-Green, "History" (cit. n. 2), 98–101.
[11] Austrian, *Herman Hollerith* (cit. n. 1), 15.
[12] Ibid.
[13] Heide, "Punched Cards" (cit. n. 3), 308.
[14] "Versammlung der Polytechnischen Gesellschaft 1896," *Polytech. Centralbl.* 57 (1896): 121–5, on 124.

to know everything statistically possible and must not misuse the census for "scientific game playing."[15] Instead, what must and would remain decisive, at least in Prussia, was whether mechanization "truly warrants additional financial investment, that is, whether it yields tangible advantages to the public at large."[16] Balancing statistical sophistication on the one hand and political and economic rationales on the other, Blenck argued that the state bore a special responsibility to its impoverished retired servicemen and their dependents in the bureau's pay. He insisted that it was the state's responsibility "to employ people rather than machines and to dispense with the added value attributed, falsely or not, to particular details and correlations. In so doing, we avoid investing a very significant sum in machinery that we may wish, in keeping with the Reich's social policy objectives, to put to use elsewhere."[17] Blenck remained an outspoken advocate of manual census compilation until his retirement in 1912. However, the reasons why the director of the Prussian census bureau resisted mechanization were far more complex than the "policy of a good heart" he repeatedly asserted, despite increasing criticism from colleagues within Imperial Germany and beyond. The Prussian census bureau's refusal to embrace mechanical innovation was not due to rejection of machines as such or to social sentiment. Rather, its system of manual data compilation was efficient, reliable, and firmly embedded in the principles of Prussian political economy.

TECHNIQUES OF MANUAL DATA PROCESSING

To put Blenck's technical as well as social concerns about mechanization in historical context, it is necessary to review how techniques of census taking in Prussia had developed since 1860. Almost everywhere in Europe, the collection of census and population statistics expanded into complex operations from the mid-nineteenth century onward, driven by efforts to render enumeration "scientific" by creating standardized procedures that would yield more than just population counts and lead to verifiable, "objective," and comparable descriptions of the population.[18] International statistical congresses hosted in different European cities starting in 1853 intensified methodological exchange and created state-of-the-art standards.[19] Prussia's statisticians were slow to adopt this new trend, but during the 1860s they radically transformed their methods and practices of census taking. The most significant conceptual as well as technical innovations in Prussian census work during this time were initiated by Blenck's predecessor, Ernst Engel, a mining engineer from the Mining Academy in Freiberg, Saxony, who was appointed director of the Prussian statistical bureau in 1860. Trained in the cameralistic tradition, Engel had developed a passionate interest in the social condi-

[15] Blenck was well known among statisticians for his emphasis on the state. Purely scientific research with the data collected must be carried out only after the duties to the state were entirely fulfilled. Georg Evert, "Nachruf auf Emil Blenck," *Bull. Int. Inst. Statist.* 19 (1912): 433–9, on 438.
[16] Ibid.
[17] "Versammlung" (cit. n. 14), 124.
[18] Ian Hacking, *The Taming of Chance* (New York, 1990); Theodore Porter, *The Rise of Statistical Thinking, 1820–1900* (1988; repr., Princeton, N.J., 1995); Alain Desrosières, *The Politics of Large Numbers: A History of Statistical Reasoning* (Cambridge, Mass., 1998).
[19] Nico Randeraad, "The International Statistical Congress (1853–1876): Knowledge Transfers and Their Limits," *Eur. Hist. Quart.* 41 (2011): 50–65. Randeraad claims that many topics discussed there were patriotically inspired and "hard to internationalize" (56), but exchange on methods and technicalities of processing census data established an international discourse of best practices.

tions created by industrialization, a concern that deepened during the 1840s, when he studied with Europe's most prominent statisticians, Frédéric Le Play in Paris and Adolphe Quetelet in Brussels.[20] As a cofounder of the International Statistical Congress, Engel was a well-known pioneer in comparative and relational statistics. At the same time, he earned a reputation as a restless reformer, advocating that census bureaus should be much more than mere accounting agencies of the state. Prior to his Berlin appointment, he was director of the Saxon Statistical Bureau, where his efforts to publish interpretations of census results—and not simply numbers—led to his dismissal in 1858.[21]

Engel's contribution in conceiving population statistics as a science in its own right to address structural problems arising from the formation of societies has been well studied.[22] Much less known, however, is his decisive role in turning the technicalities of Prussian census compilation into a data-driven operation. Less than a year into his Berlin tenure, Engel initiated a complete overhaul of the Prussian enumeration procedures. His efforts did not unfold in isolation; rather, they reflected a European discourse as Engel experimented with best practices tested in other countries and approved by the International Statistical Congress. In accordance with the latter, Engel introduced a strict distinction between "enumeration material" [*Urmaterial*] and tables, which were to contain abstractions from this information. As he stated in his seminal treatise on the methods of census taking in 1861,

> We strictly discriminate between list and table. The first one always refers to one specimen, to one particular individual. In the latter, the specimen or individual is no longer visible; the table contains already a condensed result, a summary and a grouping of information from the lists.[23]

Up to that point, this distinction between lists and tables had not been made, a fact that Engel deplored as both epistemologically unsatisfactory as well as aesthetically offensive. Prussian census information was collected in the exact same forms that were then printed: book-length lists with 625 or more rubrics.[24]

Engel's distinction between enumeration and tabulated data from 1861 can be regarded as the birth of data-driven Prussian census statistics: raw material was collected in enumeration lists, information that was then not just counted but had to be processed—that is, transformed into numbers, arranged in tables, interpreted in accom-

[20] On the mining engineer Frédéric Le Play and his career in social statistics, see Theodore Porter, "Reforming Vision: The Engineer Le Play Learns to Observe Society Sagely," in *Histories of Scientific Observation*, ed. Lorraine Daston and Elizabeth Lunbeck (Chicago, 2012), 281–302. On Adolphe Quetelet, see Jean-Guy Prévost and Jean-Pierre Beaud, *Statistics, Public Debate, and the State, 1800–1945: A Social, Political, and Intellectual History of Numbers* (London, 2012), chap. 3.

[21] H. Strecker and R. Wiegert, "Ernst Engel," in *Leading Personalities in Statistical Sciences: From the Seventeenth Century to the Present*, ed. Norman L. Johnson and Samuel Kotz (New York, 1997), 280–3, on 282.

[22] Ian Hacking, "Prussian Numbers, 1860–1882," in *The Probabilistic Revolution*, vol. 1, *Ideas in History*, ed. Lorenz Krüger, Lorraine J. Daston, and Michael Heidelberger (Cambridge, Mass., 1987), 45–68; Erik Grimmer-Solem, *The Rise of Historical Economics and Social Reform in Germany, 1864–1894* (Oxford, 2005), 62–5; Michael Schneider, *Wissensproduktion im Staat: Das Königlich-Preußische statistische Bureau, 1860–1914* (Frankfurt am Main, 2013), 223–82.

[23] Ernst Engel, "Die Methoden der Volkszählung: Mit besonderer Berücksichtigung der im Preussischen Staate Angewandten," *Z. Königl. Preuss. Stat. Bur.* 7 (1861): 149–212, on 163.

[24] Ibid. See also Schneider, *Wissensproduktion* (cit. n. 22), 228.

panying texts, and, eventually, displayed in graphs and maps.[25] As Engel put it when describing the three steps necessary to produce population statistics in mid-nineteenth-century Prussia, "Statistics are defined by methods. These methods vary depending on whether they aim at the gathering, the collection, or the utilization of data [*Daten*]."[26] It does seem more than a coincidence that he and other Prussian statisticians explicitly used the term "data" when they referred to the methodology of gathering and processing information provided by enumeration lists.[27]

Engel's description of a three-step statistical methodology based on the notion of "data" and his sharp distinction between "lists" and "tables" served as the basis for a series of far-reaching reforms that radically transformed the practices of census taking in Prussia. Within a decade, standardized enumeration procedures reshaped the gathering of raw material, while a new technique of abstracting turned the subsequent use of the gathered information into a much more refined and efficient operation. What Engel described as the "collecting of data" was omitted in this process to optimize control and accuracy: local and regional authorities, which were involved in checking and summarizing the material, were soon bypassed altogether in favor of a wholly centralized compilation effort in the Prussian census bureau in Berlin.

Equally important in reshaping both the gathering and the abstracting of census data was Engel's further adaptation of international statistical recommendations. He introduced the individual as the basic unit of Prussian census taking, doing away with households, buildings, and localities as the main categories from which census information was developed.[28] Enumeration lists drawn up for individuals rather than for households were first distributed in 1867. However, this change did not mean that Prussia's inhabitants were henceforth counted regardless of their status within a household or family. Rather, this status was one component of that person's data set, which allowed more detailed analysis of household structures than before.

The introduction of gathering data in lists in which the individual resident emerged as the ordering principle ushered in a new system of abstracting once all the data had been collected. At the heart of the Prussian census reform was a paper tool that Engel proposed in 1867: a short counting slip [*Zählblättchen*] to abstract numbers drawn from the enumeration lists. Prussia was the largest but not the only German state to introduce the counting slip that year. Statisticians from Berlin and from the Grand Duchy of Hesse also opted for this new counting slip to compile census data after returning from the International Statistical Congress in Florence held that same year. There, experts had witnessed how effectively counting slips accelerated the tedious compilation of large amounts of data. Counting slips—the Italian *cartoline*—were deployed on a grand scale in unified Italy's first national census in 1861 to streamline the many dif-

[25] For similar uses of the term "data" in the natural sciences during the late eighteenth and early nineteenth centuries, see Staffan Müller-Wille, "Names and Numbers: 'Data' in Classical Natural History, 1758–1859," and David Sepkoski, "The Database before the Computer?," both in this volume.
[26] Engel, "Die Methoden" (cit. n. 23), 162.
[27] The most explicit example may be found in Hermann Schwabe, "Bericht der städtischen Volkszählungs-Commission über die Ausführung der Zählung nebst Anlage 1–12," in *Die Resultate der Berliner Volkszählung vom 3. December 1867* (Berlin, 1869), i–xxxi. See also Engel's reply to G. H. Fabricius, "Zur Theorie und Praxis der Volkszählungen," *Z. Königl. Preuss. Stat. Bur.* 8 (1868): 198. Another example is Richard Boeckh, "Vorwort," *Berliner Städt. Jahrb. Volkswirts. Stat.* 22 (1895): iii–iv.
[28] Ernst Engel, "Die Aufgaben des Zählwerks im Jahre 1880," *Z. Königl. Preuss. Stat. Bur.* 19 (1879): 367–76.

ferent regional and local conventions of enumeration in order to produce commensurable population *dati* of the new state, in accordance with international standards.[29] The "exquisite demonstration" submitted by Italian colleagues convinced Engel and other German delegates to embrace this new technology.[30]

Like the punch card, its much younger American cousin, the counting slip was an intermediate, movable data carrier designed to facilitate and enhance the counting and sorting of data compiled in lengthy enumeration lists, praised for its ability to greatly enhance statistical complexity. As Engel noted in 1868, "This is the main advantage of the counting slip: That it allows for countless combinations of the individual data contained in it."[31] The main difference was that the information on counting slips had to be transferred from enumeration lists with a pen, not, as with punch cards, via punched holes. And whereas punch cards could be sorted and counted by machines, counting slips had to be sorted into stacks and counted by hand.[32]

The individual, movable counting slip revolutionized the work of abstracting enumeration lists and of compiling tables.[33] Until the 1870s, analysis of enumeration lists was conducted via a technique called check-listing. (This simple method entailed counting by strokes of five. The end result resembled the following: ∦ ∦ ∥ . . .). Enumeration entries were either read aloud to a group of abstractors, each responsible for marking strokes in one particular rubric of a spacious, unwieldy interim table (the so-called synchronous system), or worked through by individual abstractors many times, rubric by rubric (known as the silent system).[34] The American census of 1880 was probably the largest compilation effort conducted in this manner, with 1,495 clerks putting check marks in large tally sheets with hundreds of cells, leaving a journalist of the period to marvel, "The only wonder is . . . that many of the clerks who toiled at the irritating sheets of tally paper . . . did not go blind and crazy."[35] European statisticians agreed that these traditional methods of compilation by means of interim checklists (or tally sheets) were highly unsatisfactory, not only because of reading and writing errors, but also because hearing and counting errors hampered accuracy. Further, the tally system did not allow the tracing of these errors. Check-listing was extremely time-consuming. To make matters worse, the clumsy method proved unsuitable to create more combinations of variables, features that statisticians deemed necessary for up-to-date representations of census results.

Movable paper tools that made it easier to retrieve and sort information were common in mid-nineteenth-century library cataloging. As Staffan Müller-Wille and Markus Friedrich show in their essays, naturalists and humanists alike were using paper

[29] *Statistica d'Italia, Popolazione, Parte I: Censimento Generale (31 Dicembre 1861)* (Florence, 1867), 3. On the Italian census, see also Silvana Patriarca, *Numbers and Nationhood: Writing Statistics in Nineteenth-Century Italy* (Cambridge, 1996).
[30] Ernst Engel, "Aktenmässige Darstellung der Vorbereitungen zu den Statistischen Aufnahmen im December 1867, insbesondere der Volkszählung im Preussischen Staate und norddeutschen Bundesgebiete," *Z. Königl. Preuss. Stat. Bur.* 7 (1867): 263–321, on 304.
[31] Engel's reply to Fabricius, "Zur Theorie" (cit. n. 27), 198.
[32] A detailed description of the counting-slip technology can be found in Georg von Mayr, *Statistik und Gesellschaftslehre* (Tübingen, 1914), 122–3.
[33] Heinrich Rauchberg, "Übersicht über den Stand und die neuesten Fortschritte der Technik auf dem Gebiete der Bevölkerungsstatistik," *Allg. Stat. Arch.* 1 (1890): 99–116.
[34] Engel, "Aktenmässige Darstellung" (cit. n. 30), 303.
[35] Martin Campbell-Kelly and William Aspray, *Computer: A History of the Information Machine*, 2nd ed., Sloan Technology Series (Boulder, Colo., 2004), 15.

slips to organize and archive both notes and references flexibly.[36] But the new paper tool employed in census compilation was inspired by card games, which had informed still earlier scholarly archiving and bibliographical sorting practices as well.[37] The counting slips, as Engel explained, were to be cut from "simple sized paper" and then shaped "like a playing card."[38] The new paper tool had more in common with playing cards than size. The very process of abstracting enumeration lists via counting slips resembled card games. The size of the slips was regarded as crucially important for swift sorting, a process in which slips with matching information had to be piled into stacks according to a predetermined laying list. The limited space on each counting slip led the Prussian census office to divide the process of compilation into several distinct counts and to produce not just one, but different versions of counting slips preprinted with specific subsets of characteristics depending on the data required to compile a particular group of tables. The most common characteristics, such as the denotation of male and female, were signified by color, making them easily detectable.

The most tedious and time-consuming task of data processing via counting slips was to transfer the relevant data from the enumeration lists onto the respective counting slips by pen. Once this was achieved, the slips were robust enough to be used for numerous rounds of sorting and counting, depending on how many different characteristics or combinations of variables were needed. In each round, the slips were re-sorted anew according to different criteria and counted pile by pile, and the numerical results were noted in interim tables, a process that continued until all rubrics of the requested tables were filled. However, in place of random reshuffling after each count, the handling of the slips followed a strict, predetermined choreography. Each new round of sorting built upon the piles assembled in the previous turn; keeping the slips in order during the entire process was considered so crucial that the census bureau equipped its staff with small blocks of lead to protect the piles from unforeseen derangement.[39] Abstracting data via counting slips was a good deal less entertaining than the association with playing cards might suggest, but the process seemed germane enough to the popular leisure activity for Engel to assume that workers would quickly learn how to follow sorting, counting, and noting instructions, and to exceed, with little training, the expected piece rate average. Compared to check listing, the sorting of counting slips was considered easy work, not only faster and cheaper, but also far less prone to errors.

[36] Müller-Wille, "Names" (cit. n. 25); Markus Friedrich, "Genealogy as Archive-Driven Research Enterprise in Early Modern Europe," in this volume. On paper technologies, see also Anke te Heesen, "The Notebook: A Paper-Technology," in *Making Things Public: Atmospheres of Democracy*, ed. Bruno Latour and Peter Weibel (Cambridge, Mass., 2005), 582–9; Volker Hess and J. Andrew Mendelsohn, "Case and Series: Medical Knowledge and Paper Technology, 1600–1900," *Hist. Sci.* 48 (2010): 287–314.
[37] Card catalogs in eighteenth-century France were also inspired by playing cards, their uniform size allowing precise standardization, easy shuffling, and robust handling. Before they had a tarot pattern imprinted on the back side (around 1816), playing cards were frequently used as notepads, marriage and death announcements, *cartes de visites*, or even lottery tickets; see Markus Krajewski, *Paper Machines: About Cards and Catalogs, 1548–1929* (Cambridge, Mass., 2011), 33.
[38] Engel, "Aktenmässige Darstellung" (cit. n. 30), 303.
[39] Engel to the Prussian Ministry of the Interior, 21 June 1871, Kostenanschlag für die diesjährige Volkszählung, Geheimes Staatsarchiv Preußischer Kulturbesitz (hereafter cited as "GStA PK"), Berlin, HA I, Rep. 77, Tit. 94, No. 132, vol. 1, 44–47r.

The Italian counting slip technology that Engel introduced in 1860s Prussia soon spread across the German states; with little alteration to design and form, the technology became common practice in many European and also colonial census offices. One of the rare depictions of census workers using counting slips stems from the Indian census conducted by the British Empire in 1901 (fig. 1).[40] The movable counting slip emerged as a much-lauded technology for the same reasons that punch cards were praised a generation later.[41] In 1869, Hermann Schwabe, the director of the census commission in Berlin, demonstrated how counting slip technology could be applied to population statistics to document societal complexity.[42] Schwabe used the counting slips to establish a rich and detailed analysis of Berlin's population structure, as well the occupational and housing conditions in the Prussian (and later Imperial German) capital. The counting slips allowed Schwabe to produce tables showing the age structure (male and female) of Berlin's population subdivided by all civil status groups, a display that necessitated an additional four variables on the x-axis; variables of age, sex, and civil status were also displayed in combination with the occupational status of self-employed and salaried inhabitants and their dependent family members. Equally complex was a table showing patterns of residence by different occupational groups, male and female: whether they lived with their employer, in their own apartment, as nuclear families or with extended relatives, as subletters, or as so-called day lodgers who could only afford to rent an empty bed to sleep in after working the night shift. A set of fine-grained tables demonstrating the conditions in the city's tenements, front to rear, basement to attic, highlighted how deliberately the new statistical compilation tool was geared to reflect core social and political concerns of the time.[43]

To underscore the relevance of the complexity of scientific statistical analysis in addressing social and political challenges in a rapidly changing society, Schwabe's publication contained an additional chapter with a total of twenty-four carefully crafted graphical visualizations. These visuals included a map showing the varying degrees of population density in each city district, a population pyramid, and a bar chart showing the age structure in greater detail. Each of the graphs was accompanied by written explanations of peculiarities that might not be obvious to the untrained eye: the high rate of childhood mortality, leaving a sharp dip in the age bar chart; the large number of women (single, married, or widowed) dependent on gainful employment; or the insufficient supply of gas and water conduits in rear compared to front buildings. Several detailed area diagrams identified Berlin's trouble spots, that is, those parts of the inner city that showed the highest percentage of non-related people living under one roof, combined with the most overcrowded apartments, and the highest percentage of nonresident lodgers. Drawing on international theoretical debates on the use and benefit of graphical presentations in population statistics in which he played an active role, Schwabe sought to demonstrate what could be accomplished with the newest technical

[40] Georg von Mayr, "Das Zählblättchen und der Britisch-Indische Zensus von 1901," *Allg. Stat. Arch.* 6 (1902): 171–6.

[41] Engel's reply to Fabricius, "Zur Theorie" (cit. n. 27), 198.

[42] Schwabe, *Die Resultate* (cit. n. 27).

[43] Schwabe (1830–74) was, like most leading statisticians in Germany at the time, an active social reformer with close ties to the founders of the influential Verein für Sozialpolitik, an association founded in 1872 claiming that concern for the working class should not be a monopoly of socialism. See Grimmer-Solem, *Rise of Historical Economics* (cit. n. 22), 131, 177.

Figure 1. *Drawing of census compilers using counting slips to transfer data from bound enumeration lists during the 1901 census of India. Each of the low worktables is equipped with a pigeonhole shelf for thirty kinds of counting slips. A tablet on the floor next to one of the worktables shows fifteen different model shapes of slips that were used to signify common characteristics such as religious affiliation, civil status, caste, ethnicity, and combinations thereof. Distinctions of male and female were denoted by color.* C. E. Luard, Census of India, 1901, vol. 19, Central India, Part I, Report *(Lucknow, 1902).*

tools and graphical methods.[44] For Engel, Schwabe's publication served as compelling evidence of the potential of counting-slip technology. In his view, the richness of both tables and graphs displayed in it rested entirely on the compilation of enumeration data by means of counting slips, granting the new method "not just technical, but also ethical and pedagogical relevance."[45]

Notwithstanding his praise of Schwabe's work, Engel was not entirely satisfied with the counting slip. One remaining disadvantage, Engel lamented in 1869, was the time and effort required to transfer data from enumertions lists onto the slips before the sorting and counting could begin. He calculated that a laborer could on average transfer 800 entries per day from a list onto a slip. To transfer all the information for the state's estimated forty-eight million inhabitants onto counting slips, the Prussian census bureau would have to invest 60,000 days of work. Apart from the resulting cost of labor, finding the personnel required to carry out such a task was no mean feat.[46]

[44] Hermann Schwabe, "Theorie der graphischen Darstellungen," *Congrès international de statistique: Compte-rendu de la VIIIème session à St. Petersburg, I. Partie: Programme* (St. Petersburg, 1872), 22–5. See also Georg von Mayr, "Gutachten über die Anwendung der graphischen und geographischen Methode in der Statistik," *Z. Königl. Bayer. Stat. Bur.* 6 (1874): 36–44. Both Schwabe and Mayr were prominent figures in the "golden age of statistical graphics" from 1850 to 1910; see Michael Friendly, "The Golden Age of Statistical Graphics," *Statist. Sci.* 23 (2008): 502–35.

[45] Ernst Engel, "Die Kosten der Volkszählungen mit besonderer Rücksicht auf die im December 1870 im preussischen Staate bevorstehende Zählung," *Z. Königl. Preuss. Stat. Bur.* 10 (1870): 33–58.

[46] Engel's reply to Fabricius, "Zur Theorie" (cit. n. 27), 197–8.

Engel's plans to centralize the entire compilation process of the 1871 Prussian census in Berlin made the magnitude of the work seem all the more disconcerting.

Engel's solution to address this bottleneck was a radical shortcut modeled after the Belgian census conducted in 1846 by his mentor, Adolphe Quetelet.[47] Time, labor, and cost could be saved by omitting the process of transferring data altogether. To achieve this, Engel introduced the individual counting card, a sheet of thin cardboard almost four times larger than the counting slip (12 cm by 21 cm; fig. 2). In contrast to the counting slip, the individual counting card was large enough to display not just a portion but the entire set of characteristics requested for census taking. Each counting card contained one individual's complete data set on a single page. With the counting card, the census bureau created a "data double" of each of Prussia's inhabitants, just like the ones that Dan Bouk traces in his contribution to this volume.[48] Sent to every household in Prussia, the counting card made the time-consuming transfer of gathered data redundant. Once the enumeration effort was completed and all entries on the cards had been checked for errors, laborers could begin aggregating the data.[49] The cards were sorted into piles and counted as they were, and the resulting numbers were written down in interim tables and cross-checked, a method that also facilitated the tracing of errors.

In addition, counting cards solved other problems: not only were they much less prone to error than enumeration lists, they were also easier to handle than the heavy piles of large paper, a significant advantage in compilation efforts. From the state's fiscal perspective, the most intriguing potential benefit of counting cards was that they made census taking not only more accurate and attuned to statistical complexity but also considerably cheaper. To generate support for his grand reshaping of the Prussian census, Engel produced painstaking calculations to prove the economic potential of the new method. Wages for the workers who compiled the data were the largest expense in census taking. The use of counting cards would reduce these expenses by half.[50] At the same time, as Engel boasted in 1871, the census bureau would be able to compile tables displaying a rich combination of variables that "no other nation in the world is even close to achieving."[51] National pride and competition drove statistical innovation as the new German Empire strove for power.

From 1871 onward, Prussian authorities moved away from accumulating census data via enumeration lists, introducing individual counting cards at the point of data gathering while at the same time centralizing the compilation process in the capital. The counting card enabled census bureau statisticians in Berlin to abstract numbers and tables directly from the gathered material. Widely used in Prussia at the beginning of the twentieth century, the individual counting card, as Austrian statistician Heinrich Rauchberg acknowledged, encouraged "an expansion of what was techni-

---

[47] See Heinrich Rauchberg, *Die Bevölkerung Österreichs auf Grund der Ergebnisse der Volkszählung vom 31. Dezember 1890* (Vienna, 1895), 9.

[48] Dan Bouk, "The History and Political Economy of Personal Data over the Last Two Centuries in Three Acts," in this volume.

[49] For the sake of efficiency, the census bureau decided that the head of each household would be responsible for filling out the card, a procedure that was much debated but proved to yield reliable participation. See Schneider, *Wissensproduktion* (cit. n. 22), 236–7.

[50] Engel, "Die Kosten" (cit. n. 45), 40.

[51] Engel to the Prussian Minister of the Interior, 18 June 1871, Kostenanschlag für die Volkszählung von 1871, GStA PK, HA I, Rep. 77, Tit. 94, No. 132, vol. 1, 36.

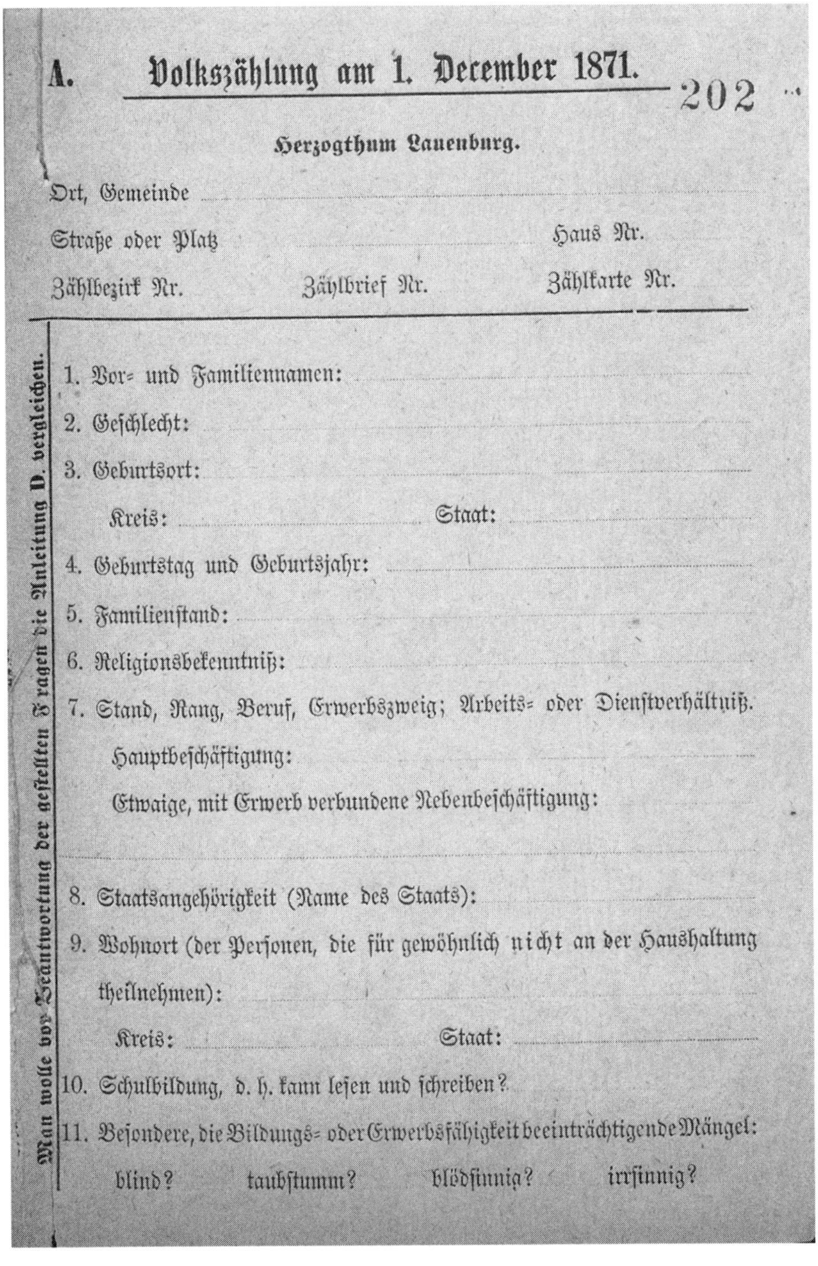

A. Volkszählung am 1. December 1871. 202

Herzogthum Lauenburg.

Ort, Gemeinde

Straße oder Platz Haus Nr.

Zählbezirk Nr. Zählbrief Nr. Zählkarte Nr.

1. Vor- und Familiennamen:
2. Geschlecht:
3. Geburtsort:
   Kreis: Staat:
4. Geburtstag und Geburtsjahr:
5. Familienstand:
6. Religionsbekenntniß:
7. Stand, Rang, Beruf, Erwerbszweig; Arbeits- oder Dienstverhältniß.
   Hauptbeschäftigung:
   Etwaige, mit Erwerb verbundene Nebenbeschäftigung:
8. Staatsangehörigkeit (Name des Staats):
9. Wohnort (der Personen, die für gewöhnlich nicht an der Haushaltung theilnehmen):
   Kreis: Staat:
10. Schulbildung, d. h. kann lesen und schreiben?
11. Besondere, die Bildungs- oder Erwerbsfähigkeit beeinträchtigende Mängel:
    blind? taubstumm? blödsinnig? irrsinnig?

Man wolle vor Beantwortung der gestellten Fragen die Anleitung D. vergleichen.

Figure 2. *Prussian counting card (21 cm by 12 cm) used for the census of 1871. GStA PK, HA I, Rep. 77, Tit. 94, No. 132, vol. 1, 202.*

cally possible in statistical terms, allowing population statistics to establish itself, once and for all, as a field of scientific research endeavor."[52] Prussian statisticians regarded the centralized processing of counting cards marked by hand as best practice in the early 1890s. Insofar as the movable counting card eliminated the error-prone

[52] Rauchberg, "Übersicht über den Stand" (cit. n. 33), 104.

and laborious process of transferring data from lists while at the same time allowing comparable statistical complexity, it was deemed technically superior to Hollerith's punch card.

MATERIALITIES, PRACTICES, LABOR

The technical, conceptual, and economic capabilities of counting cards were not the only reason why they were valued in Prussia; the cards' material and practical virtues played an equally important role. With the centralization of compilation work in 1871, the Prussian census bureau had to develop and manage workflows dictated by unprecedented waves of incoming data, strict budget constraints, and tight schedules to produce ever more refined census tables. To meet these challenges, the implementation of counting cards became embedded in a circular system of piece-rate sorting and tabulating unique to Prussia, intimately intertwined with core structures of the state and its pioneering social welfare legislation that was intended to battle the onslaught of socialism.[53]

The Prussian census bureau in Berlin became a permanent government agency in 1805 and from then on developed into an extensive apparatus. Although the census as such remained narrowly defined, Prussian officials thought that the establishment of permanent structures to collect and publish statistical evidence of the state's well-being was justified because statistical investigations were conducted at more frequent intervals than, say, the decennial operations of the U.S. government. In addition to population statistics (every three to five years), the bureau produced statistics of commerce and industry (about every five years) and determined the quantity of livestock and steam engines, the number of fires, and the annual yield of harvests in separate, annual evaluations. Furthermore, the census bureau featured departments for medical and education statistics, as well as for statistics of the civil registry. The bureau's comprehensive library represented the statistical memory of the Prussian state.[54] The census bureau's staff consisted of a small circle of senior officials who regularly attended the statistical seminar that Engel had established to discuss questions of scientific interest.[55] These statistical experts managed a modest number of permanently employed assistants. The bulk of the compilation work was performed by so-called ancillary workers paid on a daily wage basis.

Centralized census compilation increased the bureau's demand for temporary labor quite dramatically, even though the individual counting card made it possible to skip the time-consuming transfer of enumeration data from list to slip. When the 5,000 boxes containing roughly twenty-five million filled-in counting cards and various control lists from all the Prussian enumeration districts (in all 375 tons of material) arrived in Berlin in December 1871, the census bureau hired 300 extra workers to check the incoming data and count and sort the cards.[56] The three-story headquarters of the census bureau,

[53] On Prussian welfare politics, see Hermann Beck, *The Origins of the Authoritarian Welfare State in Prussia: Conservatives, Bureaucracy, and the Social Question, 1818–1870* (Ann Arbor, Mich., 1995).
[54] Emil Blenck, *Das königliche statistische Bureau im ersten Jahrhundert seines Bestehens, 1805–1905* (Berlin, 1905), 227–8 (floor plan).
[55] On the history of the statistical seminar, see Schneider, *Wissensproduktion* (cit. n. 22), 131–56.
[56] Ernst Engel, "Die Verwaltung des Königlich Preussischen Statistischen Bureaus im Jahre 1873," *Z. Königl. Preuss. Stat. Bur.* 13 (1873): 345–64, on 356.

completed in 1869, was not designed to house that many additional workers for centralized compilation, let alone for storing and moving such massive quantities of enumeration material.[57] Sorting and counting the cards required a large amount of space in order to spread the material out in piles. Thus, although an adjacent building was rented for supplementary workspace, the compilation work was not done on the bureau's premises—but distributed as piece labor to be completed in private homes.

The scope and steadiness of outsourcing to private homes suggest that it expanded not only because of a lack of office space, but also because of the need to accommodate the compilation process as such. Storing, moving, and tracking the "colossal masses" of incoming census material within the bureau's premises required much more space and time than previously assumed. Engel and his officers responded to this unforeseen challenge by altering the planned procedures. Rather than exhausting the material category by category in many successive rounds, compilers were advised to complete the work in three comprehensive counts, in order not to move each census box more than three times.[58] Additionally, choreographing the movement of the paper cards between the bureau and various private dwellings became crucial to handling the material. Just how essential this type of outsource circulation was to census compilation in Berlin became evident as early as November 1872, when a good portion of the cards from the 1871 enumeration effort had been exploited and were no longer needed for consolidating tables from interim results. As Engel explained in an urgent letter to his superiors, the used counting cards had accumulated in the census bureau into suffocating mounds of "dead data," jam-packed in every corner of the building, from basement to attic, blocking hallways and transforming the entire office space into a cramped maze. Engel asked for permission to dispose of some of this material—and fifteen million individual counting cards containing personal census data were sold as spoilage soon thereafter, setting a standard applied to (almost) all used census records for many years to come.[59]

Piecework emerged as a key component in the processing of Prussian census data over the following decades, with its volume growing considerably each year. While the Prussian census bureau temporarily employed fifty-five extra clerks on its own premises, abstracting the 1890 census records was performed mostly by as many as 264 piece-rate workers toiling from home.[60] A peak was reached when, as Engel's successor Emil Blenck reported in 1895, the census bureau employed 1,000 wage workers and about 3,000 piece-rate workers at home to compile the commerce, trade, and agricultural statistics that year, the most comprehensive statistical investigation the agency had ever undertaken, with no less than 27,760 boxes circulating all over Berlin and the city's outskirts.[61] All this was achieved, as Blenck stated, with only

[57] Blenck, *Das königliche statistische Bureau* (cit. n. 54), 226.

[58] See Engel to the Ministry of the Interior, 1 December 1872, Bericht über den Fortgang der Volkszählungsarbeiten, GStA PK, HA I, Rep. 77, Tit. 94, No. 132, vol. 2, 131–41, on 132.

[59] Engel to the Ministry of the Interior, 29 November 1872, GStA PK, HA I, Rep. 77, Tit. 94, No. 132, vol. 1, 227–39. The only exception were counting cards containing information about mentally or physically disabled individuals. Those cards were forwarded to the department of medical statistics for further use. See Kühnert to Ministry of the Interior, 3 May 1910, GStA PK, I HA, Rep. 77, Tit. 536, No. 30, vol. 2, n.p.

[60] Emil Blenck, "Die Volkszählung vom 1. December 1890 in Preussen und deren endgültige Ergebnisse," *Z. Königl. Preuss. Stat. Bur.* 32 (1892): 174–264, on 256.

[61] Emil Blenck, "Die Berufs- und Gewerbezählung vom 14. Juni 1895 und die damit verbundene landwirtschaftliche Betriebszählung," *Z. Königl. Preuss. Stat. Bur.* 37 (1897): 302–5.

five tenured civil servants working sixteen-hour shifts for weeks in crowded offices and performing a "virtually back-breaking amount of supervising and administrative tasks, controlling such a vast army of workers, and a huge amount of counting material constantly kept in circulation."[62]

The sheer mass of material required circulation beyond the census bureau. At the same time, the small size and portability of each counting card rendered such a refined system strewn all over Berlin possible. Portioned in wooden boxes by units of 5,000 or 10,000 cards, the material was shipped by horse carriage to a homeworker and his family, along with table forms and meticulous instructions explaining the sorting plan. Homeworkers were entrusted with a complex set of tasks. They had to unpack each box and double-check the material. They would then sort the cards into piles according to prescribed criteria, a process that required a fair amount of interpretation.[63] Thereafter, homeworkers had to count the cards in each pile and fill in numbers in rubrics of the respective interim table forms in black ink, draw subtotals and totals, check for errors, repack the cards in their new order, label each pack in a specific manner, close the box, reattach the lid, and return all of the material to the census office for revision. In this manner, each box was sent back and forth up to three times between the bureau and individual dwellings; each counting card was subjected to three comprehensive rounds of sorting and counting in order to establish the numbers for each rubric of the census tables.[64]

The Prussian circular system of outsourcing census work depended crucially on well-trained and perfectly reliable personnel. From 1871 onward, the census bureau painstakingly built up a skilled workforce toiling on-site or from home. The bureau used the temporary character of census compilation to optimize its male in-house staff, and to expand its reserve of homeworkers, using family members of its male employees. Long-term staffers—most of them in their late thirties and forties with a few considerably older, and all married—were authorized to earn "additional income through homework."[65] The surplus remuneration for homework supplemented each worker's income considerably; in a few cases, the piece-rate work performed at home almost doubled the worker's yearly earnings. Control was imposed via the heads of household. They were responsible for impeccable results in their own work, as well as that done by their dependents. Hefty wage deductions were put in place to punish slipshod work and to keep revisions to a minimum. Revisers—at peak times traveling from one household to the next to spare the time it took to transport the boxes—checked the material after each round of counting. Any recounts necessary were performed at the bureau and were done at the workers' expense.[66] The loss of income due to unsatisfactory

---

[62] Ibid., 302.

[63] For a more detailed analysis of the challenges that the sorting process involved, see Christine von Oertzen, "Housewifery Skills, Paper Technologies, and Labor Division in Nineteenth-Century Census Compilation," in *Working with Paper: Gendered Practices in the History of Knowledge*, ed. Carla Bittel, Elaine Leong, and Christine von Oertzen (Pittsburgh, forthcoming).

[64] Blenck, "Die Volkszählung vom 1. December 1890" (cit. n. 60), 214–5.

[65] Verzeichnis sämmtlicher Hilfsarbeiter nach dem Stand vom 1. April 1890, GStA PK, I HA, Rep. 77, Tit. 536, No. 30, vol. 1, n.p. Additionally, the bureau employed twenty-five homeworkers living on their own, most of whom were older than the on-site employees. Of these twenty-five workers, eight were female, either the widows of state officials or younger, unmarried professionals seeking additional income; ibid.

[66] Bestimmungen über die Annahme, Einstellung und Entlassung der beim Königlichen Statistischen Bureau gegen Zeit- und Stücklohn beschäftigten Hilfskräfte (n.d., ca. 1890), GStA PK, I HA, Rep. 77, Tit. 536, No. 30, vol. 1, n.p.

results averaged around 10 percent, but in some cases it amounted to up to 84 percent.[67] As in other branches of the Prussian civil service, low-ranking positions, such as the daily-wage compilation work of the census bureau, were formally reserved for those who had served in the army. During his reign, Frederick William I (1713–40), the so-called soldier-king, rewarded those who had borne arms for many years of loyalty with remunerative labor when their military service ended. By the middle of the nineteenth century, such paternalistic practices had become regulated social policy measures, ensuring that the lower ranks of the civil service were filled with former military personnel.[68] To keep the state's pension burdens to a minimum, the Prussian army functioned as a "school of the nation," providing basic schooling for lower-ranking officials who approached retirement to smooth their transition from service to clerical work: arithmetic, grammar, transcription, and list keeping were part of the curriculum. This system offered educational and social prospects to those young men of rural and artisanal background drawn to lifelong service, while at the same time infusing the civil service—and to a considerable degree society at large—with the military values of duty, rank, and obedience—key elements of what became known as Prussian militarism.[69]

The number of veterans entitled to employment in government agencies increased, especially during the decades following the Franco-Prussian War of 1870–1, and far outweighed the number of available lower-ranking positions. Particularly in Berlin, the surplus of veterans added pressure to unemployment rates.[70] This constellation proved both profitable and problematic for the Prussian census bureau. Bound to comply with the state's social policy measures to produce high-end statistical tables as quickly and cost-effectively as possible, the bureau's directors committed publicly to their mandate in the common interest. Behind the scenes, however, they embraced improvisation to operate with the most efficient, skilled, and steady workforce available.

To ensure the smooth functioning of its data-processing machinery, the bureau relied much more heavily on the professional expertise of its on-site male employees and the skills of middle-class wives, widows, and unmarried female staff members working from home than on social policy regulations granting preferred employment to retired servicemen. In 1890, only one in six compilers had actually served in the Prussian army, whereas the majority of the bureau's employees were men experienced in accounting, banking, insurance, secretarial work, and sales. A few had been high school teachers; others had been farmers.[71] Whenever demand exceeded the bureau's on-site capacities, the bureau recruited first among the families of its current staff. Only when the female labor reserve was exhausted did the bureau advertise publicly

[67] Blenck to the Ministry of the Interior, 8 May 1899, GStA PK, I HA, Rep. 77, Tit. 536, No. 30, vol. 2, n.p.
[68] Ralf Pröve, *Militär, Staat und Gesellschaft im 19. Jahrhundert*, vol. 77, Enzyklopädie Deutsche Geschichte (Munich, 2006).
[69] B. Poten, *Geschichte des Militär-Erziehungs- und Bildungswesens in den Landen deutscher Zunge*, 5 vols. (Berlin, 1895–7); Heinz Stübig, "Das Militär als Bildungsfaktor," in *Handbuch der deutschen Bildungsgeschichte, III: 1800–1870: Von der Neuordnung Deutschlands bis zur Gründung des Deutschen Reichs*, ed. Karl-Ernst Jeismann and Peter Lundgreen (Munich, 1987), 362–400, on 374.
[70] Jakob Vogel, "Der Undank der Nation: Die Veteranen der Einigungskriege und die Debatte um ihren 'Ehrensold' im Kaiserreich," *Militärgeschichtliche Z.* 60 (2001): 343–66.
[71] Verzeichnis sämmtlicher Hilfsarbeiter nach dem Stand vom 1. April 1890, GStA PK, I HA, Rep. 77, Tit. 536, No. 30, vol. 1, n.p.

to expand its temporary workforce. Increasingly, the bureau encouraged its daily wage earners to do abstracting work at home, and at peak times, it extended its mobilization efforts to include the spouses of its higher-ranking office clerks. Additionally, the bureau set aside regulations that prohibited its workers from entrusting anyone other than family members with homework. As a result, some households started to resemble busy office floors; in one of them, a bureau official was able to provide extra work and income not only for his wife, but also for her two sisters, a brother-in-law as well as two widows, an unemployed salesman, and two spinsters living nearby. This group's work was deemed so excellent that the bureau insisted on forgoing all established limits for maximum income; in fact, the official's wife hired a maid so that she would be available to help with the compilation work.[72]

Middle-class women rather than impoverished veterans proved indispensable for compilation work. Veterans, for their part, were hired in considerable numbers, though often they found themselves among the first to be laid off when demand slowed, whereas more skilled "people who have never served" as well as female homeworkers were kept busy or assigned tasks other than census work.[73] The archived files are full of petitions by dismissed and embittered servicemen claiming the bureau privileged "the fair sex." Suspecting favoritism and corruption, many veterans felt betrayed not only by the bureau itself but also by the civil service for abandoning aging soldiers, notwithstanding their entitlement to civil service work. In his response to these charges, Blenck defended the bureau's labor policy by stating repeatedly that efficiency rather than charity must rule the bureau's employment strategies—otherwise, the quality of the work would suffer and its completion would be delayed, resulting in cost overruns.[74]

During the 1890s, while Herman Hollerith traveled through Europe to advertise his punch card and electric tabulating machine, Prussian veterans' claims grew louder, as they began to organize and protest publicly against what they saw as widespread ill treatment.[75] Veterans' magazines as well as daily newspapers, most notably *Vorwärts*, the organ of social democratic radicals, castigated employment policies as well as the state's pension system for leaving retired soldiers much worse off than factory workers. With socialism on the rise, the veterans' unrest raised fears that Prussia's most loyal supporters might radicalize and join the labor movement. Until a military pension reform was passed in 1906 granting retired servicemen much higher pensions that were sufficient to forgo additional income from jobs like manual data processing, the Prussian statistical bureau seemed well advised to avoid mechanization so as not to aggravate social tensions even further.

Meanwhile, home-based census compilation done primarily by educated middle-class women proved to be the most reliable, fastest, and cheapest way to produce numbers and tables for the Prussian state. These women were not restricted to eight-hour

---

[72] Blenck to Ministry of the interior, 21 July 1898, GStA PK, I HA, Rep. 77, Tit. 536, No. 30, vol. 1, n.p. Hiring housemaids when wives were busy with compilation work was a common practice; see Blenck to the Ministry of the Interior, 8 May 1899, GStA PK, HA I, Rep. 77, Tit. 536, No. 30, vol. 2, n.p.

[73] Blenck to the Ministry of the Interior, 21 April 1898, regarding the petition of Militäranwärter Wilhelm Mittelstädt, GStA PK, I HA, Rep. 77, Tit. 536, No. 30, vol. 1, n.p.

[74] Ibid.

[75] Vogel, "Undank der Nation" (cit. n. 70); Bernd Ulrich, Jakob Vogel, and Benjamin Ziemann, eds., *Untertan in Uniform: Militär und Militarismus Im Kaiserreich, 1871–1914* (Frankfurt am Main, 2001).

shifts that were common in civil service jobs, nor did they have to refrain from work-ing on Sundays and public holidays. As a result, they produced numbers and tables for twelve to sixteen hours a day at home, invisible to the public at large.[76] By means of up-to-date paper technology and with the support of their highly skilled workers, officials, and educated female assistants working from home, Prussia's bureaucrats had consistently been able to publish fine-grained results of their population surveys faster than any other country in Europe. For these reasons as well as to address polit-ical concerns, Prussia's statistical office remained firmly committed to manual data processing, paying lip service to state paternalism rather than embracing social policy measures. Operating a technically up-to-date and cost-efficient system of manual data compilation based on trust and corporate obedience in the service of accuracy was val-ued more than mechanical scenarios, which ran the risk of surrendering state affairs to the mercy of the labor movement. As a consequence, the processing of data via punch cards, tabulating machines, and other electric sorting devices would remain unexplored in Prussia's census compilation until after World War I.

CONCLUSION

In contrast to Prussia, the Austrian Empire enthusiastically embraced punch cards and mechanical data compilation and used this technology for census compilation in 1890, less than a year after Hollerith presented his machines at the 1889 World's Fair in Paris. That Heinrich Rauchberg, the Austrian statistician in charge of that country's 1890 census, chose to act so differently than his colleagues in Prussia was mainly attributable to the fact that census data in Austria-Hungary were still gathered in enumeration lists and assessed at local and regional levels, with imperial authorities assigned the narrow task of compiling interim results based on checklists.[77] Without ef-ficient paper technologies such as counting slips or counting cards already in place, Austrian statisticians—like their counterparts in the United States—opted to central-ize the assessment of enumeration lists by using punch cards and machine process-ing in order to speed up the tabulation effort and to be able to accurately document social conditions in a quickly changing society. An early admirer of Prussian data power, Rauchberg became an ardent supporter of punch card technology because in his view, mechanical processing expanded possibilities to combine variables more readily than the data carriers and sorting practices used in manual census compilation. As efficient as the Prussian system based on counting cards and manual sorting methods had be-come, it was evident that by the end of the nineteenth century the amount of data to be processed pushed the German system to its limits and proved adverse to the greater statistical complexity he believed was needed in Austria.[78]

[76] Blenck to the Ministry of the Interior, 8 May 1899, GStA PK, HA I, Rep. 77, Tit. 536, No. 30, vol. 2, n.p.
[77] Heinrich Rauchberg, "Erfahrungen mit der elektrischen Zählmaschine," in *Allg. Stat. Arch.* 4 (1895): 131–63.
[78] Heinrich Rauchberg, "Die elektrische Zählmaschine und ihre Anwendung insbesondere bei der österreichischen Volkszählung," *Allg. Stat. Arch.* 2 (1891): 78–126, on 109. See also his detailed anal-yses of the Austrian census with regard to internal migration: Rauchberg, "Der Zug nach der Stadt," *Stat. Monatsschr.* 19 (1893): 125–71; Rauchberg, *Die Bevölkerung Österreichs* (cit. n. 47), esp. chap. 7, 90–147.

Notwithstanding Rauchberg's criticism and his enthusiasm for punch card machinery, however, both manual and mechanical data processing rested on the same principle: a movable paper tool carrying all relevant data of one person, which enabled statisticians to sort and compile census data in new ways. These technologies—manual and mechanical—had been tested against each other before the 1890 American census. That "contest"—sorting and counting St. Louis's 1880 census records—was won by Hollerith's machines, which completed the sorting task ten times faster than when the work was done manually.[79] But the holes in the punch cards and the machines to "read" the binary information were considered to make a difference much less in conceptual than in logistical and practical terms: they enhanced an established process via mechanical means.

The mechanization of census work entailed significant investments of capital and required a spatial centralization of the compilation process: Vienna's 200 hole-punchers, as well as twelve counting and sorting machines, were housed in one location, a huge, factory-like building wide enough to have thirty windows across the front. Instead of circulating census material for home-based work throughout town, statisticians in Vienna kept workers, machines, enumeration materials, as well as millions of punch cards under one roof, compiling data day and night. And instead of relying on a highly experienced long-term workforce, including trustworthy middle-class women toiling from home, Rauchberg and his colleagues counted on inexperienced, temporary working-class staff, introducing a sharp division of labor that made it possible to scale back on training. As Rauchberg boasted, the new process, implemented with draconian hire-and-fire penalties to ensure discipline and diligence, transformed the 380 individuals employed in the vast Vienna facility to operate the punching, tabulating, and sorting devices from "a motley crew of no-class unemployed persons to a well-oiled machine."[80] Few aspects of census work remained untouched by mechanization, an observation that applied to statisticians as well. As Rauchberg put it, he and his colleagues had become "factory directors . . . individuals who have only to ensure that everything proceeds according to a plan laid out well in advance."[81]

By the first decade of the twentieth century, the use of Hollerith machines in a number of European states had begun to transform government offices into environments more akin to factory floors (fig. 3). Hollerith, Rauchberg, and their counterparts in industry centralized data processing and put numbers on the assembly line and, once there, soon brought young, unmarried women out of the home and into the workforce to ensure that the monotonous labor of punching holes was performed conscientiously, quickly, and inexpensively. Investment in the capital equipment necessary to complete mass data surveys invited fresh plans to take maximum advantage of reduced costs: the machines were really only worth the investment when they were constantly kept in use.[82]

[79] See Thomas Commerford Martin, *Counting a Nation by Electricity* (New York, 1891). See also Campbell-Kelly, *ICL* (cit. n. 1), 9.

[80] The total number of employees needed to maintain a workforce of that size was more than 700. See Rauchberg, "Die elektrische Zählmaschine" (cit. n. 78), 112.

[81] Ibid. For similar dynamics and consequences for the professional division of labor in the data-driven sciences with the introduction of computers, see Sepkoski, "Database" (cit. n. 25).

[82] Hermann Julius Losch, "Die Volkszählung vom 1. Dezember 1910," *Württemberg. Jahrb. Stat. Landeskunde* (1912), 183–250, on 191.

Figure 3. *Hollerith punch keyboard operator of the U.S. Census Bureau, 1908. Courtesy of the Library of Congress Prints and Photographs Collections Division, Washington, D.C.*

However, regarding the moment of transition from manual to mechanical data compilation, it was not so much punch cards and Hollerith machines that revolutionized nineteenth-century census taking and statistical complexity, but rather the concepts, techniques, and manual sorting methods introduced since the mid-nineteenth century in Belgium, Italy, Prussia, and many other German and European states. Epistemologically, the strict distinction between "list" and "table" based on a new notion of "data" as abstracted, numerical information led statisticians to standardize the collection of census data and centralize and enhance the compilation of that data. Genealogically speaking, European playing cards, not American railroad tickets, inspired statisticians to make census data move.

A Feeling for the Algorithm:
Working Knowledge and
Big Data in Biology

by Hallam Stevens*

ABSTRACT

The term "Big Data" may serve as a useful marker for particular kinds of questions, practices, and relationships for collecting and using data. Some of the ways of talking about Big Data suggest that there might be something, if not entirely new, then at least importantly different at work in Big Data practices and problem-solving approaches. By using three examples taken from the biomedical sciences—artificial neural networks, the construction of reference genomes, and the usage of the Ensembl database—this essay shows how Big Data practices cannot be understood as mere scaling up of pen-and-paper methods but constitute qualitatively different kinds of knowledge-making practices. These practices are characterized particularly by types of human-computer interaction that are labeled "a feeling for the algorithm."

INTRODUCTION

Historians are ordinarily skeptical of any claims to newness. No doubt some of the exceptionalism claimed by proponents of "Big Data" is a product of marketing and hype. Nevertheless, Big Data may serve as a useful marker for particular kinds of questions, practices, and relationships for collecting and using data. Some of the ways of talking about Big Data described in the introduction to this volume suggest that there might be something—if not entirely new—then at least importantly different at work. The connections between older and newer ways of using data may not be straightforward, and the origins of Big Data practices may turn out to be quite independent of the kinds of practices and tools that have been applied to "small data."

The practices of Big Data, of course, are computational. Big Data are inseparable from the computers, databases, hardware, and software, in and through which they

* History Programme, School of Humanities, Nanyang Technological University, 14 Nanyang Drive, Singapore 637332; hstevens@ntu.edu.sg.

I would like to acknowledge the valuable assistance of the volume editors, Elena Aronova, Christine von Oertzen, and David Sepkoski, as well as the *Osiris* general editors and the anonymous reviewers. Work on this essay was supported by the Max Planck Institute for the History of Science (Berlin) as well as a Tier 1 Grant from the Ministry of Education (Singapore; RG56/13).

are collected, analyzed, stored, sampled, shared, and displayed. As such, these technologies exert an important influence (although not a fully determinative one) on what
Big Data are and what they can be and do. Looking at the history of these technologies
is going to be critical for finding a history of Big Data.

The aim of this essay is not to deny that some such continuities do exist, but rather
Historians have been wary about attributing an epistemically transformative role to
information technology in the sciences. Jon Agar, for instance, argues that early practices involving digital electronic computers essentially reproduced what scientists had
been doing offline.[1] Sabina Leonelli is also cautious in her claim that Big Data does
have "specific epistemic and methodological characteristics." For some fields, Leonelli
argues, Big Data gives a new prominence to data as a commodity and entails the
"emergence of a new set of methods, infrastructures and skills."[2] But Leonelli has little
more to say about what these methods, infrastructures, and skills actually are.[3] In this
volume, David Sepkoski's genealogy of "paper databases" suggests continuity between nineteenth-century "book" work and twentieth-century computerized practices
in the field of paleobiology, and Judith Kaplan demonstrates the persistence of basic
vocabulary lists in linguistics.[4]

The aim of this essay is not to deny that some such continuities do exist, but rather
to show more specifically what methods, infrastructures, and skills are associated
with Big Data and to suggest how they are different. This applies first and foremost
to recent usages of data (from the 1980s onward). Agar's argument applies to the first
generation of digital electronic computers and the first generation of scientists to use
them. But much has changed in computing since the 1950s and 1960s. It is perhaps not
surprising that the first scientists to use computers imagined few entirely novel uses
for them. However, as computers became more powerful—and users realized their
potential—different uses emerged. This happened only gradually, and certainly at different times in different disciplines. Over time, quantitative differences become qualitatively important (e.g., differences in speed). Although scientists may have initially
reproduced paper practices with computers (and sometimes still do), the ability to massively speed up some kinds of operations does make a difference to the types of questions scientists would ask and the types of solutions they could produce.[5]

To put this another way, the aim of this essay is not to draw a sharp line between
computational and noncomputational practices. Of course, many kinds of noncom-

[1] Jon Agar, "What Difference Did Computers Make?" *Soc. Stud. Sci.* 36 (2006): 869–907. Agar's
work in general emphasizes the continuities between nineteenth-century "computing" for government
bureaucracy and early electronic computing in the 1940s and 1950s. See Agar, *The Government Machine: A Revolutionary History of the Computer* (Cambridge, Mass., 2003). I find these accounts per-
suasive so far as early electronic computing is concerned, but these views do not preclude the emergence of new forms of computing practice after, say, 1970.
[2] Sabina Leonelli, "What Difference Does Quantity Make? On the Epistemology of Big Data in
Biology," *Big Data & Soc.* 1 (April–June 2014), http://journals.sagepub.com/doi/full/10.1177
/2053951714534395 (accessed 7 June 2017).
[3] Rob Kitchin, too, argues that "Big Data and new data analytics are disruptive innovations" but is
not specific about what those innovations actually consist of. See Kitchin, "Big Data, New Epistemologies and Paradigm Shifts," *Big Data & Soc.* 1 (April–June 2014), http://journals.sagepub.com/doi
/full/10.1177/2053951714528481 (accessed 7 June 2017).
[4] David Sepkoski, "The Database before the Computer?"; Judith Kaplan, "From Lexicostatistics to
Lexomics: Basic Vocabulary and the Study of Language Prehistory," both in this volume.
[5] Patrick McCray, for instance, documents changing practices and norms associated with the "born
digital" era in observational astronomy—a field that has dealt with large data volumes for a long time.
See McCray, "The Biggest Data of All: Making and Sharing a Digital Universe," in this volume.

putational practices can be reproduced on a computer, and many kinds of digital electronic operations can be reproduced without computers. Rather, the aim here is to distinguish some kinds of computational practices from others.[6] "Big Data" may be a useful label for beginning to distinguish certain kinds of computational practices that are not merely reproducing paper-based practices and that rely on new ways of thinking and working. This argument does take explicit aim at the notion that we should write a "long" and continuous history of data practices that stretches back before the twentieth century.[7] There may be, as Krajewski argues in this volume, some continuities between information processing in the Library of Alexandria and in modern online public access catalog (OPAC) library systems.[8] But putting all the world's books into Google's database, for example, opens up new possibilities for manipulating and analyzing information that were not available in the time of Callimachus.

Data-inside-computers can take on different shapes and interact with itself and with us in different ways. Information collected for science is often put inside of computers, but this is frequently done so that it can take on distinct forms and be manipulated in ways that are not possible without a computer. "The kinds of machines we use are bound up with the ways we think about nature and the ways we know it," John Tresch argues at the start of his book about nineteenth-century machines. "When our machines and our understanding of them change, so does nature, and so does our view of knowledge."[9] This must certainly be true of the digital electronic computer, perhaps more than any other machine. In what kinds of ways has the computer affected our understanding of nature? In what ways have computers reorganized the relationship between people and machines?

Here I want to examine one set of ways in which computers influence our understanding of nature by fostering particular forms of practice for producing knowledge. These practices, which involve complex interactions between human users and computers, are based on what might be called "a feeling for the algorithm." According to Evelyn Fox Keller's account, the success of Barbara McClintock's approach to biology relied on a "feeling for the organism." "Organisms have a life and order of their own that scientists can only partly fathom," McClintock argued. Understanding them requires the patience to "hear what the material has to say to you . . . how it grows, understand its parts, understand when something is going wrong with it." Above all, a "feeling for the organism" requires an appreciation for its complexity and a capacity to be surprised by it.[10]

Algorithms are not (yet) as complex as organisms, so the analogy is no doubt an imperfect one. But I argue that producing knowledge with algorithms requires the same sorts of intuition, tacit knowledge, and close attention required in working with organisms. I will use the term "algorithm" here to mean not merely the high-level or

---

[6] The line I am attempting to draw does not correspond to the line between, say, mechanical or electromechanical computers and digital electronic computers. Rather, it attempts to distinguish some kinds of digital electronic practices from others.

[7] See, e.g., Daniel Rosenberg, "Data before the Fact," in *"Raw Data" Is an Oxymoron*, ed. Lisa Gitelman (Cambridge, Mass., 2013), 15–40.

[8] Markus Krajewski, "Tell Data from Meta: Tracing the Origins of Big Data, Bibliometrics, and the OPAC," in this volume.

[9] John Tresch, *The Romantic Machine: Utopian Science and Technology after Napoleon* (Chicago, 2012), xi.

[10] Evelyn Fox Keller, *A Feeling for the Organism: The Life and Work of Barbara McClintock* (New York, 1984), 198–9.

step-by-step description of a procedure for solving a problem, but rather a fully imple-mented, working version of a program that can actually run on a computer. In other words, the algorithm is not just its description but rather involves a set of practices through which users can interact with both machines and data in specific ways.

This essay will develop three examples—all taken from the biomedical sciences—that suggest more specifically how Big Data is shaped by computers and computing practices. That is, computers (and computational practices) have generated specific ways of thinking and specific kinds of problem-solving approaches and practices that now fall under the label of "Big Data." To make this claim more precise (and to sug-gest how exactly data-inside-computers might be different), the essay will begin by introducing Davis Baird's notion of "thing knowledge."[11] Baird's claim that scientific instruments embody a distinct form of knowledge can be usefully appropriated for computing to show how it can involve new forms of knowledge and knowledge mak-ing. The aim of these examples is to suggest that algorithms and data structures contain various forms of "working knowledge" that are important in the making of new bio-logical knowledge. In other words, these algorithms and data structures do not merely store, manage, and circulate biological data; rather, they play an active role in shaping it into knowledge.

THING KNOWLEDGE, WORKING KNOWLEDGE, AND COMPUTERS

Baird argues that scientific instruments can, in themselves, constitute various forms of scientific knowledge. This can occur, he argues, in three ways. First, scientific instru-ments may serve as models of the world or of various phenomena. One of the exam-ples Baird gives is the orrery: as a model of the motions of the planets around the earth, the device embodies a kind of knowledge of the heavens. Second, Baird claims that instruments can contain something called "working knowledge." In this case, he means a device that is not intended as a model of a phenomenon but that demonstrates or instantiates a phenomenon without necessarily providing an understanding or ex-planation of it. A superconducting magnet, for example, possesses a "working knowl-edge" of superconduction, even in the absence of any theory to explain how or why it does what it does. Furthermore, tinkering with the device and improving it may lead us not only to better devices, but also to a greater understanding of what is going on. Third, there are measuring instruments. For Baird, these are a kind of hybrid of models and working knowledge.

Computers have been used in certainly the first and third ways in recent science: it is possible to use the computer to make simulations or models (e.g., of protein folding). And it is possible, by hooking a computer up to appropriate apparatus, to make it into a measuring instrument. Here, however, I will be concerned with the second category: that is, of machines instantiating "working knowledge" of some phenomenon. Ac-cording to Baird a machine itself can contain a certain kind of knowledge about the world that is not the same as a picture or words or a set of equations representing or describing it. A working superconducting magnet, again, contains a particular kind of knowledge about superconduction. That is, we can learn about the phenomenon of superconduction by building and using a working device. As Baird noted, "We

[11] Davis Baird, *Thing Knowledge: A Philosophy of Scientific Instruments* (Berkeley and Los Ange-les, 2004).

say someone knows how to ride a bicycle when he or she can consistently and success-fully accomplish the task."[12] Likewise, knowing how to build and use a super-conducting magnet means we have a "know-how" or "skill knowledge" with respect to superconduction. This knowledge does not depend, Baird argues, on any theory about superconduction or any explanation of how the phenomenon works. "We learn by interacting with bits of the world even when our words for how these bits work are inadequate," Baird claims.[13]

A successful working device, Baird continues, must have a kind of "cognitive au-tonomy." There is something epistemologically important about a working machine that literary (or mathematical) description misses. It is possible to learn things from machines, or to refine them, not by mathematical reasoning or textual description, but rather by manipulation of the device itself. Machines have a distinct "cognitive channel" that allows new knowledge to be created while remaining "ignorant of the-ory" and even while remaining unable to express in words either what the instrument teaches you or what you are doing with the knowledge so gained.[14]

In sum, there are two crucial aspects of Baird's "working knowledge." First, through a device, machine, or instrument, it is possible to do something, or make something work, without understanding in linguistic or descriptive terms everything that the de-vice or machine is doing. And second, by using and manipulating devices or instru-ments it is possible to generate new knowledge, new ideas, or new machines without understanding fully how that device works or what it is doing.[15]

Baird does not discuss computers. This is no doubt because they present a compli-cating case for his argument. At the center of Baird's claim is an argument for paying attention to, and taking seriously, materiality and material manipulations. However, many things that are important and distinct about computers are not material. Comput-ers, of course, deal with the "virtual": they virtually manipulate language, numbers, and equations. In this sense it is obvious computers are doing things with knowledge—they actually have linguistic and numerical and mathematical knowledge encoded into them. On these terms it seems like computers are simply behaving as an inscription de-vice—they are allowing us another way to write things down that is not fundamentally different from other written forms of knowledge. This corresponds to the idea that com-puters are doing the same things we can do on paper, only faster or on a larger scale.

This reduction of knowledge to the linguistic is exactly what Baird is attempting to resist. However, I want to suggest that there is something going on with computers beyond this mundane sense of them as sophisticated inscription devices. There are nonlinguistic ways in which computers are routinely used that are very much like Baird's descriptions of "working knowledge." First, algorithms or software do things,

[12] Ibid., 15.
[13] Ibid., 4.
[14] Ibid., 16.
[15] Working knowledge is related to, but not the same as, tacit knowledge and other concepts such as "gestural knowledge." On the latter, see Otto Sibum, "Reworking the Mechanical Value of Heat: In-struments of Precision and Gestures of Accuracy in Early Victorian England," *Stud. Hist. Phil. Sci.* 26 (1995): 73–106. Tacit and gestural knowledge relate mostly to the ability to operate a device or ex-perimental apparatus. Working knowledge goes beyond this to include knowledge about the world and its functions and features that is embodied in the device, rather than just knowledge about the device and its operation (in the example here, working knowledge is knowledge about the phenom-enon of superconduction, not about a specific superconducting device).

make things, and behave in ways that are not always predictable or intended or under-standable. Sufficiently complex algorithms (especially working with sufficiently large data sets) exhibit the kinds of behavior that seem similar to the cognitive autonomy Baird describes. In other words, software can do things that are often not reducible to theory or words. In working with simulations, fractals, artificial life, or artificial in-telligence, the aim is exactly to produce programs that exhibit behaviors that are not readily predictable from the code on which they are based.[16] Where large amounts of data are involved as well, data structures and databases can organize information in ways that may generate surprising results. The aim of Big Data is to organize data in particular ways that allow us to see patterns that would otherwise be impossible to discern.

The idea that computers embody working knowledge is supported by recent work in the history of mathematics. Stephanie Dick has explored how mathematicians at Ar-gonne National Laboratory in the 1970s and 1980s deployed software called the Au-tomated Reasoning Assistant (AURA) to help them search for mathematical proofs. Dick details how the computer successfully produced proofs via complex interactions and "negotiations" between the human mathematicians and their machine assistant:

> AURA was full of surprises. Although the program could only do precisely what it was programmed to do, the programmers usually could not know in advance what the con-sequences of their instructions would be. Indeed, if they could, they would have had little need for a fast computer to do this work for them. They manually studied printouts of thousands of clauses from run after run on AURA in order to understand what was going on when the engine looked for proofs. To accommodate their *a posteriori* revelations, the Argonne group committed to an experimental paradigm for ATP in which AURA was constantly improved and redesigned on the basis of results from previous runs. Those prized human insights and intuitions that the Argonne researchers reserved for the hu-man, rather than materializing from the cognitive ether, emerged from intimate and pro-longed experimental work with the engine. Experimentation determined the character, form, and relevance of human contributions to AURA proof searches. By privileging and isolating a traditional notion of human mathematical thought in their design, the Ar-gonne team in fact made possible radically new forms of intuition and insight grounded in experiential knowledge of computational behavior.[17]

Working with AURA involved a new kind of work in which the human and the ma-chine should be considered a kind of hybrid. This kind of "experimental work" with algorithms is not confined to mathematics. Rather, this kind of interaction between computers and humans is a feature of scientific computing more generally.

One way of describing a complex engineered system—such as a computer, an algo-rithm, or a piece of software—is in terms of black-boxing.[18] A software user may not

---

[16] See, e.g., Peter Galison's account of Monte Carlo simulations; Galison, "Computer Simulations in the Trading Zone," in *The Disunity of Science: Boundaries, Contexts, and Power,* ed. Peter Galison and David J. Stump (Stanford, Calif., 1996), 118–57. On artificial life, see Stefan Helmreich, *Silicon Second Nature: Culturing Artificial Life in a Digital World* (Berkeley and Los Angeles, 1998).

[17] Stephanie Dick, "AfterMath: The Work of Proof in the Age of Human-Machine Collaboration," *Isis* 102 (2011): 494–505, on 495–6.

[18] On black-boxing, see Bruno Latour, *Pandora's Hope: Essays on the Reality of Science Studies* (Cambridge, Mass., 1999). See also Langdon Winner, "Upon Opening the Black Box and Finding It Empty: Social Constructivism and the Philosophy of Technology," *Sci. Tech. Hum. Val.* 18 (1993): 365–8.

know much about how that software works; even a software designer may not know the detailed structure of a subroutine that he or she codes into a finished program. This ignorance would not stop the user (or the coder) from successfully using the software (or the subroutine) as an input-output device. In many cases, such ignorance may actually be advantageous for successful usage. What I am attempting to capture here is the opposite of simple black-boxing. Black-boxing usually refers to situations in which, in principle, the box can be opened up. What is inside the box may be hard to find out, unimportant for a given purpose, or protected by trade secrecy or some other means. Nevertheless, one can coherently imagine that someone knows what is inside the box.

On the other hand, saying that algorithms and data structures can contain working knowledge is to suggest more fundamental kinds of ignorance. It is not simply that a user or coder does not know how a program works. Rather, the user or coder cannot have access to the important knowledge embodied in and produced by interaction with the computer. The user may know in general or descriptive terms how a program works. But this is not the same as the working knowledge that can be produced through interaction with the machine. In Baird's terms, this sort of knowledge occupies a "cognitive channel" distinct from a description of the algorithm. The only way to obtain the knowledge produced by an algorithm is actually to use it.[19] We can observe the detailed properties of an algorithm (the lines of code), but this does not mean we can predict the outcome the algorithm will produce.

Even after we have run an algorithm it may be very difficult to understand precisely how it arrived at a given result. This ignorance might arise from stochastic elements of the program, from nonlinearities in its mathematics, from the complexity of the operations it performs, from the size and/or multidimensionality of the data sets with which the algorithm is working, or from some combination of these factors. This makes a significant difference to how we can use such algorithms—we use them by playing with them, running them, tweaking them, interacting with them, and so on. Through these sorts of practices we can come to develop a "feel" for their detailed behavior. It is in this sense that using algorithms seems very much like using other kinds of instruments and objects in the ways described by Baird. There is something about the complexity of algorithms and data structures that makes them more like the material instruments that Baird discusses. In the remainder of the essay, I will develop three examples that demonstrate how algorithms and data structures are put to work in biology in this way.

ARTIFICIAL NEURAL NETWORKS

Artificial neural networks (ANNs) are a form of machine learning algorithm used in many different kinds of applications. Building on the work of Warren McCulloch and Walter Pitts in the 1940s, simple ANNs were designed by early computer architects

---

[19] This has some features in common with the concept of emergence in physics, where "the micro-level interactions are interwoven in such a complicated network that the global behavior has no simple explanation." See Mark Bedau, "Downward Causation and Autonomy in Weak Emergence," in *Emergence: Contemporary Readings in Philosophy of Science*, ed. Mark A. Bedau and Paul Humphreys (Cambridge, Mass., 2008), 155–88, on 160.

(including Wesley Clark and Frank Rosenblatt) in the 1950s.[20] The basic idea of an ANN is to use a computer to simulate the way in which neurons are connected in the brain. In short, a single neuron (call it X) may have a range of input connections from other neurons. If those input neurons are activated ("fire"), they will transmit an electrical signal to X. If enough electrical signals are transmitted to X, it will reach its "threshold" and will also be activated. X may also have multiple output connections to other neurons. Once X is activated, it will transmit an electrical signal to those downstream neurons (which may, in turn, be activated). In this way, a signal is propagated along a specific path of neurons through the brain.

ANNs model this process. Conceptually, they consist of an "input layer" (the input data), one or more "hidden layers," and an "output layer" (the output data). Each layer consists of several nodes (which represent neurons). To start off, each node in the input layer is randomly connected to several nodes in the hidden layer (we will assume only one hidden layer for simplicity), and each node in the hidden layer is randomly connected to nodes in the output layer (see fig. 1). The network then requires a data set with which it can be "trained." One area in which ANNs have had considerable success is in optical character recognition.[21] Let's say the input layer consists of sixteen boxes that can be either light or dark (see fig. 2). And let's say the output layer consists of the ten numerals 0 through 9. In this case, we wish to train the ANN to successfully recognize each of the ten numerals from the patterns of light and dark that each would make in a 4 × 4 grid of boxes. Our training data would consist of light and dark patterns representing the various digits plus our own knowledge of the correct responses.

To train the ANN, we use drawings of various digits 0 through 9; these drawings are reduced to the pattern of light and dark boxes that becomes the data for the input layer. Each box can be either light or data, "on" or "off." Where the box is light, the corresponding input node is activated. Because of the connections between the input layer and the hidden layer, this activation pattern in the input layer will activate certain nodes in the hidden layer, and, in turn, the nodes in the hidden layer will activate certain nodes in the output layer. What we desire here is a pattern of connections such that, when the boxes are shown a "1," only the first output node is activated; when the sensors are shown a "2," only the second output node is activated; and so on. If that occurred, we would say that the ANN was successfully "recognizing" each of the drawn digits.

At first, because the connections between nodes are random, we expect that particular inputs will produce random output. However, the point of the training is to gradually change the behavior of the ANN, pointing it toward the "correct" responses. This gradual change is achieved by inputting the training data one instance at a time. When a certain pattern of light signals is inputted, the output is observed. If the ANN, by chance, produces "correct" outputs, the connections that led to this output are

---

[20] W. McCulloch and W. Pitts, "A Logical Calculus of the Ideas Immanent in Nervous Activity," *Bull. Math. Biophys.* 5 (1943): 115–33; B. G. Farley and Wesley A. Clark, "Simulation of Self-Organizing Systems by Digital Computer," *IRE Trans. Inform. Theory* 4 (1954): 76–84; Frank Rosenblatt, "The Perceptron: A Probabilistic Model for Information Storage and Organization in the Brain," *Psychol. Rev.* 65 (1958): 386–408. For a more detailed history of neural networks (and machine learning), see Nils J. Nilsson, *The Quest for Artificial Intelligence* (Cambridge, 2010).

[21] This example is often used in teaching ANNs. For more details, see Jianchang Mao and Anil K. Jain, "Artificial Neural Networks: A Tutorial," *Computer* 29 (1996): 31–44. For a general introduction, see C. M. Bishop, *Neural Networks for Pattern Recognition* (Oxford, 1995).

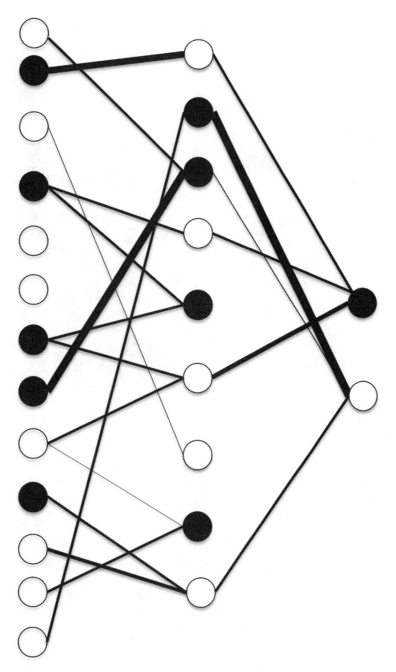

Figure 1. *A schematic diagram of an artificial neural network consisting of one input layer (left), one hidden layer, and one output layer (right). The thickness of the connections between the various nodes in the layers corresponds to the strength of the connection between them. As the network is "trained," the strength of these connections is modified to produce "correct" outputs. (Image by Hallam Stevens.)*

Figure 2. Example of inputs to an artificial neural network for recognizing Arabic numerals. Each of the sixteen squares can either be light or dark, corresponding to turning each input on or off. (Image by Hallam Stevens.)

strengthened. If, on the other hand, the network produces "incorrect" outputs, those connections are weakened.[22] In this way, the ANN is gradually disciplined toward providing "correct" responses. Eventually, the ANN can be trained to correctly "recognize" unknown input patterns.

In biology and medicine, ANNs have a variety of applications, including prediction of protein secondary structure and adverse drug reactions, gene finding, and DNA microarray analysis.[23] The example I will describe here involves using RNA expression data (from a microarray) to classify cancerous tumors.[24] This sort of prediction is important because tumors differ in multiple and complex ways (some responding to drugs and others not), and it is important not to give a patient potent, toxic drugs if their particular tumor will not respond. Here the training data consist of microarray

[22] This "updating" of the network is actually the real trick to making ANNs work. The method is called "backpropagation" and is the most mathematically involved aspect of ANNs.

[23] For a sense of the range of applications of ANNs in biology and medicine, see H. Malmgren, Magnus Borga, and L. Niklasson, eds., *Artificial Neural Networks in Biology and Medicine*, Proceedings of the ANNIMAB-1 Conference, Goteborg, Sweden, 13–16 May 2000 (New York, 2000); David J. Livingstone, *Artificial Neural Networks: Methods and Applications*, Methods in Molecular Biology (New York, 2009).

[24] For a review of the application of ANNs to cancer, see Paulo J. Lisboa and Azzam F. G. Taktak, "The Use of Artificial Neural Networks in Decision Support in Cancer: A Systematic Review," *Neur. Net.* 19 (2006): 408–15.

data that provides information about which genes in each tumor are turned "on" and "off" (this is called an "expression profile") as well as information about the true classification of each tumor. The ANN is trained on this data in the hope that it will be able to predict the proper classification of tumors in unknown cases. Such an approach has had considerable predictive success.

In 2001, Javed Khan and colleagues attempted to use an ANN to recognize "small, round blue-cell tumors" and classify them as neuroblastomas, rhabdomyosarcomas, non-Hodgkin lymphomas, or Ewing-family tumors.[25] Such tumors are difficult to distinguish via traditional methods (light microscopy), but patient prognosis differs markedly depending on diagnosis. The experiments began with sixty-three training samples (consisting of tumor tissue and tumor cell lines) and DNA microarrays containing 6,567 genes. These 6,567 inputs for each sample were reduced by principal component analysis to ten dimensions, which were used as inputs for the ANN. The outputs were the four types of tumors. The sixty-three training samples were randomly partitioned into three groups, two for training and one for validation. The training group samples were each run through the ANN one hundred times. The samples were then repartitioned and the process repeated, producing 3,750 distinct ANN models. These models were used to create a ranking of the genes that most affected the tumor classification. The entire procedure was then repeated with only the ninety-six top-ranked genes as inputs. This ANN was then tested on twenty-five different samples, correctly classifying all of them.

The point was not to try to understand how the ANN actually arrived at its predictions. Microarrays are not considered to be successful because they provide a theoretical understanding or model of what is going on inside a tumor. As Kahn et al. acknowledged, "A potential difficulty with ANN-based pattern recognition models is elucidating causal links from the outputs to the original input data."[26] Although the procedure described here attempted to mitigate this by generating a list of the "most significant genes," most of these genes were not previously known to be involved with cancer. This suggests that a great deal more experimental work would be needed to actually understand why particular genes were showing up as significant.[27] Nevertheless, the ANN results could be used to develop a lower-cost test that would distinguish different cancer types. For instance, follow-up work could design a new microarray that would test for ninety-six genes rather than 6,567 genes. The ANN produced a practical—not theoretical or mechanistic—model for diagnostically distinguishing these cancers.

The multidimensional data of the sort that comes from microarrays usually means that we have no idea how the ANN is actually arriving at its answers. Of course, it is possible to describe in words, as I have just done, how the algorithm works, or to describe it in more formal mathematical terms. But this does not provide a full description of what it is actually doing or how or why it is arriving at specific predictions. We could also write down all the connections and weightings in the network and examine

[25] Javed Khan, Jun S. Wei, Markus Ringnér, Lao H. Saal, Marc Ladanyi, Frank Westerman, Frank Berthold, et al., "Classification and Diagnostic Prediction of Cancers Using Gene Expression Profiling and Artificial Neural Networks," *Nat. Med.* 7 (2001): 673–9.

[26] Ibid., 677.

[27] And the ANN still used ninety-six genes for distinguishing just four cancer types; this is a small number compared to the 6,567 input genes, but still far too many to suggest a clear picture of how each tumor differed from the others.

them. But in most cases (and especially in cases with multidimensional data), this would not tell us much more about why the connections are the way they are or how they are producing predictions. Usually it is not apparent how ANNs are reaching a particular solution.[28]

The only way to obtain knowledge about tumors from an ANN is for the user to actually train and run it. It is a process of interaction between the user, the data, and the algorithm itself that produces working knowledge of cancerous tumors or some other system to which it is applied. The necessarily interactive aspects of this kind of knowledge building were illustrated in work done in 2015 at Google by the software engineers Alexander Mordvintsev, Christopher Olah, and Mike Tyka. Google is interested in using powerful ANNs (with between ten and thirty layers) to recognize specific objects in images (this has applications in Google's image search, for instance). "One of the challenges of neural networks is understanding what exactly goes on in each layer," Google's Research Blog reported.[29] To address this, the Google engineers began to get the ANNs to "draw" output images of what they "saw." They fed random noise to an ANN trained to find images of, for instance, starfish and observed the results. The ANNs could successfully generate images themselves (e.g., of a starfish). To experiment further, the engineers began to mix things up, feeding images of mountains to ANNs designed to find buildings or images of clouds to ANNs designed to find animals (fig. 3). Detailed but surreal images of buildings and birds began to emerge from the pictures. Finally, in a technique they called "inception" (after the 2010 Christopher Nolan movie), the engineers fed back the ANNs' own images into the algorithm over and over again, producing strange, dreamlike scenes.[30]

Although all this seems merely playful, the experiments had a serious goal. The resulting images "help us understand how neural networks are able to carry out difficult classification tasks, improve network architecture, and check what the network has learned during training."[31] In other words, it helped the Google engineers to understand their tool by revealing more about the function of specific layers in the ANN. But the experiments also revealed things about the objects to which the algorithms were applied. In this case, engineers learned something about the essence of the images they were feeding to their algorithms (e.g., they learned that images of dumbbells were inseparable from images of muscled arms). This kind of tinkering with the algorithm allowed the users to develop a kind of working knowledge of their images. If "inception" were applied to ANNs trained on biological data, it might reveal useful working knowledge of biological systems (such as what features of tumors or gene networks are important or are active in a particular process).

ANNs can make accurate and reliable predictions about the world. And they capture things that we cannot fully reduce to linguistic descriptions. Moreover, we can certainly gain knowledge from an ANN: they can lead us to new and important knowledge about

[28] J. J. Hopfield, "Neural Networks and Physical Systems with Emergent, Collective Computational Abilities," *Proc. Natl. Acad. Sci. USA* 79 (1982): 2554–8.

[29] Alexander Mordvintsev, Christopher Olah, and Mike Tyka, "Inceptionism: Going Deeper into Neural Networks," *Google Research Blog*, 17 June 2015, http://googleresearch.blogspot.sg/2015/06/inceptionism-going-deeper-into-neural.html (accessed 16 July 2016).

[30] See https://photos.google.com/share/AF1QipPX0SCl7OzWilt9LnuQliattX4OUCj_8EP65_cTVnBmS1jnYgsGQAieQUc1VQWdgQ?key=aVBxWjhwSzg2RjJWLWRuVFBBZEN1d205bUdEMnhB (accessed 16 July 2016).

[31] Mordvintsev, Olah, and Tyka, "Inceptionism" (cit. n. 29).

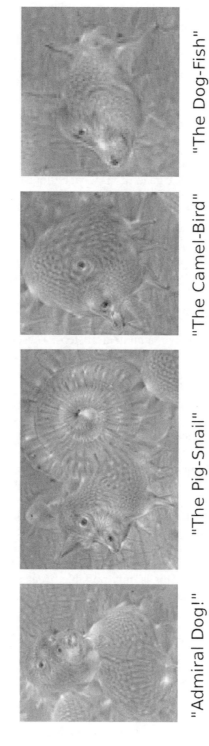

"Admiral Dog!"   "The Pig-Snail"   "The Camel-Bird"   "The Dog-Fish"

Figure 3. *"Funny Animals" made by Google Research using artificial neural networks and the "inceptionism" techniques. (Reproduced under CCA 4.0 license; images by Alexander Mordvintsev, Christopher Olah, and Mike Tyka. See Mordvintsev, Olah, and Tyka, "Inceptionism" [cit. n. 29]. Images have been modified to black and white.)*

cancerous tumors or other complex systems. Moreover, we can improve on the algorithm itself, refining it by using additional training data or interacting with it and tweaking the details of the network. This sort of work involves careful interaction with the algorithm. The knowledge derived here relies on tinkering and playing, as well as intuition about how the algorithm works. The point is not to wholly comprehend what the algorithm is actually doing in any given instance, but interacting with it allows us to learn from it or to improve on it.

REFERENCE GENOMES

My second example also involves a complex algorithm working on a large data set. In this case, however, working knowledge emerged in the interaction between the algorithm and its user. The term "reference genome" is what biologists usually mean when they talk about sequencing genomes. These are the finished sequences of most parts of a species' chromosome (human, elephant, platypus, etc.).[32] The Human Genome Project (HGP) aimed to build a reference genome for humans. Importantly, the reference genome is not derived from a single individual. Rather, it is ordinarily constructed from samples taken from many (or at least several) individuals of the same species; it thus becomes a representative example of the genome for that species. Reference genomes act as a "reference" since once a reference genome is complete it can be used as a scaffold for assembling the genome sequences of other individuals of the same species much more quickly.

Making a reference genome involves a number of steps. These steps have changed over time as sequencing technology and bioinformatic software has advanced. Because I am interested here in the algorithm that was used for the assembly of the reference genome for the HGP, I will describe the process as it was carried out around the year 2000. First, a single chromosome from the genome is broken up into many longer fragments in order to prepare it for insertion into bacteria or yeast. Sequencing technology requires millions of identical copies of a piece of DNA in order to produce a strong signal. Second, this copying was performed by splicing the DNA into bacterial or yeast DNA "vectors" and allowing those organisms to do the copying. Third, the copies of the long fragments are then extracted from the bacteria (or yeast) and broken randomly into many smaller fragments. These fragments are small enough to be sequenced by the sequencing machines (which can handle sequences of only up to 1,000 base pairs). Since the breakage is random, each of the copies may not be broken up in the same way. Next, each of these fragments is sequenced in a sequencing machine and stored in a computer database. And finally, computers are used to reconstruct the original long fragments. This is only possible because the random breaks mean that there will be overlapping segments of sequence that can be used to match up the pieces from end to end.

It is this last step that I am concerned with here. The "assembly problem" turned out to be a critical challenge for the HGP. This task involved 30,000 clones and 400,000 fragments, which together covered 88 percent of the human genome, or 2.7 billion base pairs. This has sometimes been compared to solving a jigsaw puzzle with 400,000 pieces.

---

[32] "Finished" is a technical term in genomics that refers to agreed-upon levels of completeness and accuracy. See, e.g., https://www.genome.gov/10001812 (accessed 16 July 2016).

In fact, it is much more difficult: this is a jigsaw puzzle with 400,000 sometimes re-dundant, always overlapping pieces, some of which are missing and many of which are blurry around the edges.

For the HGP, this massive computational task was tackled by Jim Kent, a graduate student at the University of California, Santa Cruz. Kent had started his career as a software designer in the 1980s, founding a computer animation business. In the late 1990s, Kent decided to return to school to train as a biologist. His work initially focused on the worm genome. In mid-2000, the HGP was struggling to find a way to assemble its data into a complete genome. David Haussler, who was leading HGP's assembly efforts, asked Kent, with his previous programming experience, to give it a try. Kent's software, called GigAssembler, produced the first complete assembly of the human genome, just managing to beat out the efforts of a private company, Celera Genomics. The most computationally intensive step took about three days to run on a cluster of one hundred 800-MHz Pentium CPUs.[33]

How did GigAssembler work? The blurriness of the data (arising from inaccuracies in sequencing) meant that it was necessary for Kent to make the maximum and best use of all the data at his disposal. The aim here was not getting the "right" answer (because the data came from no individual's actual genome, there was not a single "right" answer) but rather getting the "best" answer. In practice, this meant using a set of heuristics to score the various ways of lining up the pieces of the genome, which required careful judgment calls about what was most important in constructing the alignment (e.g., is it better to have an error or a gap?). Researchers valued the relative weight of matches, mismatches, and tails (areas adjacent to matching areas that overlapped but did not match). Here is how Kent and Haussler describe the process:

> The scoring function is crucial here. It is not unusual for the data to conflict. It is important that especially the first joins be based on the strongest matches. The current scoring function strongly favors overlaps that are unique, weakly favors overlaps that are repeat masked, strongly discourages sequence mismatches and inserts within the aligning blocks, and moderately discourages tails. Alignments below a certain threshold of the scoring function are not used to build rafts.[34]

We can see how these decisions are built into the algorithm itself by examining a part of the code for the scoring function:

```
int fragOverlapScore(struct psl *alignment)

/* Return score from roughly 500 to -500 for
fragment/fragment overlap based on alignment. */

{
int milliBad;  /* Misalignments in rough parts
    per thousand.*/
```

[33] W. James Kent and David Haussler, "GigAssembler: An Algorithm for the Initial Assembly of the Human Genome Working Draft," UCSC-CRC-00-17, 27 December 2000, https://cbse.soe.ucsc.edu/sites/default/files/gig_asmtech.pdf (accessed 14 June 2017).

[34] W. James Kent and David Haussler, "Assembly of the Working Draft of the Human Genome with GigAssembler," *Genome Res.* 11 (2001): 1541–8, on 1544.

```
int startTail;  /* Size of starting tail. */
int endTail;    /* Size of ending tail. */

/* Calculate basic score. */

milliBad = calcMilliBad(alignment);

findTails(alignment, &startTail, &endTail);

score = -30*milliBad - (startTail+endTail)/2 +
        round(25*log(alignment->match+1) +
        log(alignment->repMatch+1));

/* Add some penalties for not having many unique
matches or being part of a small fragment. */

if (!enclosingPsl(alignment))

        {
        if (alignment->match < 20)
            score -= (20-alignment->match)*25
        if (alignment->qSize < 5000)
            score -= (5000-alignment->qSize)/40;
        }

return score;
}
```
[35]

This subroutine in the scoring code calculates how to "score" two overlapping fragments. The higher the score, the more likely these fragments will be joined in the finished alignment. The score is based on the fraction of mismatches in the overlap, the
size of the overlapping segment overall, and the size of the tails on each end. Notably,
the scoring function includes various integers (in bold) that place relative weights on
these different factors.

   There are other aspects of GigAssembler that exhibit similar features to the scoring
subroutine. Building and using such software is not simply a matter of entering the
data and waiting for the computer to finish its job: the software needs to be actively
manipulated to make it work in each case. Determining the numbers in the scoring
function requires judgment and intuition about how to make the best possible alignment. It also requires finding out what combinations of parameters are likely to "work"
by accumulating experience with the program and adjusting parameters accordingly.
The results of adjusting particular parameters are not predictable in advance—such
work requires a feeling for the algorithm and a kind of tinkering to make it produce
the best possible result (i.e., a genome with the least gaps, most matches, etc.).

---

[35] Kent and Haussler, "GigAssembler" (cit. n. 33).

Kent's description of the process of writing his program suggests how this was not simply a matter of running the program once and letting it spit out the answer:

> And so I decided that I would just write something quick and simple that would do it. And it did do it. It was pretty quick to come up. It was harder than I thought at first. It took, I guess, it took about a week before the very first—from when I started to the very, very first thing that worked at all did. And for me that's actually a long time. Usually when I write a piece of program, most programs will have the very first skeletal thing that will take about a day or two because I like to sort of build it so that you've got the skeleton first, and then you kind of layer stuff on top of it. But it's much easier to test the program if you always have a little something working. And then sort of add a little bit more to do and a little bit more to it. And so I always try and get the first thing that's kind of working and close very quickly. So it took longer than I thought. And then we just kept adding stuff to it.[36]

It was not only the fact that more data was becoming available that necessitated multiple runs of the algorithm. Rather, more and different types of data meant adjusting the algorithm, making improvements, and learning more about the behavior of the algorithm from the output it produced.[37] Describing Kent's achievement to the *New York Times*, Haussler emphasized the rapid pace of his work: "This program represents an amount of work that would have taken a team of 5 or 10 programmers at least six months or a year. Jim in four weeks created the GigAssembler by working night and day. . . . He had to ice his wrists at night because of the fury with which he created this extraordinarily complex piece of code."[38] This detail suggests the intensity of the interaction between computer and human required for this kind of work. Kent needed a deep and intuitive connection with the algorithm and his data that allowed him to do work that could ordinarily only be performed by a large team.

This kind of work resembles the kind of work that is done with instruments. That is, there is a kind of working knowledge in the algorithm that it is impossible to capture in a mere description of it. Even though it is literally possible to read through the algorithm itself and even to understand what it does in general terms, in practice this does not tell us everything about what the algorithm does, what it can make, or how it does it. This is true even for the programmer, Jim Kent himself: successfully making a reference genome involved a specific kind of interaction between the programmer (or user), the data, and the machine. It is not merely that Kent did not know all the details of how the algorithm would behave (as in the case of black-boxing), but rather that he cannot practically know how it will behave without actually running it and observing the results.

GigAssembler produced not only new biological knowledge but also a new biological object (a reference genome). Yet this production did not rely on necessarily un-

---

[36] Interview with James Kent, "On Involvement in Genomics: Developing Assembly Software," 31 May 2003, Cold Spring Harbor Oral History Collection, 2012, http://library.cshl.edu/oralhistory/interview/genome-research/involvement-genomics/involvement-genomics-developing-assembly-software/ (accessed 16 July 2016).
[37] See Nicolas Wade, "Reading the Book of Life; Grad Student Becomes Gene Effort's Unlikely Hero," *New York Times*, 13 February 2001, http://www.nytimes.com/2001/02/13/science/reading-the-book-of-life-grad-student-becomes-gene-effort-s-unlikely-hero.html (accessed 16 July 2016); Roger Smith and Alexandra Weber Morales, "Some Assembly Required," *Dr. Dobb's*, 1 June 2001, http://www.drdobbs.com/some-assembly-required/184414738 (accessed 16 July 2016).
[38] Wade, "Reading" (cit. n. 37).

derstanding everything the algorithm did, and it certainly did not rely on a complete accounting of how it arrived at its specific solution. One could know generally how GigAssembler weighed different kinds of alignments, and so on, but the user did not know the detailed process through which fragment A was aligned with fragment B, and fragment B with C, and so on. The user remained ignorant of the detailed process through which a reference genome was built. In principle, because a computer is a deterministic machine, it is possible that a human (or group of humans) could follow the exact instructions in the algorithm and make a reference genome themselves. But replicating the work done with GigAssembler would mean not just running the program once but running it over and over again with slightly different code and parameters each time. Even if this were conceivable, doing this work with human computers or on paper would engender a fundamentally different set of interactions between the "user" and the "program." Actually making a reference genome requires tinkering, playing with the algorithm, adjusting parameters, running it again, and so on; it means interacting with the algorithm to get the "best" result out of it.

ENSEMBL AND HYPERTEXT

My third example is a database system rather than an algorithm. This case suggests that working knowledge can inhere not only in algorithms but also in the complex ways in which data is ordered and presented. Ensembl is the preeminent database of the European Bioinformatics Institute (EBI; based in Hinxton, Cambridgeshire).[39] The project was initiated in 1999 in an effort to organize and systematize the data from the HGP. The database is most commonly accessed and used as the Ensembl Genome Browser: this is a Web interface that presents genome data in a visual format, allowing biologists to scroll along the lengths of various genomes, zooming in on various features (fig. 4). This is similar to other "Genome Browsers" maintained by the National Center for Biotechnology Information (at the National Library of Medicine in Bethesda, Md.) and the University of California, Santa Cruz.[40]

Ensembl itself produces no data. Rather, it collects data from a wide variety of sources and makes it available in a common, consistent, and coherent format. Moreover, the Ensembl Genome Browser is the front end of a much more complex and interconnected system of databases that store data and allow access to it through a variety of tools (e.g., the data-mining tool BioMart). Ensembl's history can be traced to a database to manage information for the worm genome project, ACeDB (A *C. elegans* Data Base).[41] This database was initiated in 1989 as a joint project between Richard Durbin (Laboratory for Molecular Biology, Cambridge) and Jean Thierry-Mieg (Centre National de la Recherche Scientifique, Montpellier). Durbin and Thierry-Mieg wanted to find a way to store sequence information, physical and genetic maps, and literature citations within the same database. To do this, they realized that they required a very flexible system in order to manage not only the quantity of data but also

[39] Ensembl can be accessed here: http://www.ensembl.org/index.html (accessed 16 July 2016). The original publication describing the project is T. Hubbard, D. Barker, E. Birney, G. Cameron, Y. Chen, L. Clark, T. Cox, et al., "The Ensembl Genome Database Project," *Nucl. Acids Res.* 30 (2002): 38–41.

[40] The NCBI Genome Browser: http://www.ncbi.nlm.nih.gov/genome/ (accessed 16 July 2016). The UCSC Genome Browser: https://genome.ucsc.edu/ (accessed 16 July 2016).

[41] For more information on ACeDB, see http://www.acedb.org/ (accessed 16 July 2016).

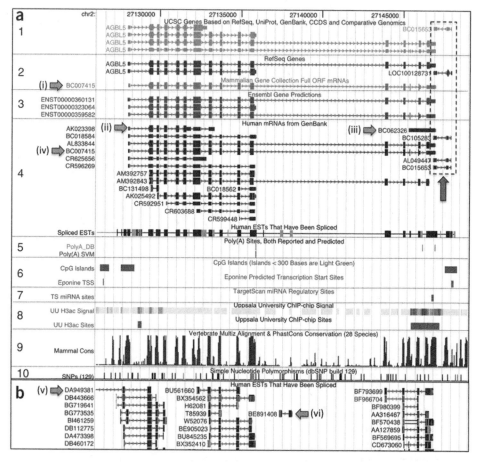

Figure 4. *Screenshot of the Ensembl Genome Browser with various genome "tracks" running across the screen from left to right. (Reproduced with permission from Melissa S. Cline and W. James Kent, "Understanding Genome Browsing," Nat. Biotechnol. 27 [2009]: 153–5, fig. 1.)*

its variety. They wanted to help biologists organize and manipulate data in order to find the important relationships within it.

The result was a system that was not merely a database but also a visualization system (fig. 5). ACeDB was being designed at the same time that Tim Berners-Lee was developing the tools of the future World Wide Web: HTML and HTTP. Durbin and Thierry-Mieg made use of these tools, linking their data using hypertext. ACeDB was not only solving a storage problem but also demonstrating how to make data useful and valuable. To do so, its designers had to find ways to organize and link the data in accessible and findable ways. For this purpose, hypertext proved particularly useful.[42]

ACeDB proved extremely successful. It was quickly picked up and used beyond the worm genome project. Both the Wellcome Trust Sanger Institute and the Genome Institute at Washington University adapted ACeDB to be used as their main database for

[42] Richard Durbin and Jean Thierry-Mieg, "The ACEDB Genome Database," in *Computational Methods in Genome Research*, ed. S. Suhai (New York, 1994), 45–55.

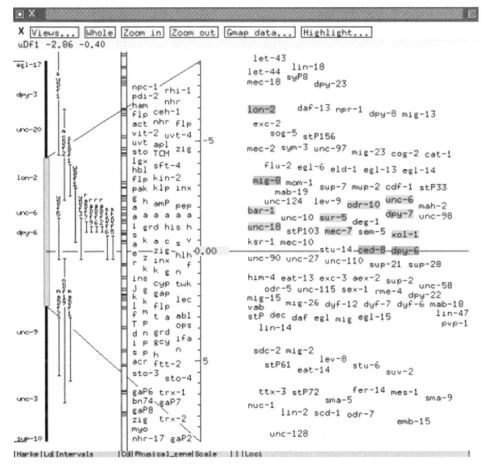

Figure 5. *Screenshot of ACeDB genome browser. The chromosome or sequence element runs vertically down the screen with genes and other features shown on the right. (Reproduced with permission from http://www.acedb.org/Tutorial/brief-tutorial.shtml [accessed 16 July 2016].)*

storing and organizing human genome sequences for the HGP. Beginning in 1998, Durbin realized that the HGP required an even more powerful tool for representing genomic sequence data. He put some of his graduate students to work on this problem: Ewan Birney, Tim Hubbard, and Michele Clamp developed the code for Ensembl using ACeDB as a model. The eventual name adopted for this database system—a "browser"—echoed the increasingly important Web browsers being developed by Netscape and others at the same time.[43]

The way that biologists use Ensembl also reflects how it is linked to the technologies of the Web. In Ensembl (or other genome browsers), users navigate by clicking through links, which allows them to call up or drill down to the specific data that they might need, generating and seeing relationships between various data. In most cases, specific Ensembl Web pages and links are dynamically generated from the database by the Ensembl code. That is, selecting certain options or clicking certain links will cause

[43] See Hallam Stevens, "The Politics of Sequence: The Bermuda Principles and Open Source Biology," *Inform. & Cult.* 50 (2015): 465–503.

the data to be displayed in new arrangements. This sort of work can reveal novel patterns or relationships in data that can tell us about how organisms evolve or function.

Ensembl's programmers at the EBI give examples of how Ensembl's Web tools can be used to find the function of a gene or to discover genomic elements likely to be regulating a gene.[44] For the human gene *MYO6*, for instance, the browser first allows the user to look for similar genes (homologues) in other organisms ("Do so by clicking on the *orthologues* link at the left of the gene tab"). This reveals an extensively studied mouse gene called *Myo6*. Functions for this gene can be revealed by looking at the associated gene ontology (GO) terms: "Clicking on the mouse protein identifier ENSMUSP00000108893, then on the *gene ontology* link at the left shows the GO terms associated with the mouse protein." The known functions of the mouse gene can be used as a basis for predicting the analogous function of the human gene. To predict regulatory sequences associated with the human gene, the authors suggest looking for histone modification and methylation sites near the gene:

> Regulatory features are drawn alongside the human genome along with *MYO6* transcripts, which are on the forward strand. Variations are also drawn and colour-coded indicating the position of the variation with respect to the gene (ie. intronic, upstream, coding). Clicking on variations opens a pop-up box with specific information, and a link to the variation tab.[45]

Navigating around near the human gene can lead to new knowledge about what this gene does and how it is regulated. But it can also foster more tacit forms of knowledge about how regulatory elements act, what evolutionary relationships between genes look like, and so on. This is based on a visual, interactive, and hypertextual activity. It does not require knowing how the Ensembl system itself works, nor does it require understanding the structure of the underlying data. Instead, it is through hypertextual interaction with the data that it is possible to gain working knowledge of genes and genomes.

This is not simply a question of black-boxing. It is not just that the users of Ensembl do not need to know how it works. Rather, it is that no one (not even Ensembl's designers) can describe or predict all the possible states that the data structure may produce. There is no practical way of reasoning through the large sets of data, operations, and constraints. Here, this has as much to do with the amount and multidimensionality of the data as with the complexity of the algorithms at work.[46] The size and complexity of the data mean that data structures play a critical role in shaping the interaction between data and users; a user's understanding of the data, his or her ability to manipulate it or work with it, and his or her "feel" for it depend on data structures.

The use of Ensembl is similar to how we use many parts of the Web. Web technologies allow us to navigate through data in specific ways. The Web imposes a structure on the data that constrains what we do with it. Tinkering with the Web ("surfing") allows us to discover certain kinds of patterns and relationships that otherwise might not

---

[44] Giulietta M. Spudich and Xosé M. Fernández-Suárez, "Touring Ensembl: A Practical Guide to Genome Browsing," *BMC Genomics* 11 (2010): 295–304.

[45] Ibid.

[46] On multidimensional data in genomics, see Adrian Mackenzie, "Machine Learning and Genomic Dimensionality: From Features to Landscapes," in *Postgenomics: Perspectives on Biology after the Genome*, ed. Sarah S. Richardson and Hallam Stevens (Durham, N.C., 2015), 73–102.

be visible. For instance, websites such as Kayak (www.kayak.com) aggregate information from different databases (in this case, airline flight data) based on user input and display it in novel ways. This provides specific knowledge about the cheapest flights to particular destinations on particular days. But interacting with Kayak also fosters a kind of working knowledge of how to search for the cheapest flights, when the best deals can be found, what kinds of parameters to input, how flexible to make one's search, and even perhaps a more intimate knowledge of how the arcane system of airline ticket pricing works.

The constraints and categories that are created by the Web itself constitute a structure within which we can operate on and interact with the data. Working knowledge emerges in the interaction between the data, the structure, and its users. Users are able to generate new knowledge through interacting with the data, even though they cannot precisely describe what that knowledge is or how it arises. The possibilities of the Web structure coupled with massive amounts of data make it impossible, in practice, to predict the kinds of relationships the data might reveal, except by actually interacting with it via tools such as Ensembl. The underlying data of the human genome or the Web elude linguistic or mathematical description. But data structures and tools provide ways of tinkering with data that allow users to "see" particular features.[47] Using Web browsers or genome browsers allows users to develop a working knowledge of the underlying data through interaction with it.

CONCLUSIONS

The three examples here all show different ways in which large quantities of data have been utilized to make biological knowledge. Whether we call these "Big Data" or not, I have argued that the kinds of computational work involved here entail different practices and methods from older ways of working with pens, paper, and people. Some of these practices and methods predate the recent fascination with Big Data. But they are not necessarily related to or derived from conventional statistical or scientific data-processing methods. Instead, they have emerged from fields such as artificial intelligence and machine learning to play a more prominent role in a range of fields. "Big Data"—because it draws attention to size rather than practice—may be an inappropriate label for this kind of work, but the wide application of these methods does nevertheless mark an important shift.

This shift represents not merely a scaling up of work done with pen and paper or books and tables. Rather, the quantities of data processed by computers, and the algorithms needed to deal with them, make a qualitatively different kind of knowledge. In other words, this is a "more is different" argument. Philip Anderson in the 1970s made this argument for the importance of solid state physics vis-à-vis particle physics: having more molecules to deal with not only involved a scaling up but generated situations in which qualitatively new phenomena (such as superconductivity) emerged:

> The behavior of large and complex aggregates of elementary particles, it turns out, is not to be understood in terms of a simple extrapolation of the properties of a few particles.

[47] Software allows biologists to "see" their objects in particular ways that open it up for certain forms of understanding and manipulation. See Hallam Stevens, *Life Out of Sequence: A Data-Driven History of Bioinformatics* (Chicago, 2013), chap. 6.

Instead, at each level of complexity entirely new properties appear, and the understanding of behaviors requires research which I think is as fundamental in its nature as any other.[48]

This is an emergence argument: that the properties of the higher level cannot be practically derived from the properties of the lower level. By analogy, I am suggesting that large amounts of data, subjected to complex algorithms and data structures, present a similar situation. The behavior of such systems cannot be predicted in advance, and the objects and solutions that they produce are not reducible in any simple way to paper-based manipulations of text or numbers. At some point, more data is different.

Although I have taken my examples here from biology, the phenomena I have described extend well beyond that field into other domains of knowledge. Big Data has generated the most interest as a set of tools with huge commercial potential. As Dan Bouk argues in this volume, various forms of personal data have recently been imbued with added value and power.[49] This value depends not only on an increased ability to aggregate data but also on new kinds of algorithms and ways of using them. And indeed, we can see versions of working knowledge present in some of the most commercially important algorithms. Google's "Adsense" and "Adwords," for instance, are algorithms that place targeted advertisements on websites.[50] These algorithms strive to match the numerous businesses that wish to advertise online with the many websites that, for a fee, are available to host ads (e.g., online news sources, social media sites, blogs). The aim of Adsense/Adwords is to optimally connect advertisers to those who wish to display advertisements. Because Google is paid only when a Web surfer clicks on an ad, it has an interest in placing ads on sites where they are most likely to receive clicks. Usually, Google assumes, such clicks will occur where the advertisement is relevant (in some way or another) to the content of the Web page being viewed.

Posed in this way, Google faces a very big data problem. Given all the varied content of the web, how is it possible to decide whether a given website is relevant to a given topic? In other words, Google needed to find a way of determining the meaning of a website without actually having anyone read it (there are far too many websites for that to be practical). One of the most important parts of the system that Google uses to achieve this is called PHIL: Probabilistic Hierarchical Inferential Learner.[51] It is possible to describe in broad and conceptual terms how PHIL (and Adsense/Adwords) works. However, such an understanding does not make it possible to predict which ads will be assigned to which websites. This is only practically knowable by letting the algorithm loose on the Web and seeing what it does. Like an ANN, even knowing the specific weightings that PHIL uses to link words into clusters and meta-clusters is

[48] P. W. Anderson, "More Is Different: Broken Symmetry and the Nature of the Hierarchical Structure of Science," *Science* 177 (1972): 393–6, on 393.

[49] Dan Bouk, "The History and Political Economy of Personal Data over the Last Two Centuries in Three Acts," in this volume.

[50] Google's description of Adsense can be found at http://www.google.com/adsense/start/ (accessed 16 July 2016). See also Joel Comm, *The Adsense Code: What Google Never Told You about Making Money with Adsense*, 2nd ed. (New York, 2006).

[51] For more information, see Ruchira S. Datta, "PHIL: The Probabilistic Hierarchical Inferential Learner," Tenth Annual Bay Area Discrete Mathematics Day, 2005, http://math.berkeley.edu/~datta/philtalk.pdf (accessed 16 July 2016). See also G. Harik and N. Shazeer, "Method and Apparatus for Learning a Probabilistic Generative Model for Text," U.S. Patent 7,231,393 B1, filed 26 February 2004, and published 12 June 2007.

not going to help in understanding why certain ads appear in certain places online. Indeed, Google's business implicitly relies on this unpredictability—if it was possible to know how to attract particularly valuable ads, bloggers and websites would adjust their content in order to gain advertising revenue from their site. Instead, some advertisers and websites attempt to "game" Google's algorithms by experimenting with them—by adjusting their content to attract more ads and more money.[52] In practice, this is the only way it is possible to learn about the behavior of PHIL/Adsense/Adwords algorithms. Gaining a working knowledge of Google's algorithms in this way can be a financially valuable undertaking.[53]

In each case discussed here, algorithms and data structures provide ways of thinking about and solving biological problems that have now become part of the woodwork not only of biology but also of many parts of the Web and other information technologies we interact with every day. Interacting with algorithms and data structures allows us to gain knowledge of genomes or the Web or other large and complex data sets. The significance of Big Data lies not so much in its size or speed or scope, but rather in the particular ways in which it produces knowledge through what I have described here in terms of "working knowledge" or a "feeling for the algorithm."

Arguing that technologies such as algorithms or data structures are important in shaping the kind of knowledge that emerges from them does not amount to technological determinism. Someone designed such algorithms and data structures. These coders and designers no doubt had specific goals and interests and styles of reasoning that made their software the way it is. This essay, as much as anything else, is an attempt to call attention to these (often hidden) processes of design and how they shape subsequent knowledge production. This shaping takes place through an interaction between human designers, coders, and users and their machine-based tools. If new biological knowledge is based on computers—if it somehow "belongs to" computers—then one important task is to begin to interrogate these computational structures in detail, including the histories of their production by humans. Who built them? For what purpose? Why were they built in one particular way and not another?

Understanding the history of Big Data requires insight into the histories of these algorithms and data structures—the history of machine learning, the history of algorithms like GigAssembler and Adsense, the history of hypertext. The history of Big Data is tied to the history of these algorithms and data structures. Big Data practices are not entirely novel—rather, they have a history that is rooted in the kinds of stories about algorithms and data structures that I have sketched here. These histories overlap with, but are importantly different from, histories of noncomputational, paper-based, "small data" practices. To understand the political, social, and cultural meanings of Big Data, we need to apply the kinds of sociological, historical, and political analysis that we have applied to labs and scientists to the insides of algorithms and data structures.

[52] In fact, there is an industry based on these practices called "search engine optimization."
[53] Frank Pasquale, *The Black Box Society: The Secret Algorithms That Control Money and Information* (Cambridge, Mass., 2015).

The Database before the Computer?

by David Sepkoski*

ABSTRACT

Are characteristic practices of modern data-driven science—the compiling of databases, quantitative analysis of large data sets, standard graphical representations of data patterns—a product of the computer era? I explore this question through a comparative analysis of nineteenth- and twentieth-century data approaches in paleontology. Drawing on examples of large-scale quantitative data collection and analysis of both paper-based taxonomic compendia and eventual electronic databases, I argue that, in fact, paleontologists engaged in what we might call "databasing" long before computers arrived on the scene. The arrival of computers in paleontology, I argue, fits closely into a pattern that Jon Agar has described whereby preexisting practices and epistemologies are adapted to new technologies. However, I also attend to the ways in which changes in the technology and material culture of data between the nineteenth and later twentieth centuries affected the "moral economy" of data. Each era faced particular challenges for coping with "data friction," and new technologies and materialities of data had consequences for the professional division of labor in data-driven taxonomic sciences.

Can the history of databases be separated from the history of electronic digital computers? At first blush this might seem like an anachronistic question: after all, electronic databases are perhaps *the* exemplary tool of the modern era of "data-driven" science and the precondition for a world—both in science and in broader political and economic culture—dominated by "Big Data."[1] The *Oxford English Dictionary* appears to endorse this notion, defining a database as "a structured set of data held in computer storage and typically accessed or manipulated by means of specialized software."[2] This definition seems to assume that there is something irreducibly electronic about databases, or at least that some fundamental shift in the collection, storage, and analysis of data arrived with the advent of electronic computers—a notion widely reflected in the recent literature about Big Data. It is certainly the case that while the term "data" was in wide cir-

* Max Planck Institute for the History of Science, Boltzmannstrasse 22, 14195 Berlin, Germany; dsepkoski@mpiwg-berlin.mpg.de.

I am grateful to all of the members of the "Historicizing Big Data" working group at the Max Planck Institute for the History of Science for many stimulating conversations and for very helpful feedback on earlier versions of this essay. I would like to thank Hallam Stevens in particular for a number of very productive conversations through which what initially appeared to be differences ended up looking much more like common ground.

[1] The literature on Big Data and data-driven science has become vast. For citations, see Elena Aronova, Christine von Oertzen, and David Sepkoski, "Introduction: Historicizing Big Data," in this volume.

[2] *Oxford English Dictionary*, s.v. "database," http://www.oed.com/view/Entry/47411 (accessed 16 September 2014).

culation for hundreds of years before computers arrived on the scene, the term "data-base" (or "data base") was first coined in the 1950s, explicitly in the context of computer data storage and processing. Technically speaking, then, one might even argue that the notion of a preelectronic database is a contradiction in terms.

And yet. Surely earlier kinds of data collections—Hollerith punch cards, for example—contain many of the central attributes assigned to even the more narrow definitions of the term "database" (internally stored programs, structured organization of data, the ability to sort, automated quantitative analysis).[3] If we expand our definition to include one preelectronic technology, is there any reason not to add other "paper technologies," such as paper card storage collections, tables, lists, and printed catalogs and compendia?[4] Viewed from this perspective, one could argue that human beings have been collecting large volumes of abstracted, organized data for quite a long time—whether in the form of weather information, astronomical tables, lists of plants and animals, bibliographic indices, and so on—that might well be considered, if not identical to, then at least homologous with modern electronic digital databases. In fact, even the *OED* entry contains a hidden ambivalence. While eight of the nine illustrative quotations provided support the narrowly electronic definition, one seems to baldly contradict it: a quotation from a 1985 issue of *Ashmolean Magazine* that reads, "A museum and its records are one vast database."

Data and databases have been very much on the minds of historians, philosophers, and sociologists of science in recent years, and various authors have responded quite differently to the technological reducibility of data. On one end of the spectrum is what we might call a "digital determinist" view, which restricts all use of the term "data-base"—and even "data"—to electronic computerized contexts. That approach is characterized most provocatively and ably by Hallam Stevens, who has argued in a recent book that "data belong only to computers; they are part of a set of practices that make sense only with and through computers."[5] Perhaps the opposite view has been defended by Geoffrey Bowker, who has argued that far from being an invention of the computer era, "if anything the computer revolution is a product of the drive to database."[6] In between these extremes, scholars including Paul Edwards, Bruno Strasser, Sabina Leonelli, Dan Rosenberg, Markus Krajewski, and others have emphasized that on the one hand there are continuities in both practice and epistemology of data across historically specific material cultures, but that on the other new technologies have brought new social arrangements and tools to data-oriented science.[7]

---

[3] See Christine von Oertzen, "Machineries of Data Power: Manual versus Mechanical Census Compilation in Nineteenth-Century Europe," in this volume.

[4] On paper technologies, see, e.g., Markus Krajewski, *Paper Machines: About Cards and Catalogs, 1548–1929* (Cambridge, Mass., 2011); Hallam Stevens, *Life Out of Sequence: A Data-Driven History of Bioinformatics* (Chicago, 2013).

[5] Stevens, *Life Out of Sequence* (cit. n. 4), 7.

[6] Geoffrey C. Bowker, *Memory Practices in the Sciences* (Cambridge, Mass., 2005), 109.

[7] Sabina Leonelli, "Making Sense of Data-Driven Research in the Biological and Biomedical Sciences," *Stud. Hist. Phil. Biol. Biomed. Sci.* 43 (2012): 1–3; Lisa Gitelman, ed., *"Raw Data" Is an Oxymoron* (Cambridge, Mass., 2013); Paul N. Edwards, *A Vast Machine: Computer Models, Climate Data, and the Politics of Global Warming* (Cambridge, Mass., 2010); Bruno J. Strasser, "Data-Driven Sciences: From Wonder Cabinets to Electronic Databases," *Stud. Hist. Phil. Biol. Biomed. Sci.* 43 (2012): 85–7; David Sepkoski, *Rereading the Fossil Record: The Growth of Paleobiology as an Evolutionary Discipline* (Chicago, 2012); Sepkoski, "Towards 'A Natural History of Data': Evolving Practices and Epistemologies of Data in Paleontology, 1800–2000," *J. Hist. Biol.* 46 (2013): 401–44.

In this essay, I am going to enter this debate by arguing that so-called data-driven science has a long and continuous genealogy that stretches back well before the introduction of computers to science. I will argue further that even the most distinctive feature of modern data-driven science—the database—need not be conceived as the product of a specific technology or material culture. There has indeed been a transformation of what Bruno Strasser has called the "moral economy" of data since the introduction of electronic computing, but this change should be understood neither as an epistemic rupture nor as a revolution in practice.[8] But while I broadly agree with what Strasser and others have concluded about the continuities between pre- and postelectronic digital practices in the life sciences, I take an even more extreme opposition to the "digital determinist" view. For example, Strasser has argued that while modern databases are analogous to earlier "organized assemblages of standardized objects" like collections and catalogs, he maintains that they are not homologous—in other words, there is no direct "historical connection, whether social, intellectual, or cultural, between collections and databases."[9] I will argue rather that in some instances preelectronic assemblages of data—printed compendia, tables, and catalogs—are precisely homologous with modern electronic databases, and that the history of electronic databases in the natural sciences is continuous with a history of earlier paper data practices.

I will make these arguments primarily through the example of paleontology, a discipline with a long history of data accumulation and analysis. What the history of paleontology shows is that from the standpoint of both epistemology (the goals of data-oriented research) and practice (the methods and tools used to collect and analyze data) there was a fairly remarkable continuity from the mid-nineteenth to the late twentieth centuries. The printed compendia in which nineteenth-century paleontologists organized their data about the fossil record functioned (as I will show) quite similarly to much later electronic databases, at least as databases were employed during the discipline's first engagement with computers from the late 1960s to the early 1980s. The lesson here is that in paleontology (and I argue that this example is extendable at least to other natural historical disciplines like botany and zoology), similar cultures of systematic collection, abstraction, commensuration, and quantitative analysis of data persisted over almost 200 years and in a variety of distinct material cultures. Furthermore—and perhaps even more remarkably—the kind of quantitative analysis performed by paleontologists on these preelectronic data collections in the mid-1800s did not differ significantly from approaches in the 1980s and beyond. These comparisons can be extended to the basic format of data collections (the structure and the "metadata" of the databases), the central questions being asked (generally interpretations of the patterns of diversification over the history of life), and the format of the (usually graphical) summaries of the analysis performed (distinctive charts and graphs that offer summaries of data analysis).

At the same time, I will be attentive to what did change with the introduction of computers to paleontology. Computers had a profound impact on the disciplinary arrangement of many natural historical disciplines, and paleontology was no exception.

---

[8] Bruno J. Strasser, "The Experimenter's Museum: GenBank, Natural History, and the Moral Economies of Biomedicine," *Isis* 102 (2011): 60–96.

[9] Bruno J. Strasser, "Collecting Nature: Practices, Styles, and Narratives," *Osiris* 27 (2012): 303–40, on 336–7.

In fact, as I have documented elsewhere, computerized approaches to data collection and analysis helped bring about what has been called a "paleobiological revolution" during the 1970s and 1980s, whereby a small group of organized actors successfully brought an approach to the study of the history of life that was informed by theoretical models and oriented toward quantitative evolutionary analysis from the margins of pale-ontology to the center of the discipline.[10] However, while computers helped spread the popularity of quantitative data analysis in paleontology in a variety of ways, they did so mostly by streamlining, accelerating, or black-boxing existing, preelectronic data prac-tices, and not (in the first instance, at least) by creating a wholly new data culture. This development was consistent with the general observation made by Jon Agar that, in most cases, "Computerization, using electronic stored-program computers, has only been attempted in settings where there *already existed* material and theoretical compu-tational practices and technologies," and where, furthermore, "existing practices and technologies were still capable of the computational task at hand."[11]

What computers did allow in paleontology was to make many of those existing prac-tices easier or more accessible, enormously increasing their popularity and eventually greatly expanding the analytical power and epistemic authority of data-driven practices in the discipline. To use a term introduced by Paul Edwards, computers greatly reduced "data friction" in paleontology, making it much easier to collect, store, rearrange, share, and analyze data than had been the case in the era of paper databases and hand calcu-lations. Mainframe computers—and ultimately PCs and the World Wide Web—also offered new possibilities for collective empiricism that are still expanding today in the form of open-access electronic data collectives and projects that seek to link data-bases of distinctly heterogeneous content (e.g., fossil taxonomy with geochemistry).[12] At the same time, computers created quite considerable "disciplinary friction" (to mod-ify Edwards's term) in paleontology, both by intensifying older epistemic divisions be-tween analytical and descriptive approaches and by introducing new divisions of labor between the producers and the manipulators of data that have caused significant ten-sion. In this last regard, the history of computerization in paleontology tracks closely with disciplinary changes that have been described by Strasser, Stevens, Leonelli, and others in the emergence of molecular genetics and bioinformatics, whereby a class of worker has emerged in the field trained almost wholly in computer science or database analysis with little expertise or interest in more traditional biology.[13]

Ultimately, though, as Agar has observed, arguments about the historical relation-ship between technology and scientific practice call into question assumptions about the nature of historical change that are empirically testable: for example, "the claim that certain scientific activities would be 'impossible' without computers . . . presumes that historical change is discontinuous."[14] This is precisely the kind of assumption that

---

[10] Sepkoski, *Rereading the Fossil Record* (cit. n. 7).
[11] Jon Agar, "What Difference Did Computers Make?," *Soc. Stud. Sci.* 36 (2006): 869–907, on 872–3; emphasis in the original.
[12] See, e.g., the collaborative database "Macrostrat," which combines "macrostratigraphic, litholog-ical, and environmental data"; https://macrostrat.org (accessed 16 September 2014).
[13] Strasser, "Data-Driven Sciences" (cit. n. 7); J. K. Dolven and H. Skjerpen, "Paleoinformatics: Past, Present and Future Perspectives," in *Computational Paleontology*, ed. Ashraf M. T. Elewa (Berlin, 2011), 45–52. See also Patrick McCray, "The Biggest Data of All: Making and Sharing a Digital Uni-verse," in this volume, for an examination of similar divisions in late twentieth-century astronomy.
[14] Agar, "Difference" (cit. n. 11), 872.

I am challenging in this essay by historicizing data practices in paleontology. Paraphrasing the title of Agar's article, we could ask, "What difference did computers make to data-driven science?" In asking this question, I will resist the teleological impulse to adopt modern electronic databases as the exemplar against which preelectronic technologies should be compared. Rather, by introducing the term "data collection" as a genus-level term capable of describing assemblages of data in any format, I will suggest that the modern electronic database is a stage in a contingent genealogy of practices around data collections that ultimately shaped how modern data-driven science operates. In other words, the point is not to show that paper data collections in nineteenth-century paleontology "look like" modern databases, but rather to make the deeper historical point that late twentieth- and twenty-first-century data practices were strongly constrained by epistemologies, structures, and techniques that emerged long before the computer—a set of practices that eventually came to be named "database." This essay, then, traces the historical emergence of the conditions that allowed the very notion of the database.

NINETEENTH-CENTURY DATA COLLECTIONS

One way to think about the relationships between data collections across distinct material cultures is to introduce a level of abstraction by considering what are known to computer programmers as "data structures." Data structures are frameworks that allow collections of data to be organized into manageable relationships—they are schemes for organizing information. A list, for example, is a very simple data structure in which "objects" are organized sequentially and assigned consecutive integers for retrieval. In computer science, data structures require programming languages capable of supporting particular kinds of objects and structures, and algorithms designed to access the information stored therein. One limitation of simple lists, for example, is that they are not randomly accessible or automatically re-sortable without re-creating the entire list itself. But the earliest higher-level computer languages like FORTRAN and COBOL supported multidimensional "arrays" in which data can be structured (and accessed) nonsequentially. There now exist a great many complex data structures—arrays, stacks, binary trees, and so on—supported by equally complex algorithms designed to optimize specific tasks, like sorting or retrieving objects that exist in complex relationships with other objects in the data structure. This makes possible the complex relational databases used today in Big Data applications from commerce to the natural sciences. But in theory, data structures are independent of any specific material platform in which they are implemented—they can exist in any medium capable of supporting them. It is hard to imagine re-creating a complex multidimensional array with paper technologies, but simpler data structures like lists or tables can certainly exist on paper. Algorithms now search and sort and analyze data automatically, but some of the most simple tasks they perform can be—and were—performed by humans by hand (albeit more slowly and cumbersomely).

We can say, then, that a database has four major requirements: data, a data structure, a medium (mechanical or electronic) in which to exist, and routines (algorithms) to sort and access that data. None of these features are exclusive to electronic technology. In a Hollerith punch card machine, for example, the data structure is physically imposed by the arrangement of cards stored in the machine (and by the architecture of the machine itself), and algorithmic functions are performed by mechanical manipulation of that machine. The earliest electronic computers simply translated this arrangement to binary electronic signals. But even a printed list or table of data can also be utilized to perform these operations. Another way of avoiding anachronism or presentism, then, is to conceive of the history of data as a history of evolving technologies onto which data structures have been imposed to allow access to data collections.

Bruno Strasser has convincingly argued that there are important similarities between collections of objects in natural history disciplines and electronic databases in the experimental life sciences. He regards this similarity as mostly analogical but considers the analogy nonetheless powerful: like databases, "collections isolate something in nature (say a bird or a gene), strip it of its relations to its surroundings (the forest or the genome), leave behind most of its properties (such as being alive or chemically modified) and turn it into a specimen embedded in a new system of relations with other specimens in a collection."[15] It is this relational aspect that is most significant, since both natural history collections and experimental databases are "composed not only of individual things, but also of the many relations among the things they contained."[16] Here Strasser is arguing that a database may be thought of as a kind of collection—"the same category of *collection* as collections of plants, animals, fossils, and so on."[17] I think there is great value in conceiving of such "organized assemblages of standardized objects," whether found in museum specimen drawers or electronic databases, as species of collections. But I want to make a slightly different move here: instead of comparing collections of physical objects themselves with databases, I am comparing organized collections of data about such natural specimens—which from the nineteenth century through the later twentieth took the form of printed catalogs and compendia—with electronic databases. Such compendia were organized collections of abstracted data—sometimes pictorial, sometimes descriptive, and sometimes numerical—that functioned as what Strasser calls a "second nature" for further abstraction and analysis in producing knowledge about the natural world.[18]

Collections of things have always been the starting point for knowledge production in natural history, but in themselves these collections are not very useful unless they can be converted to information—data—that is abstracted, standardized, and recontextualized. As long as naturalists have been amassing organized collections, they have been producing catalogs, which are an indispensable corollary to collecting. Whether one examines an eighteenth-century catalog like John Woodward's *An Attempt Towards a Natural History of Fossils in England*, a nineteenth-century atlas like Adolphe Brongniart's *Prodrome d'une histoire des végétaux fossiles*, or a massive, multivolume compilation from the twentieth century like the *Treatise on Invertebrate Paleontology*, catalogs and compendia convert things to data in a way that opens up a variety of epistemic,

[15] Bruno J. Strasser, "Collections," in *Eine Naturgeschichte für das 21. Jarhundert: In honor of Hans-Jörg Rheinberger*, ed. Safia Azzouni, Christina Brandt, Bernd Gausemeier, Julia Kursell, Henning Schmidgen, and Barbara Wittmann (Berlin, 2011), 25–7, on 26.
[16] Strasser, "Collecting Nature" (cit. n. 9), 321.
[17] Ibid., 311.
[18] One could make an analogy between Strasser's notion of "second nature" and what Dan Bouk (borrowing from sociologists of bureaucracy) terms a "data double." Both are attempts to re-create some vital aspects of natural phenomena (in Strasser's case, natural history, and in Bouk's, human lives) by abstraction and reconfiguration through data. See Bouk, "The History and Political Economy of Personal Data over the Last Two Centuries in Three Acts," in this volume.

analytical, and communal practices.[19] While many densely verbal or illustrated compendia of fossil information bear little obvious similarity to modern databases, other examples—reaching well back into the early nineteenth century—were organized into data structures that were functionally identical to early electronic databases in paleontology. For example, Samuel Woodward's 1830 *A Synoptical Table of British Organic Remains* compiled a catalog of British fossils drawn from earlier, regional fossil catalogs, published descriptions, and private collections, which Woodward (no relation to John Woodward) arranged as a 100-page table. The primary data in this arrangement are the taxonomic groups—class, family, genus, and species—while the headings or identifiers (what would today be called the "metadata") are the strata and locality where the specimens are found, as well as the bibliographic references to original descriptive literature.[20] The middle decades of the nineteenth century saw a proliferation of such tabular catalogs in Britain, France, and Germany, all of which followed this general format. They did not replace illustrated atlases or dense verbal taxonomic encyclopedias, but over the next century—and in fact up to the present day—compendia have constituted a distinctive genre in paleontology and other naturalist disciplines (zoology, botany, etc.). In fact, the earliest electronic databases in paleontology—such as J. John (Jack) Sepkoski Jr.'s *A Compendium of Marine Fossil Invertebrates* or Michael Benton's *The Fossil Record II*—followed a nearly identical structure and format to these older catalogs (and indeed, both Benton and Sepkoski have referred to such earlier paper collections as "databases").[21]

Some early tabular catalogs might have looked superficially like modern databases, but the important question is whether these compendia and catalogs were used in the same way that scientists later used databases. Databases are, at root, tools used for the production of knowledge, and what distinguishes modern electronic databases from other types of collections are the kinds of uses to which they can be put—namely, particular kinds of recontextualizations, abstractions, and quantitative analyses. Some authors who have argued against a continuity between pre- and postcomputer data practices have touched on this very issue of use: Stevens, for example, has argued that while many nineteenth-century naturalists "dealt with large numbers of object and paper records," calling such material "data" is anachronistic because it is only in the modern (i.e., electronic) sense that data is understood as "symbols on which operations are performed."[22] However, even if one accepts Stevens's narrow modern definition of data,

---

[19] John Woodward, *An Attempt Towards a Natural History of the Fossils of England* (London, 1729); Adolphe Brongniart, *Prodrome d'une histoire des végétaux fossilles* (Paris, 1828). The *Treatise on Invertebrate Paleontology* is a definitive compendium of taxonomic invertebrate fossils begun during the 1950s by Raymond C. Moore. It has been continuously expanded and updated ever since, involving the contributions of several hundred paleontologists and spanning more than fifty volumes, which have been published in print by the University of Kansas and, more recently, as an electronic resource. On the epistemic and communal functions of such compendia, see Sepkoski, "'Natural History of Data'" (cit. n. 7).

[20] Samuel Woodward, *A Synoptical Table of British Organic Remains* (London, 1830). The term "metadata" is, in this case, an anachronism, since it was not introduced until the 1960s. However, according to the most basic definition, metadata are simply identifiers or descriptors—"data about data"—that help organize data structures. The term, then, describes a practice much older than electronic computers, and I use the anachronism for simplicity's sake.

[21] Michael Benton, "The History of Life: Large Databases in Palaeontology," in *Numerical Paleobiology*, ed. D. A. T. Harper (New York, 1999), 249–83; J. John Sepkoski Jr., "Ten Years in the Library: New Data Confirm Paleontological Patterns," *Paleobiology* 19 (1993): 43–51.

[22] Stevens, *Life Out of Sequence* (cit. n. 4), 242n46.

the rest of his claim is not borne out. One reason for this, as I will show in the next section, is that some nineteenth-century paleontologists used printed compendia of data in almost precisely the same way that twentieth-century paleontologists used electronic databases—that is, as repositories of abstracted, commensurated data to be quantitatively analyzed in order to produce general arguments about the history of life.[23] According to Stevens, a crucial characteristic that distinguishes bioinformatics from "pre-informatic biology" is that informatic practices "allow biologists to pose and answer *general* questions" and to employ "statistical approaches." I argue that nineteenth-century paleontologists did precisely this. But I also draw on support in this claim from many of the scientists most closely associated with the emergence of computerized data analysis in late twentieth-century paleontology. Jack Sepkoski, who compiled the first electronic fossil database and pioneered statistical analysis of the data from the late 1970s onward, defined taxonomic databases as "compilations of taxa of various ranks, relationships, and regions accompanied by estimates of their times of first appearances and final disappearances in the fossil record" and noted that "from Phillips (1860) onward, such data bases have been used to investigate changes in diversity, biotic composition, and origination and extinction through life's history."[24] This suggests that what—in paleontology, at least—came to be known as a "database" was so strongly constrained by earlier data collections that no meaningful distinction exists between data collections of 150 years ago and today.

H. G. BRONN'S PAPER DATA COLLECTION

To test my argument about the homology between nineteenth-century paper data collections and later electronic databases, we need to examine a case in detail to see what a nineteenth-century paleontologist actually did with his data. The example I will use is the German naturalist H. G. Bronn, who between the 1830s and his death in the early 1860s devoted his career to systematic compilation and analysis of global fossil data. Numerical approaches to studying the distribution of plant, animal, and fossil species were fairly common in early nineteenth-century natural history; this kind of analysis, popular in the circle around Alexander von Humboldt, has been described by Janet Browne as "botanical arithmetic."[25] This approach applied fairly elementary statistics (e.g., computing ratios of taxonomic groups in a particular region), and results were often displayed in tabular format. Browne argues, however, that such attempts were basically exercises in numeracy for numeracy's sake and did little to advance understanding of biogeography, since they "were rarely used to substantiate specific hypotheses, nor did they generate any important new questions about geographical phenomena."[26]

As I have argued elsewhere, I think Browne potentially misses a deeper and more interesting strain of quantitative analysis that ran through paleontology from the 1830s through the 1850s, especially in England, France, and Germany.[27] Unlike slightly earlier counterparts in botany and zoology, paleontologists of this time did genuinely attempt to

---

[23] Ibid., 44–5.
[24] Sepkoski, "Ten Years" (cit. n. 21), 43.
[25] Janet Browne, *The Secular Ark: Studies in the History of Biogeography* (New Haven, Conn., 1983).
[26] Ibid. Martin Rudwick makes much the same point when discussing numerical analysis in paleontology during this period, e.g., in the work of H. G. Bronn and Charles Lyell. See Martin J. S. Rudwick, "Charles Lyell's Dream of a Statistical Palaeontology," *Palaeontology* 21 (1978): 225–44.
[27] Sepkoski, "'Natural History of Data'" (cit. n. 7).

answer broad questions about the history of life using numerical tools, and their data collections were the catalogs and compendia that were increasing in popularity at the time. In a broad sense, Bronn's approach was to adopt the existing practice of cataloging fossil genera and species toward a global, quantitative analysis of patterns of diversity in the history of life. This was a career-long project: his first important publication, the 1831 *Italiens Tertiär-Gebilde und deren organische Einschlüsse* (Italian Tertiary formations and their organic contents), presented a 140-page taxonomic catalog of fossils found in Italian Tertiary rocks (now known to be roughly 66 to 2.5 million years old), compiled from a bibliographic survey of works by more than forty authors.[28] Appended to this catalog were twenty-six foldout tables of numerical data and fifty pages of analysis in which Bronn computed a variety of ratios and relationships: relative proportions of species and genera found in the Italian Tertiary, ratios of still living versus extinct groups in this period, relative "richness" of species at particular times (i.e., diversity of species within individual genera), and changes in these relationships over time. In his last major work, the 1858 *Untersuchungen über die Entwicklungsgesetze der organischen Welt* (Investigation of the laws of development in the organic world), Bronn gave a broad theoretical account of the history of life, including hundreds of pages of tables, graphs, and numerical analysis based on a massive data collection of the known global fossil record.[29]

Bronn's data collection was the massive *Index Palaeontologicus* he assembled during the 1840s as a comprehensive compendium of fossil data. The *Index* is divided into two parts: the first, which Bronn subtitled the "Nomenclator palaeontologicus," is a two-volume alphabetical listing of the taxonomic names of plant, invertebrate, and vertebrate fossils drawn from hundreds of bibliographic sources and spanning more than 2,000 pages.[30] The second, subtitled "Enumerator palaeontologicus," recontextualizes the taxonomic information as a "systematic compilation" of fossil taxa, arranged taxonomically in tables that display the geological periods in which fossil genera and species appear. If the first part of the *Index* (the Nomenclator) presents information about fossils as a fairly standard catalog, it is the second part—the Enumerator—that Bronn used as his actual data collection.

I propose, then, to examine the Enumerator palaeontologicus of Bronn's *Index* in some detail in order to illustrate his approach to data compilation and analysis. The standard data format in the Enumerator is depicted in figure 1, which is one of some 700 pages of such tables of data in the volume. The taxonomic entries are arranged systematically according to conventions of the day, with genera and species organized within classes and subclasses of plants and animals. This forms a list running down the left side of the page; within classes and subclasses, genera and species are arranged chronologically—that is, they appear in order of the appearance of the group in the fossil record. At the top of the page are the fields for the horizontal axis of the table. This represents the stratigraphic location of the taxa, from early to recent, and it is divided into geological period from "Kohlen" (e.g., "coal," or in modern designation the Carboniferous period, about 358 mya), through "Saltz" (Permian), "Oolithic" (Jurassic and Triassic),

[28] H. G. Bronn, *Italiens Tertiär-Gebilde und deren organische Einschlüsse* (Heidelberg, 1831).

[29] H. G. Bronn, *Untersuchungen über die Entwicklungsgesetze der organischen Welt während der Bildungs-Zeit unserer Erdoberfläche* (Stuttgart, 1858).

[30] H. G. Bronn, *Index Palaeontologicus, oder, Übersicht der bis jetzt bekannten Fossilen Organismen* (Stuttgart, 1848).

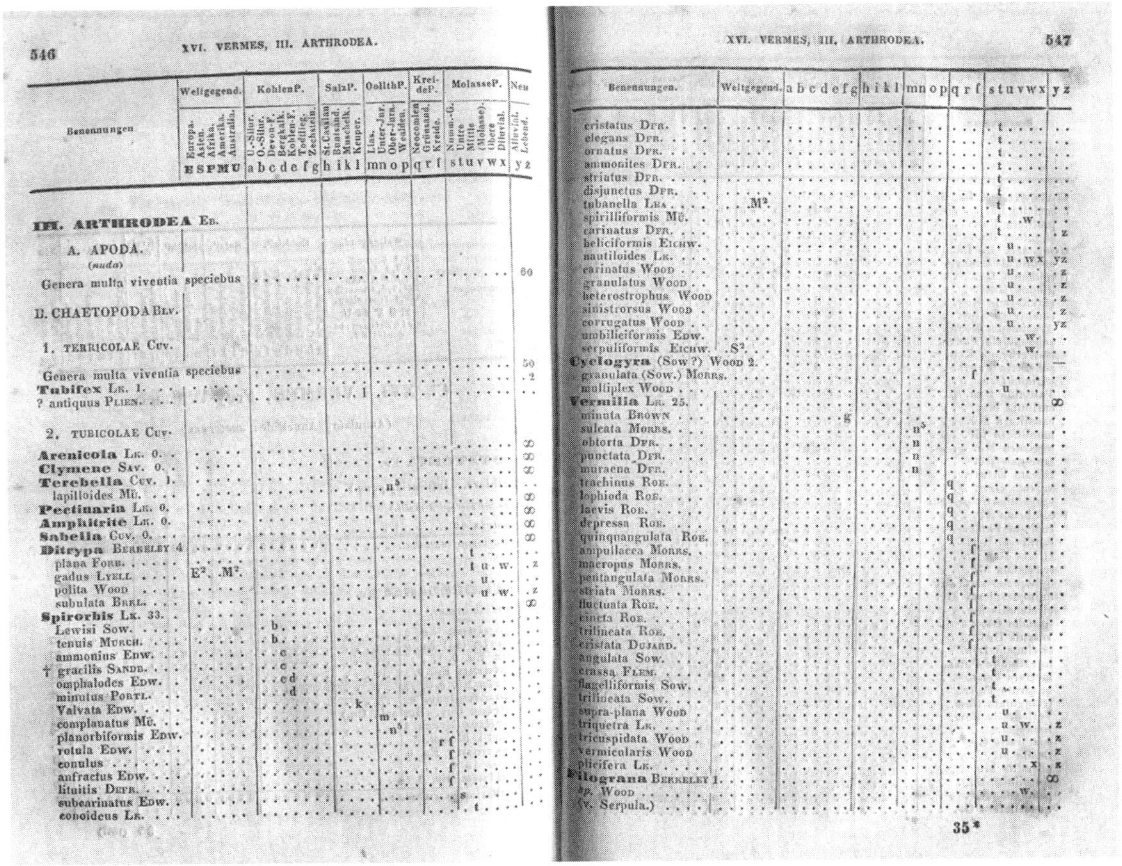

546 XVI. VERMES, III. ARTHRODEA.

Benennungen	Weltgegend. (Europa. Asien. Afrika. Amerika. Australia.) E S P M U	KohlenP. (U.-Silur. O.-Silur. Devon-F. Bergkalk. Kohlen-F. Todtlieg. Zechstein.) a b c d e f g	SalzP. (St.Cassian Buntsand. Muschelk. Keuper.) h i k l	OolithP. (Lias. Unter-Jur. Ober-Jura. Wealden.) m n o p	Krei-deP. (Neocomien Grünsand. Kreide.) q r f	MolasseP. (Numm.-G. Untre Mittle (Molasse). Obere Diluvial.) s t u v w x	Neu (Alluvial. Lebend.) y z
III. ARTHRODEA Eb.							
A. APODA. (*nuda*)							
Genera multa viventia speciebus							60
B. CHAETOPODA Blv.							
1. TERRICOLAE Cuv.							
Genera multa viventia speciebus							50
Tubifex Lk. 1.			l				2
? antiquus Plien.							
2. TUBICOLAE Cuv.							
Arenicola Lk. 0.							∞
Clymene Sav. 0.							∞
Terebella Cuv. 1.							∞
lapilloides Mü.				n^5			
Pectinaria Lk. 0.							∞
Amphitrite Lk. 0.							∞
Sabella Cuv. 0.							∞
Ditrypa Berkeley 4							∞
plana Forb.						t	
gadus Lyell	E^2 M^2					t u w	z
polita Wood						u	
subulata Brkl.						u w	z
Spirorbis Lk. 33.							∞
Lewisi Sow.		b					
tenuis Murch.		b					
ammonius Edw.		c					
† gracilis Sandb.		c					
omphalodes Edw.		c d					
minutus Portl.		d					
Valvata Edw.			k				
complanatus Mü.				m			
planorbiformis Edw.				n^5			
rotula Edw.					r f		
conulus					f		
anfractus Edw.					f		
lituitis Defr.					f		
subcarinatus Edw.						s	
conoideus Lk.						t	

XVI. VERMES, III. ARTHRODEA. 547

Benennungen.	Weltgegend.	a b c d e f g	h i k l	m n o p	q r f	s t u v w x	y z
cristatus Dfr.						t	
elegans Dfr.						t	
ornatus Dfr.						t	
ammonites Dfr.						t	
striatus Dfr.						t	
disjunctus Dfr.						t	
tubanella Lea	M^2					t	
spirilliformis Mü.						t w	
carinatus Dfr.						t	z
heliciformis Eichw.						u	
nautiloides Lk.						u w x	y z
carinatus Wood						u	z
granulatus Wood						u	z
heterostrophus Wood						u	z
sinistrorsus Wood						u	z
corrugatus Wood						u	y z
umbiliciformis Edw.						w	
serpuliformis Eichw.	S^2					w	
Cyclogyra (Sow?) Wood 2.							—
granulata (Sow.) Morrs.					f		
multiplex Wood						u	
Vermilia Lk. 25.							∞
minuta Brown		g					
sulcata Morrs.				n^5			
obtorta Dfr.				n			
punctata Dfr.				n			
muraena Dfr.				n			
trachinus Roe.					q		
lophioda Roe.					q		
laevis Roe.					q		
depressa Roe.					q		
quinquangulata Roe.					q		
ampullacea Morrs.					f		
macropus Morrs.					f		
pentangulata Morrs.					f		
striata Morrs.					f		
fluctuata Roe.					f		
cincta Roe.					f		
trilineata Roe.					f		
cristata Dujard.					f		
angulata Sow.						t	
crassa Flem.						t	
flagelliformis Sow.						t	
trilineata Sow.						t	
supra-plana Wood						u	
triquetra Lk.						u w	z
tricuspidata Wood						u	z
vermicularis Wood						u	z
plicifera Lk.						x	z
Filograna Berkeley 1.							∞
sp. Wood						w	
(v. Serpula.)							

35*

Figure 1. *A data table from H. G. Bronn's* Index Palaeontologicus *(cit. n. 30), 546–7.*

"Kreide" (Cretaceous), and so on. Within the periods, Bronn has designated subperiods with lowercase letters, which loosely correspond to epochs and ages in the modern geological timescale (although in Bronn's usage they refer more to the particular formation of which the fossils are characteristic). Finally, Bronn also noted with the uppercase letters E, S, F, M, and U whether the fossils are found in Europe, Asia, Africa, the Americas, or Australia, respectively—this is indeed a global compendium. Reading down the table, then, a user could locate a particular species within a class or subclass and then learn when the species appears in the fossil record (Bronn helpfully marks this with the lowercase letter corresponding to the geological subperiod so the reader does not have to trace up the page).

Why did Bronn arrange his data this way? In the first place, it gives an at-a-glance overview of the number of fossil species that make up a particular genus, which is a very general indication of the diversity of the genus. Similarly, it offers a snapshot of the temporal distribution of species; one can see, for example, that the earliest species of genus *Spirorbis* (a type of polychaete worm) appeared during the Lower Silurian of the "Coal" period ("b" on Bronn's timescale) and was replaced by a succession of related forms spaced fairly evenly across subperiods through the Upper "Molasse," when a proliferation of spirorbids appeared, including some species with living representatives ("y" and "z"). As useful as this format is—and in many ways, it is virtually identical to the much later format used in Sepkoski's early electronic database—for Bronn this was just the beginning. Following more than 700 pages of these taxonomic tables, the Nomenclator continues with another dozen numerical tables under the heading "Compilation of the Previous Sums," which include tables summarizing a variety of relationships in the taxonomic data. These tables include the absolute numbers of species and genera of plants and animals, which are the "raw" data compiled by hand from the entire compendium. But Bronn did not stop there, including several additional tables computing relationships such as the ratio of fossil to living genera, representative proportions of genera and species within each period and subperiod, and the relative geographical distribution of genera and species (fig. 2).

These tables are where the real action is, and they show us what Bronn was after with his quantitative approach to fossil data. Far from approaching this as an arbitrary exercise in mere *Tabellenstatistik*, Bronn converted his massive taxonomic catalog to a paper data collection in order to try to answer broad, theoretical questions about the history of life. In previous decades paleontologists had made the qualitative observation that certain geological periods seem to have been populated by distinctive plants and animals, and that there appears to be some kind of regular succession of related forms as one traces the fossil record through the stratigraphic column. But Bronn was not satisfied with qualitative observation. He wanted to base paleontology on rigorous, quantitative analysis that could potentially reveal rules or laws of distribution and development; this is why the subtitle of the Nomenclator is "Systematic Compilation and Geological Laws of Development in the Organic Realm." The quantitative tables of ratios found in the *Index* (and also in the earlier *Italiens Tertiär-Gebilde*) are an intermediate step: regularities and patterns that are hinted at in the data collection come into sharper focus, successively, as the tables progress. The tables of genera and species numbers are only partially "cooked"—they still contain absolute numbers, which do not yet fully express the kinds of relationships and regularities he was seeking. One problem was that Bronn did not know how complete his data were: because of sampling bias (disproportionate collecting in certain areas and stratigraphic intervals),

III. TABELLE: VERHÄLTNISSE DER FOSSILEN GENERA ZU DEN LEBENDEN.

(Lebende Genera, welche eine Periode überspringen, sind in dieser nicht mit gezählt; sonst würde ihre Zahl grösser ausfallen; aber auch die überspringenden fossilen Genera würden gezählt werden müssen. — Viele Genera sind eigentlich noch ..., aber im Fossil-Zustande durch andere Namen angedeutet bei Pflanzen, Krustern etc.

Perioden: Zahl der darin vorkommenden Genera:	I. Kohlen-P. aller	der lebenden absolut	Quote	II. Trias-P. aller	der lebenden absolut	Quote	III. Oolith-P. aller	der lebenden absolut	Quote	IV. Kreide-P. aller	der lebenden absolut	Quote	V. Molassen-P. aller	der lebenden absolut	Quote	I–V. Periode aller	der lebenden absolut	Quote	VI. jetzige Per. aller lebenden	Quote d. fossilen davon
PLANTAE.																				
Cellulares	8	0	0	2	0	0	18	0	0	12	0	0	21	4	0,19	38	4	0,10	718	0,005
Vasculares	116	0	0	37	0	0	37	0	0	24	0	0	168	56	0,33	312	56	0,18	5811	0,010
Monocotyledones	101	0	0	27	0	0	29	0	0	9	0	0	27	5	0,19	152	5	0,03	1172	0,004
A. Cryptogamae	84	0	0	22	0	0	24	0	0	4	0	0	7	1	0,14	105	1	0,01	89	0,011
B. Phanerogamae	17	0	0	5	0	0	5	0	0	5	0	0	20	4	0,20	47	4	0,09	1083	0,004
Dicotyledones	15	0	0	10	0	0	18	0	0	15	0	0	141	51	0,36	160	51	0,33	4639	0,001
A. Monochlamydae	12	0	0	9	0	0	15	0	0	9	0	0	57	17	0,30	70	17	0,24	300	0,057
B. Corolliflorae	—	0	0	—	—	—	—	—	—	—	0	0	13	6	0,46	14	6	0,43	2280	0,003
C. Choristopetalae	1	0	0	1	0	0	1	0	0	3	0	0	57	28	0,49	59	28	0,48	2059	0,013
D. Dubiae	2	0	0	2	—	—	2	—	—	2	—	—	14	—	—	17	—	—	—	—
Summa	124	0	0	39	0	0	74	0	0	36	0	0	189	60	0,32	350	60	0,17	6529	0,009
ANIMALIA.																				
I. PHYTOZOA	146	37	0,25	34	17	0,50	125	69	0,55	199	111	0,55	307	215	0,70	524	242	0,48	652	0,37
I. Pseudozoa	—	—	—	—	—	—	—	—	—	(1)	1	1,00	(1)	1	1,00	(2)	2	1,00	13	0,15
II. Amorphozoa	11	3	0,27	7	4	0,59	10	6	0,60	26	9	0,35	17	12	0,76	42	15	0,32	15	1,00
III. Polygastrica	(1)	1	1,00	(1)	1	1,00	—	—	—	7	4	0,57	80	68	0,85	84	69	0,82	168	0,41
IV. Polypi	82	30	0,38	16	8	0,50	70	49	0,70	105	77	0,73	164	113	0,68	251	138	0,55	245	0,56
A. Polythalami	7	4	0,57	7	4	0,57	15	15	1,00	38	31	0,82	67	55	0,82	81	59	0,73	77	0,75
B. Bryozoa	38	11	0,30	7	2	0,28	24	11	0,46	40	22	0,55	56	27	0,48	97	33	0,34	75	0,44
C. Anthozoa	37	15	0,40	9	6	0,67	32	23	0,72	29	24	0,89	41	31	0,76	73	46	0,63	93	0,50
V. Entozoa	—	—	—	—	—	—	—	—	—	1	0	0	3	0	0	3	0	0	60	0,00
VI. Acalephae	52	3	0,06	11	5	0,45	45	14	0,31	59	20	0,34	42	21	0,50	142	28	0,20	75	0,00
VII. Echinodermata	52	3	0,06	9	3	0,33	15	5	0,24	15	6	0,40	6	5	0,83	77	8	0,10	76	0,37
Stelleridae	50	1	0,02	4	1	0,25	11	2	0,07	15	—	0,28	3	2	0,67	63	2	0,03	36	0,22
Crinoidea	14	1	0,02	3	1	0,33	4	2	0,50	1	—	1,00	1	1	1,00	5	2	0,40	4	0,50
Ophiulidae	1	1	1,00	1	—	0,67	3	2	0,67	7	3	0,43	1	1	1,00	9	4	0,44	14	0,14
Asteriadae	1	1	1,00	2	1	0,50	2	—	0,67	2	2	1,00	2	2	1,00	62	18	0,29	18	0,22
Echinidae	—	—	—	(2)	2	1,00	12	7	0,34	44	14	0,32	35	16	0,46	3	2	0,67	29	0,62
Fistulidae	—	—	—	—	—	—	(2)	2	1,00	—	—	—	1	0	0	—	—	—	11	0,18

Figure 2. *An analytical table from Bronn's Index showing the proportion of living versus extinct genera of plants and animals by geological period (Bronn, Index Palaeontologicus [cit. n. 30], 738–9).*

absolute numbers of genera or species give only a partial picture of changes in diversity over time. This is why he then computed various ratios of, for example, the average number of species per genus for all of the groups included in his database, arranged by geological period (fig. 3). This table allows both comparisons of relative diversity between groups (*Animalia*, say, tend to have more diverse genera than *Vegetabilia*), and also within groups over time (particular groups have higher "clade diversity" at certain periods than at others).

But even here Bronn was not finished. One feature we can observe in Bronn's approach is a successive streamlining of his presentation of the analysis of his data. In the first instance, even in the "raw" tabular data collection, it is possible for an experienced eye to pick out general patterns—such as the succession of species within genera over time mentioned above. However, in Bronn's era, most readers would not have been familiar or comfortable with this form of data presentation, so he also provided the intermediate steps showing how he computed his ratios and averages. But even the data tables would not tell the whole story, so as a final step Bronn utilized a number of innovative diagrams to provide a quantitative visual summary of his analysis. For instance, in the final section of the *Index* and also in the more elaborate *Untersuchungen*, Bronn introduced what are now called "spindle diagrams," which are horizontal bar charts representing the relative diversity of genera or classes by the thickness of a horizontal line spanning the longevity of that group in geological time (fig. 4).

In the example presented here, one can see how genera of marine invertebrates begin with few members, expand over time to reach what he calls a "culmination point" of maximal diversity, and then slowly wane (except in the case of groups that have not yet become extinct). When placed together, the individual spindles also allow a quick visual summary of the relative dominance of particular groups at particular times; here we see, for example, that the Spirifer had their heyday early on, while other groups emerged later and diversified slowly over time. In other diagrams, Bronn visually summarized the duration ("range data") of higher taxonomic groups, along with calculations of their absolute diversity over time. Two points are important here. First, these visual summaries are the product of considerable data refinement—"cooking"—that converts the messy information in his data collection and the numerical tables into a format where the quantitative arguments can be seen "at a glance." Second, this approach—the visual style, the underlying methodology, and the epistemic goal—is virtually identical to the practice of paleontologists in the 1980s working with the first electronic databases. What paleontologists are now able to do quickly with the aid of computers Bronn did painstakingly by hand, but the end product—graphical representations of changes in taxonomic diversity over time that stand in for the underlying numerical data—is essentially the same.

Bronn's approach was in fact a fairly widespread practice among his contemporaries, in Germany as well as in France and Britain. The French paleontologist Alcide d'Orbigny, for example, compiled elaborate range diagrams that were similar to Bronn's in a companion atlas to his 1849 *Cours élémentaire de paléontologie et de géologie stratigraphiques*, and the British geologists Charles Lyell and John Phillips attempted broad quantitative summaries of historical changes in diversification based on analysis of fossil data.[31] Of these other attempts, Phillips's was perhaps the most noteworthy: in

---

[31] See Sepkoski, "'Natural History of Data'" (cit. n. 7); Rudwick, "Lyell's Dream" (cit. n. 26).

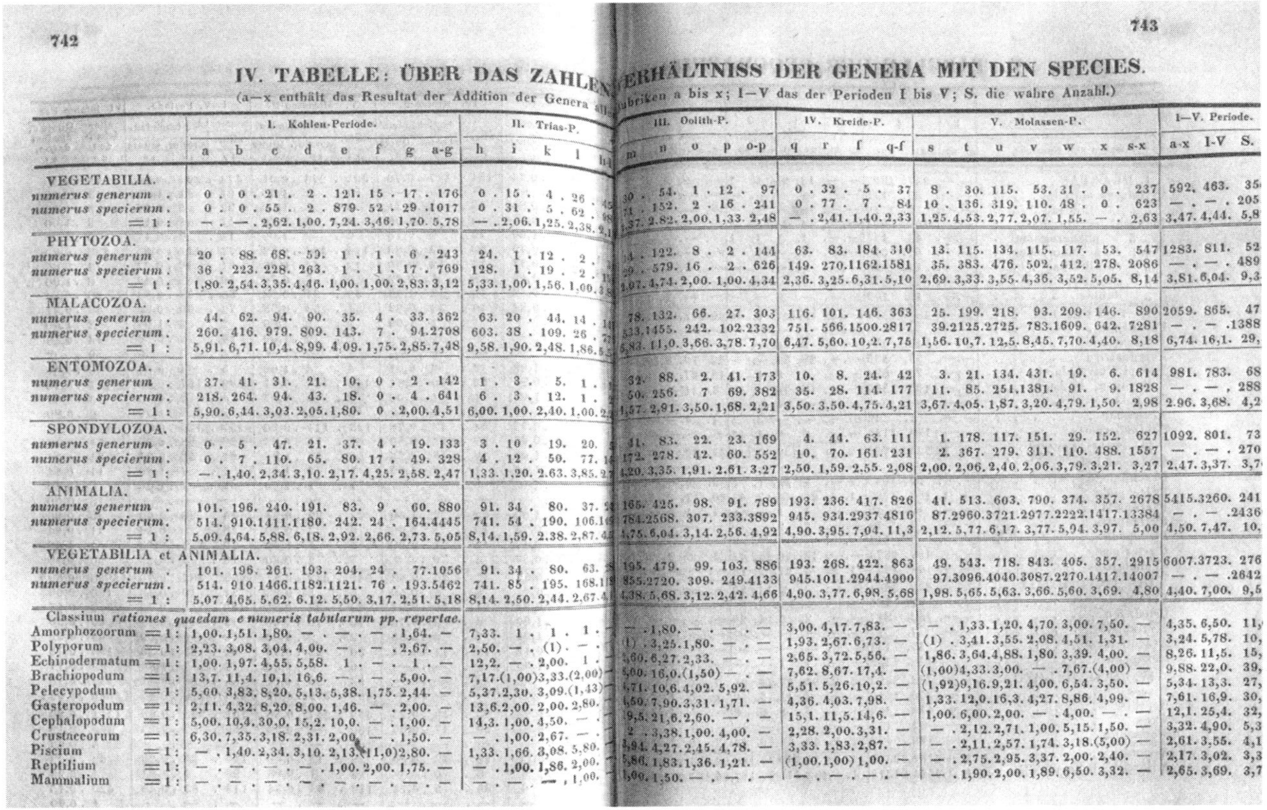
742 743

IV. TABELLE: ÜBER DAS ZAHLENVERHÄLTNISS DER GENERA MIT DEN SPECIES.

(a—x enthält das Resultat der Addition der Genera aller Rubriken a bis x; I—V das der Perioden I bis V; S. die wahre Anzahl.)

I. Kohlen-Periode. II. Trias-P. III. Oolith-P. IV. Kreide-P. V. Molassen-P. I—V. Periode.

VEGETABILIA.
numerus generum
numerus specierum
= 1 :
PHYTOZOA.
MALACOZOA.
ENTOMOZOA.
SPONDYLOZOA.
ANIMALIA.
VEGETABILIA et ANIMALIA.
Classium rationes quaedam e numeris tabularum pp. repertae.
Amorphozoorum
Polyporum
Echinodermatum
Brachiopodum
Pelecypodum
Gasteropodum
Cephalopodum
Crustaceorum
Piscium
Reptilium
Mammalium

Figure 3. Bronn's calculation of the number of species and genera within major taxonomic groups, again arranged in tabular format by geological period (Bronn, Index Palaeontologicus *[cit. n. 30], 742–3).*

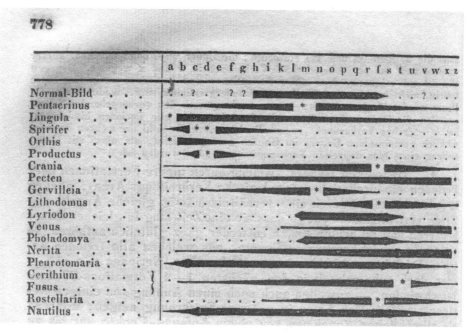

778

Figure 4. One of Bronn's first "spindle diagrams," depicting the diversification (and extinction) of groups of marine organisms. This figure is also keyed to the geological periods (alphabetically at the top) found in his tabular diagrams (Bronn, Index Palaeontologicus [cit. n. 30], 778).

his short book *Life on the Earth* (1860), Phillips analyzed John Morris's *Catalogue of English Fossils* to produce a variety of visual summaries, including spindle diagrams and a line graph depicting successive changes in diversity that illustrates his argument that geological eras are populated by "characteristic" fauna that "begins at a minimum, rises to a maximum, and dies away to a final minimum, to be followed by another system having similar phases" (fig. 5).[32] Phillips's 1860 graph is almost identical to a similar one produced by Jack Sepkoski depicting the same phenomenon in the late 1970s, a fact that was not lost on Sepkoski (who, recall, happily considered Phillips's resource a "data base").

The emergence of these distinctive tables and visual summaries of data arguments in the nineteenth century points to a feature of data-driven science that Staffan Müller-Wille and Isabelle Charmantier have observed in the context of earlier, Linnaean information management: the production of "new entities and relationships" that result from the ontological commitments implied by the "infrastructure" of a data-driven research program.[33] The innovative visual analytic summaries presented by Bronn, Phillips, and others descended from the numerical tables used by naturalists like Bronn to identify patterns and relationships in their "raw" data. It seems likely, in Bronn's case, that this approach was strongly influenced by the early nineteenth-century tradition of

[32] John Phillips, *Life on the Earth; Its Origin and Succession* (Cambridge, 1860), 64.
[33] Staffan Müller-Wille and Isabelle Charmantier, "Natural History and Information Overload: The Case of Linnaeus," *Stud. Hist. Phil. Biol. Biomed. Sci.* 43 (2012): 4–15.

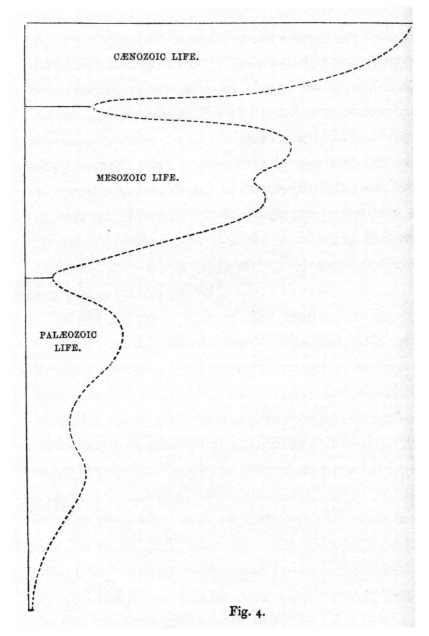

CÆNOZOIC LIFE.

MESOZOIC LIFE.

PALÆOZOIC
LIFE.

Fig. 4.

Figure 5. *John Phillips's line graph illustrating successive waves of diversification of life over time. Note the vertical orientation (life begins at the bottom and flows toward the present at the top): this orientation is the standard geological perspective, following the vertical orientation of stratigraphic diagrams since the time of Cuvier. In contrast, Bronn's horizontal diagrams were quite unusual for their time—and are an indication of the influence of the horizontal tabular structure of his data tables on his pictorial representations. Phillips,* Life on the Earth *(cit. n. 32), 64.*

cameralism or *Staatswirtschaft* in which he initially trained, which similarly adapted tables (which were themselves relatively new epistemic objects) to the analysis of hidden ratios and relationships in social and economic data.[34] According to Desrosières, the form of the table introduced a new epistemic convention in which the "diversity" of heterogeneous entities (i.e., states) could be taken in "at a glance," which he considers a new form of "graphic reasoning."[35] Bronn's approach essentially applied this idea to the history of life and in doing so introduced another new epistemic object—in this case, the concept of "diversification" as the fluctuating quantity of lower taxonomic groups within higher ones (e.g., species within genera or genera within families), and their relative durations. That these relationships (and their accompanying visual representations) would remain central objects of paleontological analysis for more than a hundred years underlines the considerable continuity between pre- and postcomputer approaches to data in paleontology. The fact that Bronn and others may have incorporated their paleontological methods from a preexisting tradition in quantitative state administration suggests continuities in the other direction as well: Bronn's method did not emerge from nowhere—it is continuous with other practices and traditions of expertise and is suggestive not only of temporal continuities in data practices, but also of continuities between the natural and human sciences that were deep and ongoing.

THE DATABASE AFTER THE COMPUTER

I hope that I have succeeded in at least demonstrating that nineteenth-century data practices in paleontology were sophisticated enough to warrant the name. What remains to be shown, though, is that those earlier practices differed little enough from postelectronic approaches that they may be considered "database analysis." And even if my broad continuity arguments about data collections are accepted, I still want to draw attention to the very real impact that computers had on paleontology, particularly from the perspective of disciplinary identity and organization. In Edwards's terms, there was very considerable computational and data friction involved in analyzing paper databases; nowhere is this more obvious than in the succession of formats in which Bronn presented his data in the *Index Palaeontologicus*, each of which involved dozens if not hundreds of additional printed pages, to say nothing of the massive expenditure of time and effort required to generate them by hand. Perhaps in part because of the great amount of labor required, Bronn's and Phillips's approach to broad-scale analysis of fossil data did not become mainstream practice in paleontology between the 1870s and the 1940s.[36] Computers do ease data and computational friction, and it is undeniable that the eventual introduction of computers to paleontology greatly increased the

[34] Bronn's formal training was in cameralism at the University of Heidelberg, and his initial appointment at the same university in 1821 was in "applied natural history," where he taught courses on forestry, mining, agriculture, and other "applied" topics in addition to paleontology and zoology. See Sepkoski, "'Natural History of Data'" (cit. n. 7); David Sepkoski and Marco Tamborini, "'An Image of Science': Cameralism, Statistics, and the Visual Language of Natural History in the Nineteenth Century," *Hist. Stud. Nat. Sci.* (forthcoming).

[35] Alain Desrosières, *The Politics of Large Numbers: A History of Statistical Reasoning* (Cambridge, Mass., 1998), 21.

[36] For a discussion of late nineteenth-century debates about quantitative methods in German paleontology, see Marco Tamborini, "Paleontology and Darwin's Theory of Evolution: The Subversive Role of Statistics at the End of the 19th Century," *J. Hist. Biol.* 48 (2015): 575–612.

ease and attractiveness of such methods. However, the impact of computers on data practices in paleontology was neither immediate nor obvious.

Like most of the natural sciences, paleontology entered the computer age in the late 1950s and early 1960s, thanks to the wider availability of mainframe electronic digital computers on university campuses in the United States during the post-*Sputnik* era. This period did coincide with a flourishing of data-intensive approaches in paleontology, particularly in the United States and exemplified by the vertebrate paleontologist G. G. Simpson and the invertebrate specialist Norman Newell. However, while the work of paleontologists like Newell and Simpson had a profound effect on the subsequent development of quantitative analysis in paleobiology, few paleontologists used computers in any significant way before the 1970s, nor did computers play a necessary role in the kinds of quantitative analysis that most paleontologists favored. In fact, the approach taken to the storage and analysis of data taken by paleontologists in the postcomputer era—eventually with the assistance of computers—did not differ greatly either methodologically or epistemically from precomputer approaches, at least in the first decade or two of paleontology's involvement with electronic computing.

An example was Newell's pioneering quantitative study of "Periodicity in Invertebrate Evolution," published in 1952, an attempt to quantify patterns of successive waves of evolutionary diversification of invertebrates in the fossil record and to illustrate them using simple graphs. Newell drew his data from existing published sources, as well as from a number of colleagues whom he thanked for generously contributing private "lists of raw data" on various groups for use in his analysis.[37] Despite the availability of more sophisticated statistical tools to assess rates of evolution from fossil data, he observed that "by far the most practical is a simple measurement of rate of differentiation of genera."[38] The graphs that accompanied the paper showed both the average total number of genera plus the rate of appearance of new genera over time, and also the percentage increase of genera during the same intervals. These are very simple calculations, obtained by counting up the first and last appearance of marine invertebrate genera in the fossil record, and are quite similar to those produced by Bronn, although instead of summarizing his analysis using spindle diagrams, Newell employed line graphs (similar to Phillips). Newell did not use a computer to perform this study, nor was there any technical aspect that could not have been accomplished with (and indeed was not anticipated by) nineteenth-century methods. In a 1959 paper addressing the "adequacy" of data in the fossil record, Newell drew attention to "the recent application of electronic IBM computers," which he noted meant that "we may have at our disposal the means for more or less routine quantitative solution of all sorts of paleontological problems involving complex interrelationships of many variables."[39] However, while he explicitly viewed the computer as a powerful calculating machine, he made no reference to its use in the collection and storage of data.

Indeed, probably the greatest innovation to paleontological data analysis in the 1960s and 1970s was not tied to the emergence of any particular technology but rather involved the incorporation of sophisticated multivariate statistics for analyzing fossil samples and populations. Improved statistical techniques (such as regression, analysis

---

[37] Norman D. Newell, "Periodicity in Invertebrate Evolution," *J. Paleontol.* 26 (1952): 371–85, on 371.
[38] Ibid., 373.
[39] Norman D. Newell, "Adequacy of the Fossil Record," *J. Paleontol.* 33 (1959): 488–99, on 490.

of variance, and rarefaction) gave paleontologists greater confidence in the analysis of fossil data (which were notoriously incomplete) and helped popularize data-intensive approaches within the discipline. There is no doubt that the introduction of computers made these kinds of studies easier and more attractive to perform—and hence more popular—but this development is quite consistent with Agar's model in which computers are generally introduced to assist with problems and techniques that are already tractable.[40]

Generally speaking, paleontologists were rather slow adopters of computers in part because initially computers did not appear to offer obvious advantages over traditional methods of data storage and analysis. Eventually, however, paleontologists did begin to imagine the advantages of compiling an electronic collection of fossil data. The first major attempt at such a project was the result of a collaboration between the Geological Society of London and the British Palaeontological Society, which resulted in a publication in 1967 titled simply *The Fossil Record*. In a short history of databases in paleontology, paleontologist Michael Benton describes *The Fossil Record* as "the first published comprehensive data base designed specifically for studies of the nature of the history of life."[41] I would argue that Bronn's work (of which Benton was apparently unaware) fits this description as well, but it is certainly the case that *The Fossil Record* can be considered one of the first modern published global fossil data collections. And yet *The Fossil Record* is not an electronic database—it is a print volume. The introduction to the volume explains that "in planning so large an assembly of data, the Editorial Sub-Committee had in mind the possibility of processing the data by computer," but this was limited to analysis of data extracted by hand from the printed collection and did not involve compiling a true electronic database.[42] Nor was this printed database truly comprehensive: the editors described it as "an abstract" of published documentation of fossil groups, and the taxonomic resolution was generally at the level of order or suborder, meaning that the included data was quite general and fairly sparse. The presentation of the data was fairly simple: each taxonomic listing provides the first and last appearance of the taxon in the fossil record, along with the bibliographic reference(s) to descriptive literature. This format is essentially identical to that used in Bronn's Nomenclator section of the *Index Palaeontologicus* (with the distinction that Bronn did not provide the last instance). Chapters in *The Fossil Record* are arranged by class, and for each chapter a chart is provided plotting the duration of the included taxa, where "each taxon selected is represented by one vertical line only, designed to show in visual form the total known stratigraphic range as documented in the text."[43] This is exactly the format used by Bronn to visually summarize the same information.

While *The Fossil Record* was not an electronic database, it did gesture toward the use of computers for fossil analysis. The final chapter of the volume is a paper titled "Numerical Analysis of *The Fossil Record*," which describes the manual entry of the data onto magnetic tape for numerical analysis on a mainframe computer. The results are a series of line graphs showing changes in diversity over time. The authors discussed the

[40] A detailed history of this era of the introduction of computers to paleontology is presented in Sepkoski, *Rereading the Fossil Record* (cit. n. 7).

[41] Benton, "History of Life" (cit. n. 21), 256.

[42] W. B. Harland, C. H. Holland, M. R. House, N. F. Hughes, A. B. Reynolds, M. J. S. Rudwick, G. E. Satterthwaite, L. B. H. Tarlo, and E. C. Willey, eds., *The Fossil Record* (London, 1967), 2.

[43] Ibid., 138.

utility of this method in positive terms, but it should be stressed that their data entry was only for the purpose of analysis—that is, it was not a "translation" of *The Fossil Record* onto an electronic database—and also that the authors noted that the database itself "can in no sense be regarded as a complete documentation of the fossil record."[44] Within a decade, however, paleontologists were discussing producing a much more complete database of marine invertebrate fossils, based on the range of existing catalogs, compendia, and monographic literature amassed since the early nineteenth century. In the early 1970s, Stephen Jay Gould and a group of colleagues began using computers to simulate hypothetical evolutionary lineages and wanted to test these simulations against patterns in the actual fossil record.[45] Since existing data collections were either unwieldy (like the *Treatise on Invertebrate Paleontology*) or incomplete (like *The Fossil Record*), Gould asked his graduate student, Jack Sepkoski, to compile a database at the resolution of families and genera from existing compilations. Over the next several years (and after its original purpose was essentially abandoned), this project ballooned to become *A Compendium of Fossil Marine Families*, a resource that eventually came to be known simply as the "Sepkoski database" (fig. 6).[46] This was the ancestor of most subsequent global taxonomic electronic databases used in paleontology to this day.

However, the Sepkoski database was not an electronic database either. It was a print volume comprising a list of fossil marine families indicating the stratigraphic range of each taxon (first and last appearance) plus bibliographic citations. It was, essentially, an updated version of the Nomenclator section of the *Index Palaeontologicus*. In fact, like Bronn, Sepkoski initially arranged his database in systematic (taxonomic) order but was asked by his publisher to change it to alphabetical listing at the last minute. This occasioned "several nights with razor blade and tape rearranging the wretched manuscript lines and wondering why I had not typed the thing into a computer file."[47] Most likely, this was because at the time typing the data onto a computer was not a trivial task; the first fully electronic version of the database was produced around the time of publication by Sepkoski's colleague Ted Foin, who translated the data to computer tape that could be mounted on a mainframe and accessed on a remote terminal.[48] Versions of this first truly electronic database circulated privately over the next decade between Sepkoski and his colleagues, but it was not until 1992, when the second edition of the *Compendium* was published (again, as a book), that "a diskette in ASCII format (DOS) containing the data in this publication" was offered for purchase (for $25!).[49] The third and final version of the *Compendium*, which provided genera-level data and was published posthumously after Sepkoski's death in 1999, was also published as a printed book, although this version came with a free CD containing the data.[50]

    [44] J. L. Cutbill and B. M. Funnell, "Numerical Analysis of the Fossil Record," in Harland et al., *The Fossil Record* (cit. n. 42), 791–820, on 793.
    [45] Sepkoski, *Rereading the Fossil Record* (cit. n. 7), chap. 7.
    [46] J. John Sepkoski Jr., *A Compendium of Fossil Marine Families* (Milwaukee, 1982). On the history of this database, see J. John Sepkoski Jr., "What I Did with My Research Career; or How Research on Biodiversity Yielded Data on Extinction," in *The Mass-Extinction Debates: How Science Works in a Crisis.*, ed. William Glen (Stanford, Calif., 1994), 132–44; and Sepkoski, *Rereading the Fossil Record* (cit. n. 7), chap. 8.
    [47] Sepkoski, "My Research Career" (cit. n. 46), 140.
    [48] Ibid.; Arnold I. Miller, e-mail communication with David Sepkoski, 29 August 2014.
    [49] J. John Sepkoski Jr., *A Compendium of Fossil Marine Animal Families*, Contributions in Biology and Geology, vol. 83 (Milwaukee, 1992).
    [50] J. John Sepkoski, *A Compendium of Marine Fossil Genera*, Bulletins of American Paleontology, no. 363 (Ithaca, N.Y., 2002).

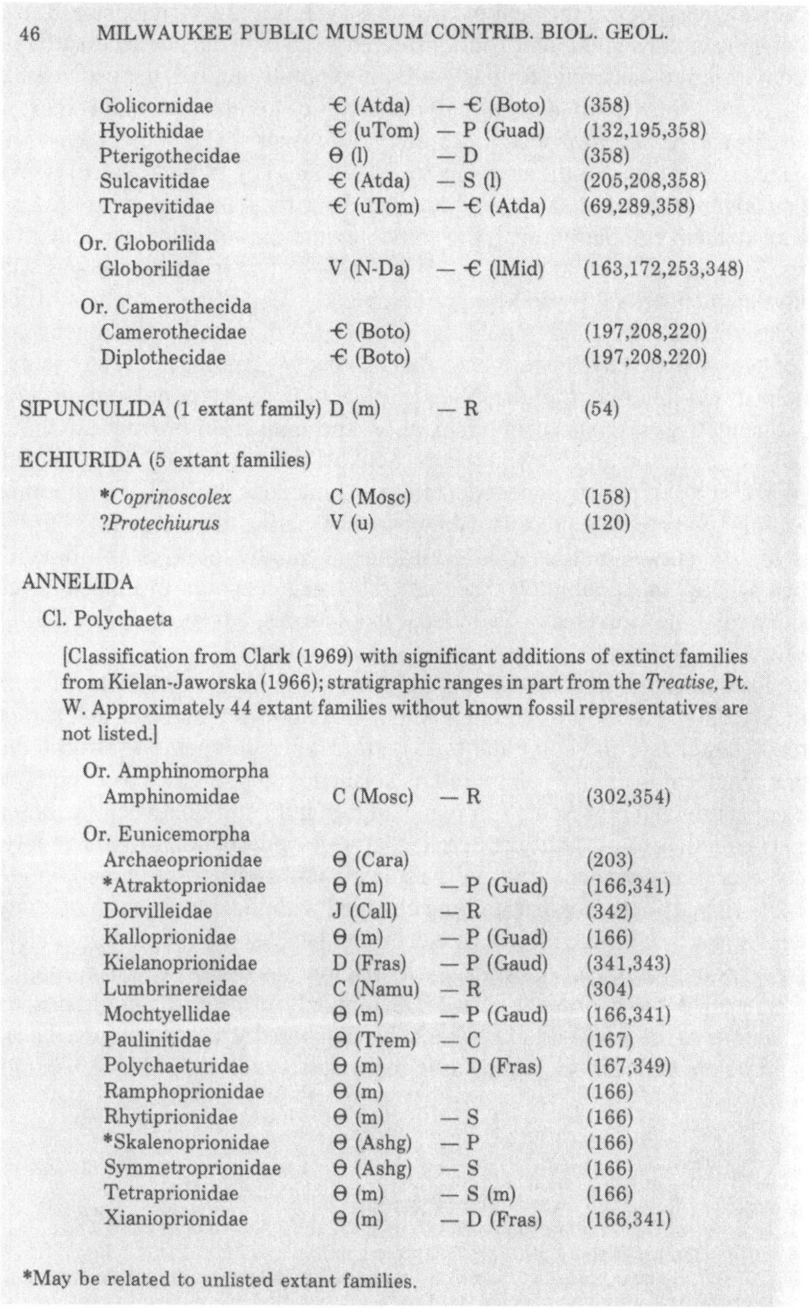
46 MILWAUKEE PUBLIC MUSEUM CONTRIB. BIOL. GEOL.

Golicornidae	Ꞓ (Atda)	— Ꞓ (Boto)	(358)
Hyolithidae	Ꞓ (uTom)	— P (Guad)	(132,195,358)
Pterigothecidae	Θ (l)	— D	(358)
Sulcavitidae	Ꞓ (Atda)	— S (l)	(205,208,358)
Trapevitidae	Ꞓ (uTom)	— Ꞓ (Atda)	(69,289,358)
Or. Globorilida			
Globorilidae	V (N-Da)	— Ꞓ (lMid)	(163,172,253,348)
Or. Camerothecida			
Camerothecidae	Ꞓ (Boto)		(197,208,220)
Diplothecidae	Ꞓ (Boto)		(197,208,220)
SIPUNCULIDA (1 extant family)	D (m)	— R	(54)
ECHIURIDA (5 extant families)			
**Coprinoscolex*	C (Mosc)		(158)
?*Protechiurus*	V (u)		(120)

ANNELIDA

Cl. Polychaeta

[Classification from Clark (1969) with significant additions of extinct families from Kielan-Jaworska (1966); stratigraphic ranges in part from the *Treatise*, Pt. W. Approximately 44 extant families without known fossil representatives are not listed.]

Or. Amphinomorpha			
Amphinomidae	C (Mosc)	— R	(302,354)
Or. Eunicemorpha			
Archaeoprionidae	Θ (Cara)		(203)
*Atraktoprionidae	Θ (m)	— P (Guad)	(166,341)
Dorvilleidae	J (Call)	— R	(342)
Kalloprionidae	Θ (m)	— P (Guad)	(166)
Kielanoprionidae	D (Fras)	— P (Gaud)	(341,343)
Lumbrinereidae	C (Namu)	— R	(304)
Mochtyellidae	Θ (m)	— P (Gaud)	(166,341)
Paulinitidae	Θ (Trem)	— C	(167)
Polychaeturidae	Θ (m)	— D (Fras)	(167,349)
Ramphoprionidae	Θ (m)		(166)
Rhytiprionidae	Θ (m)	— S	(166)
*Skalenoprionidae	Θ (Ashg)	— P	(166)
Symmetroprionidae	Θ (Ashg)	— S	(166)
Tetraprionidae	Θ (m)	— S (m)	(166)
Xianioprionidae	Θ (m)	— D (Fras)	(166,341)

*May be related to unlisted extant families.

Figure 6. *A page from Jack Sepkoski's "database" of marine fossil families. The data were published in a print volume, although an electronic version was eventually made available and circulated privately on diskette. Sepkoski,* Compendium of Fossil Marine Families *(cit. n. 46), 46.*

The point here is not that the Sepkoski database was not really a database. Rather, it is that what it meant for something to be considered a database did not necessarily depend on its conversion to electronic format. As Benton points out, it is the use to which the database is put—that is, broad quantitative analysis of fossil data—that is most important, but this can be accomplished in a variety of formats: "The book at least provides a permanent reference, but the user has to extract and key in data. [Jack] Sepkoski's disk is an advance in that it saves retyping, but disks must often be translated in some way to make them compatible with the computer and the software available to the researcher."[51] Indeed, Sepkoski spent much of 1978—when his database was still a set of handwritten entries in a spiral-bound notebook—"count[ing] numbers of families within classes and orders and punch[ing] them onto computer cards to generate long strips (of fanfold paper) of Phanerozoic 'clade diversity' diagrams."[52] The results, however, helped revolutionize the discipline of paleobiology by popularizing the use of broad, quantitative analysis of diversification and extinction beyond anything seen in Bronn's or Simpson and Newell's eras. Sepkoski's initial analyses of his database resulted in a series of classic papers depicting a "kinematic model" of evolutionary diversification: ultimately, he presented the history of marine life as a series of three overlapping logistic curves, and his work contributed centrally to the establishment of the "Chicago school" of paleobiology (so named because its major proponents were and are still based at the University of Chicago) that has pushed paleontology into the age of Big Data.[53]

There is no question that Sepkoski and later paleobiologists had access to greater quantities of data, more sophisticated mathematical tools for analysis, and technology that greatly eased data and computational friction when compared to Bronn, Phillips, and other nineteenth-century paleontologists. But not only did the basic methodology of data collection and representation not change significantly between the nineteenth and late twentieth centuries, the epistemic goals—the production of broad analyses of diversity over time represented visually as charts and graphs—remained remarkably consistent. This is not to say that nothing changed with the introduction of computers to paleontology, or with the arrival of electronic databases. For one thing, electronic databases made data extraction much easier. Before his database was computerized, Sepkoski recalled that in order to perform factor analyses he had to count originations and extinctions of the 3,500-odd families in his database by hand to build a data matrix that could be entered into the mainframe for analysis with a FORTRAN program.[54]

[51] Benton, "History of Life" (cit. n. 21).
[52] Sepkoski, "My Research Career" (cit. n. 46), 138.
[53] Sepkoski, *Rereading the Fossil Record* (cit. n. 7); J. John Sepkoski, Jr., "A Kinetic Model of Phanerozoic Taxonomic Diversity. III. Post-Paleozoic Families and Mass Extinctions," *Paleobiology* 10 (1984): 246–67. If one objects that the modest ambitions of paleontological data collection do not meet the standards of twenty-first century Big Data initiatives, it is instructive to consider the Paleo Deep Dive initiative, spearheaded by a product of the Chicago school, Shanan Peters. As the initial article describing the project explains, the goal is "a machine reading system, PaleoDeepDive, that automatically locates and extracts data from heterogeneous text, tables, and figures in publications," and furthermore which "performs comparably to humans in several complex data extraction and inference tasks and generates congruent synthetic results that describe the geological history of taxonomic diversity and genus-level rates of origination and extinction." S. Peters, C. Zhang, M. Livny, and C. Ré, "A Machine Reading System for Assembling Synthetic Paleontological Databases," *PLoS ONE* 9:12 (2014).
[54] Sepkoski, "My Research Career" (cit. n. 46), 142.

Arnie Miller, one of Sepkoski's first PhD students in the early 1980s, recalled performing similar tasks with preelectronic databases, something he called "a real pain in the ass." Conversion to electronic format streamlined all of these efforts: as Miller put it, "Digitization, including the original digitization of Jack's compendium, made it a hell of a lot easier to work with these data in different ways."[55] Dave Raup, who along with Sepkoski was one of the major figures in the Chicago school and the quantitative turn in paleobiology during the 1970s and 1980s, agrees with Miller's assessment and also points out that digitization opens the way to sharing and integrating databases.[56]

Integration and collective empiricism have been the most fruitful results of the electronic turn in paleontology databases: the Sepkoski database was eventually absorbed into a much larger electronic database project in the late 1990s that joined Sepkoski's marine invertebrate data with mammalian and terrestrial fossil taxonomic and ecosystem databases, which in the early 2000s became the web-based Paleobiology Database. The "PaleoBioDB" now archives twenty "major data sets" developed independently by individuals or groups and, according to its website, boasts some 52,811 references for 306,244 taxa, comprising 1,207,841 occurrences drawn from 162,275 collections.[57] The database describes itself as "a non-governmental, non-profit public resource for paleontological data," whose "purpose is to provide global, collection-based occurrence and taxonomic data for organisms of all geological ages, as well [as] data services to allow easy access to data for independent development of analytical tools, visualization software, and applications of all types."[58] Users can access the database remotely from anywhere in the world and are able not only to download existing data archives but also to contribute data (subject to authorization by an executive committee comprising paleontologists from a variety of institutions). In addition to accessing the data, the website allows users to automatically perform a variety of analyses of the data in the collection, including counting taxa, generating data summary tables, generating diversity curves, and analyzing abundance and taxonomic ranges. And in 2015, an article in *Nature* announced that the PaleoBioDB project would begin experimenting with automatic text recognition to "mine" paleontological data directly from empirical research articles.[59]

The great advantage of resources like the PaleoBioDB, beyond obvious issues of data access, is that many of the main techniques pioneered first by Bronn and others in the nineteenth century and later by Simpson, Newell, Raup, Sepkoski, and others in the twentieth have been almost completely automated and black-boxed. An end user need know very little about how the techniques work or where the data come from; this has dramatically increased the appeal of quantitative paleontology in recent decades. On the other hand, this has brought changes to the discipline that not every paleontologist—even the most data oriented—is entirely comfortable with. As early as 1979, when quantitative analysis was being promoted as part of a "paleobiological revolution," Raup warned his colleague Tom Schopf that "if you and I and Steve [Gould] really succeed in selling our current brand of nomothetic paleontology . . . scores of

---

[55] Miller, e-mail communication (cit. n. 48).
[56] David M. Raup, e-mail communication with David Sepkoski, 6 September 2014.
[57] Paleobiology Database, paleobiodb.org (accessed 26 September 2014).
[58] Ibid., FAQs, http://paleobiodb.org/#/faq (accessed 26 September 2014).
[59] E. Callaway, "Computers Read the Fossil Record," *Nature* 523 (2015): 115–6.

young paleontologists will be plotting survivorship curves (or whatever) in a slavish and unthinking manner."[60] More than three decades later, Raup sees his fears as having been confirmed: he told me in 2014 that "as databases grow ever larger, involving mixtures of data drawn from different disciplines, we will inevitably see a new breed of data-driven scientists, populated by people who have more invested in the technology than the content. And along with this is the risk of seeing a lot of terrible research, when the end-user has lost the ability to see exactly what is being done to what."[61] This, of course, is a problem faced by many data-driven disciplines, as the "informatics" approach has become ascendant, often at the expense of traditional modes of expertise. But the tensions introduced into the discipline by this trend are also balanced by some fairly obvious benefits, many of which were outlined by Benton, writing at the dawn of the era of web-based databases in paleontology:

> The advantages of the system are that (1) it provides an open form of scientific communication to enable testing of published hypotheses: (2) the researcher has control of the arrangement of his/ her own data sets, and can find out who is using the material; (3) data transfer is instant, with no need to send disks through the post, and it allows easy transfer between widely different computer systems and software; (4) copying errors are minimised since the data set may be cut and pasted directly from the spreadsheet in which the data were assembled during the research phase into the web site, and then to any other person's computer; (5) this is a way to make data available that journals are often reluctant to publish.[62]

While it may be the case that electronic databases have made data sharing, integration, publication, and analysis easier for paleontologists, the question remains whether they have introduced a genuinely new kind of science. From a practical perspective, the first electronic databases (e.g., the Sepkoski database, and also Benton's slightly later *The Fossil Record II*) were flat-file databases, meaning that they were essentially tables of data where each entry occupies one line and the associated metadata is repeated for each occurrence of a particular datum. Technically speaking, any such list, whether written on paper or stored in electronic form, shares the same data structure; Bronn's *Index Palaeontologicus* and Sepkoski's digitized *Compendium* are both flat files. While more recent databases like the PaleoBioDB have a more complex, relational structure for organizing the raw data, the standard practice when using such databases is to download a subset of the data as a spreadsheet (the most common kind of electronic flat-file database) for analysis. The main advantage of computerization, then, is speed and automation of data extraction and analysis, but the computer simply automates what might otherwise be done by hand.

However, digitization has certainly opened up new research possibilities that might not have been conceived or implemented prior to computers. One area is Monte Carlo simulation, which some authors have suggested produced a genuine epistemic shift when these techniques were introduced to physics in the 1940s and early 1950s. Peter Galison, for example, has argued that Monte Carlo methods in particle physics led to a conceptualization of an "alternate reality" inside the computer that allowed for sim-

---

[60] David M. Raup to Thomas J. M. Schopf, 28 January 1979, Thomas J. M. Schopf Papers, Smithsonian Institution Archives, Box 3, Folder 30.
[61] Raup, e-mail communication (cit. n. 56).
[62] Benton, "History of Life" (cit. n. 21), 279.

ulated "experiments" on entities that were not directly physically observable.[63] According to Galison, the wide popularity of Monte Carlo simulations helped produce a broad culture of "stochasticism," in which "the simulation was of a piece with the natural event," that spread into disciplines beyond physics (in fact, he directly cites work by Sepkoski and Raup in the early 1980s on random clustering of extinction events as one such example).[64] Stevens takes up this argument in the context of the development of bioinformatics, arguing that thanks to Monte Carlo methods, "The computer itself became . . . a simulation machine."[65] Putting aside the fact that other historians have questioned whether Monte Carlo simulations were a unique product of the computer age, it is definitely the case that simulations became an important tool during the 1980s and beyond for data analysis in paleontology, especially in Raup's and Sepkoski's early analyses of fossil databases.[66] I am fairly sympathetic to Galison's and Stevens's arguments, especially since a major early thrust of quantitative paleobiology during the 1970s was the development—by Gould, Schopf, and Raup—of a "stochastic view" of evolutionary processes that depended heavily on simulation experiments performed on computers.[67] On the other hand, these early stochastic explorations had little to do with actual fossil data and were not integrated into database analysis. In the early 1980s, Sepkoski made considerable use of Monte Carlo techniques when analyzing his database, but these were primarily simulations that explored what particular patterns might look like under certain assumptions that were then compared to actual data patterns. As Raup later put it, "some of these applications steered us away from databases and toward purely theoretical questions . . . such as 'What would the world (or this organism) look like if it were controlled by this or that set of conditions?'"[68] In other words, while it is clear that computers opened up powerful methods for simulation and statistical analysis that were applied to the analysis of data, it is less clear that the electronic database was the crucial development, or that data-driven science is possible only with computers.

CONCLUSION

In the end, what I hope I have shown is not that computers made no difference to the practice of paleontology, or even that electronic databases do not carry advantages over paper data collections. Rather, my point is that a model of sharp discontinuities in data practices, epistemologies, and cultures does not hold up to close historical scrutiny, at least in the case of paleontology. I suspect that this applies to other natural history disciplines as well, and perhaps even more broadly in the natural and human sciences. Technologies—like the computer—are not unimportant, but nor do they necessarily determine how approaches to data evolve to the extent that some authors have suggested. While, for example, Stevens has made a compelling case that in molecular genetics

[63] Peter Galison, *Image and Logic: A Material Culture of Microphysics* (Chicago, 1997), 691.
[64] Ibid., 743, 748.
[65] Stevens, *Life Out of Sequence* (cit. n. 4), 17.
[66] Agar, e.g., points to Enrico Fermi's hand calculations of Monte Carlo simulations in the 1930s as evidence that the method was not coextensive with the arrival of computers, while Steven Stigler has traced the roots of stochastic simulation back into the nineteenth century. Agar, "Difference" (cit. n. 11), 892; Stigler, "Stochastic Simulation in the Nineteenth Century," *Statist. Sci.* 6 (1991): 89–97, on 89.
[67] Sepkoski, *Rereading the Fossil Record* (cit. n. 7), chap. 7.
[68] Raup, e-mail communication (cit. n. 56).

"biology adapted itself to the computer, not the computer to biology," I think it can be fairly confidently stated that paleontologists did adapt computers to existing data practices, at least initially.[69] One reason for this difference—and I believe it is a legitimate difference—is that, unlike bioinformatics, paleontology is an old discipline with a fairly long-established data tradition. Bioinformatics, on the other hand, emerged simultaneously with computers, and its central practices, objects of study, and conceptual framework were much more closely related to the architecture of electronic digital computers, as Stevens has shown.

At the same time, I would urge caution in making arguments about data that reduce epistemologies and practices of data to any particular material culture. The charge that referring to a preelectronic practice using a term—"database"—explicitly coined in a recent context is anachronistic can equally be met with the reverse argument, that the true anachronism lies in the assumption—by modern practitioners of digital computing as well as by historians—that there was something necessarily special about converting information to electronic format. Technology does influence practice, and Stevens, for example, has convincingly shown that the particular architecture of early electronic digital computers influenced not just what could be done with data in bioinformatics, but also how the very entities of that discipline were conceived. Stevens's story, where function follows form (to use an analogy from evolutionary morphology), is a lesson in the path dependency of particular outcomes (the "datafication" of molecular genetics) on contingent events (the arrival of computers at a crucial moment in the conceptual understanding of genetic mechanisms). But that one case cannot automatically be extended to others. Paleontology had its own unique tradition of data practices, tied to its own contingent technological and conceptual developments: it may well be the case that the particular notion of "data" developed in nineteenth-century paleontology—as information stored on a list, counted and analyzed as a table, and summarized in particular visual styles—strongly influenced eventual electronic formats and methods used by paleontologists. Perhaps it was no accident, in other words, that the flat-file database was and continues to be the dominant and most useful database genre in paleontology, despite the availability of more sophisticated options.

Nonetheless, I will conclude with one suggestive generalization about the implications of my story. Despite their apparently exemplary status, some of the more emblematic examples of Big Data science—like molecular genetics—may actually be fairly anomalous because of their recent emergence. Paleontology, like physics, chemistry, botany, zoology, economics, and sociology, was part of the nineteenth-century professionalization of scientific disciplines, and its current practices have thus been constrained by a much longer history of data and material cultures. At the same time, even more recent cases—as Stevens has shown—reveal how the technological conditions of a particular moment can constrain later conceptual and methodological developments. More recently (in this volume), Stevens has clarified his stance, acknowledging that while "many kinds of digital electronic operations can be reproduced without computers," his goal has been to "distinguish certain kinds of computational practices that are not merely reproducing paper-based practices and that rely on new ways of thinking and working."[70] I think this is an excellent attitude to take. As other essays in this vol-

---

[69] Stevens, *Life Out of Sequence* (cit. n. 4), 41.

[70] Hallam Stevens, "A Feeling for the Algorithm: Working Knowledge and Big Data in Biology," in this volume.

ume show, there is no single unified history of data that covers all possible examples; our stories must, of necessity, be specific to individual disciplines and contexts and attentive to the different relationships between technologies, practices, and cultures that each history demonstrates. The history of technology is vitally important to the history of data, but practices around the collection, storage, analysis, and representation of data are underdetermined by the material culture context in which they have been deployed.

From Lexicostatistics to Lexomics:
Basic Vocabulary and the Study of Language Prehistory

by Judith Kaplan*

ABSTRACT

This essay analyzes the data of prehistorical work in comparative linguistics. These data typically derive from the sector of "basic vocabulary"—words thought to be especially frequent, universal, and resistant to change over time. I show how basic vocabulary data have facilitated transfer between languages, methods, and disciplinary groups. The essay focuses on the standardization of these wordlists by the anthropological linguist Morris Swadesh (1909–67). It argues that the history of basic vocabulary exemplifies "data drag" rather than data-driven change: the labor-intensiveness of wordlist compilation and calibration has only reinforced the use of basic wordlists despite foundational criticisms straddling the move to electronic computing.

> One should work with well selected materials for each language. . . . If such boiled-down material is not available, it has to be prepared.
>
> —Morris Swadesh, 1963[1]

In March 2014, the Max Planck Institute for Economics in Jena was realigned, and, by November of that year, a new Institute for the Science of Human History (Max-Planck-Institut für Menschheitsgeschichte) was up and running under the directorship of "archaeogeneticist" Johannes Krause and "evolutionist" Russell Gray.[2] Bringing specialists in such fields as genetics, bioinformatics, documentary linguistics, and phylogeography together under one roof, organizers promised from the outset a "thoroughly integrated, interdisciplinary approach." But while looking forward to new insights from "state-of-the-art" experimental and computational methods, institute members were simultaneously looking back. After some 150 years, one press release heralded, "evolutionary research returns to its origins in Germany"—to the very hometown, in fact, of evolutionary heavyweights August Schleicher, Ernst Haeckel, and Matthias

---

* University of Pennsylvania; juka@sas.upenn.edu.
[1] Morris Swadesh, "A Punchcard System of Cognate Hunting," *Int. J. Amer. Linguist.* 29 (1963): 283–8, on 284.
[2] The new institute has already undergone a name change, having first been named the MPI for the study of "history and science." The society initially announced the reorganization here: https://idw-online.de/de/news593799 (accessed 5 July 2015).

Schleiden.[3] Significantly, this homecoming brought with it a renewed willingness to conjoin linguistically and biologically defined groups: an association made not just possible, but palatable, by widespread enthusiasm for Big Data.[4]

Indeed, Gray's Department of Linguistic and Cultural Evolution is explicit about the centrality of bigness and data to its scientific mission. "Big picture questions" are its stated focus—questions to be tackled by "novel language documentation methods, global linguistic and cultural databases and analyses using evolutionary theories and computational methods." Taken together, Gray and colleagues believe, "these new computational tools" will open up "research questions that were previously deemed difficult or even completely intractable by traditional linguists."[5] Though Gray's work remains controversial in the eyes of those traditionalists, it represents a considerable and growing alliance—one that looks optimistically to computerized phylogenetic methods as a way to crack open the riddles of language prehistory.[6]

Curiosity about the origins, chronology, prehistory, and naturalistic patterning of human language has motivated Gray's career in linguistics—or, "lexomics," as he has termed it—since at least 2000.[7] In a *Nature* article published that year, he and coauthor Fiona Jordan urged historical linguists to adopt the kind of statistical methods that "revolutionized" evolutionary biology decades before.[8] For those who have followed their lead, confidence in the possibility of scientific knowledge about language prehistory has grown in tandem with the use of phylogenetic methods since the turn of the millennium.[9]

The literature in this subfield is full of declarations about the novelty of computing technologies and the practices involved.[10] Such rhetoric pervades the science of human prehistory more generally, which has heard repeated assertions of a "new synthe-

[3] For more about the Max Planck Institute for the Science of Human History, see http://www.shh.mpg.de/en (accessed 5 July 2015).

[4] On the linkage of linguistic and biological affinity, see Simon J. Greenhill, Robert Blust, and Russell D. Gray, "The Austronesian Basic Vocabulary Database: From Bioinformatics to Lexomics," *Evol. Bioinform.* 4 (2008): 271–83. Here the authors conclude that a better understanding of linguistic relationship is a worthwhile goal because it can help geneticists improve their sampling designs (280). The same article foregrounds the desirability of Big Data, suggesting that linguists ought to have their own version of GenBank (271).

[5] For the research outline of the Department of Linguistic and Cultural Evolution, see http://www.shh.mpg.de/48816/research_outline (accessed 5 July 2015).

[6] See Colin Renfrew, ed., *Time Depth in Historical Linguistics: Papers in the Prehistory of Languages*, 2 vols. (Cambridge, 2000); Peter Forster and Colin Renfrew, eds., *Phylogenetic Methods and the Prehistory of Languages* (Cambridge, 2006). Significantly, Renfrew and Forster cite Gray. For a list of phylogenetic programs, see http://evolution.genetics.washington.edu/phylip/software.html (accessed 6 July 2015).

[7] Greenhill, Blust, and Gray, "Austronesian Basic Vocabulary Database" (cit. n. 4), 273.

[8] Russell D. Gray and Fiona M. Jordan, "Language Trees Support the Express-Train Sequence of Austronesian Expansion," *Nature* 405 (2000): 1052–5. On this history, see Joel Hagen, "The Introduction of Computers into Systematic Research in the United States during the 1960s," *Stud. Hist. Phil. Biol. Biomed. Sci.* 32 (2001): 291–314. For further reference to the adoption of multivariate statistics in the life sciences during the 1960s and 1970s, see David Sepkoski, "The Database before the Computer?," in this volume.

[9] For example, see Andrew Kitchen, Christopher Ehret, Shiferaw Assefa, and Connie J. Mulligan, "Bayesian Phylogenetic Analysis of Semitic Languages Identifies an Early Bronze Age Origin of Semitic in the Near East," *Proc. Roy. Soc. B* 276 (2009): 2703–10.

[10] A good example is Stephen C. Levinson and Russell D. Gray, "Tools from Evolutionary Biology Shed New Light on the Diversification of Languages," *Trends Cognit. Sci.* 16 (2012): 167–73.

sis" since the 1950s.[11] Black-boxed in these accounts and decidedly not new, however, are the actual data in question—typically, wordlists of "basic" vocabulary. Words like *all, louse, seed, blood, claw, belly, bite, know, sun, yellow, night, new,* and *round* have been selected for their apparent frequency, universality, and resistance to change over time. Defined in opposition to "cultural" vocabulary, this lexicon has been rigorously sampled in an effort to construct "diagnostic" lists, typically some 100 or 200 items in length. While embedded assumptions about historical process and translatability have been challenged since these wordlists were first proposed in the mid-twentieth century, they have become such a staple of routine fieldwork and comparative practice that even their loudest critics can be found endorsing their expediency on record.[12] Simply put, basic wordlists are one of the most ubiquitous "little tools" of historical linguistics and its growing network of allied disciplines.[13]

Like Mirjam Brusius's emphasis in this volume on the "thing itself," my essay traces the intuition, standardization, and operationalization of basic words over time.[14] With such emphasis on the very givens, it challenges claims that computing technology has caused a radical break in the development of the language sciences. To the contrary, researchers have consistently gone back to their philological roots in an effort to improve wordlists on the basis of what is already known. Echoing the strong continuity thesis David Sepkoski advances in his discussion of paleontology, the transition from paper to electronic formats in historical linguistics is depicted here as one of direct transcription.[15] I view lists of basic vocabulary as immanent to specific practices and technologies in linguistics—not only of computing, but of salvage anthropology, linguistic fieldwork, and the Comparative Method as well. These considerations point to the significance of the linguistic case for histories of data more broadly: designed to sustain comparisons on a global scale reaching back thousands of years, basic vocabulary highlights the way in which data forge equivalences across systems (disciplinary, technological, and linguistic) that would otherwise be incommensura-

---

[11] In a widely cited article, H. A. Gleason Jr. attributed the "new" convergence between anthropology and linguistics to advances in Africanist studies, machine data processing, and statistics; Gleason, "Counting and Calculating for Historical Reconstruction," *Anthropol. Linguist.* 1 (1959): 22–32. Some forty years later, population geneticist Luigi Luca Cavalli-Sforza wrote of the "synthesis" then underway between archaeology, genetics, and historical linguistics; Cavalli-Sforza, *Genes, Peoples and Languages* (1996; repr., London, 2000). The view has since been repeated by Renfrew and Forster, who refer to a "new synthesis" joining the three fields; Forster and Renfrew, *Phylogenetic Methods* (cit. n. 6), 3.

[12] Compare, for instance, Harry Hoijer's strong position against universality in "Lexicostatistics: A Critique," *Language* 32 (1956): 49–60, with his reluctant admission that at least "we can avoid items which are obviously culture-bound . . . and which have reference to geographical and climatic factors of restricted distribution" in a paper published the same year; Hoijer, "Chronology of the Athapaskan Languages," *Int. J. Amer. Linguist.* 22 (1956): 219–32. For a survey of workaday implementations, see Annette Dobson, Joseph B. Kruskal, David Sankoff, and Leonard J. Savage, "The Mathematics of Glottochronology Revisited," *Anthropol. Linguist.* 14 (1972): 205–12, n. 1.

[13] Peter Becker and William Clark, eds., *Little Tools of Knowledge: Historical Essays on Academic and Bureaucratic Practices* (Ann Arbor, Mich., 2001). See also Staffan Müller-Wille and Isabelle Charmantier, "Lists as Research Technologies," *Isis* 103 (2012): 743–52.

[14] Mirjam Brusius, "The Field in the Museum: Puzzling Out Babylon in Berlin," in this volume.

[15] Sepkoski, "Database" (cit. n. 8). Cf. Janet Martin-Nielsen, "'It Was All Connected': Computers and Linguistics in Early Cold War America," in *Cold War Social Science: Knowledge Production, Liberal Democracy, and Human Nature,* ed. Mark Solovey and Hamilton Cravens (New York, 2012), 63–78. Martin-Nielsen argues that computers were essential to the shift from Bloomfieldian structuralism to transformational grammar.

ble.[16] Investments in the universality of basic vocabulary have trumped other attributes (of frequency and stability), supporting translations both literally and figuratively among dialects, methods, and disciplinary communities.[17] Because so much hard work has gone into calibrating the lists for statistical validity, and because they conform so neatly to the efficiency demands on fieldwork practice, there has been little impetus to throw them out, despite heavy critiques over time.

My story then, developed in five sections, is really one of "data drag" rather than data-driven change. In this sense, it conforms to the way in which Geoffrey Bowker and Susan Leigh Star have characterized other sorts of scientific standards, insofar as these have "significant inertia and can be very difficult and expensive to change."[18] Section I looks at the intuitive grasp of the category of basic vocabulary before it was standardized in the mid-twentieth century. Section II shows how that intuition aligned with other linguistic priorities and commitments in the first half of the twentieth century. Section III focuses on the immediate postwar context, examining the efforts of Morris Swadesh (1909–67) to forge a rigorous definition of the diagnostic list.[19] Section IV addresses the crystallization of basic vocabulary through testing and revision. Section V examines the uptake of basic vocabulary data in computerized research. Taken as a whole, the essay adds another brick to the antideterminist wall built by Jon Agar and others.[20] At the same time, it has the potential to open up new avenues of research on the linguistic underpinnings of data structures and ontologies, as they are manifest in a wide variety of scientific traditions.[21]

I. "THE MOST INTIMATE FUND OF SPEECH"

Those interested in the historical patterns and processes of language change have three categories of evidence at their disposal: morphology (concerning structure), phonology (concerning systems of sound), and vocabulary (concerning units paired for form and meaning). While evidentiary preferences with respect to these three categories have fluctuated over time, linguists have consistently tried to pick out characters that will insulate historical information from chance or other noninherited explanations of similarity (namely, borrowing and onomatopoeia). When it comes to questions of a broadly anthropological nature—especially concerning relationships among unwritten languages and prehistory—the third category, vocabulary, has been prioritized. What is more, linguists have had remarkably stable intuitions about which kinds

---

[16] Another way of putting this is to say that they reduce "friction" in the sense developed by Paul Edwards in, e.g., *A Vast Machine: Computer Models, Climate Data, and the Politics of Global Warming* (Cambridge, Mass., 2010). Similarly, Sabina Leonelli's notion of "packaging" for travel fits here; see Leonelli, "Packaging Small Facts for Re-Use: Databases in Model Organism Biology," in *How Well Do Facts Travel? The Dissemination of Reliable Knowledge*, ed. P. Howlett and M. S. Morgan (Cambridge, 2010), 325–48.

[17] D. H. Hymes, "Lexicostatistics So Far," *Curr. Anthropol.* 1 (1960): 3–44, on 7. Here Hymes contends, "the development of the test lists so far has given priority to the attribute of universality, for obvious practical reasons."

[18] Geoffrey Bowker and Susan Leigh Star, *Sorting Things Out: Classification and Its Consequences* (Cambridge, Mass., 1999), 14.

[19] Morris Swadesh, "Lexico-Statistic Dating of Prehistoric Ethnic Contacts: With Special Reference to North American Indians and Eskimos," *Proc. Amer. Phil. Soc.* 96 (1952): 452–63.

[20] Jon Agar, "What Difference Did Computers Make?," *Soc. Stud. Sci.* 36 (2006): 869–907.

[21] On "data structures," see Sepkoski, "Database" (cit. n. 8); on ontologies, see Hallam Stevens, *Life Out of Sequence: A Data-Driven History of Bioinformatics* (Chicago, 2013), 117–27.

of words are likely to retain inherited features best over time. These have been words for things like body parts, counting, colors, and attributes of the natural world. With meanings thus supposedly held constant, similarities observed between collections of words have been taken as reasonable evidence—beyond chance—of common inheritance.[22]

Linguist Lyle Campbell points to a lively tradition, born during the late Renaissance, of compiling "large-scale word collections for language comparisons." Konrad Gesner (1555), Johan Christoph Adelung (1782, 1806), Lorenzo Hervas y Panduro (1784, 1800), and Peter Simon Pallas (1786) are each known to have made notable contributions in this respect.[23] But efforts to collect words from "a sector of vocabulary, common to all languages . . . especially criterial of relationship" has a more specific history of its own.[24] As early as 1643, the Dutch geographer Jean De Laet (1581–1649), for instance, gestured toward a distinct sector of vocabulary including the names of body parts, numerals, kinship relations, and geographical concepts. De Laet emphasized the importance of seeking out "the names of those things which are domestic and most common to [a] nation." He suggested that only these could rule against something like the idea of chance as "it is not difficult to find words in all languages agreeing to some extent with other languages" through an unlimited search.[25]

De Laet was certainly not the only early modern thinker to appeal to a concept of basic vocabulary. In his expansive etymologizing, for example, Leibniz also advocated the use of basic words. In a brief 1698 essay on how to study the history of the Slavic languages, he put forward the following list:

> When the Lord's Prayer is not easy to obtain, we could still ask for words that are in it. . . . We could also ask for the names of other things, as, for instance, for some limbs of the human body and their use, such as: eyes, ears, to hear, to see, etc. And the same for relations and next of kin, as father, mother, daughter, brother, sister, man, woman, child, etc. The same for foods, drinks, clothes, arms: also for the names of some familiar animals, for the four elements and what pertains to them.[26]

This intuitive grasp of basic vocabulary persisted through the nineteenth century, a period generally regarded as the heyday of comparative-historical linguistics.[27] Standing at the vanguard of this tradition, the Danish linguist Rasmus Rask (1787–1832) advocated the comparative analysis of structural evidence in questions of historical relationship. That said, he also allowed that lexical similarities could be counted toward

---

[22] Markus Friedrich's discussion of "bare data" in the service of genealogical "conjunctions" or "junctures" offers an intriguing methodological analogy; see Friedrich, "Genealogy as Archive-Driven Research Enterprise in Early Modern Europe," in this volume.

[23] Lyle Campbell, "The History of Linguistics," in *The Handbook of Linguistics*, ed. Mark Aronoff and Janie Rees-Miller (Oxford, 2002), 81–104, on 86. See also James Turner, *Philology: The Forgotten Origins of the Modern Humanities* (Princeton, 2014), on 101.

[24] Dell Hymes, "Lexicostatistics and Glottochronology in the Nineteenth Century (with Notes toward a General History)," in *Lexicostatistics in Genetic Linguistics: Proceedings of the Yale Conference, Yale University, April 3–4, 1971*, ed. Isidore Dyen (The Hague, 1973), 122–76, on 128.

[25] De Laet, quoted in ibid., 128.

[26] Leibniz, quoted in Hans Aarsleff, "The Study and Use of Etymology in Leibniz," in *Akten des Internationalen Leibniz-Kongresses (Hannover 14–19 November 1966): Erkenntnislehre, Logik, Sprachphilosophie, Editionsberichte*, vol. 3 (Wiesbaden, 1969), 173–89, on 187–8.

[27] On comparative-historical linguistics in the nineteenth century, see Anna Morpurgo Davies, *History of Linguistics*, vol. 4, *Nineteenth-Century Linguistics* (London, 1998).

kinship provided samples were limited to "the most essential, concrete, indispensable words," in his estimation, "the foundation of the language."[28] This idea was reinforced by those engaged in fieldwork, as linguist Dell Hymes (1927–2009) has shown. Expressing what appears to have been a common idea in and around the Société de Géographie in Paris, for instance, the American diplomat David Bailie Warden (1772–1845) argued,

> For there to be agreement between two languages, it is necessary that it be found among the common words, such as those of number, the sun, moon, land, water, wind, thunder, to eat, to drink, good, bad, the names of parts of the body and of animals, etc.; for in supposing that these words should have undergone great change with regard to pronunciation, there would exist nevertheless enough resemblances to discover their origin.[29]

For those with religious motivations, the quintessentially human category of basic vocabulary also held out considerable appeal. Arthur James Johnes (1809–71), for instance, issued two propositions in his 1846 *Philological Proofs of the Original Unity and Recent Origin of the Human Race*: first, "that the various nations of our Globe are descended from one Parent Tribe," and second, "that the introduction of the Human Species into the system to which it belongs, cannot be referred to an epoch more ancient than the era indicated . . . by our received systems of chronology."[30] In an effort to establish this monogenist prehistory, he then turned to basic vocabulary, defined as follows:

> With respect to the particular words selected for comparison, I have chosen the names for the following objects: "Fire, Sun, Day, Eye, Moon, Heaven, a Human Being, Man, and Woman. . . . The most important parts of the Human Frame, (viz. "The Hand, Arm, Foot, Leg, Ear, Tongue, Head.") "Water." . . . That the selected specimens of the languages of Africa are sufficiently *numerous* for this end is plain. It only remains to be shown that their *nature* is such as to render them eminently suitable and conclusive.[31]

While it is not my focus here, these examples call for an analysis of the politics of sourcing and reuse, such as the one Joanna Radin offers in her contribution to this volume.[32] For now, I want simply to point out that basic wordlists have been instrumental to the anthropological investigation of human origins and relationships over the *longue durée*. Linguists have roughly adhered to this collection protocol for generations, upholding an intuitive theory of language that trumpets the universality of basic vocabulary. The eminent late nineteenth-century philologist William Dwight Whitney (1827–94) gave voice to this underlying theory, linking basic vocabulary to the very essence of language. For him it was no less than "the most intimate fund of speech, the words significant of those ideas without whose designation no spoken tongue would

---

[28] Rask, quoted in Rulon Wells, "Lexicostatistics in the Regency Period," in Dyen, *Lexicostatistics* (cit. n. 24), 21.

[29] Warden, quoted in Hymes, "Lexicostatistics and Glottochronology" (cit. n. 24), 142.

[30] Arthur James Johnes, *Philological Proofs of the Original Unity and Recent Origin of the Human Race Derived from a Comparison of the Languages of Asia, Europe, Africa, and America: Being an Inquiry How Far the Differences in the Languages of the Globe Are Referrible to Causes Now in Operation* (London, 1846), xv.

[31] Ibid., 8; emphasis in the original.

[32] Joanna Radin, " 'Digital Natives': How Medical and Indigenous Histories Matter for Big Data," in this volume.

be worthy of the name."[33] As I hope to show in this essay, efforts to commensurate old and new data in linguistics entrenched the lists in practice, well beyond Whitney's theoretical intuition.

II. "PROGRESS MUST TAKE THE FORM OF CONSTANTLY IMPROVING TENTATIVE RESULTS"

In his 1933 landmark work, *Language*, the pioneering American structuralist Leonard Bloomfield (1887–1949) straightforwardly asserted that basic words were especially informative, as "neither accident nor borrowing will explain them."[34] This insight anticipated controversies that later surrounded the role of the wordlists in "lexicostatistics," the quantitative analysis of vocabulary data. While proponents of historical lexicostatistics have argued since the 1950s that the very definition of basic wordlists reduces the possibility of noise due to chance and borrowing, critics have repeatedly opened up intuitions about the frequency, universality, and stability of basic vocabulary to empirical scrutiny. These critical assessments, examined below, have paradoxically reinforced the category, contributing to the phenomenon of data drag.

This section of the essay contextualizes the efforts of Morris Swadesh to standardize the category of basic vocabulary. Swadesh was the inventor of "lexico-statistic glottochronology," a midcentury approach to historical linguistics that made the intuitions about frequency, universality, and stability just surveyed explicit, and therefore testable.[35] To understand what motivated and sustained the undertaking, I begin with a look at his education and early career.

Born in the planned industrial community of Holyoke, Massachusetts, in 1909, Morris Swadesh was multilingual from a young age.[36] His parents were Jewish immigrants who spoke Russian and Yiddish at home, a foundation he would later build upon as a student of modern languages at the University of Chicago. Upon the completion of his bachelor's degree in 1930, Swadesh began working with the eminent linguistic anthropologist Edward Sapir (1844–1939), one of the founders of American linguistics, and himself a student of Franz Boas (1858–1942). Under Sapir's direction, and commissioned by the International Auxiliary Language Association, Swadesh undertook a master's project on comparative semantic analysis. He undoubtedly drew upon that work years later in a series of published responses to C. K. Ogden's "Basic English," an international auxiliary language that attracted considerable publicity after the Allied victory in World War II.[37] Sensitivity to the cross-cultural translation of semantic units, interest in the possibility of language universals, and familiarity with de-

---

[33] William Dwight Whitney, *Language and the Study of Language* (1867; repr., London, 1997), 197–8.

[34] Leonard Bloomfield, *Language* (Chicago, 1984), 298–9. There are many more such examples from the early twentieth century. See Sidney Herbert Ray, "The Polynesian Languages in Melanesia," *Anthropos* 14–15 (1919): 46–96, on 53. Instead of "basic," Roland G. Kent simply referred to the categories under discussion as the "words of family relationship"; Kent, *The Sounds of Latin, a Descriptive and Historical Phonology*, 3rd ed. (Baltimore, 1945), 15.

[35] Morris Swadesh, "Glottochronology," in *A Dictionary of the Social Sciences* (Paris, 1964), 289.

[36] Biographical material on Swadesh has been compiled from Dell Hymes, "Morris Swadesh: From the First Yale School to World Prehistory," in *The Origins and Diversification of Language* (Chicago, 1971), 228–70; Anthony Grant, "Swadesh's Life and Place in Linguistics," *Diachronica* 27 (2010): 191–6.

[37] Morris Swadesh, "The Future of Basic English," *Bulletin of the Linguistic Circle of New York*, 1943; Swadesh, "Scientific Linguistics and Basic English," *Basic Engl.* 17 (1944): 194–206; Swadesh, "Basic English," in *Collier's Encyclopedia* (New York, 1950), 220. He was not in favor of Basic En-

bates about "basic" language were thus part of his toolkit from an early age. They form a crucial context for understanding Swadesh's subsequent work on the standardization of basic vocabulary.

Sapir accepted his professorship at Yale University in 1931 on the strict condition that Swadesh be allowed to travel with him. Thus began the twenty-two-year-old's immersion in the study of the indigenous languages of North America. Swadesh took yearly field trips over the next decade, "salvaging" languages from British Columbia to the Gulf of Mexico, and completed his doctoral dissertation on *The Internal Economy of the Nootka Word* in 1933.[38] Significantly, Swadesh devoted this project to word-level analyses, as described in these prefatory remarks:

> I am now engaged, as research assistant to Professor Sapir, in the work of cataloguing the lexical data contained in the material in preparation for compiling a dictionary of the language. Inasmuch as Nootka is a polysynthetic language, problems of the structure of the word, phonetically and semantically, are of prime importance. The present paper is an attempt to discover the principles underlying the semantic construction of the word.[39]

The fundamental problem Swadesh confronted in this project was how to understand the fundamental units of unwritten languages with highly composite words that had not previously been written down. "On the one hand," Swadesh wrote,

> a complete understanding of the internal economy of the word is possible only from the study of the data that would be contained in a rather complete lexicon. . . . On the other . . . a lexicon cannot be intelligently compiled without a thorough understanding of the structure of words. *Progress must take the form of constantly improving tentative results.*[40]

Both the emphasis on words as the starting point for work on lesser-known languages and the description of research as a process of tinkering with "tentative results" were characteristic of Swadesh's later attitude toward diagnostic lists of basic vocabulary. To Swadesh, like others, word-level characters were especially well suited to mapping out new frontiers in anthropological linguistics.

Swadesh elaborated upon this fieldwork approach in a 1937 paper for *American Anthropologist.*[41] Counseling novice fieldworkers, he made practical suggestions, like "avoid making your informant speak so slowly as to make his speech unnatural." But the first step in his protocol was as follows:

> *Record a few hundred short utterances.* . . . It saves time to have a list of English words made out in advance. Any list will do. If the informant cannot give you the equivalent of one of the words in your list without long thinking, skip it.[42]

---

glish (largely because of a lack of standardized orthography), and thought the Latinate auxiliary of the International Auxiliary Language Association was better. See James McElvenny, "Meaning in the Age of Modernism: C. K. Ogden and His Contemporaries" (PhD diss., Univ. of Sydney, 2013).

[38] Morris Swadesh, "The Internal Economy of the Nootka Word" (PhD diss., Yale Univ., 1933).

[39] Ibid., 2.

[40] Ibid., 2–3; emphasis added.

[41] Morris Swadesh, "A Method for Phonetic Accuracy and Speed," *Amer. Anthropol.* 39 (1937): 728–32.

[42] Ibid., 729. The expectation here was that the interviews would be recorded on tape, the cost of which partially explains the urgency of these instructions. On the media history of such salvage work, see Brian Hochman, *Savage Preservation: The Ethnographic Origins of Modern Media Technology* (Saint Paul, Minn., 2014).

While the arbitrariness of word collection here appears to be at odds with attempts to rigorously define the category of basic vocabulary, this directive nevertheless shows how standardized lists fit into fieldwork practice. These are steps Swadesh himself would have taken in his own efforts to record and revitalize the Tarascan language in a state-funded project commissioned by the Mexican government during the late 1930s.[43] Days after the attack on Pearl Harbor, however, he returned to the United States, where he began work on Japanese with the foreign language program of the American Council of Learned Societies. Participating in a movement that would transform American linguistics during and after World War II, Swadesh then became a translator with the Language Section of the Information and Education Division of the U.S. Army under Henry Lee Smith.[44] He remained on the East Coast after the war, joining Roman Jakobson and other émigrés in the Linguistic Circle of New York—"the place for unorthodox thoughts and concern with semantics"—assuming the editorship of the Circle's journal, *Word*, in 1946. These interests in semantics and translation, I believe, encouraged Swadesh in his efforts to standardize basic vocabulary, the goal being to identify equivalences common to all languages. In this way, basic vocabulary data was ready-made to commensurate across systems.

According to biographer Stanley Newman, Swadesh was "labeled unambiguously as a 'leftist' during the noisiest period of the McCarthy Era," following his participation in a 1948 student strike at the City College of New York.[45] Banned from the American professoriate, he was a self-funded independent scholar when he laid the groundwork for "lexico-statistical glottochronology" with its standardized wordlists of basic vocabulary. With a grant from the Phillips and Penrose Funds of the American Philosophical Society (APS), he began work as a cataloger in the vast Boas Collection of the APS Library the following year. His efforts were part of a larger project in comparative Iroquoian languages, and he took advantage of special access to the collection for his own work, developing a punch card system of analysis using vocabulary data from the Salish language.[46]

Punch cards arguably helped build confidence in (or at least tolerance for) the idea of semantic equivalence underpinning the use of basic vocabulary data in historical reconstruction. As Swadesh described his technique, "Boas' vocabulary was coded by myself in pencil with letters a, b, c etc. used to identify cognates." Note here the expectation that vocabulary data should be shared and reused: similar to the way in which Staffan Müller-Wille characterizes museums in this volume (as "collections of collections"), lexicostatistics has always depended on samples taken from samples.[47] "For each semantic item," he continued, "all forms marked a are cognate with

[43] For more about Tarascan, see https://www.ethnologue.com/subgroups/tarascan (accessed 29 September 2016).

[44] Swadesh to Fred Eggan, 17 August 1950, Box 44, Folder 14, Eggan Papers, Special Collections Research Center, University of Chicago Library. Boyd Davis and Raymond O'Cain note, "World War II drew many [linguists] into wartime language work . . . in the postwar years, they fought for and secured the autonomy of linguistics proclaimed by their mentors"; see Davis and O'Cain, *First Person Singular: Papers from the Conference on an Oral Archive for the History of American Linguistics (Charlotte, N.C., 9–10 March 1979)* (Amsterdam, 1980), on xi.

[45] Stanley Newman, "Morris Swadesh," *Language* 48 (1967): 948–57, on 949.

[46] Swadesh to Hymes, 19 February 1955, Dell H. Hymes Papers, No. 55, Series I, Morris Swadesh correspondence, American Philosophical Society Manuscript Collections, Philadelphia (hereafter cited as "APS MS Coll."). On Salish, see https://www.ethnologue.com/subgroups/salish (accessed 14 July 2015).

[47] See Staffan Müller-Wille, "Names and Numbers: 'Data' in Classical Natural History, 1758–1859," in this volume.

each other, those marked b are cognate with each other, etc." Swadesh then created a punch card for each semantic unit and assigned holes to each language with numbers corresponding to his lettered judgments of cognancy.[48] The size of the cards would have imposed an upper limit on the possible total number of semantic matches, and holes would have reduced ambiguity as to the degree that words were actually forced to match cross-linguistically. This analysis was an essential first step in the preparation of Swadesh's 1950 publication on "Salish Internal Relationships"—his earliest print foray into lexicostatistics and glottochronology.[49] In addition to the engagement with semantic and quantitative analysis just suggested, the formulation of that method also required an explicit consideration of time—both as it correlated with the observed diversity among languages, and with respect to rates of change.

III. "HISTORIC KNOWLEDGE SERVES TO CLARIFY PREHISTORY"

Swadesh had already explored these ideas in a 1948 Viking Fund Supper Conference presentation on "The Time Value of Linguistic Diversity." The Viking Fund context is significant: there was widespread enthusiasm and debate surrounding human prehistory in this forerunner of the Wenner-Gren Foundation, and Willard Libby introduced his method of radiocarbon dating to the same audience that very year.[50] Libby's work was a crucial stimulus for Swadesh, who became increasingly committed to the idea of calculating absolute linguistic chronologies using differential rates of lexical replacement in standardized wordlists of basic vocabulary.

Swadesh devoted the bulk of the Salish paper to a discussion of "linguistic distances determined by basic vocabulary correspondences." Engaging the smallest units of meaning in language (morphemes), he wrote:

> The method used is a simple one based on the percentage of common elements in a selected basic vocabulary in 30 Salishan languages and dialects. This approach should prove fairly dependable, that is it should correlate well with other lines of linguistic and ethnographic evidence *because it is a well known fact that certain types of morphemes are relatively stable.*[51]

Like his nineteenth-century predecessors, Swadesh evidently started with an intuitive grasp of the category of basic vocabulary, something that could be sampled ("a selected basic vocabulary") for the purposes of quantitative analysis. With the help of Columbia University's Gordon Gould—another unambiguous "leftist" from Yale, and the disputed inventor of laser technology—Swadesh was able "to reduce originally cumbersome computations to . . . simple formulas."[52] Correspondences discovered for

---

[48] Swadesh to Hymes, 19 February 1955 (cit. n. 46). See also Morris Swadesh, "Machine-Aided Linguistic Research in the National University of Mexico," in *The Use of Computers in Anthropology*, ed. Dell Hymes (The Hague, 1965), 523–35.

[49] Morris Swadesh, "Salish Internal Relationships," *Int. J. Amer. Linguist.* 16 (1950): 157–67.

[50] Willard Libby, "Archaeological Ages by Natural Radiocarbon Content," 9 January 1948. See the complete list of recorded Supper Conference presentations, www.wennergren.org/history/conferences-seminars-symposia/supper-conferences (accessed 6 June 2017). See Susan Lindee and Joanna Radin, "Patrons of the Human Experience: A History of the Wenner-Gren Foundation for Anthropological Research, 1941–2016," *Curr. Anthropol.* 57 (2016): S218–S301.

[51] Swadesh, "Salish Internal Relationships" (cit. n. 49), 157; emphasis added.

[52] Ibid., 158.

165 "suitable basic items" were scored on the presence or absence of cognancy, and these counts were converted into percentages. Gould gave Swadesh a formula by which he could measure the time depth of separation between any two languages using these percentages and a constant (r), designed to correspond to a universal "percentage of basic vocabulary retained after one standard period of time."

This constant was basically a numerical expression of the long-standing intuition, surveyed above, that basic vocabulary is partly defined by its stability. But how did Swadesh and Gould derive an explicit number for this idea? This involved reference to existing philological knowledge. "The standard value of r used was 85%," Swadesh reported, "which is the amount found to persist in modern English by reference to Old English . . . a little over 1000 years, in a vocabulary list similar to that used in Salish." Though he allowed that this unit of time was "arbitrary" in the Salish case, he nevertheless concluded that "relative values obtained can be used with considerable confidence."[53]

In other words, Swadesh used evidence from related languages of known chronology to generate a constant rate of change that could then be applied to experimental cases for which relationship and time depth were not yet known. In this early pilot of the method, the list was not standardized (being merely "similar") and the rate of change was not calibrated. Nevertheless, this example shows how basic vocabulary data continued earlier research traditions: proceeding from the known to the unknown, the proposed method extrapolated from written records to address questions relating to prehistory. The validity of lexicostatistics was understood to depend on tried and true philological scholarship, which did not require replication.

Swadesh was cautiously optimistic about this method of lexicostatistical glottochronology. The following year, he applied it to a "test" of the most theoretical dispute between Boas and Sapir concerning linguists' ability to scientifically distinguish "diffusional cumulation" (horizontal transfer, borrowing) from "archaic residue" (vertical transfer, inheritance). For this, he used a test list "consisting of 215 items empirically chosen for their relatively stable character." He elaborated further:

> The test list, designed as a general measure of affinity between related languages, is not fully satisfactory in its selection of stable items, but will serve our purposes. The procedure in using any such technique has to be standardized so that one may always get approximately the same results from the same material.[54]

Swadesh derived the constant in this study from a consideration of the basic vocabulary in English, French, and German—again, building established comparative-philological knowledge into the validity of the method itself. When it came to the experimental case in question, however, the going got somewhat tougher. "If suitable dictionaries were available it would be a simple matter to measure basic vocabulary correspondences," he wrote of the Na-Dene languages, allowing for an easy comparison with the Indo-European results. Absent such references, he urged researchers to avail themselves of private field notes (belonging in this case to one Pliny Earle Goddard)—an imperfect practice made necessary by the labor-intensiveness of descriptive linguis-

[53] Ibid.
[54] Morris Swadesh, "Diffusional Cumulation and Archaic Residue as Historical Explanations," *Southwest. J. Anthropol.* 7 (1951): 1–23, on 12–3.

tics. His comments on the matter shed light on the hybrid sourcing of basic vocabulary data, which has depended on single observations as well as recorded media, published reference works, and specialist research outputs. The literature on lexicostatistics was rife with rhetoric about the laudable "objectivity" of a quantitative approach, echoing midcentury discussions of anthropological method as a whole, it was also vulnerable time and again to the charge of bad inputs.[55]

Swadesh successfully standardized the test list of basic vocabulary for his 1952 publication on the "Lexico-Statistic Dating of Prehistoric Ethnic Contacts."[56] As the title suggests, this paper was centrally concerned with matters of prehistory—that "great obscure depth which science seeks to penetrate"—and it championed lexicostatistical methods as the best way in.[57] Data and discipline, the science of prehistory in this case, were co-constitutive, as Swadesh indicated:

> Powerful means have been found for illuminating the unrecorded past, including the evidence of archaeological finds and that of the geographic distribution of cultural facts in the earliest known periods. Much depends on the painstaking analysis and comparison of data, and on the effective reading of their implications. *Very important is the combined use of all the evidence*, linguistic and ethnographic as well as archaeological, biological, and geological.[58]

With a view to the "comparison of data" across disciplines, he stressed the need for absolute—rather than relative—chronologies, to reduce "friction." These dates, it was argued, would help commensurate evidence from the various traditions in prehistorical research. In his words,

> One of the most significant recent trends in the field of prehistory has been the development of objective methods for measuring elapsed time. Where vague estimates and subjective judgments formerly had to serve, today we are often able to determine prehistoric time within a relatively narrow margin of accuracy. This development is important especially because it adds greatly to the *possibility of interrelating* the separate reconstructions.[59]

Swadesh had Libby's radiocarbon method in mind when he wrote these words, recalling the broader anthropological context.[60] He concluded that while the words in-

---

[55] Overcoming subjectivity was the stated aim of another foundational paper in the statistical analysis of historical language relationship: A. L. Kroeber and C. D. Chrétien, "Quantitative Classification of Indo-European Languages," *Language* 13 (1937): 83–113, on 85. The authors wrote, "What statistical analysis can do is to validate and correct insight, or, where insight judgments are in conflict, help to decide between them. In short, it increases objectivity, sharpens findings, and sometimes forces new problems." This paper initiated its own research tradition (and was a resource for Swadesh); for more about this, see Sheila Embleton, *Statistics in Historical Linguistics* (Bochum, 1986), 5–15. The "elements" Kroeber and Chrétien engaged, however, were not strictly lexical—they did not work with basic vocabulary data—and so, despite overlaps (e.g., Dobson et al., "Mathematics" [cit. n. 12]; Dyen, *Lexicostatistics* [cit. n. 24]), I will not directly address this tradition here.

[56] Swadesh, "Lexico-Statistic Dating" (cit. n. 19).

[57] Ibid., 452.

[58] Ibid.; emphasis added.

[59] Ibid.; emphasis added.

[60] See E. C. Anderson, W. F. Libby, S. Weinhouse, A. F. Reid, A. D. Kirschenbaum, and A. V. Grosse, "Radiocarbon from Cosmic Radiation," *Science* 105 (1947): 576–7, on 576.

volved in lexicostatistic study differed materially from radioactive carbon in organic substances, the "broad theoretical principle" nevertheless carried over. Thus,

> the fundamental everyday vocabulary of any language—as against the specialized or "cultural" vocabulary—changes at a relatively constant rate. The percentage of retained elements in a suitable test vocabulary therefore indicates the elapsed time.[61]

This passage made the fundamental ideas of lexicostatistics and glottochronology explicit. First, linguistic differentiation was conceived as a function of time. Second, owing to the "relatively constant rate" of change in basic vocabulary, absolute time depths could be known scientifically. Finally, Swadesh set basic vocabulary data apart from culture, holding out a universal conception of human communicative needs that facilitated intercultural and linguistic comparison.

As he had done previously, the "constant rate" proposed in this paper came from "controlled material," a 200-item list, sourced from Indo-European, Semitic, and Sino-Tibetan for which "historical materials are abundant and go back as far as 6,000 years."[62] Established reconstructions and textual evidence, again, were instrumental for calibrating a rate of replacement that could be scaled up through time and space. "In this way," Swadesh proclaimed, "historic knowledge serves to clarify prehistory"—repurposing old data.[63] It was only later that questions about the relative stability of written versus unwritten languages would emerge.

IV. "LEXICOSTATISTICS HAS THUS FAR OPERATED . . . ON A SHOE-STRING OF BASIC RESEARCH"[64]

It is a principal contention of this essay that the avalanche of research touched off by doubts about the universality and stability of basic words paradoxically served to fix that category in linguistic research practice—this being the phenomenon of data drag. Swadesh stood at the forefront of this critical response, hoping to ever improve the method he had proposed. Through collaboration with Robert Lees (1922–96), Joseph Greenberg (1915–2001), and several other colleagues, Swadesh launched a coordinated effort during the 1950s to test and validate the numerical constant on which lexicostatistical glottochronology relied.[65] Lees published his own account of the method the following year, discussing a universal "Basic-Root-Morpheme-Inventory" based on the 200-item Swadesh list. Lees's test engaged thirteen languages and revealed "on the average about 81% of the basic-root-morphemes of a language will survive as cognates after 1000 years, *for all languages, at all times*."[66]

---

[61] Swadesh, "Lexico-Statistic Dating" (cit. n. 19), 452.

[62] Swadesh, "The Time Value of Linguistic Diversity," paper presented at the Viking Fund Supper Conference for Anthropologists, 12 March 1948, cited in ibid., 454.

[63] Ibid.

[64] Morris Swadesh, "Towards Greater Accuracy in Lexicostatistic Dating," *Int. J. Amer. Linguist.* 21 (1955): 121–37, on 122.

[65] Greenberg, controversial in his own right, contributed African language data and helped Swadesh secure funding from the Columbia University Social Science Research Fund. Swadesh reported on phonographic field work using 200-item diagnostic lists made possible by this collaboration in Swadesh, "On the Penutian Vocabulary Survey," *Int. J. Amer. Linguist.* 20 (1954): 123–33.

[66] Robert Lees, "The Basis of Glottochronology," *Language* 29 (1953): 113–27, on 119; emphasis added.

This claim cried out for empirical reinforcement. With a grant from the APS, Swadesh set out to collect as many vocabularies as possible, uniting his methodological interests with a concern to salvage evidence of languages at risk of extinction.[67] He described the work in a report to the society, which merits quotation at length:

> At the outset the grantee had a collection of about 30 diagnostic vocabularies. . . . It was necessary to find a way of rapidly assembling similar material from other Amerindian languages, ideally from all of the 1,500 or so languages of the New World. He undertook to accomplish this by communicating, by mail or personally, with linguists, ethnologists, teachers and missionaries, wherever he knew of anyone in a position to provide the needed material. Each was requested to furnish a list of 200 selected words. . . . About 100 vocabularies have been submitted and many more have been promised. They are being published by the grantee in a series called Amerindian Non-Cultural Vocabularies (in mimeographic form), which is being distributed to individuals, educational institutions, and libraries. This method of obtaining material is evidently capable of eventual success in gathering together a complete collection of diagnostic vocabularies for the entire hemisphere.[68]

Swadesh proved to be a diligent correspondent. In a handwritten March 1954 postscript, for instance, he invited Hymes to contribute data: "Carl [Voeglin] mentions you among his students who have had field experience and who might be willing to fill out vocabularies for my collection," the senior linguist wrote, "so I enclose blanks [*sic*] sheets with my request that you give me whatever vocabulary or vocabularies you may have. . . . *I can repay your kindness in vocabularies.*"[69] Through such exchanges, he collected wordlists from original research and printed sources, an early and informal kind of open access rooted in a disciplinary logic of collective and distributed effort.[70] Conferences became opportunities to make contacts "for getting additional vocabularies."[71] Swadesh became a kind of one-man clearinghouse, running the Amerindian vocabulary exchange out of the basement of his home in Denver. As figure 1 shows, he cultivated the observational, interview, and notational skills of contributors through correspondence: this involved the diagnostic list itself, instructions on how to use it, and considerable editorial labor. All of these efforts were undertaken with a view to standardizing linguistic data "for the sake of uniformity . . . to facilitate comparative study," reinforcing Jon Agar's point that computerization has only been attempted (as we shall see) in those fields where significant routinization "*already existed.*"[72] Swadesh was equally frank when it came to the rules governing this economy of data exchange—each wordlist entitled its contributor as many as ten others in return, with additional lists provided for only 5¢ each. This pricing encouraged a spirit of collaboration among researchers, arguably necessary for any such large-scale com-

---

[67] Grant no. 1685 (1954), $900. Morris Swadesh, "Linguistic Time Depths of Prehistoric America: Penutian" (grant report, American Philosophical Society, 1954), 375–6.

[68] Ibid.

[69] Swadesh to Hymes, 26 March 1954, Dell H. Hymes Papers, No. 55, Series I, Morris Swadesh correspondence, APS MS Coll.; emphasis added.

[70] The contrast with Sabina Leonelli's characterization of open access in contemporary biology here is intriguing; Leonelli, "Why the Current Insistence on Open Access to Scientific Data? Big Data, Knowledge Production, and the Political Economy of Contemporary Biology," *Bull. Sci. Tech. Soc.* 33 (2013): 6–11. On this division of labor in linguistics, see Judith Kaplan, "Archiving Descriptive Language Data: A New GOLD Standard?," *Limn* 6, http://limn.it/issue/06/ (accessed 6 June 2017).

[71] Swadesh to Hymes, 21 July 1954, Dell H. Hymes Papers, No. 55, Series I, Morris Swadesh correspondence, APS MS Coll.

[72] Agar, "What Difference Did Computers Make?" (cit. n. 20), 872; emphasis in the original.

parative undertaking.[73] Still, there were problems, for example, with the very definition of basic vocabulary:

> The terminology of "cultural" versus "non-cultural" is not good. The following ideas have
> been suggested. (1) distinguish universal versus local, with intermediate types like regional,
> widespread, fairly universal. (2) distinguish between natural objects and cultural objects,
> also culturized natural objects. (3) to some extent one can speak of relatively stable and rel-
> atively unstable areas of vocabulary. (4) this is onomatopoeia-influenced vocabulary and its
> opposite. etc. etc.[74]

Such reflections followed an outpouring of critical engagement with the method after
its first formal articulation. Charles Hockett exemplified this response, which held the
door open to further research.[75] As Hockett saw things, the "validity of the procedure"
was "still in doubt, but the possible yield is so great that extremely diligent study is
imperative."[76] While the majority of his critique focused on the tenuous relationship
between history and statistics, he also addressed concerns about the selection of basic
vocabulary data:

> By the basic lexicon is meant a semantically-defined stock of forms which we can be sure
> will be found in every human language. . . . For purposes of statistical treatment it is de-
> sirable for the basic lexicon to be as large as possible. But when one begins to look for
> meanings that are universally or almost universally given lexical expression, the reliable
> list becomes smaller and smaller. *Here . . . a great deal of experimentation is needed. . . .*
> Although much of the necessary experimental and checking work has already been car-
> ried out, the answers are not yet sure; *more work is still needed.*[77]

A growing number of these challenges led to a significant revision of the diagnostic
list.[78] Like Hockett, Swadesh had initially expressed a desire to expand the test list beyond 200 items in order to boost statistical significance, but he quickly discovered that
this compromised the universal validity of individual items, undermining practitioners' ability to interpret observed similarities historically.[79] Rather than expansion, he
thus called for "a drastic weeding out of the list, in the realization that quality is at least
as important as quantity."[80] Thus bigness in lexicostatistics had more to do with a capacity to move beyond the local circumstances of linguistic production than it did with
sheer size.[81]

---

[73] On collecting, see Bruno Strasser, "Collecting Nature: Practices, Styles, and Narratives," *Osiris* 27
(2012): 303–40.

[74] Swadesh to Hymes, 21 July 1954 (cit. n. 71).

[75] Charles F. Hockett, "Linguistic Time-Perspective and Its Anthropological Uses," *Int. J. Amer. Linguist.* 19 (1953): 146–52.

[76] Ibid., 148.

[77] Ibid.; emphasis added.

[78] For a useful overview, see Embleton, *Statistics* (cit. n. 55).

[79] He provided the theoretical justification for this move in a letter to Fred Eggan the year before,
writing: "Lees' conception that there are thousands of words, of which our list is a sample, is wrong.
This is not a sampling, but a diagnostic test. The theory of a diagnostic test is scientifically sound, even
tho the conception is not so widely disseminated as the theory of sampling." Swadesh to Eggan, 25 May
1954, Box 44, Folder 13, Eggan Papers, Special Collections Research Center, University of Chicago
Library.

[80] Swadesh, "Towards Greater Accuracy" (cit. n. 64), 124.

[81] Note the analogy here with Joanna Radin's characterization of bigness in the case of the Pima
Indian Diabetes Dataset (PIDD); Radin, "'Digital Natives': How Medical and Indigenous Histories
Matter for Big Data," in this volume.

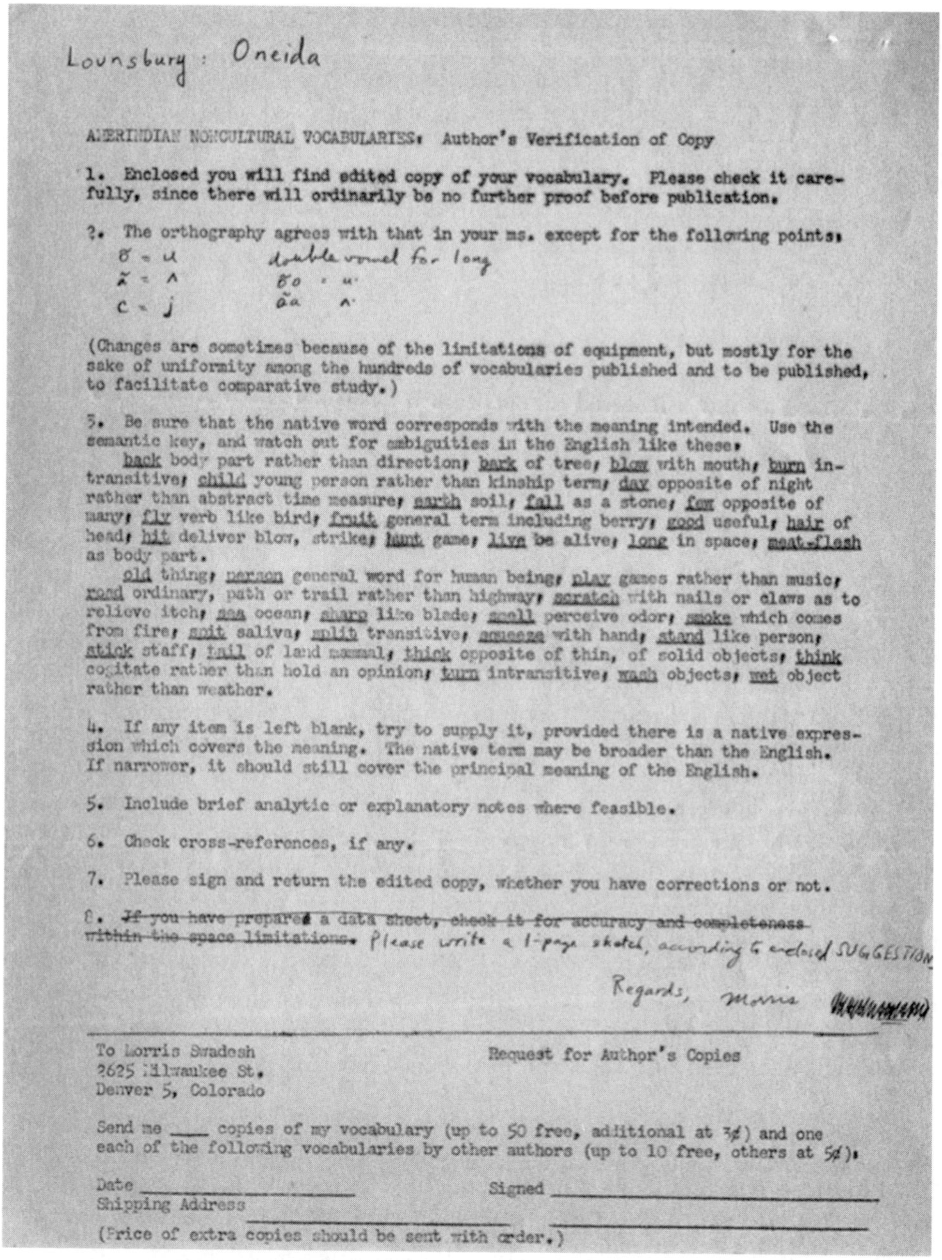

Lounsbury: Oneida

AMERINDIAN NONCULTURAL VOCABULARIES: Author's Verification of Copy

1. Enclosed you will find edited copy of your vocabulary. Please check it carefully, since there will ordinarily be no further proof before publication.

2. The orthography agrees with that in your ms. except for the following points:

δ = u   double vowel for long
ž = ʌ   δo = u·
c = j   ãa = ʌ·

(Changes are sometimes because of the limitations of equipment, but mostly for the sake of uniformity among the hundreds of vocabularies published and to be published, to facilitate comparative study.)

3. Be sure that the native word corresponds with the meaning intended. Use the semantic key, and watch out for ambiguities in the English like these:
    back body part rather than direction; bark of tree; blow with mouth; burn intransitive; child young person rather than kinship term; day opposite of night rather than abstract time measure; earth soil; fall as a stone; few opposite of many; fly verb like bird; fruit general term including berry; good useful; hair of head; hit deliver blow, strike; hunt game; live be alive; long in space; meat-flesh as body part.
    old thing; person general word for human being; play games rather than music; road ordinary, path or trail rather than highway; scratch with nails or claws as to relieve itch; sea ocean; sharp like blade; smell perceive odor; smoke which comes from fire; spit saliva; split transitive; squeeze with hand; stand like person; stick staff; tail of land mammal; thick opposite of thin, of solid objects; think cogitate rather than hold an opinion; turn intransitive; wash objects; wet object rather than weather.

4. If any item is left blank, try to supply it, provided there is a native expression which covers the meaning. The native term may be broader than the English. If narrower, it should still cover the principal meaning of the English.

5. Include brief analytic or explanatory notes where feasible.

6. Check cross-references, if any.

7. Please sign and return the edited copy, whether you have corrections or not.

8. If you have prepared a data sheet, check it for accuracy and completeness within the space limitations. Please write a 1-page sketch, according to enclosed SUGGESTION

Regards, Morris

To Morris Swadesh   Request for Author's Copies
2625 Milwaukee St.
Denver 5, Colorado

Send me ____ copies of my vocabulary (up to 50 free, additional at 3¢) and one each of the following vocabularies by other authors (up to 10 free, others at 5¢):

Date _____   Signed _____
Shipping Address _____
(Price of extra copies should be sent with order.)

Figure 1. Floyd Lounsbury's verification copy for the Oneida wordlist of Swadesh's Amerindian Non-Cultural Vocabulary project. Floyd Lounsbury Papers, No. 95, Folder 15, Morris Swadesh correspondence, APS MS Coll. Reprinted courtesy of the American Philosophical Society.

The new list included just 100 items and boosted the rate of vocabulary retention from 81 percent to 86 percent per millennium. As one reviewer opined, "The closer we get to universal validity, the fewer items we have . . . there will be no items at all on the perfect list."[82] Reflecting contemporary debate, Swadesh acknowledged that universality was an attribute of basic vocabulary subject to empirical testing. Items bound to particular areas of the world ("snow," "snake," "sea") were dropped to facilitate comparison, as were those unacceptably "cultural" in their behavior ("salt" and "rope"). For Swadesh, culture terms were useless in the quest for "archaic residue" because of the "readiness with which they may be borrowed along with the objects" to which they were thought to refer.[83] Swadesh addressed additional technical concerns with redundancy in the list (e.g., "wife": "woman"), sound symbolism (e.g., "blow"), and structurally embeddable words (e.g., "at" and "in"). Finally, he responded to critics who were able to discern heterogeneous rates of change within the test list. This led to further testing, modification, and "pruning,"[84] which generated more and more list-governed data over time.

V. "WE ARE NOW CALLING ON COMPUTERS MORE AND MORE"

The problem of basic vocabulary was taken up in countless publications, international congresses, and collection initiatives in the years that followed.[85] Much of this work addressed the purported universality of the diagnostic lists, spurring pilot studies in parts of the world not yet touched by the Americanists, their Indo-Europeanist counterparts, or the published record.[86] As Harry Hoijer observed, for many the idea was that "further application of the method, to more and more languages and language families" would "make it into an efficient and precise technique of dating."[87] A chief interest in doing so was to bring together distinct lines of evidence, reducing disciplinary friction in the joint pursuit of prehistory.

Despite the coincident rise of machine translation, linguists began to despair that they would never find appropriate matches using the Swadesh "semantic key." In this regard, many questioned the decision to halve the diagnostic list, when this was already said to be just a fraction of the total basic lexicon. Sarah Gudschinsky posed the question then on several minds: "If only 100 items can be postulated as clearly members of the basic root morpheme inventory, is it probable that even these will prove to be in any real sense universal?"[88] Criticism mounted further during the 1960s,

---

[82] Karl Teeter, "Lexicostatistics and Genetic Relationship," *Language* 39 (1963): 638–48, on 642; Knut Bergsland and Hans Vogt, "On the Validity of Glottochronology," *Curr. Anthropol.* 3 (1962): 115–53.

[83] Swadesh, "Towards Greater Accuracy" (cit. n. 64), 125.

[84] "Pruning" of the Swadesh wordlists has marched on into the new millennium. Notable recent revisions are the Leipzig-Jakarta list, which has 62 percent overlap with the 1955 100-item Swadesh list, and George Starostin's selection of fifty "ultra stable" words.

[85] For an overview of this literature, see Embleton, *Statistics* (cit. n. 55), chap. 2.

[86] See, e.g., Arnold C. Satterthwait, "Rate of Morphemic Decay in Meccan Arabic," *Int. J. Amer. Linguist.* 26 (1960): 256–61; David D. Thomas, "Basic Vocabulary in Some Mon-Khmer Languages," *Anthropol. Linguist.* 2 (1960): 7–11; George W. Grace, "Lexicostatistical Comparison of Six Eastern Austronesian Languages," *Anthropol. Linguist.* 3 (1961): 1–22; Leo F. McNamara, "Morpheme Retention in Irish," *Anthropol. Linguist.* 3 (1961): 23–30.

[87] Hoijer, "Lexicostatistics: A Critique" (cit. n. 12), 49.

[88] Sarah Gudschinsky, "Three Disturbing Questions Concerning Lexicostatistics," *Int. J. Amer. Linguist.* 22 (1956): 212–3.

spanning the introduction of computers and paralleling debates within biological tax-onomy. In 1962, just a year before Robert Sokal and Peter Sneath published *Principles of Numerical Taxonomy*, Knut Bergsland and Hans Vogt initiated a conversation in the pages of *Current Anthropology* that ravaged one of the key presuppositions of glotto-chronology—the idea that "fundamental vocabulary changes at a constant rate."[89] With evidence sourced from printed reference works and living correspondents, Bergsland and Vogt demonstrated highly variable retention rates in Old Norse and Armenian, undermining this original claim. Their conclusion focused on inherent problems with "control" study design, as they wrote,

> it seems more important to gather extensive lexical material and to study in detail its re-lation to morphology and syntax and to such extralinguistic factors as social and natural background, than to compile short word lists in ever increasing number, in the extrava-gant hope that this will shed new light on the rate of vocabulary change in human lan-guage in general.[90]

As their comments suggest, "short word lists" had indeed been piling up "in ever in-creasing number." While these were not enough to quiet concerns about the validity of lexicostatistics and glottochronology, they did become a valuable cache of data in their own right, constraining future collecting activities and facilitating work in the interdis-ciplinary study of prehistory.

Energized by the availability of electronic computing, the debate swung from em-pirical questions back to foundational mathematics during the second half of the 1960s. In 1966, for example, archaeologist Nikolaas J. van der Merwe authored a study focused on the problem of heterogeneous retention—not among languages, but instead among items within the diagnostic list. This threatened its cohesion, forcing him to the seemingly ridiculous conclusion that "the test list can be subdivided further and further, until there are as many subdivisions and rates of retention as there are words on the list."[91] Surprising as it was to some of his interlocutors, this proposal seemed entirely feasible to van der Merwe, who could now refer the Sisyphean task to an electronic com-puter. University of Nevada professor William H. Jacobson Jr. concurred. Marveling at the eagerness with which linguists repeatedly took up the dead horse of glottochronol-ogy, Jacobson discerned something new:

> How is it . . . that [van der Merwe] thought it worthwhile to take up again this widely rec-ognized problem and to take the trouble to work out the equations involved? The answer, I think, is to be found in the fact that we are now calling on computers more and more to carry out highly repetitive or intricate operations in linguistics, as in other fields, that sim-ply could not be attempted without their aid.[92]

Here Jacobson was thinking especially of work by Robert Oswalt and Isidore Dyen. Oswalt's willingness to discuss the nuts and bolts of his methodology, in particular,

---

[89] Bergsland and Vogt, "Glottochronology" (cit. n. 82), 125. *Principles of Numerical Taxonomy* launched a parallel approach to classification in biological systematics based on the numerical anal-ysis of character states.

[90] Ibid., 129.

[91] Nikolaas J. van der Merwe, "New Mathematics for Glottochronology," *Curr. Anthropol.* 7 (1966): 485–500.

[92] William H. Jacobson Jr., "Comment," ibid., on 493.

sheds important light on how the Swadesh wordlists met with computerization.[93] Taking aim at the common preoccupation that chance resemblances would mislead lexicostatisticians, Oswalt developed a computer program to determine how many similarities might be expected between two languages under various formal and semantic similarity criteria. His publication of this research in 1970 significantly marked a probabilistic application of electronic computing to problems of long-range linguistic relationship.[94]

Oswalt was fundamentally skeptical of the notion that basic wordlist construction ruled out the noise of chance similarities. To illustrate this danger—particularly for cases involving long-range relationship—he experimented with "shift tests," where matches tallied among basic words were compared to a "background mean." This mean was derived from re-scoring formal similarities when list items were bumped one slot down the line up to the nth position (in this case, 90). Subtracting this mean from the initial judgment of cognancy, Oswalt hoped to arrive at a measure of statistical significance, something still eluding linguists today.[95] Working with the IBM 7090 at Berkeley's Computer Center, Oswalt transferred premechanical routines of comparative-historical linguistics to instruction cards that would determine how the computer subsequently sorted the introduction of basic vocabulary data. "The phonetic range allowed in . . . sets of correspondences is in keeping with what many linguists allow intuitively, though they do not usually explicitly formalize their criteria of similarity," Oswalt contended. "According to the theories of sound change developed in the past, mainly with Indo-European languages, variations in sound correspondences should be explainable."[96] These phonological conditioning factors were directly transcribed onto the instruction cards using hyphens to initiate subroutines where, for example, the t of Italian *vita* would be counted as a match with the d of Spanish *vida* according to further punched information about the fully voiced vocalic surroundings of these consonants.

So far, so continuous. But what became of the 100-item Swadesh wordlist data that Oswalt employed for this study? These were sourced from eight well-known languages using a variety of different collection strategies: Oswalt elicited the Greek, Russian, German, Persian, and Finnish words directly from ESL students he was teaching at the time; borrowed the Hindi wordlist directly from colleague John Gumperz; and indicated that the

> 100-word list in Italian is from John A. Rea, "Concerning the validity of lexicostatistics" . . . which also itemizes the Swadesh 100-word list. The English forms are as given in the Rea article except that *flesh*, *hot*, and *you* were used instead of *meat*, *warm*, and *thou*.[97]

---

[93] Robert L. Oswalt, "The Detection of Remote Linguistic Relationships," *Comput. Stud. Human. Verb. Behav.* 3 (1970): 117–29.

[94] On probabilistic uses of computers, see Agar, "What Difference Did Computers Make?" (cit. n. 20), 892–7; Stevens, *Life Out of Sequence* (cit. n. 21), chap. 2.

[95] See William H. Baxter and Alexis Manaster Ramer, "Beyond Lumping and Splitting: Probabilistic Issues in Historical Linguistics," in *Time Depth in Historical Linguistics*, ed. Colin Renfrew, April McMahon, and Larry Trask (Cambridge, 2000), 167–88.

[96] Oswalt, "Detection" (cit. n. 93), 119.

[97] Ibid., 120.

This citation exemplifies the phenomenon I am calling data drag. Though Oswalt was keen to discredit aspects of lexicostatistics, he marshaled basic vocabulary data defined by lexicostatistical protocols in order to do so, serving to entrench the list further, now in the computerized context of historical linguistics.

Punch cards were capacious but not unlimited; computers saved time but could still be tedious. Basic vocabulary data was simplified considerably to conform to the IBM keyboard and to Oswalt's limited field of inquiry. "Inflectional and other *affixes not critical to the desired meaning of the vocabulary items were segmented off* whenever there was enough information to do so with a comparatively good degree of surety."[98] This simplification conforms to Jon Agar's characterization of mechanization—with increased data processing power, researchers tended to shift their gaze from rare to common phenomena. It further illustrates his point that early computerization was more easily applied to analytic than descriptive research.[99]

Isidore Dyen has also been willing to simplify the information associated with basic words in his long-standing engagement with lexicostatistics and historical linguistics. His description of the Comparative Indo-European Database, published in collaboration with Joseph Kruskal and Paul Black in 1997, highlights the continuity of language data through successive processes of transcription and mechanized handling:

> data were placed on punched cards in the 1960s, and transferred to disc circa 1990. [The database] gives cognation data among 95 Indoeuropean speech varieties. For each meaning in the list of 200 basic meanings (chosen by Morris Swadesh in 1952), this file contains the forms used in the 95 speech varieties collected by Isidore Dyen and the cognation decisions among the speech varieties made by Isidore Dyen in the 1960s. Note, however, that the speech forms are represented only as upper case English letters that were punched long ago into IBM cards, and do NOT contain the diacritical marks that were handwritten onto the surface of the cards. Thus the cognation decisions are provided in full detail, and the forms are described well enough for an expert to check which forms were chosen, BUT . . . THE FORMS ARE NOT DESCRIBED WELL ENOUGH FOR PHONOLOGICAL OR PHONETIC ANALYSIS![100]

Stripped of their diacritical markings, these data are not fit for traditional comparative-historical analysis. Nevertheless, the incorporation of written historical records allows the authors to promote this database as "an important baseline" in the analysis of "language groups for which there is no historical record"—work based on lexicostatistical judgments of overall similarity.[101]

Russell Gray has done just that, sourcing data from Dyen, Kruskal, and Black for some of his most important and controversial papers on prehistorical "phylolinguistics."[102] While most of his discussion has focused on the novelty of the computational phylogenetic methods involved, with any luck this essay has shown that the data in use are much older and thoroughly cooked. Decades of controversy with respect

---

[98] Ibid., 121; emphasis added.

[99] Agar, "What Difference Did Computers Make?" (cit. n. 20), 877, 898.

[100] See http://www.wordgumbo.com/ie/cmp/iedata.txt (accessed 17 July 2015). See also Isidore Dyen, Joseph B. Kruskal, and Paul Black, "An Indoeuropean Classification: A Lexicostatistical Experiment," *Trans. Amer. Phil. Soc.* 82 (1992): 1–132.

[101] See http://www.wordgumbo.com/ie/cmp/ (accessed 15 July 2015).

[102] Russell D. Gray and Quentin D. Atkinson, "Language-Tree Divergence Times Support the Anatolian Theory of Indo-European Origin," *Nature* 426 (2003): 435–9.

to the key claims that basic vocabulary words are especially frequent, universal, and stable over time have reified basic vocabulary as a tool that can be picked up and applied to new stochastic models of human prehistory. If absolute time depths held promise to reduce disciplinary friction in the 1950s, phylogenetic networks enjoy similar attention today.

VI. CONCLUSION

This essay has demonstrated significant continuity in the history of basic vocabulary. The intuition that basic vocabulary hangs together and is especially "criterial" of genetic relationship has been remarkably stable over time. Moreover, the philological foundations on which diagnostic lists of basic vocabulary were first developed go back hundreds of years and have been calibrated over decades of testing and modification. Digitization in recent years has been a fairly straightforward process of direct transcription, give or take a diacritic. Far from a radical break in the history of comparative linguistics, the advent of computing can be seen as the next step in a long line of attempts to make intuitive linguistic knowledge explicit—assumptions about the constancy and homogeneity of the rate of change; the universality of basic terms; and, lately, model parameters and significance criteria.

What threatens to undermine the whole project is the understanding that semantic content cannot be made similarly explicit or objective.[103] This is one of the peculiarities of linguistic data—insofar as it irreducibly concerns meaning, there seems to be an upper limit to the possibility of valid abstraction. Though, as we have seen, basic vocabulary encodes assumptions about the universality of human communicative needs, fundamentally this data is used in studies of the contingencies and particularities of human (pre)history. There would "be no items at all on the perfect list."

But Swadesh and his followers have never shown much interest in perfection. Rather, linguists have presented basic vocabulary data as a "tentative" and preliminary path forward, a way to bring previously intractable questions about very remote time depths into the domain of science. Significantly, this has involved coordinating linguistic results with those from other disciplines—most prominently biology and archaeology. Non-linguists in this new "new synthesis" have greeted linguistic data with enthusiasm because basic vocabulary gives a strong signal when run through phylogenetic models. It is important to temper this enthusiasm with the sober recognition that "such boiled-down material" signals clearly, at least in part, because it has been "prepared" to do so.[104] The evidentiary status of basic vocabulary data, in other words, is considerably lower within linguistics than it is in the study of human prehistory more generally.

As these interdisciplinary encounters suggest, "following the data" often allows us to see historical connections that would otherwise be missed.[105] Through its emphasis on the conceptualization and automatic handling of basic vocabulary, this essay has begun to bridge the gap between the anthropological tradition in linguistics and those

---

[103] Cf. Alexi Kassian, George Starostin, Anna Dybo, and Vasiliy Chernov, "The Swadesh Wordlist: An Attempt at Semantic Specification," *J. Lang. Relation.* 4 (2010): 46–89.

[104] Josh Berson (personal communication with Judith Kaplan, Berlin, 1 October 2015) argues that Austronesian, which has received considerable attention from Gray and colleagues (Greenhill, Blust, and Gray, "Austronesian Basic Vocabulary Database" [cit. n. 4]), behaves well in phylogenetic models because of features that are specific and inherent to languages in that family.

[105] Stevens, *Life Out of Sequence* (cit. n. 21), chap. 4.

concerned with the "inner nature" of language itself.[106] I hope the integrated picture of descriptive and theoretical linguistics that emerges from this account of the "most intimate fund of speech" will justify further study of the relationship between linguistics and the language of data practices going forward.

The simple fact is that language learning is demanding, and linguistic fieldwork can be extremely labor-intensive. At the same time, the reconstruction of linguistic prehistory has been understood to depend on the availability of a maximally broad comparison of known human languages. These realities define a field of collaborative and conservative inquiry where progress is achieved through data accretion, "borrowing," and "second sourcing."[107] Because of these logistics, the sixty-year arc from lexicostatistics to lexomics is characterized by data drag, not data-driven change. As basic vocabulary has been highly conserved, new disciplinary constellations are starting to be formed around it.

---

[106] Hymes, "Lexicostatistics and Glottochronology" (cit. n. 24), 123. As others have argued, the historiography of linguistics has suffered "from being 'blinded by the light' that is Noam Chomsky." See Martin-Nielsen, "'It Was All Connected'" (cit. n. 15), 75.

[107] W. Lewis, S. Farrar, and T. Langendoen, "Linguistics in the Internet Age: Tools and Fair Use," in *Proceedings of the EMELD' 06 Workshop on Digital Language Documentation: Tools and Standards: The State of the Art* (Lansing, Mich., 2006), 2, http://staff.washington.edu/farrar/documents/review/LewFarLang_review.pdf (accessed 6 June 2017).

Tell Data from Meta:

Tracing the Origins of Big Data, Bibliometrics, and the OPAC

by Markus Krajewski*

ABSTRACT

This essay discusses "Big Data" as a historical phenomenon with a long trajectory, dating back to the ancient library of Alexandria and its methods to counter information overload. The essay looks at different situations in which the proliferation of data posed problems and focuses on the development of systems and techniques— for example, bibliometrics—to cope with this abundance. After a brief inquiry into the origins of Big Data as a phenomenon, I will explore a distinction that is necessary to grasp the concept of Big Data, and which proves imperative in order to cope with the consequences of information overload. The distinction between data and metadata will be discussed and developed by dint of the history of the electronic library catalog called OPAC, which fuses two strands of Big Data: quantification and automatization.

THE BEGINNING OF BIG DATA

The literal meaning of the word "data," derived from the Latin *dare*, is "the given," in contrast to the term "fact," which stems from the Latin *facere* and stands for something manufactured, fabricated, generated, and so on.[1] Today, semantics usually work the other way around, since data signifies something generic, while the meaning of "fact" has shifted almost to the realm of something naturally given. What, then, does the epithet "big" denote in conjunction with data? First, it is nothing but a qualifier of a quantity: a huge amount of generated facts, somehow homogenized in the sense that they come in a common yet specific shape usually called a "format"—like this text, a homogenized collection of different facts comes in a certain format, in Latin characters and Arabic numerals (this UTF-8 encoding is, of course, a format) arranged and stored either on a sheet of paper with a specific size (DIN A vs. letter format) or in a portable document format. There are no data without a format delimiting and, at the

* University of Basel, Department Arts, Media, Philosophy, Holbeinstrasse 12, CH-4051 Basel, Switzerland; markus.krajewski@unibas.ch, gtm.mewi.unibas.ch.
[1] For a closer examination of the consequences of this semantic shift, see Hans-Jörg Rheinberger, "Wie werden aus Spuren Daten, und wie verhalten sich Daten zu Fakten?," *Nach Feierabend: Zürch. Jahrb. Wissensgesch.* 3 (2007): 117–25.

same time, enabling any further processing of its content—but if data are just something generated, homogenized, and squeezed into a format, then this term is still extremely wide-ranging and therefore quite vague. And if the basic term, data, is already that broad, how could a catchphrase like "Big Data" gather such attention that it could become a symbol characterizing a new paradigm in the sciences, or even a new era in the way of being digital? In this article, I do not discuss Big Data as a historical phenomenon of the twenty-first century.[2] Instead, I will argue for a longer trajectory for the concept of Big Data by examining the historical situations in which data, in its sheer abundance, became big and played a significant role or posed a major problem, which was then tackled by developing certain measures and techniques in order to cope with this abundance. After a brief inquiry into the origins of Big Data as a phenomenon, I will explore a distinction that is necessary to grasp the concept of Big Data, and which proves imperative in order to cope with the consequences of information overload. The distinction between data and metadata helps avoid the pitfalls of the sorcerer's apprentice drowning in too much data.

Who, since Herodotus and the beginning of history itself, has dealt with and reflected upon Big Data? Candidates for the position include the first polymaths, such as Lucius Cornelius Alexander Polyhistor, known as Alexander of Miletus, who lived in the first century before Christ, and anonymous civil servants in the harbor of Alexandria in ancient Egypt, who dealt with large amounts of economic data in their bureaucratic work. However, at least for oriental-occidental history, this question of who was the first will inevitably lead to an equally simple answer: the foremost experts in Big Data, now as then, are, of course, librarians. The simple reason is that libraries are the most prominent collections and databases of big (and structured) data, in ancient times as well as today. The more complex reason is that over time, librarians had to develop tools and systems to manage increasing amounts of data. In addition, they were eager to use cutting-edge technology to deal with this proliferation of data. These tools include the groundbreaking concept of alphabetical order and the endeavor to store references for everything, which ancient librarians attempted to accomplish as well. Thus, Callimachus, known for his work in the Library of Alexandria in the third century BC, might be another candidate. He lived from about 310 to 240 BC and was more a scholar than a librarian. He may have been the first to confront the problem of Big Data as something too large to be grasped by a single mind and therefore in need of description, abstraction, and indexing in order to be accessed by others.

When Caliph Omar had the famous library of Alexandria set on fire (as the legend goes),[3] the almost equally famous *pinákes* were supposed to be destroyed as well. Callimachus created the *pinákes*, a catalog of about 120 "books" that listed, described, and critically organized the Greek writings of the library while also providing brief biographies of their authors. In this case, the *pinax*, the term for a votive tablet made from painted wood, marble, bronze, or terracotta, or for a wax-covered writing tablet,

[2] When can a set of data be considered "big"? For a critique on the scalability of data into bigness, see Hallam Stevens, "A Feeling for the Algorithm: Working Knowledge and Big Data in Biology," in this volume. Christine L. Borgman, in *Big Data, Little Data, No Data: Scholarship in the Networked World* (Cambridge, Mass., 2015), discusses different categories of data scales and their characteristics. On the emergence of the use of the term "Big Data" since the late 1970s, first as a challenge and more recently also as a promise, see Rebecca Lemov, "Anthropology's Most Documented Man, Ca. 1947: A Prefiguration of Big Data from the Big Social Science Era," in this volume.
[3] For arguments that the library was not burned, see Luciano Canfora, *Die verschwundene Bibliothek* (1986; repr., Berlin, 1988).

was made of parchment or papyrus containing a table. It was placed above library shelves to list the authors' names and to summarize the respective literary genres. The tragedians and the lyricists appeared in different columns. Thus, *pinákes* brought a kind of systematic order to the library's holdings. Although we do not know exactly how Callimachus produced his catalog—it is assumed that he received help from the library's staff—his work provided at least one thing: a schema for organizing long lists. On the one hand, Callimachus organized the texts according to the authors' names, which may be the first extended use of alphabetical order; on the other hand, each *pinax* also creates a systematic order that might be considered a model for a library's overall classification, though it remains fragmentary by nature.[4] The word "catalog," according to the Greek origin of the term, means "I am enumerating," or "I am specifying." This meaning of enumeration underlines the specific relevance of numbers, counting, and computing from the very beginning of librarianship. The most important feature or invention of Callimachus, however, is the fact that he distinguished between data and metadata, when he jotted down the authors' names, their biographies, their other works, their work's title, and the first lines of its content. He even provided a stichometry of the papyruses he found on the shelves, a computation of the lines included in the papyrus rolls. Callimachus ultimately provided an extensive inventory of the Greek literature of his time, ordered by genre, in addition to his comprehensive alphabetical list of authors. In short, Callimachus is not simply the first known cataloger in a long line of famous scholars that includes Leibniz and Lessing. With his stichometry and his cataloged and numbered rows of manuscripts, he undertook the first bibliometric exploration in the history of knowledge.

Callimachus seems to be one of the first scholars who knowingly distinguished between data and metadata. This distinction is important because the role of metadata cannot be overstated in the context of huge amounts of information. Metadata enables users and developers to control and navigate the sea of information. Thus, this creation of metadata implicitly recognizes the library not only as the place of knowledge production but also as the most prominent site where Big Data resides for future processing. Without this processing, the collection of information would be an idle task since nothing could be retrieved. Only categorization by headwords, only indexing of the vast writings turns them into formation on a second level, beyond the formation of letters and words in a unidirectional, linear sentence. The processing of the words literally forces them "in formation" and transforms them into information on a different level. The description of the data's content, its abstraction into metadata, makes the information accessible to other readers who can then use it as input in various informational processes.

The library serves as the classic place where Big Data is at issue. But what makes the library, apart from its mere accumulation of material, its abundance of words and letters, a hot spot of Big Data? The answer is rather simple: over the course of centuries, the library is where strategies have been developed to get synopses of the overwhelming amount of written material;[5] at the library, tactics for organizing data with

---

[4] See Friedrich Schmidt, *Die Pinakes des Kallimachos*, vol. 1 of *Klassisch-Philologische Studien* (Berlin, 1922), 49–50, 57–8. See also Rudolf Blum, "Kallimachos und die Literaturverzeichnung bei den Griechen," *Arch. Gesch. Buchwes.* 18 (1977): 1–360.

[5] See Ann Blair, *Too Much to Know: Managing Scholarly Information before the Modern Age* (New Haven, Conn., 2010).

metadata have been proven and applied,[6] perhaps even invented. Where, if not here, can the key to Big Data be found? And what are the specifications of library information processing in detail?

First, it is one of the oldest techniques of knowledge classification, again dating back to Callimachus and his successors in medieval and early modern libraries: setting up the collection of books according to the order and classification on the shelves; in this case, the location of the books generates relevant data itself. The spatial proximity between two books is a crucial bit of information because titles are grouped by subject. The arrangement of books serves as metadata that allow a specific orientation in the library halls. In German, the rich term *Aufstellung* is used for this technique.[7] Thus, if Big Data is all about correlation and less about the individual item, like a book, a sentence, a word, or even a syllable, but rather all about the context, then it is this spatial proximity, the close relationship that makes similar hits retrievable while not only searching for a book but also seeing its immediate surroundings.

Second, it is the catalog, divided in alphabetical and systematic versions, both of which originated with Callimachus, as we have seen. For centuries, the library's catalog served only as an inventory. As late as the end of the eighteenth and the beginning of the nineteenth century, this medium became a document in its own right, entailing meticulous preparation as well as careful exegesis.[8] In order to build a catalog, it is critical to examine the books closely, analyze them, perform an autopsy, to use the correct librarian's term, which means to write a phenomenological description of the exterior of the book, and then open it with a surgeon's fingers, in order to describe the interior and the internal relations to be found in the content.

Finally, it is the subject cataloging, a specific subdiscipline of the inevitable analytics of cataloging, where the formal characteristics like the size of the book, author's name, publication date, year, and so on are recorded. Also, the content of the text is mirrored by a selection of headwords. Traditionally, this is done manually, or, in the terms of librarianship, "intellectually." However, during the last few decades, this process has been delegated to machines. In other words, the crucial words of a text, the main terms of Heidegger's *Sein und Zeit*, besides "time" and "being"—for instance, the "call of consciousness" or *Geworfenheit*—are supposed to be located, analyzed, and listed automatically by a machine rather than a skillful reader.

Thus, the problem of subject cataloging, that is, how to derive crucial information from large amounts of data *without* reading all of the contents "intellectually," has been delegated to computers in libraries. And those machines offer many more services, of course, than simply computing the numbers of lines on papyrus rolls as Callimachus's stichometry did. From the end of the eighteenth century until the middle of the twentieth century, one central instrument ruled the organization of knowledge, and not simply in the library: the index or catalog, either bound or made of ephemeral cards. It dominated the realm of knowledge administration, not only in the context of what is known today as information science, founded by the Benedictine monk Martin Schrettinger in 1808: *Bibliothek-Wissenschaft*. During the fin de siècle, the catalog was also gradually disseminated into offices and incorporated itself into accounting

[6] For another example of the use of metadata, see Patrick McCray, "The Biggest Data of All: Making and Sharing a Digital Universe," in this volume.
[7] Uwe Jochum, *Bibliotheken und Bibliothekare, 1800–1900* (Würzburg, 1991), 14–5.
[8] Ibid., 24–5.

techniques; it subsequently influenced early electronic data processing by mainframe computers. From the world of business accounting, this technology, then, flows back into the library.

The following section will briefly outline this different tradition of working with books by means of statistics rather than reading, after libraries adopted computers initially for administrative purposes and later for everyday users. This was a twofold process that happened during the 1960s and 1970s—though it had its predecessors— mainly in the United States, eventually resulting in the Online Public Access Catalog (OPAC).[9]

A (TRANSLUCENT) HISTORY OF THE OPAC

From the end of the eighteenth century until the mid-twentieth century, the classical card catalog with its movable elements dominated the realm of cataloging, bringing the flexibility of dynamic information processing into the context of librarianship.[10] Early electronic experiments concerning online information retrieval in libraries can be traced back to 1954, though serious attempts to automate the library with the help of electronic systems did not occur before the early 1960s. It was not until 1974 that the first OPACs were mounted, when early users were staring at pale cathode tubes at the recently founded Online Computer Library Center (OCLC) of Ohio State University Library in cooperation with the Research Libraries Group/Research Libraries Network (RLG/RLIN), an association of the East Coast's largest libraries.[11]

In this early experimental stage, the skepticism of both librarians and engineers concerning electronic equipment was considerable. And they expected similar attitudes from library patrons and readers, ranging from noble reserve to open disapproval. On the one hand, library staff assumed that readers would prefer the world of the analog, that is, the well-established card catalogs, instead of unfamiliar and clumsy new devices. However, this turned out to be a conservative prejudice, a librarian's fantasy, since the first surveys of users showed that readers were willing to try new methods of

---

[9] For various reasons, I cannot attempt a detailed investigation of the roots of the OPAC in this essay. An important methodological concern is the difficulty of pursuing the cultural or social history of software. This subdiscipline is still in its infancy and has yet to be developed. For first approaches, see Susan Elliott Sim, "A Small Social History of Software Architecture," in *Proceedings of the 13th International Workshop on Program Comprehension (IWPC 2005), 15–16 May*, ed. IEEE Computer Society (St. Louis, Mo., 2015), 341–4; Capers Jones, *The Technical and Social History of Software Engineering* (Upper Saddle River, N.J., 2014). A promising concept, because it follows a rather theory-laden paradigm, is the work of Jörg Pflüger, "Writing, Building, Growing: Leitvorstellungen der Programmiergeschichte," in *Geschichten der Informatik. Visionen, Paradigmen, Leitmotive*, ed. Hans Dieter Hellige (Heidelberg, 2004), 277–319; see also Pflüger, "Konversation, Manipulation, Delegation: Zur Ideengeschichte der Interaktivität," in ibid., 367–410. As a metadata-driven analysis of this history, see also Martin Campbell-Kelly, "The History of the History of Software," *IEE Ann. Hist. Comput.* 29 (2007): 40–51. Though a comprehensive social history of software has yet to be written, the OPAC movement could serve as a model for a collaborative software development experience and the necessary communications structures that are laying the groundwork for later patterns, e.g., in the open software movement, since the OPAC started during an early stage of cooperative software projects.

[10] See Markus Krajewski, *Paper Machines: About Cards and Catalogs, 1548–1929* (Cambridge, 2011). For the advantages of mobile elements, exemplified by punch cards and their predecessors for manual use, see also Christine von Oertzen, "Machineries of Data Power: Manual versus Mechanical Census Compilation in Nineteenth-Century Europe," in this volume.

[11] Charles R. Hildreth, *Online Public Access Catalogs: The User Interface* (Dublin, Ohio, 1982), 2–3.

research like "ducks accept the water."[12] On the other hand, the complex electronic devices made a strong impression on the staff; the age of the PC was still ten years in the future. A mainframe computer would occupy a significant amount of space and would have to be in an area of the library not open to the public, space that at one time would have been used for cataloging or storing books. It is not surprising that librarians thought computers were quite complicated to operate (in addition to requiring a large amount of space) and questioned their usefulness, especially considering the costs of purchasing and maintaining them.

As a very skeptical librarian pointed out in 1971, the "electronification," or digitization, of the metadata of books, especially of small and local book collections, looked like "an action tantamount to renting a Boeing 747 to deliver a bonbon across town."[13] In rhetoric, this comparison simply means that the *aptum*, the appropriate goal, has been missed: according to this librarian, it was simply not appropriate to work with computers in libraries. The library users seemed unimpressed by this prophecy of doom, since access to the OPACs increased steadily and without the anticipated reservations. In fact, the opposite proved to be true, and, therefore, the data basis of the OPACs was also steadily enhanced and expanded by other collections and new bibliographical data, gradually becoming the "true local information center."[14]

Following the classification of Charles Hildreth, the protagonist behind this cataloging system, there are three generations of OPACs.[15] The first generation consisted of nothing more than a clumsy, nongraphical simulation of card catalogs in the virtual realm. It was not until the 1990s that a second generation became more flexible by including multimedia elements like more intuitive graphical user interfaces, or more differentiated search queries. Only with those features did the paradigm and the era of the card catalog gradually disappear, even in electronic contexts.

However, it took until the third generation, the so-called E'OPACs of today—the enhanced, expanded, and extended OPACs[16]—for the system to become a substantially different provider of information, even changing its ontological status. Inasmuch as the former catalog integrated whole bibliographies that themselves contained digital full texts, indices, reviews, articles, images, audio files, and other media of all kinds, the catalog's former distinction of data and metadata collapsed. The system that others called the catalog resembled more and more "the universe which others call the library," to quote Borges's famous phrase,[17] because it evolved into an all-encompassing information center that led from metadata to data, from bibliographic entry to full text with a simple click. The catalog began turning into the universal library itself.

The final step in this development consisted of linking those local information centers together into one universally connected system. In short, the whole knowledge of

---

[12] Shiao-Feng Su, "Dialogue with an OPAC: How Visionary was Swanson in 1964?," *Libr. Quart.* 64 (1994): 130–61, on 136–7.

[13] Ellsworth Mason, "The Great Gas Bubble Prick't; or, Computer Revealed—by a Gentleman of Quality," *College Res. Libr.* 32 (1971): 183–96, on 183.

[14] Su, "Dialogue" (cit. n. 12), 144.

[15] Hildreth, "Advancing toward the E'OPAC: The Imperative and the Path," in *Think Tank on the Present and Future of the Online Catalog: Proceedings*, ed. American Library Association (Chicago, 1991), 17–38, on 21–3, 37–8.

[16] Ibid., 17. Any resemblance to military terminology may be purely coincidental. Or are command, control, communications, and intelligence at stake at the OPAC as well?

[17] Luis Borges, *Die Bibliothek von Babel: Erzählungen*, trans. Karl August Horst and Curt Meyer-Clason (1941; repr., Stuttgart, 1974), 47.

the libraries is interconnected with, and institutionalized in, an all-encompassing network structure offering a maximum range operating under descriptive addresses like worldcat.org. It would be inappropriate to mention another famous site that might come to mind in this context. No, books.google.com should not be mentioned here. Not because it resembles a jumbo jet but because it evades even the most basic librarian's schema like a reliable differentiated field search. Anyone who has searched for the second edition of Johann Jacob Moser's autobiography through the interface of books.google.com will understand the deep gulf that still persists between an OPAC and a common search engine. One might object that this problem is singular and idiosyncratic; however, Geoffrey Nunberg's careful general testing of this search engine for scholarly needs proves this objection wrong. In his systematic search for errors and flaws in Google Books' search feature, Nunberg found a wide range of examples. Among the errors he found were incorrect publication dates (e.g., 325 books about Woody Allen before he was born, eighty-one hits for Rudyard Kipling before 1865, etc.); odd attributions ("*Madame Bovary* By Henry James"); questionable classifications: "An edition of *Moby Dick* is labeled Computers; *The Cat Lover's Book of Fascinating Facts* falls under Technology & Engineering. And a catalog of copyright entries from the Library of Congress is listed under Drama (for a moment I wondered if maybe that one was just Google's little joke)"; and alternative titles of well-known works ("*Moby Dick: or the White Wall*").[18] Obviously, it leads to different views if a web search engine (technology) is mistaken for (or used as) a library catalog. In other words, Google Books is more a mess than it is the promise of a finely ordered pile of books with a meticulously refined set of metadata that enables most sophisticated research tasks with its database.

As the opening statement of this essay already claimed, in history, it is quite hard to find a truly unprecedented idea, especially as it concerns the fantasy of universally connected information centers. This reservation also applies to electronic search engines' claim to gather all the information of the world in one (virtual) place. Therefore, a brief historical digression is essential, in order to widen the perspective of the origin of the concept and overcome the US-centric, Silicon Valley–narrowed view of historical inventions in information technology.

The history of innovation demonstrates that most "new" inventions or ideas are modifications of an older insight or a fusion of two or three existing ideas. The trick is to skillfully conceal the giants on whose shoulders every inventor, entrepreneur, R&D department, or researcher apparently stands. Most of these people, however, do not stand on the shoulders of giants but rather on those of forgotten dwarves,

---

[18] All quotes from Geoffrey Nunberg, "Google's Book Search: A Disaster for Scholars," *Chronicle of Higher Education*, 31 August 2009, http://chronicle.com/article/Googles-Book-Search-A/48245/ (accessed 20 September 2016). Google, ironically, hearkens back to the old system by keeping digitized versions of historical library catalog facsimiles. However, since the electronic indexers do not read or understand the sense of the arrangement, it is neither analyzed as a model nor adapted to their system. The unorthodox use or limited value of Google Books' extended field search is also criticized by Dirk Lewandowski, "Der OPAC als Suchmaschine," in *Handbuch Bibliothek 2.0*, ed. Julia Bergmann and Patrick Danowski (Berlin, 2010), 86–108, on 91. A more recent, empirical account of the fallacy and failures of Google's Book Search is provided by Klaus Graf, "Dramatische Verschlechterung bei der Arbeit mit Google Books," *Archivalia*, 13 October 2016, https://archivalia .hypotheses.org/59938 (accessed 28 October 2016), as well as Graf, "Wie Google Books uns derzeit gewaltig in die Irre führt," *Archivalia*, 15 October 2016, https://archivalia.hypotheses.org/60007 (accessed 28 October 2016).

who not only take on the necessary drudgery but also often provide unique insights. Therefore, it can hardly be surprising that the promise of representing all knowledge in one place is about as new as yesterday's papers.

Today, the aim of worldcat.org or books.google.com is to collect the totality of information in the world of texts, but at least three smaller-scale projects could claim to be the forerunners of these enterprises.[19] In 1897, an imperial librarian in Berlin, Christlieb Gotthold Hottinger, announced a project that he called "Book Slip Catalogue and Bio-Icono-Bibliographical Collection," which was meant to be an archival storehouse of total world knowledge composed of paper slips containing bibliographic information. At about the same time, another endeavor was starting up in Brussels—with exactly the same purpose, although it used ready-made index cards instead of homemade paper slips—when the Institut International de Bibliographie was founded by Paul Otlet and Henry La Fontaine as the "memory of the world."[20] Later known as the Mundaneum, its goal was to copy and unify the titles from large libraries and to catalog every new publication on index cards. In 1911, while Hottinger was starting a second attempt to unite the world's data at his home in Berlin, another project, called the Bridge, was initiated by Karl Wilhelm Bührer. His main idea was to establish an "information center of information centers."[21] Also described as the "World Brain" (at least by its initiators), the Bridge aimed to collect all the world's information in its printed form. If the Mundaneum collected bibliographic data from all the books in the world, the Bridge sought to gather together all other intellectual products, particularly sculptures, images, sagas, and popular sayings, an all-encompassing survey of all thoughts, classified and categorized with the help of Dewey's Decimal Classification System, but also a comprehensive directory of the addresses of all intellectual laborers. Munich would become the home for Bührer's center of information and meta-information, providing answers to questions from all over the world. As in Otlet's and Hottinger's projects, the core aim was to be a huge card catalog containing all information, including a complete set of bibliographic data of all books ever printed, as well as a universal dictionary containing all the languages of the world—the "Weltwörterbuch."[22] One could address any kind of question to the Bridge and expect to receive more information, a list of valuable references for further research, or, if they were fortunate, the direct answer to the inquiry—"42," for example. However, in June 1914, after three years of cataloging and collecting, the Bridge went bankrupt, and when the bailiff closed its office, he found the World Brain in a state of incompletion.

---

[19] Others could be added, e.g., Aby Warburg's library; see Markus Krajewski, "Mobility on Slips, or: How to Invest in Paper: The Aby Warburg Style," *J. Phil. Photogr.* 9 (forthcoming, 2017). See also Emily Levine, *Dreamland of Humanists: Warburg, Cassirer, Panofsky, and the Hamburg School* (Chicago, 2013); or the work of Herbert Haviland Field described by Colin Burke, *Information and Intrigue: From Index Cards to Dewey Decimals to Alger Hiss* (Cambridge, Mass., 2014).

[20] See, e.g., Paul Otlet and Ernest Vandeveld, *La Réforme des bibliographies nationales et leur utilisation pour la bibliographie universelle: Rapport présenté au Vième Congrès international des éditeurs* (Brussels, 1906); for a historical analysis of Paul Otlet, see W. Boyd Rayward, *The Universe of Information: The Work of Paul Otlet for Documentation and International Organization* (Moscow, 1975); Paul Otlet, *International Organisation and Dissemination of Knowledge: Selected Essays of Paul Otlet,* trans. and ed. with an introduction by W. Boyd Rayward (New York, 1990).

[21] Karl Wilhelm Bührer and Adolf Saager, *Die Organisierung der geistigen Arbeit durch "Die Brücke"* (Ansbach, 1911), 78 (my translation).

[22] Ibid., 107.

Nothing remains of Hottinger's project. Meanwhile, the Mundaneum increased its collection to 14 million index cards that remain more or less intact—this long outdated collection is now housed in Mons, near Brussels, a monument to the necessary failure of well-intentioned if hubristic attempts to aggregate all information. In the case of the Bridge, the contents of the World Brain are, like the Mundaneum, still extant. Unlike the Mundaneum, however, its legacy could fit inside a shoebox. It amounts to a couple of stamps issued by companies for advertising purposes and a complete collection of picture postcards of the Bavarian town of Ansbach. This still-extant, material evidence raises the question even more urgently: What will be the remnants of books. google.com in 2117, 100 years from now? Perhaps the racks stuffed with electronic gadgetry on those sites called server "farms" (as data cattle are tended there) will be hazardous waste or they will resemble the car dumps in the deserts of New Mexico.

Why did all three of these early attempts at creating global search engines fail? Not because Hottinger and the others did not have access to contemporary technologies (blogs, social media, etc.) for directing more attention to their ideas. (Like the Bio-Icono-Bibliographical Collection? Thumbs-up.) One reason for the lack of success of these hubristic projects lies in the definitions of the projects themselves, and the important distinction between the person who conceives the scheme and the staff who are in charge of realizing the idea. It is one thing to build castles in the air but quite another thing to implement such schemes, and the planners simply did not consider the hard work involved in collecting all the information in the world. The promises of the technical media of the time—the telegraph, the radio, the new global transportation system of ocean liners synchronized with railroads and coaches—suggested an ever-shrinking, connected world in which such grand projects seemed plausible.[23] But if the World Brain consisted of nothing but a complete collection of postcards of Ansbach, it was because Bührer took the idea of completeness to be certain knowledge rather than information whose validity had to be proven through the success of his project.

Furthermore, the promise of an all-encompassing catalog usually proves to be misguided because there is a great difference between a thing and a thought, and between a thought and a reference. A catalog is nothing but a necessarily limited collection of references or pointers that cannot represent a totality. It is only a collection of references that point to thoughts or real entities. Something is always missing. A catalog, therefore, always remains a virtual order of things. But this very fundamental premise seems to have been forgotten by both the inventors of the turn-of-the-century cataloging projects and by those currently scheming to organize the world's knowledge as a whole. What to do with books.google.com and its information if it does not even meet the basic standards of bibliographic cataloging? How to deal with its often false or unverified metadata? Furthermore, search engines can only provide disordered information in bits and pieces, but no answers at all—those must be generated by the users themselves, who have to turn the brittle fragments of information into a new narrative, something that actually constitutes knowledge.

What might be the advantage of OPACs over the usual search engines, if the collection—unlike the historical projects mentioned above—is aware of its principal limi-

---

[23] For details about this line of thinking, see Markus Krajewski, *World Projects: Global Information before World War I* (Minneapolis, 2014), chap. 1.

tations? To know the boundaries is always better than to claim to reference everything, especially if everything comes in large numbers (and the second page of results is ignored by the users). In particular, OPAC's ability to access selected articles or even whole books emancipates the catalog not only from the individual library,[24] but also from the concrete, sometimes fetishized book as a paper copy. Thus, the reader has no need to care about a specific call number, interlibrary loan, or a book's glocal (= global + local) position, whether on the shelves where he or she was supposed to find this book and others or on another continent, because the real place of storage is removed from the user's attention or knowledge. Each access to the OPAC is by definition a remote access to standardized data, the actual location of which is known to (almost) no one. And, in addition to this spatial presence, yet another entity seems to be outdated: the "sub-sub-librarian" (as Herman Melville referred to this species)[25] or, in other words, the staff member formerly known as the library servant, that is, the assistant who carried the books from the outer stacks or most remote shelves to the reader's desk, where they were arranged by his superiors. The future of those people seems to be inevitable idleness, if no one requests "real" books anymore. This could be the end of both brick-and-mortar libraries and the history of the origins of Big Data.

FROM BIBLIOMETRICS TO INFORMATION PROGNOSTICS

Almost anything can be data. As long as it is abstracted, in words, an apple can be measured, quantified, and categorized, and in that way it becomes data. For example, take an object that weighs 143 grams, with a spherical shape and a green, gleaming, waxy surface. The analysis, measurements, description ("autopsy"), and quantification suggest that the object might be an apple. One takes a bite, to derive information about its taste, fruitiness (limited), resistance to biting (strong), sweetness (nonexistent), the sound it makes when breaking apart (crumpling), juiciness (rather dry), the variety of flavors that ranges from echoes of early brambles to the strong and sharp smell of seagrass in the sun. Alas, the first bite reveals that it is actually an unripe orange. Because of confusion between data and metadata, one now has to endure a furry taste on the tongue. Metadata only combine specific elements of an object in order to create a seemingly better classification. After all, it was not entirely wrong to assign the object to the "fruit" category. The fact that the "ripe/unripe" criterion was not part of the current analysis is a temporary problem that is easily rectified through a system update. However, the difficulties of accessing content/data through metadata/descriptors pose a much more serious problem, no matter how comprehensive the descriptors (the so-called ontology of the object) are. Any categorization can only be an approximation. One should not easily confuse apples with oranges.

But what exactly are metadata? What purpose do they serve? An established definition summarizes metadata as "data about data," thus implementing the common usage of the Greek preposition μετά.[26] Therefore, metadata is information that makes

---

[24] The copyright issues raised by this accessibility are not discussed here.

[25] Herman Melville, *Moby-Dick; or, The Whale*, 2nd ed. of the 150th anniversary ed. (1851; repr., New York, 2002), 8.

[26] For a critique of the "meta" and its philosophical genealogy, see Tim Boellstorff, "Die Konstruktion von Big Data in der Theorie," in *Big Data: Analysen zum digitalen Wandel von Wissen, Macht und Ökonomie*, ed. Ramón Reichert (Bielefeld, 2014), 105–31, on 113.

it easier to structure, process, regulate, and control—in short, handle—other information. Normally, metadata include information about who compiled which data in which format, for example, in a file. They do not directly relate to the content; nevertheless, they play an important role in further processing that content.

What are metadata for books or, more generally, for texts and other complex repositories of knowledge? It is not surprising that libraries in particular have a long tradition of collecting, processing, and servicing metadata. Again, Callimachus's endeavors in categorization, including stichometry, can be pinpointed as the beginning of this tradition. As a matter of fact, all kinds of paratexts can be understood as metadata of the text. This includes the whole apparatus of textual practices: from the year of publication, a convention since the Gutenberg age, to the place of printing, the publishing house, the table of contents, indices, registers, page references, marginalia, and, in modern monographs at least, the jacket blurb. Complemented with keywords that characterize the content, these metadata constitute an essential part of the library catalog record. Furthermore, they serve as a starting point for bibliometrics, a method that quantifies texts and measures their circulation.

According to Alan Pritchard's definition from 1969, bibliometrics means "to quantify the processes of written communication."[27] Pritchard first traced this term back to none other than himself. However, he did not consider that the now-famous Belgian philanthropist and librarian Paul Otlet used this term as early as 1934, when he and his aforementioned Institut International de Bibliographie in Brussels and their collection of 14 million index cards proposed to measure books in this modern sense, analyzing the bibliography of each text, tracking down cross-references, and so on—in short, analyzing the web of science citations. Of course, the emerging idea of the science citation index has an even older history, dating back to judicial reports and journals: *Raymond's Report* of 1743, *Douglas' Reports* of 1783, or *Shepard's Citation* of 1873 can be seen as the starting point of this concept.[28]

What is gained by measuring books instead of simply describing them? Is it not sufficient to have only a title, the author's name, which is—at least since the eighteenth century—a brand of its own, a few headwords, and a publisher and its location? The most prominent feature of metadata is the ability to control—in the broadest sense of this process. The most interesting feature of metadata is that they can serve as a tool of control or surveillance. An example where metadata have been used to get information about information offers insights into a strange control-dispositive, which was intended to monitor accountants' use of pencils in a 1930s office:

> The control of the pencils which of all the stationery are the most precious could be handled in such a way that the blunt or consumed pencils are thrown into the mail outbox attached to a flag made of strong cardboard taken from the material supply. This flag, which bears the same information as the previously described wrapper, is transferred by in-house mail to the supply department where the pencil is either sharpened or placed in the reserve material box or replaced by a new one. In the latter case the recipient will

[27] Alan Pritchard, "Statistical Bibliography or Bibliometrics?," *J. Document.* 25 (1969): 348–9, on 349.
[28] For a brief history of this project, see the remarks by the Citation Index's "inventor," Eugene Garfield, "The Evolution of the Science Citation Index," *Int. Microbiol.* 10 (2007): 65–9, and his "invention," outlined in Eugene Garfield, "Citation Indexes for Science: A New Dimension in Documentation through Association of Ideas," *Science* 123 (1956): 108–11.

be charged in the "list of pencils issued." This list documents the pencil consumption of each accountant and each department, and it will be tabulated on a monthly basis in order to transfer the results to the competent authorities.[29]

Statistics of individual pencil use are generated in order to control the clerks. It is quite easy to translate this example to the present: if you know how much (or how little) one suspicious subject (a lazy clerk, e.g.) is writing, you know a lot about his or her productivity—even if you do not know the content, what exactly is written. And if the authorities know with whom he or she is communicating, how long and how often, even more metadata is analyzed, which ultimately renders a more detailed picture of the suspect's behavior. Additionally, this pencil-control-dispositive proves the fact that metadata are used to control both data and users of data.

Are there any statistical advantages when library computers control the readers? What is gained when readings are not only listed but also measured and analyzed in electronic contexts? When using metadata as data for correlations between the books, their reading statistics, their citation statistics, their word statistics, it becomes technically easy not only to count how often texts are retrieved, but also to know which passages are read (Amazon's Kindle already proved this feature), to what extent, close or far, intensively or fast; this information may provide new and different paths for future consultations because the catalog already knows what you are interested in. The catalog may have suggestions for you about what to read next. However, not in the Amazon sense of suggestions like: "others who read this book also read," but with a more subtle algorithm that correlates your actual turning of the pages, your mouse movements or fingertips on the track pad (telling name!), the time span you have spent reading this passage—all those parameters can be measured already, they are brought to you, with compliments, by your catalog. The next step, then, correlates the terms to be found on these pages with the whole set of Big Data where similar groups of words can be found, in order to present those texts with the highest resemblance or statistical proximity. In short, the catalog becomes your "smart" research partner while suggesting new texts. And this selection is based on nothing more than pure statistical correlation, not on causality. The catalog has the power not only to navigate to your future readings, it has the technology to control and influence the selection. The list itself becomes smart.

So, why is the E³OPAC such a landmark in comparison to former catalogs? It is not only because it changed its media basis from paper to bits. Due to this media and format change, it also offers a much richer basis for connected and accumulated information. It exceeds its ontological status as a catalog while itself becoming a text. Due to its extensibility on the fly, its endless records, but most of all because of its capacity for swapping the use of data and metadata, the OPAC and its entries allow a new kind of connectivity between the references. The catalog operates as a huge table and table of contents, with browsable pages. Leaving all copyright issues aside, the OPAC theoretically becomes the embodiment of the "absolute book," the text that contains all possible information, a fantasy for which the early Romanticists were already

---

[29] Irene Margarete Witte, "Amerikanische Büro-Organisation," in *Psychotechnik der Organisation in Fertigung, (Büro-)Verwaltung, Werbung*, ed. Irene Margarete Witte, Johannes Wiedenmüller, and Hans Piorkowski (Halle, 1930), 55–88, on 74 (my translation).

searching.[30] It is one of the OPAC's strengths that it has command over both data (references of the text) and metadata (information as to its usage) that can be taken into account while the search process is active.[31] Once an entry from the hit list is selected after issuing a query, the metadata enables the client to move to the full-text pdf file. So, the OPAC works as a machine that converts metadata to data again, while it collapses the vital difference between metadata and data on the fly.

The first generation of OPACs had clumsy interfaces that bore little resemblance to card catalogs, the sophisticated tools used for more than 150 years, which electronic tools were intended to replace. The second generation was more user-friendly, but still a highly mimetic digital arrangement of what functioned perfectly in the analog world. Only the third electronic catalog generation, the E³OPACs, offered something that an analog card catalog could never achieve: they became a rather rich text while not only integrating a plethora of data and metadata but also directly linking full texts, whole books, and articles into their texture. The OPAC evolved into "new media" in its own right: today, the OPAC includes more and more details, tables of contents, subject headings, indexes, reviews of books; it can be read almost instead of the content itself. During this substitution the information about the content turns back into the content itself, metadata are converted to data again. This ability to constantly raise and collapse the distinction between metadata and data, by seamlessly paving the user's path from references to full texts and rich information and back again, turns out to be the OPAC's crucial feature. What the OPAC demonstrates, then, is that the distinction between data and metadata is always fragile, view dependent, and highly dynamic. However, what are the practical as well as epistemological consequences of this collapsing difference?

The general history of (big) data is the story of storing and retrieving, of shuffling, transferring, and, above all, processing information in order to know more about the usage and behavior behind the data.[32] In this sense, the OPAC, the "opaque," is a perfect term for hiding and also revealing this usage. To generate further knowledge of clients' or users' behavior, more or less sophisticated algorithms are applied, either on paper, as with bibliometrics in the late eighteenth century, or with computer programming languages. Big Data's whole arrangement of information processing has one main goal, generating and analyzing correlations between heterogeneous data. For example, the death rates caused by cholera distributed over a territory, in Soho, London, for example, when the British physician John Snow counted and related the deaths that occurred near a specific water pump on Broad(wick) Street in 1854. As Steven Shapin notes, "The map Snow produced, in 1854, plotted cholera mortality house by house in the affected area, with bars at each address that showed the number of dead. The closer you lived to the Broad Street pump, the higher the pile of bars."[33] The bars on the map indicated the danger itself, but, more important, they allowed

[30] See Jens Schreiber, *Das Symptom des Schreibens: Roman und absolutes Buch in der Frühromantik (Novalis/Schlegel)* (Frankfurt am Main, 1983), 129–41.

[31] See also Lewandowski, "Der OPAC als Suchmaschine" (cit. n. 18), 91.

[32] See Viktor Mayer-Schönberger and Kenneth Cukier, *Big Data: A Revolution That Will Transform How We Live, Work, and Think* (Boston, 2013).

[33] Steven Shapin, "Sick City: Maps and Mortality in the Time of Cholera," *New Yorker*, 6 November 2006, http://www.newyorker.com/magazine/2006/11/06/sick-city (accessed 14 August 2015). The whole story is told in greater detail by Steven Johnson, *The Ghost Map: The Story of London's Most Terrifying Epidemic and How It Changed Science, Cities, and the Modern World* (New York, 2006).

investigators to identify the pump that was the source of the disease, which they then stopped using. Though the absolute numbers were quite small in comparison to present data volumes, this example shows the significance of bigger data samples, which allow a hypothesis to be proven, while turning correlation (death rates and territorial distribution) into causality (the infectious pump as the source of the disease).[34]

Big Data is a promise to tackle (and master) one of the three basic functions of media (storage, transmission, and processing). By now, transmission itself poses no significant problems; in most cases, bandwidth and interference determine success. Storage has likewise mostly become unproblematic thanks to server farms and storage media with petabyte capacity. It is processing, the third and—according to Hartmut Winkler—neglected category of media that plays the crucial role.[35] Brief intervention: one could reply that the same description applies not only to Big but also to little data, which seems both justified at first glance and yet not entirely accurate. If the task, for example, consists of harvesting a plum tree in a year with rich treasures of fruit, consider the plums as data (the "given" fruit on the tree): a small young tree—the connection to the "trees of knowledge" is not arbitrary here—with little data raises different problems than a large, mature tree with many branches and fruit with a wide range of qualities (some of them already rotten on the tree, some of them carrying the whole range of flavors of an aged plant).[36] Storage and conveyance would not be the largest problems if one could rely on (mechanical) aids like buckets, conveyor belts, and air-conditioned warehouses during and after the harvest. However, it is clear that the processing of Big Data (many plums with different qualities) in comparison to little data (fewer plums with more homogeneous qualities) poses a problem: a season with many blossoms requires more effort with a large tree, even in spring, because blossoms have to be pruned to maximize the fruit's flavor. Furthermore, the large number of plums makes classification of the fruit a challenging task, for it takes time to analyze each plum—to assess its smell (pleasant or not), taste (sour or sweet), ripeness (ripe or rotten), and overall quality (premium or substandard) and to determine whether it has been eaten by worms. Bigness demands more effort, care, and maintenance not only because of size but because of different properties such as varying quality, heterogeneity, and complexity. In short, in the context of little data, processing does not seem to be much of a problem; in the context of large-scale data sets, however, processing is the decisive factor in order to cope with the multitude as well as with the vast quantity of items that must be qualified, in other words, measured, assessed, estimated, and categorized in a more complex way than with manageable quantities. The trajectory between "little" and "big" is not linear, and increases require different steps of control and additional measures in mastering the data.

If processing is crucial to managing Big Data, no organization—neither the National Security Agency (NSA) nor anyone else—seems to be able to keep pace. Despite the tightest means of data surveillance, no one predicted the Paris attack on *Charlie Hebdo*—as far as the public knows. One might also ask why no alarm is

---

[34] Staffan Müller-Wille provides similar evidence within early nineteenth-century taxonomy; see Müller-Wille, "Names and Numbers: 'Data' in Classical Natural History, 1758–1859," in this volume.

[35] See Hartmut Winkler, "Processing: The Third and Neglected Media Function," paper presented at the "Media Theory in North America and German-Speaking Europe" conference, University of British Columbia, Vancouver, 8–10 April 2010, http://homepages.uni-paderborn.de/winkler/proc_e.pdf (accessed 11 June 2017).

[36] See, e.g., Barbara Bader, ed., *Einfach komplex: Bildbäume und Baumbilder in der Wissenschaft* (Zurich, 2005).

set off when a seventeen-year-old American male dropout buys weapons and the tracking data of his mobile phone point in the direction of his former school. It is because the processing cannot possibly keep up. Therefore, the distinction between data and metadata merely offers some consolation, that by the simplification achieved, while accessing the structure of the content instead of the content itself, one might happen upon significant signals. However, this conclusion is elusive since every "datum" can become a "metadatum" and vice versa. Thus, the value of metadata is always limited because they do not represent privileged information. On the contrary, they are regular data, which only momentarily acquire the special status of metadata and lose it when the task, research question, or query changes. Likewise, the rhetoric of government institutions and Internet providers proves futile, when they argue that they collect "only" metadata (which, as Edward Snowden has proven, is only partly correct). The metadata of Internet providers (i.e., who called whom and the duration of the call) are just as instructive as the contents of the respective phone calls, maybe even more conclusive. The subterfuge often used by notorious data collectors (NSA, F8, Apple, and various telecommunication companies), claiming not to be interested in the contents, which are supposedly left untouched, is once again proven wrong. The difference between "meta" and "data" always collapses easily and repeatedly, just as the distinction between signal and noise depends on the parameters of the search.

The library knows it all. But librarians do not. Nor do readers. The eternal problem of Big Data will not be the step of gathering and collecting heterogeneous information at virtual sites. It will be the next step of analyzing and, more important, processing the information. But who knows where the "good stuff" resides? What is relevant at all, and what makes the distinction as to what might be considered data or metadata? Is it the algorithm? The developer of the software? Or, rather, the data operator, an anonymous monitoring subject in Fort Meade staring at big screens?

Finally, the most important aspect of why Big Data can be regarded as something comparable to the oil of the twenty-first century, is not only its ability to help navigate the flood of information and to control the user's behavior, but, first and foremost, the promise of prognostics. The example of a large reseller who already knew about the pregnancy of a teenage customer before the father of the client has been notoriously cited in order to illustrate the (frightening) power of prognostics derived from Big Data.[37] For purposes of surveillance and economic "improvement," this issue is the subject of intense debate. But what is gained by predicting future readings? Is information prophecy really so innocent? To adapt a quotation from Jean Anthelme Brillat-Savarin, one of the Enlightenment fathers of modern gastronomy (coding and cooking are closely related practices), "Tell me what you read, and I will tell you who you are."

But how can this approach of quantification, correlation, and prognostics be applied to scholarly disciplines? These methods are used in history and historiography—not to mention the heirs of the former Bibliothek-Wissenschaft. Perhaps "cliodynamics" is the new method that proves that Big Data might serve as an epistemologically challenging tool for historiography as well.[38] The hope of Peter Turchin and his colleagues seems to be that, after quantifying all kinds of historical data, history itself will become predictable, at least in (very) short future time spans. But again, what would

---

[37] Mayer-Schönberger and Cukier, *Big Data* (cit. n. 32), 57–9.
[38] See Peter Turchin, "Arise Cliodynamics," *Nature* 454 (2008): 34–5.

be gained if behavior in general and history (of science) were to be computable for future reasoning? A rather short answer would be that surprise would be minimized to a great extent. Laplace's and other demons would garner attention once again.[39] Whether this would be an advantage or an obstacle for historiography depends on one's perspective and one's individual belief in contingency versus determination.

What are data and meta again? It is muddled as in Denis Diderot's novel *Jacques, le Fataliste, et son maître* of 1773. Jacques's master is a confessed supporter of free will even as he remains idle, fatefully, in order to let Jacques act for him. The servant, on the other hand, believes in unlimited determination and underlines this presumption with his motto that "tout était écrit là-haut" (everything was written up there).[40] As a consequence, he lives a carefree and autonomous life. He acts as he wants to, because everything is written "up there" already. From the perspective up there, data are always contingent; only metadata count as a determining factor. It seems that the source code is compiled, and the program is running, controlled only by some opaque agency. The code is not open and will always remain restricted, in defiance of cliodynamics, no matter how big data may be.

CONCLUSION

For several years, Big Data has been rising as a new paradigm of applying statistical algorithms to huge collections of heterogeneous data. According to its proponents, it is nothing less than a "revolution that will transform how we live, work, and think."[41] Big data, however, as is typical of heterogeneous collections, seems to be based on a somewhat obscure concept when it incorporates various notions of bigness—for example, the abundance of data—and the promise of prognostics based on algorithms that analyze large collections. In this essay, I differentiated the umbrella term "Big Data" into two historical strands that converge into one entity in the twentieth century, the electronic library catalog called OPAC.

The OPAC deals with Big Data in two senses. First, it counteracts information overload by structuring the contents and serving as a tool to locate and retrieve information. Second, it generates new information by organizing the content of different entries by common headwords or similar arguments taken from the blurbs, abstracts, and other paratexts of the text. The OPAC provides the opportunity to fuse the two strands of Big Data, the history of quantification and automatization, with measures to counter the flood of information.

No catalog can be prepared without quantification. Every entry in the long, electronic list of books, whether it is an inventory, a source for bibliographies, or a virtual collection of many local catalogs, is the sum of quantified data, since it always includes—as in the case of Callimachus—a more or less detailed stichometry. Besides the title and the author, each record lists the number of pages the text contains as well as a time stamp (year of publication). Cataloging means counting. Records of monographs might also list the number of illustrations as well as a sequence of headwords abstracting the content, in order to provide potential readers with a condensed impres-

---

[39] See Jimena Canales and Markus Krajewski, "Little Helpers: About Demons, Angels and Other Servants," *Interdiscipl. Sci. Rev.* 37 (2012): 314–31.

[40] Denis Diderot, *Jacques der Fatalist und sein Herr*, trans. Ernst Sander (1776; repr., Stuttgart 1972), 6, 9, 53, 91, and so on throughout the remainder of the text.

[41] Mayer-Schönberger and Cukier, *Big Data* (cit. n. 32), 7–10.

sion of what they can expect from examining the book itself. The connections between different entries are drawn, automatically, because of semantic similarities.
In other words, every catalog is based on bibliometric measures. Furthermore, the catalog is not merely a list of books; it holds a plethora of additional information such as
the table of contents, abstracts, blurbs, reviews, rankings, or even the full text itself, in
order to provide the reader with a more detailed glimpse into the content, perhaps
enough to make the reader decide not to read the text. The OPAC, therefore, serves
as a tool to manage information overload; it structures the abundance and makes it
processable. At the same time, it offers new paths to other works by showing links
to comparable and related texts. So the OPAC situates a reference within the network
of knowledge.

The only way to cope with Big Data is by distinguishing between data and metadata. Some special information is necessary to control the quantity of information.
This difference, however, is artificial, because it depends on one's point of view,
as I have tried to exemplify by way of the OPAC and its entries: data can be regarded
as metadata while metadata can be understood as data itself. The distinction is purely
a matter of perspective.

What did Callimachus do at the shelves in the Library of Alexandria? With his
pinakes, he developed a tool to counter information overload, while structuring the
paths potential readers might explore, orienting them with information about the authors and their other works, and also situating the text within an order, designating
each one as a certain discipline or genre. At the same time, however, he organized heterogeneous information by grouping texts that dealt with similar issues and placing
them on the same shelves because of their descriptions, thus generating proximity of
the texts by moving them into the same niche, forming a systematic place for them
in the same area. Callimachus performed a correlation of heterogeneous data, while
linking texts with probable proximities. By the way, when reordering the texts that
share a common topic or refer to each other and moving them next to each other, Callimachus did what the original meaning of *meta* in ancient Greek signified: it does not
imply a somehow transcendental layer of semantics, since it only defines a succession
of order, something placed beside or after another item.

While organizing the library and producing new data by cataloging, Callimachus
generated new facts—the metadata of the *pinakes*, which describe references and authors, combining them into both a catalog and a brief history of literature.[42] He enhanced the given data and made them "bigger," while at the same time reducing the
text to a few words describing the author, content, and relations to other texts in order
to provide new tools that could be used to manage information overload.

Perhaps Big Data will be an episode rather than a "revolution." Within the long
history of data practices, the fundamental concern of this episode is the distinction
between data and metadata, which is artificial (maybe even politically motivated in
order to disguise interest in the real content, especially when authorities proclaim that
they use only metadata instead of the content itself). However, the long history of data
proves that becoming "big" always brings up issues of control, processing, and counting. In the case of the OPAC, coping with the information overload of Big Data is related to (and correlated with) bibliometrics. These concepts, though quite distinct from
one another, are thoroughly intertwined.

[42] Schmidt, *Die Pinakes des Kallimachos* (cit. n. 4), 59, 99.

ECONOMIES OF DATA

The Biggest Data of All:
Making and Sharing a Digital Universe

by W. Patrick McCray*

ABSTRACT

Throughout the twentieth century, astronomers moved from and between different data-collection regimes, from the photographic to the electronic and, finally, to the born-digital era. At the same time, the focus of scientific discovery shifted away from the telescope itself to the hard drive, the database, and digital archives. This essay builds on the assumption that the sharing and circulation of astronomical data—as with other kinds of scientific data—have become core research activities that demand an increasing fraction of researchers' time, money, and expertise. The examples presented here give insights into the larger and gradual digitization process that unfolded throughout the entire international astronomy community. Although the examples chosen here depict local processes, the importance of sharing digital data transcended specific institutions, individual research questions, and national boundaries.

> As if an astronomical observatory should be made without any
> windows and the astronomer within should arrange the starry
> universe solely by pen, ink, and paper.
> —Charles Dickens, *Hard Times*, 1854[1]

To neophytes, doing astronomical research might appear relatively straightforward. When a science writer asked Wallace Sargent what he would do with better and bigger instruments, he responded that he would "point the fucking telescope at the sky and see what's out there."[2] If taken at face value, the noted observational astronomer's pro-

---

* Department of History, University of California, Santa Barbara, CA 93106-9410; pmccray@history .ucsb.edu.

Thanks to the Huntington Library and the California Institute of Technology for providing opportunities to do the initial research for this essay. The Max Planck Institute for the History of Science contributed a short-term residency in 2014 that enabled ideas here to coalesce further. David Brock, David DeVorkin, Alison Doane, Josh Grindlay, Richard S. Ellis, George Djorgovski, Robert Hanisch, Robert Kirshner, and Donald C. Wells generously offered thoughts and recollections. Finally, I wish to thank the other contributors to this volume and *Osiris*'s anonymous reviewers for their helpful suggestions.

[1] Dickens, *Hard Times* (London, 1854), 289.

[2] Ann Finkbeiner, *A Grand and Bold Thing: An Extraordinary New Map of the Universe Ushering in a New Era of Discovery* (New York, 2010), 1. I interviewed Wallace Sargent twice myself, once in February 2003 (https://www.aip.org/history-programs/niels-bohr-library/oral-histories/31826-1 [accessed 15 September 2014]) and again in February 2012, about eight months before his death on 29 October 2012. Based on our conversations, his colorful statement reflected his personal vision of observational astronomical practice.

---

fane reply implies a traditional mode of research anchored to the telescope. At about the same time, another (younger) researcher presented a different picture. "I do not believe," he said, "discoveries are made at the telescope riding the weather variations like a cowboy riding a bucking bronco." Instead, they emerge "after long nights at a terminal trying to reconcile an awkward data set with preconceived models."[3]

Over the last four decades, observational astronomers' locus for discovery has expanded to include sites other than the telescope; these include the hard drive, the database, and the digital data archive. This essay builds on the assumption that sharing and circulating astronomical data—as opposed to collecting it—have become core research activities that demand an increasing fraction of researchers' time, money, and expertise. Astronomers, like their counterparts in biology, meteorology, and other fields, have increasingly been obliged to become data scientists as well.

In the past half century, astronomers encountered different and often overlapping data regimes. In the photographic era—its origins in the mid-nineteenth century were roughly contemporaneous with those of astrophysics—raw data was collected via photographic means, and it remained photographic. Adjoining and overlapping this, starting shortly before World War II, was an electronic era. Here, devices such as photomultipliers and image tubes augmented and complemented established photographic techniques. However, data was still recorded in analog fashion on strip charts or punch cards that a person would later analyze. Note that "computerization" is not necessarily equivalent to "digitization"; that is, data produced via electronic instruments did not have to terminate in a digital format.[4] Eventually, astronomical practice gradually moved into a born-digital era where the raw data—itself a contested term—was collected, manipulated, and stored in a digital format.[5] The boundaries between these three eras were blurred and indistinct as older technologies and techniques endured and complemented newer modes.

And what of this digital astronomical data? Some experts have claimed that astronomical data differs from other kinds of scientific or commercial data. James N. Gray, a computer software scientist and data management expert, often joked with colleagues that he and other data experts "liked working with astronomers because their data is worthless."[6] By this, Gray meant that astronomical data has little commercial value (a point that can be compared with Dan Bouk's comments on the value of data in this volume).[7] And, unlike the massive databases maintained by corporations, health enterprises, or government agencies, astronomers' data poses no legal or ethical implications or privacy constraints. And although astronomers' data has little relevance

---

[3] Matt Mountain, "New Observing Modes for the 21st Century: A Summary," in *New Observing Modes for the Next Century*, ed. Todd Boroson, John Davies, and Ian Robson (San Francisco, 1996), 235–44, on 240.

[4] See Nathan Ensmenger, "The Digital Construction of Technology: Rethinking the History of Computers in Society," *Tech. & Cult.* 53 (2012): 753–76, in which Ensmenger highlights the distinction between computerization and digitization.

[5] Lisa Gitelman, ed., *"Raw Data" Is an Oxymoron* (Cambridge, 2013).

[6] Gray, whose 2007 disappearance at sea prompted a massive high-tech search, often deployed this anecdote in his public presentations. It appears in a number of places, including a tribute to Gray from a colleague: Alexander S. Szalay, "Jim Gray, Astronomer," *Comm. ACM* 51 (2008): 58–65. Gray's talks and papers are available at http://research.microsoft.com/en-us/um/people/gray/ (accessed 15 September 2014).

[7] Dan Bouk, "The History and Political Economy of Personal Data over the Last Two Centuries in Three Acts," in this volume.

for national security (unlike census or climate data), it correspondingly has few political or policy implications.[8]

Besides possessing the ability to be shared with relative ease, astronomical data is
derived from actual observations. It is, in other words, both "real" and "well-documented,
spatially and temporally."[9] As data goes, however, it isn't perfect. Cosmic rays, airplanes
passing overhead, and poor weather can make astronomical data "dirty," which, to some
computer experts, presents intriguing challenges. Moreover, the questions one can ask of
the data are scientifically interesting. These differences in degree, if not kind, are something to consider when comparing astronomical data to other kinds of collected data—
genomic, biological, environmental—as well as data generated in a lab, differences that
other authors in this volume consider.

Astronomers work in a profoundly data-rich world. For example, since its 1990
launch, the Hubble Space Telescope (HST) alone has transmitted dozens of terabytes
of observational data back to earth. The overabundance of data, in fact, presented scientific communities with tremendous challenges.[10] But, during the 1970s, astronomers began noting an especially significant discontinuity. It is here that one begins
to find astronomers' desperate-sounding references to floods, deluges, and explosions
of data. A British committee reported, for example, that "data generated by powerful
new detectors . . . are overwhelming," while American astronomers grudgingly accepted the "potential disruption" of computers in the "quiet austerity of a telescope
dome," because scientists simply "cannot cope with the large amount of raw data produced by electronic detection systems."[11] Where once astronomers complained that
they lacked sufficient data, they now started to worry about drowning in it.

There were, however, considerable obstacles associated with converting, sharing,
processing, and archiving scientific data. Paul Edwards uses the metaphor of "data
friction"—a term adopted in this volume by contributors like David Sepkoski and
Etienne Benson—to explain the "costs in time, energy, and attention" needed to "collect, check, store, move, receive, and access data."[12]

---

[8] Gray's comment only applies to certain types of contemporary astronomical data. In the past, the
military services were keenly interested in particular types of astronomical information. This, of
course, connects to broader discussions of the nature of scientific objectivity itself; Lorraine Daston
and Peter Galison, *Objectivity* (New York, 2007).

[9] Alexander S. Szalay, "Publishing Large Datasets in Astronomy—the Virtual Observatory," in
*Electronic Scientific, Technical, and Medical Journal Publishing and Its Implications: Proceedings
of a Symposium*, ed. Technical Committee on Electronic Scientific, and Medical Journal Publishing
(Washington, D.C., 2004), 83–6, on 85.

[10] This sense of crisis was by no means a new phenomenon; see Daniel Rosenberg, "Early Modern
Information Overload," *J. Hist. Ideas* 64 (2003): 1–9; as well as Daniel R. Headrick, *When Information Came of Age: Technologies of Knowledge in the Age of Reason and Revolution, 1700–1850* (Oxford, 2000); Lars Heide, *Punched-Card Systems and the Early Information Explosion, 1880–1945*
(Baltimore, 2009).

[11] "Report of the Panel on Astronomical Image and Data Processing," March 1979, personal papers
of Richard S. Ellis (hereafter cited as "RSE"); copies in W. Patrick McCray's possession. Lloyd B.
Robinson, "On-Line Computers for Telescope Control and Data Handling," *Annu. Rev. Astron. Astrophys.* 13 (1975): 165–85, on 175. See also Stephen E. Strom to Peter Boyce, 6 July 1977, personal
papers of Donald C. Wells (hereafter cited as "DCW"); copies in McCray's possession.

[12] Paul N. Edwards, *A Vast Machine: Computer Models, Climate Data, and the Politics of Global
Warming* (Cambridge, 2010), 84. See also Edwards, Matthew S. Mayernik, Archer L. Batcheller,
Geoffrey C. Bowker, and Christine L. Borgman, "Science Friction: Data, Metadata, and Collabora-
tion," *Soc. Stud. Sci.* 41 (2011): 667–90; David Sepkoski, "The Database before the Computer?";
and Etienne Benson, "A Centrifuge of Calculation: Managing Data and Enthusiasm in Early Twentieth-
Century Bird Banding," both in this volume.

In addition to data friction, we also encounter social friction. Whether it was analog or digital, collecting, analyzing, and sharing astronomical data required that considerable work be performed. Different communities of technical and scientific experts were implicated in the transnational project of constructing a digital facsimile of the universe. Goals and methods were not always aligned, and not all researchers were willing to readily cede primacy to the data archive instead of the telescope. At its core, this essay examines ways in which these many and varied points of friction were successfully greased or remained stubbornly sticky.

This essay extends our understanding of modern astronomical practice beyond the telescope itself. Much of the history of astronomy, at least where it intersects with technology, has focused on the building of institutions and instruments along with the politics and patronage that made this possible.[13] Less attention has been paid to astronomy's "knowledge infrastructure" in which the production and circulation of images is central.[14] But the data—once analog in form but now almost always digital—is indispensable for producing new knowledge about the universe. This essay, in other words, seeks to better understand what happens at the other end of the telescope.

Finally, the practices and activities associated with the world of astronomical data offer an opportunity to think more directly about the economies associated with astrophysics and modern science in general. There are, of course, issues of political economy as scientists maneuver to secure the resources necessary for building increasingly costly and complex instruments. But there is also the moral economy of astronomy to consider.[15] When interviewing astronomers over the age of fifty or so, it is not hard to elicit stories about colleagues with offices full of exposed photographic plates or data tapes that remained unanalyzed and unpublished—what one scientist referred to as the "mine, mine, mine syndrome."[16] While perhaps apocryphal, such tales are instructive in that they are often presented as a critique of scientists' behaviors. These two differ-

---

[13] See, e.g., Richard Hirsh, *Glimpsing an Invisible Universe: The Emergence of X-Ray Astronomy* (Cambridge, 1983); David H. DeVorkin, *Science with a Vengeance: How the Military Created the US Space Sciences after World War II* (New York, 1992); W. Patrick McCray, *Giant Telescopes: Astronomical Ambition and the Promise of Technology* (Cambridge, Mass., 2004).

[14] Astronomers, perhaps reflecting their long visual tradition, routinely refer to their data as images. While this term might in some cases refer to familiar pictures of stars and galaxies, more often these data are astronomical spectra that yield information about an object's composition, temperature, and other physical conditions. The term "knowledge infrastructure" is adapted from Christine L. Borgman's *Big Data, Little Data, No Data: Scholarship in the Networked World* (Cambridge, 2015), 4.

[15] In my usage, a moral economy refers to the values and beliefs associated with the production of knowledge and the circulation of resources. These values are not fixed but instead change over time and vary from place to place. New technologies (or changes to existing ones) can also modify the moral economy in which scientists and engineers do their work. The classic articulation of the concept of a moral economy remains E. P. (Edward Palmer) Thompson's "The Moral Economy of the English Crowd in the Eighteenth Century," *Past and Present*, 50 (1971): 76–136, as well as his earlier book, *The Making of the English Working Class* (Vintage, 1963). Historians of science have deployed the idea in several different ways. See, e.g., Steven Shapin, "The House of Experiment in Seventeenth-Century England," *Isis* 79 (1988): 373–404; Robert E. Kohler, *Lords of the Fly:* Drosophila *Genetics and the Experimental Life* (Chicago, 1994); Lorraine Daston, "The Moral Economy of Science," *Osiris* 10 (1995): 2–24; W. Patrick McCray, "Large Telescopes and the Moral Economy of Recent Astronomy," *Soc. Stud. Sci.* 30 (2000): 685–711. More recently, moral economy has been used as a framework for exploring the behavior of scientific communities in terms of credit and authorship; see, e.g., Bruno J. Strasser, "The Experimenter's Museum: GenBank, Natural History, and the Moral Economies of Biomedicine," *Isis* 102 (2011): 60–96.

[16] Finkbeiner, *A Grand and Bold Thing* (cit. n. 2), 44.

ent economies are not separate spheres but rather overlapping regimes in which the circulation and ownership of resources—money, data, credit—are central.

Compared with relatively new subfields like radio or X-ray astronomy—fields that emerged after World War II and carried less historical baggage—the traditional optical astronomy community had more deeply ingrained data practices. Astronomers observing in optical wavelengths had used photographic methods to collect data and record images and spectra since the mid-nineteenth century. But in a relatively short span of time, between roughly 1960 and 1980, observational astronomers' view of the sky transformed to a digital one. Not surprisingly, changes in the moral economy were most contested and the "coefficient of friction" highest in this community. Accordingly, this is where I have focused most of my attention.[17]

This essay uses several examples as probes to explore the social life of astronomical data. Instead of focusing on one specific institution or national context, I have opted instead to present illustrative snapshots that represent a diverse community of actors and institutions. However, a common meta-theme connects these activities: circulation. One could argue that astronomers—even if they did not always articulate it directly as such—directed their efforts toward the broader goal of reducing friction so that data could move about and between researchers.

The digitization of astronomy and its effects on practices like data sharing was not restricted to one country or disciplinary subfield. Rather, it was a process that instrument builders, observers, and theoreticians alike experienced in some way. Nonetheless, the examples presented here give insights into larger data-centric processes that unfolded throughout the entire international astronomy community. This approach allows us to see the transformation at several different scales, from the local context of individual laboratories to the transnational circulation of data. Although the examples chosen here depict local processes, the importance of sharing digital data transcended specific institutions, individual research questions, and national boundaries. For all astronomers, it was, in both senses of the phrase, a universal concern.

CONVERSION EXPERIENCES

In August 1970, astronomers and engineers convened at the Royal Observatory in Edinburgh to discuss best practices for applying automation techniques to astronomical data. The Edinburgh meeting occurred at a time when there was a good deal of introspection concerning the relationship between scientists and the data they collected. Astronomers' anxiety and enthusiasm mirrored feelings expressed earlier by high-energy physicists when particle accelerators and bubble chambers grew enormously in complexity and cost as well as the amount of data generated.

Despite one scientist's warning—if "we rely on automats to do everything . . . there will be no more men in the full sense"—at the Edinburgh meeting the mood was gen-

---

[17] Data collected at optical wavelengths with ground-based or space-based telescopes, alone or combined with data from other wavelength regimes such as radio or X-ray, still results in the majority of research publications; Helmut Abt, "The Most Frequently Cited Astronomical Papers Published during the Past Decade," *Bull. Amer. Astron. Soc.* 32 (2000): 937–41. Since 1970, astronomers have become less parochial in terms of the wavelength regimes (optical, infrared, radio, etc.) in which they worked; Abt, "The Growth of Multi-Wavelength Astrophysics," *Publ. Astron. Soc. Pacif.* 105 (1993): 437–9.

erally optimistic.[18] According to Jaap Tinbergen, a Dutch scientist who had recently migrated from radio astronomy to optical wavelengths, one device appeared as "an extremely adult sort of marriage" between computer and mechanical techniques.[19] Tinbergen directed his compliment toward an ambitious prototype machine, built and promoted by Edward J. Kibblewhite, which could automatically scan and measure astronomical photographs.

In 1970, Kibblewhite was a twenty-six-year-old graduate student just a few months away from filing his dissertation at Cambridge University. Like many in his professional cohort, Kibblewhite had moved into astronomy from another field (in his case, it was electrical sciences). Driving this demographic shift was the growing sophistication of electronic instrumentation used to collect data. As in other fields, research using optical telescopes—as was already the case for emerging fields like radio astronomy—became increasingly dependent on instrumentation developed by "gadgeteers" and "electronickers."[20]

In 1966, Kibblewhite proposed building an "automatic Schmidt reduction engine" for his dissertation.[21] Its prime application would be analyzing images taken with Schmidt telescopes, instruments whose optics are designed to take in much wider fields of view compared with conventional telescopes. To understand the data challenge posed by these large-scale survey telescopes, consider their output. The photograph itself is a negative—bright objects like nearby stars appear as dark black spots while galaxies are fainter and fuzzier. A typical exposure might contain as many as one million astronomical objects. About half of these might be stars, the other half galaxies. In addition, the chemical emulsion that records the images has an inherent graininess. Thus, distinguishing between very faint stars, distant galaxies, and the emulsion background itself posed a challenge. Another factor was the sheer amount of data each photographic plate contained. Some two billion "individual picture points" required sifting in order to find "data of interest."[22]

For the next five years, Kibblewhite designed and built what eventually became the Automated Photographic Measuring facility (or APM).[23] The initial cost was just under £33,000, a considerable sum in the late 1960s (roughly US$800,000 in 2016).[24] In developing his design, Kibblewhite looked to previous machines as something to improve upon. For example, astronomers had routinely used "measuring engines" since the 1950s. These instruments scanned photographs and electronically recorded information such as the coordinates of stars and galaxies.[25] Commercial firms like Perkin-

[18] Quote from Jean Rösch, "Introductory Address," in *Automation in Optical Astrophysics*, ed. H. Seddon and M. J. Smyth (Edinburgh, 1971), 3–7, on 7.

[19] Discussion following Edward J. Kibblewhite, "The Cambridge Automatic Plate-Measuring Project," in Seddon and Smyth, *Automation* (cit. n. 18), 122–3, on 123.

[20] Leo Goldberg to Jesse L. Greenstein, 15 October 1958, Box 12, "Goldberg" folder, and Greenstein to Alec Boksenberg, 17 September 1975, both from Jesse L. Greenstein papers, Archives of the California Institute of Technology, Pasadena.

[21] Letter and memo from Kibblewhite to Roderick Redman, 20 June 1966, personal papers of Edward J. Kibblewhite (hereafter cited as "EJK"); copies in W. Patrick McCray's possession.

[22] These data challenges are described in Ed Kibblewhite, "Counting the Stars by Computer," *New Scientist* 99 (1983): 478–82.

[23] There are various names ascribed to the acronym in the literature. In addition to Automatic Photographic Measuring facility, there is the Advanced Plate Measuring facility.

[24] Edward Kibblewhite, "The Design of Automatic Systems for the Analysis of Astronomical Photographic Data," unpublished report, 11 July 1967, EJK.

[25] John Lentz and Richard Bennett, "Automatic Measurement of Star Positions," *Electronics* 27 (1954): 158–63; S. Vasilevskis, "Automatic Measurement of Astrographic Plates," *Astron. J.* 62 (1957): 208–12.

Elmer eventually made "microphotometers" that allowed researchers to manually map the location of a star or galaxy on a photographic plate, measure its optical density, and convert the signal into a value representing the object's brightness.[26]

In 1970, the most sophisticated scanning machine for astronomy to date was located at the Royal Observatory, Edinburgh. The GALAXY machine—a tortured acronym for "General Automatic Luminosity and X-Y" measuring engine—was first proposed in the late 1950s by Peter B. Fellgett, a professor of Cybernetics and Instrument Physics at the University of Reading.[27] Development and testing took more than a decade, and the finished machine did not start operation until 1969. Fellgett's idea was to shine a light beam through a photographic plate. The transmitted signal, recorded on the other side, was diminished when the light passed through an image of a star or galaxy. The automated machine would then register the object and record its position and magnitude.

However, the GALAXY machine projected a relatively faint spot from a cathode ray tube and could only record about 1,000 objects per hour from a photographic plate. In contrast, Kibblewhite decided to use a bright laser beam as the light source for his APM. When rapidly moved, the scanner could process an entire Schmidt plate about a hundred times faster and do so automatically.[28]

Besides the optomechanical parts of the scanner, Kibblewhite and the small team working on the APM built a system to handle the large amount of data their machine generated. For the actual image analysis, Kibblewhite found assistance from an unexpected source. In 1967, he met a cancer researcher at Cambridge's Pathology Department who was developing software to process images of cell nuclei. When stained black, biological cells, Kibblewhite wrote, "looked just like star images."[29] The ability to subtract the background as well as delineate the edges of "fuzzy" objects was essential for researchers in both biology and astronomy, and Kibblewhite adopted a variation of this biology software for his APM.

The APM was interactive in the sense that a user could monitor the data conversion process in real time. It could also combine data from a number of different photographic plates. This was important because astronomers often observe the same patch of sky using different filters. As a result, only certain wavelengths were recorded on each photographic plate. Comparing plates made with different "colors" allowed one to distinguish, for example, between stars and other objects such as quasars.

This feature enabled Kibblewhite and a small group of scientists in 1987 to publish the discovery of the first quasar with a redshift of 4, making it the most distant object seen in the universe at the time.[30] How the APM was used by astronomers is also telling. The 1987 quasar paper, for example, used data—in this case, photographs from the UK's Schmidt Telescope facility in Australia—collected by another observer. The actual discovery, in other words, occurred after the data was reanalyzed using the APM, a point that reinforced the potential of working with archived data.

---

[26] For a good overview of these machines' history as well as technical details, see http://www.astro .virginia.edu/~rjp0i/museum/index.html (accessed 30 June 2014).

[27] G. S. Walker, "The Design and Development of the GALAXY Machine," in Seddon and Smyth, *Automation* (cit. n. 18), 103–8.

[28] The entire system is described in Edward J. Kibblewhite, "The Automatic Measurement of Astronomical Photographs" (PhD thesis, Univ. of Cambridge, 1971).

[29] Kibblewhite, "Counting the Stars" (cit. n. 22).

[30] S. J. Warren, P. C. Hewett, M. J. Irwin, R. G. McMahon, M. T. Bridgeland, P. S. Bunclark, and E. J. Kibblewhite, "First Observation of a Quasar with a Redshift of 4," *Nature* 325 (1987): 131–3.

Kibblewhite described the APM as a "national facility" available to "astronomers from all over the world" who wanted to convert and analyze their photographic data.[31] Scientists could come to Cambridge with their own astronomical photographs, and once the conversion was done—it took about seven hours to convert a typical plate into about two billion digital pixels—the astronomer could then "walk away with his data and start working out what it all means."[32] If two people wanted to study the same astronomical objects, "we would try to persuade them to collaborate," Kibblewhite recalled. But, if they were reluctant, "we would scan the same plates with the same selection criteria and provide them each with their own, but different, magnetic tape."[33] Established rules of sharing and ownership prevailed as converted data still belonged to the individual scientist rather than going to a common repository for later use by another person. To be fair, a major reason was technical, not cultural. There simply was not enough computer memory available for the APM project to retain copies, so the data "archive" still resided in the original photographic plates.

By the end of the 1970s, astronomical data could be rendered digital, either via conversion or at the moment of its creation via a veritable zoo of increasingly sophisticated instruments. These technological developments compelled astronomers, electrical engineers, and software writers to collaborate with one another more often.[34] Besides fostering increased need for collaboration and an expanded professional skill set, the digital nature of astronomical data raised an increasingly important issue. As opposed to the physical artifacts that characterized the photographic era, once data was digital, it became—at least in principle—more portable. As Joanna Radin's essay in this volume notes, one attribute of "Big Data" is its ability to "radically transcend" the "locality of its production."[35] Scientific data that could circulate more easily had the potential to more thoroughly disrupt long-standing community traditions and norms about ownership and access. But in order for astronomers to circulate their data more easily, they had to be able to share it.

"*ANY* WELL DEFINED FORMAT IS INFINITELY PREFERABLE TO NONE"

Imagine it is 1976 and you are an observational astronomer. Regardless of what kind of telescope you used—optical or radio, public or private, orbiting in space or sitting on a mountaintop—if you wanted to share your data, it was hard to do. In the older analog tradition, astronomers might loan photographic plates to colleagues, and observatories maintained physical libraries of the same. But, as more data was born-digital or converted to a digital format, the ability to share it posed an increasingly problematic issue.

Several factors contributed to astronomers' growing concern about their data. The prime driver was the "swelling flood of data" that astronomers' nightly observing runs produced.[36] Scientists also lacked appropriate tools to tackle the "daunting task" of

---

[31] Kibblewhite, "Counting the Stars" (cit. n. 22).
[32] Ros Herman, "Starmap Calibration Goes Automatic," *New Scientist* 84 (1979): 522–3.
[33] Edward J. Kibblewhite, personal correspondence with W. Patrick McCray, 25 June 2014.
[34] W. Patrick McCray, "How the Astronomers' Sky Became Digital," *Tech. & Cult.* 55 (2014): 908–44.
[35] Joanna Radin, "'Digital Natives': How Medical and Indigenous Histories Matter for Big Data," in this volume.
[36] M. J. Disney and P. T. Wallace, "STARLINK," *Quart. J. Roy. Astron. Soc.* 23 (1982): 485–504.

turning data into an "astrophysically useful form."[37] So, despite the growing capabilities of new digital detectors and instruments, astronomers lagged in their ability to "extract and study the relevant bits from this mass of data."[38] Greater challenges lay ahead. In the late 1970s, scientists in the United States, the United Kingdom, and Europe awaited the launch of what became the HST. The "immense amount of data" generated by Hubble and other space-based facilities meant the existing data glut would be "greatly aggravated" while a "lot of valuable science could be lost" because of poor data-handling capabilities.[39]

At the same time, astronomers' research practices were changing. More scientists wanted to combine data collected at different parts of the electromagnetic spectrum. However, digital data recorded by scientists using a radio telescope in Australia was rarely compatible with that collected, for instance, by optical astronomers in California. "The data transport problem," a scientist at the Netherlands Foundation for Radio Astronomy noted, "is getting larger each year as more people seek to combine data from different instruments."[40] Moreover, each institution typically used its own software packages to read the often unique data formats its instruments produced. In other words, considerable data friction inhibited astronomers' ability to share research with colleagues or combine data collected at different instruments or telescopes. One way to grease this friction was to adopt a common format for astronomical data that the whole community used.

Starting in late 1976, a small group of astronomers at national observatories in the United States and Europe began to address the problem. At Kitt Peak National Observatory, for example, Donald C. Wells took a lead role. Wells started his career as a research astronomer but also taught himself how to program in FORTRAN and ALGOL.[41] After he moved to Kitt Peak's Tucson headquarters in 1972, Wells's interests shifted from astronomical research to information management and data handling. Because the national observatory's telescopes were accessible to any astronomer who successfully submitted a peer-reviewed proposal, Wells wanted to likewise build tools for data handling that a broader community could use.

In December 1976, Ronald Harten, an American-born radio astronomer working in the Netherlands, visited Wells at Kitt Peak. Radio astronomers, because their data is inherently "born electronic," had previously faced many of the challenges confronting their optical counterparts. Harten disliked the difficulty of moving data between radio telescopes in the Netherlands where it was collected and the offices where scientists later analyzed it. He told Wells about his experiments with a "magic record size" that might offer an initial step toward a solution.[42] At this point, different computer systems read data files in basic units of information interchange called "record lengths" that varied in size. If the chunk of data used was a common multiple of the various record

[37] Michael J. Disney, "Centre for Optical Data Analysis," unpublished report, n.d. (but likely 1978), RSE.

[38] Strom to Boyce, 6 July 1977 (cit. n. 11).

[39] "Minutes of the 1st Meeting, Panel on Astronomical Image and Data Processing," 27 October 1978, RSE. The overall situation closely resembles a "presumptive anomaly," as described by Edward Constant, *The Origins of the Turbojet Revolution* (Baltimore, 1980), 15.

[40] Ronald H. Harten to colleagues, 9 June 1978, DCW.

[41] Donald C. Wells, oral history interview by W. Patrick McCray, 16 July 2012.

[42] Donald C. Wells, "Happy Birthday, FITS!," 29 March 1992, e-mail message, http://www.cv.nrao.edu/fits/documents/overviews/history.news (accessed 15 October 2012).

lengths that commercially available computer systems could read, then this "universal commensurability" would enable the "packing and unpacking" of files on "a wide variety of computers."[43]

Wells and Harten devoted considerable time to engineering the "header" of the data record. Akin to what today is called "metadata," the header gives crucial information—where a picture was taken and with what instrument, celestial coordinates, observing conditions, and so forth—that precedes the data of the actual astronomical image. Because Wells and Harten represented the optical and radio astronomy communities, respectively, they needed to create headers general enough to apply to data collected in either wave band. They also wanted to create a header system that would be "flexible and self-defining" yet open to "indefinite expansion" in the future.[44]

As they developed their respective data interchange formats, neither was especially committed to the formats they had personally designed. As Wells wrote to Harten, "I believe that *any* well defined and widely accepted format is infinitely preferable to none." Harten agreed, noting that a "general purpose scheme" could attract the interest of as many scientists as possible. Through their respective efforts, Wells wrote, "the community is being exposed to our ideas," but the time was quickly coming for an "attempt to meld the opinions of a number of people to try to reach a compromise that can be accepted by all." Securing support from researchers at the major national observatories was highly desirable because, as Harten predicted, if scientists at these publicly funded institutions got on board, "then most of the battle is won."[45]

In January 1979, the National Science Foundation (NSF) arranged a meeting for representatives from the major national observatories in the United States to discuss digital image analysis. Given general agreement that a "tape interchange standard is important," a small committee representing Kitt Peak, the National Radio Astronomy Observatory, and NASA was set up to "facilitate the communication of digital data."[46] Three months later, Wells and his counterpart in the radio astronomy community, Eric Greisen, drafted an informal agreement based on a data format developed by Harten. The design for what they called the Flexible Image Transport System (FITS) would "implement the transfer of images between observatories" in a "general format" that was "flexible and contains virtually unlimited room for growth."[47] Their mutual acceptance of a common record length meant that data standardized into the FITS format could be read "on all computers commercially available in the U.S. today." Wells and Greisen successfully tested their system with a trial exchange of data, and the results were presented at an international meeting in June 1979.[48]

[43] D. C. Wells, E. W. Griesen, and R. H. Harten, "FITS: A Flexible Image Transport System," *Astron. Astrophys. Suppl. Ser.* 44 (1981): 363–70. In this era, an alphanumeric character was represented by a byte, which generally was 8 bits, where a bit is a single information element with a value of 0 or 1. The number ultimately chosen was 23,040 bits, equivalent to 2,880 8-bit bytes or 3,840 6-bit bytes. Moreover, it was evenly divisible by the byte lengths of computers on the market then; i.e., it is divisible by 6, 8, 12, 16, 18, 24, 32, 64, etc.

[44] Ibid.

[45] Letters between Wells and Harten, 17 May 1978 and 5 June 1978, DCW; emphasis in the original.

[46] The founding of the Space Telescope Science Institute (STScI), which manages the Hubble Space Telescope, was not formally announced until January 1981.

[47] "Draft of Flexible Image Transport System," 29 March 1979, DCW.

[48] Donald C. Wells, "FITS: A Flexible Image Transport System," in *International Workshop on Image Processing in Astronomy: Proceeding of the 5th Colloquium on Astrophysics, Trieste, June 4–8, 1979*, ed. G. Sedmak, M. Capaccioli, and R. J. Allen (Trieste, 1979), 445–71.

FITS offered astronomers a "syntax" for sharing data with each other or between their respective institutions. Greisen, Wells, and Harten saw that FITS also had value as an archival format. Their goal was that information preserved with FITS should be able to be read by all computer systems, old or new, in the future. Wells later claimed he saw this as analogous to James Madison's goal of protecting minority interests in the drafting of the U.S. Constitution.[49] Therefore, a policy of "once FITS, always FITS" was adopted to ensure backward compatibility. Moreover, aware of the potential value of astronomical data collected decades earlier in older formats such as photographic plates, they came to see FITS "not only as a way to talk to remote astronomers in the here and now" but also as a tool "to talk to future astronomers."[50]

Producing a common data exchange format, however, would be fairly worthless if other institutions didn't adopt it. This made promoting FITS a political as well as a technical activity. For Wells, this meant "trying to mobilize an opinion in the community of sharing data, of always using the same formats. I was trying to stamp out the heretics, people with alternative data formats."[51] To Wells's relief, astronomers quickly recognized the value of data standardization. By the end of 1980, national observatories in Sweden and Australia, in addition to those in the Netherlands and the United States, had adopted FITS as their basic data format.[52] In 1982, the International Astronomical Union officially sanctioned this by recommending that "all astronomical computer facilities recognize and support" FITS as the standard global interchange format for digital data.[53]

Of course, advocates of FITS could not compel astronomers to share their data. But for scientists inclined to do so, the process was now simpler and smoother. FITS presented astronomers with a lingua franca to foster easier sharing and archiving of digital data. Of course, the more scientists and institutions adopted FITS, the more essential it became for other scientists to enlist as well.[54] Once astronomers adopted FITS as an international data standard, it provided a potent oil to reduce friction inherent in the interinstitutional and transnational circulation of data. And, once digital data began to move and circulate more freely, some astronomers began to imagine a working world in which the digital tools to interact with it could also be shared.

---

[49] Donald Wells to W. Patrick McCray, e-mail message, 22 May 2011. It is worth noting that Wells sometimes used the famous quote from Benjamin Franklin to John Hancock—"We must indeed all hang together or, most assuredly, we shall all hang separately"—to encourage the astronomy community to unite behind FITS or another suitable standard.

[50] Donald C. Wells, "Speculations of the Future of FITS," in *Astronomical Data Analysis Software and Systems VI*, ed. Gareth Hunt and Harry Payne (San Francisco, 1997), 257–60. Lorraine Daston makes a similar point about collections created with the needs of future researchers in mind; see Daston, "The Sciences of the Archive," *Osiris* 27 (2012): 156–87.

[51] Donald C. Wells, oral history interview by W. Patrick McCray, 17 July 2012.

[52] See, e.g., Denis Warne of the Mount Stromlo and Siding Spring Observatories in Australia to Donald Wells, 2 August 1979, DCW.

[53] "Resolution C1," in *Proceedings of the 18th General Assembly, Transactions of the International Astronomical Union, Volume XVIIB*, ed. Richard M. West (Dordrecht, 1982), 46–7. One expert noted that, seen in the broader sense, FITS was perhaps the "first digital standard for data exchange," preceding formats like PDF, GIF, etc.; Robert Hanisch, personal communication with W. Patrick McCray, 1 September 2014.

[54] The process of getting other scientists to adopt FITS resembles, of course, the process of enlisting the participation of other actors described in Michel Callon and John Law, "On Interests and Their Transformation: Enrolment and Counter-Enrolment," *Soc. Stud. Sci.* 12 (1982): 615–25.

SHARING SCIENCE BY SHARING TOOLS

After retiring from Cardiff University, Michael Disney composed a roman à clef about his career in science. Originally trained as a theorist, in the 1970s Disney began to shift his research attention to observational astronomy. At the end of an observing run, astronomers like Disney would leave the telescope with "one or more inscrutable magnetic tapes" filled with data.[55] Before it could be "turned into useful astrophysics," this "crude and dirty" data had to be "calibrated, corrected, and cleaned."[56] Disney, according to his fictionalized account, found that "almost all of his time was going into writing and testing trivial but necessary computer programs . . . to carry out mundane but unavoidable housekeeping tasks."[57] Moreover, Disney observed that his colleagues were also writing their own algorithms and routines for processing data. "So," Disney (speaking via a fictional protagonist named "Cotteridge") asked, "why couldn't they share?"

One problem was technical. Before FITS became the community's data standard, different machines and programming languages created barriers to sharing. But, even with a common data format, astronomers faced a bewildering assortment of image processing programs. There was little in the way of standardization as researchers came up with fragmented solutions that were disorganized and ad hoc.[58] Software development efforts at one site were often duplicated at another place. Although astronomers often devoted considerable time to programming computers, Disney claimed they were "mostly incompetent" at this task, or they simply did not like doing it.[59] British astronomers described their situation as especially serious. The national investment was substantial, with the astronomy community receiving almost 20 percent of the research monies that came from the United Kingdom's Science Research Council. Data reduction, Disney and other scientists saw, was "beginning to create a bottleneck" where "a lot of valuable science" that their government had paid for "could be lost," placing them at a disadvantage compared to other members of their "highly competitive community."[60] Although building an adequate data processing infrastructure would not come cheaply, to "ignore or starve it" would "make as much sound sense as building a telescope in a cloudy site."[61]

The other hurdle was cultural. "Morgan," one of the scientists in Disney's fictional account, describes astronomers as "ambitious, competitive, selfish egoists. We all want to be the next Galileo, the next Isaac Newton." So why would a scientist with a better tool for data processing share it with a "more cunning rival who might use it to overtake us in the race for glory?" Cotteridge replies, "Because if we don't, we'll

---

[55] M. J. Disney and P. T. Wallace, "STARLINK," *Quart. J. Roy. Astron. Soc.* 23 (1982): 485–504.

[56] Ibid.

[57] Michael J. Disney, unpublished/untitled work of fiction, from chap. 4, "STARLINK, 1977"; electronic copy shared with W. Patrick McCray by Disney in 2013 (in McCray's working files).

[58] Richard S. Ellis, oral history interview with W. Patrick McCray, 28 November 2011.

[59] M. J. Disney, "Centre for Optical Data Analysis," unpublished report, February 1979, RSE.

[60] "Minutes of the 1st Meeting, Panel on Astronomical Image and Data Processing," (cit. n. 39); "Report of the Panel on Astronomical Image and Data Processing" (cit. n. 11), 6–7.

[61] "Report of the Panel on Astronomical Image and Data Processing" (cit. n. 11), 31. The reference to "cloudy site" was a barbed reference to the questionable decision made a decade earlier by British politicians to locate the 100-inch Isaac Newton Telescope at a poor location in the United Kingdom and a jab at the prioritizing of politics over scientific requirements; John Irvine and Ben Martin, "Assessing Basic Research: The Case of the Isaac Newton Telescope," *Soc. Stud. Sci.* 13 (1983): 49–86.

waste our entire lives writing trivial computer programs, leaving no time to do astronomy, vain-glorious or otherwise." Ah, true, Morgan retorts, but how do you "turn shits into saints?"[62] How, in other words, could one transform astronomy's moral economy in such a way that everyone is encouraged to give "at least a modest push to the common wheel"?[63]

In late 1978, with approval from the Science Research Council, a small Panel on Astronomical Image and Data Processing (PAIDP) chaired by Disney began to chart a new course.[64] Panel members perceived three basic options. One was to preserve the existing "laissez-faire" system. With "no central co-ordination," each institution would be "free to propose its own computer system and configuration" so as to "give astronomers the maximum choice."[65] However, this ran counter to the "spirit of co-operation" panel members wanted to foster.[66]

A distinct alternative to the existing system was to establish a single national center for all U.K. astronomers to use. This option offered streamlined management and funding as well as the elimination of redundancy as all data processing programs would be developed at a central facility. However, a single location meant that most users would have to travel to it, bypassing the "informally interactive system required by the very nature of the research."[67] One can see, in both cases, how perceptions of their discipline's moral economy were part of the committee's thinking.

In the best Goldilocks fashion, the PAIDP steered between these two extremes. It recommended a linked minicomputer network that the committee christened STAR-LINK.[68] Like other "star networks," STARLINK had a central node to which other networks' points would be linked. Besides acting as the switching center for communications, the central node would also service the network, update the computer systems, and make sure software was adequately documented. System software developed and shared between sites would ensure compatibility and prevent devolution back to the existing situation. Indeed, software sharing between scientists at different sites was the "laudable, and indeed compulsory goal" the PAIDP wished to achieve.[69]

In choosing the networked option, the PAIDP looked within its own national borders for an example. The Rutherford Laboratory, located near Oxford, hosted the Interactive Computing Facility. Set up in 1978, this facility was a general purpose computer network that scientists used to share software for applications like circuit design

---

[62] Disney, unpublished/untitled work of fiction, from chap. 4, "STARLINK, 1977" (cit. n. 57).

[63] Disney and Wallace, "STARLINK," (cit. n. 55), 501.

[64] The committee was chaired by Michael J. Disney. Joining him were three optical astronomers (Alec Boksenberg, Richard S. Ellis, and Robert Fosbury) as well as two computer experts (James Alty and Igor Alexsander).

[65] "Report of the Panel on Astronomical Image and Data Processing" (cit. n. 11), 13. Although no evidence exists to support this, one might speculate on the prevalence of laissez-faire in the context of British Thatcherite politics ca. 1980.

[66] "Discussion of Some Image Processing Alternatives for U.K. Astronomers," March 1979 Annex to "Minutes of the 5th Meeting of the Panel on Astronomical Image and Data Processing," 22 March 1979, RSE.

[67] "Report of the Panel on Astronomical Image and Data Processing" (cit. n. 11), 13.

[68] The recommendation is in ibid. A technical description of such a network is given in Lawrence G. Roberts and Barry D. Wessler, "Computer Network Development to Achieve Resource Sharing," in *AFIPS '70 (Spring) Proceedings of the May 5–7, 1970, Spring Joint Computer Conference*, ed. Harry L. Cooke (New York, 1970), 543–9.

[69] "Report of the Panel on Astronomical Image and Data Processing" (cit. n. 11), 10.

and fluid mechanics.[70] The PAIDP's recommendation to base STARLINK at Rutherford, rather than at one of the royal observatories or at a university with a large and/or eminent astronomy department, reflected the priority given to software sharing over astronomical research per se. At the same time, the PAIDP insisted that "STARLINK must at all times respond to astronomical needs."[71] However, there was an obvious tension between these two goals that became more pronounced over time.

In its initial configuration, STARLINK was based on six VAX-11/780 minicomputer machines. Besides the central node in Chilton, other machines went to places such as Cambridge, University College London, and the Royal Greenwich Observatory, choices made based on estimates that 80 percent of U.K. astronomers worked at or within twenty miles of these sites. Leased telephone lines from Britain's Post Office connected the nodes. At each of STARLINK's sites, astronomers could access two image-display systems that allowed them to interact with their data in real time. The STARLINK project also adopted FITS as its data interchange format. The whole system—with an initial cost of £1.8 million—was inaugurated in October 1980.[72]

Implementing STARLINK as a tool for sharing software proved more difficult than astronomers originally expected. There were several reasons for this mismatch between aspiration and actualization. First, the science community expressed "conflicting requirements" as to how STARLINK should function. On the one hand, scientists wanted the "immediate no-nonsense development of a large number of application programs" they could access via STARLINK.[73] These "ultra-pragmatists" believed that "little if any supporting software over and above application routines were needed."[74] They were opposed by "idealists" who "cannot brook the slightest departure from complete portability across machine types."[75] In short, some scientists wanted to start using STARLINK in a quick and dirty fashion while others wanted to wait for a "comprehensive, soundly architectured, easy to use, and efficient" system. A fault line also ran between professional disciplines. Astronomers, Disney observed, wanted immediate results "no matter how inefficient, *ad hoc*, and inelegant" the data processing techniques were that yielded them. Computer scientists, in contrast, were as much "concerned with methods as with a particular astronomer's results."[76]

STARLINK advocates also had to contend with what Disney's fictional character Morgan had indecorously called the "saints versus shits" problem. Even before STARLINK was officially launched, astronomers anticipated that scientists at the various network nodes might "implement only software that was locally in demand."[77]

---

[70] Information on the Interactive Computing Facility is at http://www.chilton-computing.org.uk/acd/literature/annual_reports/p004.htm (accessed 10 July 2014). In 1979, the Appleton Laboratory moved to Chilton and was combined with the Rutherford Laboratory to create the Rutherford Appleton Laboratory.

[71] "Report of the Panel on Astronomical Image and Data Processing" (cit. n. 11), 16, 20.

[72] The inauguration featured a computer program that allowed astronomical images to be retrieved over the network from the various STARLINK nodes and then assembled on an image display terminal at Chilton. Described in "Inauguration of STARLINK," *Enterprise: STARLINK Information Bulletin*, no. 3 (November 1980): 1. See also http://www.chilton-computing.org.uk/acd/starlink/p004.htm (accessed 29 December 2012).

[73] "STARLINK Applications Software," *Enterprise: STARLINK Information Bulletin*, no. 3 (November 1980): 2.

[74] Disney and Wallace, "STARLINK" (cit. n. 55), 493.

[75] Ibid.

[76] Mike Disney, "Concluding Remarks," in *International Workshop on Image Processing in Astronomy*, ed. G. Sedmak, M. Capaccioli, and R. J. Allen (Trieste, 1979), 495–500, on 498.

[77] "Discussion of Some Image Processing Alternatives" (cit. n. 66).

This, of course, ran counter to STARLINK's original purpose—avoiding wasted effort inherent in the original laissez-faire model by "coordinating much of the software common to many requirements into a few universal and centrally supported packages."[78] Even more worrisome was the possibility that some centers might develop especially innovative software and "be reluctant to share it."[79] STARLINK's premise was that users would develop data processing software and share it over the network. However, once an institution received its VAX machine and interactive terminals, there was no way to compel its astronomers to share programs. In fact, astronomers actually had incentive not to share software via STARLINK.

Consider the differences between FITS and STARLINK. All scientists worked with data; FITS ensured its portability between researchers and institutions, benefiting all. Researchers interacted with STARLINK, however, after they already had their rough data. At this point, the impetus was to convert data into career-enhancing scientific results and publications. If one already had superior image processing software—perhaps written personally or by a colleague—then there might be reduced incentive to share it via STARLINK. Soon after the system was inaugurated, Disney restated his belief that there was "*no alternative* to sharing software development." Doing so, however, demanded "alertness, openness, generosity, and a strong spirit of compromise."[80] Fortunately, STARLINK could draw on an expanding user community, which included enough people possessing a combination of altruism and self-interest—there were over 1,000 users by 1988—to help oil Disney's "common wheel."

While not wanting to oversimplify, one can situate the aspirations Disney and other advocates had for STARLINK and other community-developed data processing tools in a broader context. Although it was years before the open-source software movement, there was an ethos of sharing in the 1970s-era computer culture, typified by the members of the Homebrew Computer Club and other hobbyist groups.[81] Of course, the groovy world of Bay Area hackers was considerably removed from research programs at Cambridge or Durham. Nonetheless, one can detect a common focus on sharing that reflects larger community aspirations and norms.

Likewise, Disney and other STARLINK advocates expressed a certain sense of idealism, perhaps even technological utopianism, about what their digital systems might accomplish. As Disney later recalled in his fictionalized account, Cotteridge "was designing The Future." Likewise, Disney and his colleagues speculated on how systems like STARLINK might affect the "shape of astronomy in the 21st century."[82] To look forward, Disney looked back in time, noting how the "cheap plane ticket" had caused the "backyard telescope and the staff astronomer" to give way to the "remote National facility and the guest observer."[83] Software systems and data sets accessed via high-

---

[78] R. J. Dickens, R. A. E. Fosbury, and P. T. Wallace, "The Development of STARLINK Applications Software," unpublished report, 6 October 1980, RSE. Early ARPANET developers faced similar issues as the reality of resource sharing often fell below expectations; Janet Abbate, *Inventing the Internet* (Cambridge, Mass., 1999), 96–7.

[79] "Discussion of Some Image Processing Alternatives" (cit. n. 66).

[80] Disney and Wallace, "STARLINK" (cit. n. 55), 501; emphasis in the original.

[81] Steven Levy, *Hackers: Heroes of the Computer Revolution* (New York, 1984); Christopher M. Kelty, *Two Bits: The Cultural Significance of Free Software* (Durham, N.C., 2008).

[82] I. Elliott, "Starlink," *Irish Astron. J.* 14 (1980): 197; this was elaborated on in Disney and Wallace, "STARLINK" (cit. n. 55).

[83] Disney and Wallace, "STARLINK" (cit. n. 55), 503.

bandwidth links would cause similar changes in astronomy's working world. In time, "authors on different continents could collaborate," and eventually, the "community of locations" would yield to the "international team based on community of interest."[84] And, as the network grew, Disney imagined using systems like STARLINK to share not only data processing tools but "equations, drawings, papers, data, and pictures." It will be, he predicted, as if all astronomers "live in the same electronic corridor," where data circulated in a more frictionless fashion.[85]

By 1990, much of the technical friction in the way of achieving this ideal had been smoothed over. Data were almost always "born digital," while older photographic forms of data could be routinely converted to zeros and ones. The astronomical community had agreed-upon data standards, and it could share tools for processing data. Nonetheless, at least one major barrier remained to achieving Disney's dream. A National Academy of Sciences report described the situation in the United States thus: "most ground-based astronomical data obtained outside the national observatories are treated as the private property of the observer," and, consequently, there was "no imperative" to share it.[86] Overcoming this particular point of friction—part technical and part social—would require significant additional changes in how astronomers thought about intellectual property, sharing, proprietary rights, and so forth—that is, the norms and behaviors underpinning the community's moral economy.

ARCHIVING THE UNIVERSE

In 2003, information managers at the Space Telescope Science Institute reviewing data usage statistics noticed something interesting. The total number of publications astronomers produced using the HST the previous year was just under 600. For the first time, half of these referred papers used archived Hubble data, a percentage that continued to climb.[87] Even more notable was the fact that roughly 40 percent of HST publications had used only archived data.

Modern astronomy's emergence as an archival science can be detected in other ways. Starting in the mid-1960s, the astronomy community regularly conducted a disciplinary review. These "decadal reports" set national research priorities for the next ten years. It was not until the third such survey, concluded in 1980, that the word "data" appeared in association with "archives."[88] The online SAO/NASA Astrophysics Data System tells a similar story. Searches of papers published between 1970 and 1980 show that the phrase "digital archive" was not used at all and that "data archive" appeared in paper titles on only four occasions. Jumping ahead a decade, one finds nineteen titles containing "digital archive" and 244 instances of "data archive." The

---

[84] Ibid.

[85] Ibid. Historians, of course, will recognize the familiar trope here as new communication tools were imbued with all sorts of utopian desires, including that of the "paperless office." See, e.g., "Office of the Future," *Business Week*, 30 June 1975, 48–70.

[86] John N. Bahcall, ed., *The Decade of Discovery in Astronomy and Astrophysics* (Washington, D.C., 1991), 96.

[87] This statement is based on statistics at http://archive.stsci.edu/hst/bibliography/pubstat.html (accessed 10 July 2014). I am using HST as an example because, among major public observatories, it maintains the most accessible records of this sort. Similar patterns would be found at other observatories.

[88] George Field, ed., *Astronomy and Astrophysics for the 1980's: Reports of the Panels* (Washington, D.C., 1983).

word "archive" alone shows a more dramatic jump, from thirty-one instances to almost 800. Such numbers confirm that the data archive—once containing tangible photographic plates but now composed of digital bits and pixels—had become an important research site.

Two developments helped catalyze this shift. In the 1990s, the astronomy community carried out several large survey projects in a variety of wavelengths. Often conducted with specially designed instruments, these surveys were planned from the outset to produce coherent data sets recorded with well-defined standards. The flurry in survey activity coincided with changes in computing technology and astronomical instrumentation. Computer processing speeds and memory storage capacity continued to march in step with Moore's law. While it took a quarter century for the light-collecting area of telescopes to double, the number of pixels in astronomers' digital detectors doubled every few years. Just as critical was the emergence of the Internet and the World Wide Web as legitimate research tools. The zeitgeist of the dot-com boom and the desire of astronomers to take advantage of new digital tools merged, for instance, when Johns Hopkins University astronomer Alexander Szalay and Microsoft computer scientist Jim Gray described astronomers' plans to "make the Internet act as the world's best telescope—a World-Wide Telescope."[89]

As astronomers in the United States and Europe began to consider building so-called virtual observatories—the term first appears in the literature around 1997—they also reevaluated the fundamental nature of their data. As Szalay and Gray, two of the most prominent advocates for virtual observatories, described it, astronomers' raw digital data was a complex assortment of "fluxes . . . spectra . . . individual photon events." Moreover, unlike data in other disciplines that "can be frozen and distributed to other locations," astronomical data often needs reprocessing and recalibration such that it "stays 'live' much longer . . . [and] needs an active 'curation.'" However, because each research group had its "own historical reasons" and methods for saving its data "one way or another," a single centralized repository, like the molecular biologists' GenBank, didn't seem feasible. They concluded that a "federated" system that would unite existing databases seemed more realizable.[90]

From the inception of the idea, rhetoric around virtual observatories was imbued with utopian aspirations that accompanied other Internet-related endeavors at the turn of the century. Virtual observatory advocates gushed about the liberating possibilities of "mining the sky." For the "clever people who don't have access to a big telescope," said Caltech's George Djorgovski, a virtual observatory "will allow them to do first-rate observational astronomy."[91] Claims made on both sides of the Atlantic hinted at the possibility of political, not just scientific, revolution. Virtual observatories could "lead to a true democratization of astronomy" and represented a "fresh wind blowing through the graveyard of old and unused data."[92] Riffing on journalist Thomas Friedman's best-selling 2005 book *The World Is Flat*, Matt Mountain suggested that

---

[89] Alexander Szalay and Jim Gray, "The World-Wide Telescope," *Science* 293 (2001): 2037–40.
[90] Ibid., 2038.
[91] Ron Cowen, "Mining the Sky: Taking Some Big Bytes of the Universe," *Sci. News* 159 (2001): 124–5.
[92] "Democratization" quote from Caltech's Robert Brunner in Govert Schilling, "The Virtual Observatory Moves Closer to Reality," *Science* 289 (2000): 238–9; "graveyard" quote from German astronomer Wolfgang Voges in Toni Feder, "Astronomers Envision Linking World Data Archives," *Phys. Today* 55 (2002): 20–2, on 20.

publicly accessible astronomical databases might "make the sky flat" by increasing access to data resources.[93] Not everyone was so sanguine, however. One astronomer predicted that virtual observatories only "breed a generation of astronomers who sift through data without knowing about instruments."[94] Others questioned data archives' costs, technical challenges, and perceived banality compared to building a giant new telescope.

Astronomers' imaginings of virtual observatories were more far-reaching in scope than other data-focused efforts described in this essay. Kibblewhite's APM and STARLINK were locally bound machines or systems. FITS gradually became a community standard and a necessary first step in facilitating digital data sharing, but it addressed a specific technical problem. Ambitions for virtual observatories, in contrast, were fully transnational in scope, bringing together databases from disparate countries and observations made across the spectrum so that they might constitute "one uniform, consistent data set."[95]

In the United States, a blue-ribbon panel of astronomers gathered by the National Academy of Sciences nudged the "National Virtual Observatory" (NVO) forward by making it a top priority for the early twenty-first century.[96] Researchers from observatories and computer science centers put together an implementation plan, and the NSF awarded $10 million toward their efforts.[97] In the United States, proponents described it as a "new research environment for astronomy with massive datasets," the creation of which would be "technology-enabled but science-driven."[98] Similar efforts emerged in the United Kingdom and the European Union, resulting in the creation of the International Virtual Observatory Alliance. In 2010, the NVO transitioned to become the Virtual Astronomical Observatory, while Elsevier's launch of the journal *Astronomy and Computing* provided a forum for peer-reviewed publications. By 2014, when the effort concluded, some $16 million from the NSF and NASA had helped dozens of scientists and computer programmers begin to build a more robust data infrastructure for astronomy.[99]

Digital archives retained the potential for slowly eroding established data-sharing conventions. Many of these changes accelerated as more scientists embraced a multiwavelength astronomy that relied on data collected at multiple facilities. Taxpayer-funded "Big Science" has also helped to drive the process. In August 1982, for example, Riccardo Giacconi, director of the recently formed Space Telescope Science Institute, which would handle science operations for the as-yet-to-be-launched HST, explained how individual researchers' data would have a proprietary period of just

[93] Matt Mountain, "Flattening the Astronomy World," *Phys. Today* 67 (2014): 8–10.
[94] Feder, "Astronomers" (cit. n. 92).
[95] Jim Gray in Cowen, "Mining the Sky" (cit. n. 91).
[96] Christopher McKee and Joseph Taylor, eds., *Astronomy and Astrophysics in the New Millennium* (Washington, D.C., 2000); the decadal report recommended that $60 million be directed to the NVO.
[97] "Building the Framework for the National Virtual Observatory," 23 April 2001 proposal to the National Science Foundation, electronic copy archived at http://www.us-vo.org/pubs/index.cfm (accessed 30 June 2014).
[98] "Towards a National Virtual Observatory: Science Goals, Technical Challenges, and Implementation Plan," white paper prepared for National Academy of Sciences, 8 June 2000, electronic copy archived at http://www.us-vo.org/pubs/index.cfm (accessed 30 June 2014).
[99] Robert Hanisch, "A Brief History of the US VO Effort," 10 July 2014 presentation, Pasadena, Calif.; copies of slides in W. Patrick McCray's working papers. Also, Robert Hanisch, personal communication with McCray, 8 September 2014.

one year. After that, the data would "be made available to the community at large," becoming a public good.[100]

Jump ahead thirteen years. For ten consecutive days in late 1995, Hubble took 342 exposures of a small region of the sky in the constellation Ursa Major. The resulting data, once released to astronomers, caused great public wonderment.[101] Since then, the Hubble Deep Field has become an iconic scientific image. The thousands of jewel-like galaxies it revealed have been reproduced on calendars, coffee cups, and screen savers, while scientists used the shared data set to produce hundreds of refereed papers.[102] Far from being the property of a single investigator, the Hubble Deep Field (and similar surveys that followed) presented astronomical data as a shared community resource. One wonders what Disney's dyspeptic character Morgan might have thought.

* * *

In 1945, Henry Norris Russell advised a colleague about to assume the directorship of a major observatory that he should hire a "good man who knows modern electronic instrumentation."[103] The subsequent postwar "de-astronomization of astronomy" brought many new experts into the field.[104] This trend has continued with a redefinition of who and what an astronomer is. In the decades after Russell's suggestion, astronomers migrated away from their analog data traditions, taking first an electronic and then a digital turn. Events of the past decade suggest another turn in process. Starting in 2008, a new term—"astroinformatics"—began to appear in the international online repository of astronomy papers. Proponents described it as a "new data-oriented paradigm for astronomy."[105] The neologism reflects the tendency, seen in many scientific fields, to embrace, or at least come to terms with, a broader data turn—that is, bio-

---

[100] Riccardo Giacconi, "Science Operations with Space Telescope (NASA CP-2244)," in *The Space Telescope Observatory*, ed. Donald N. B. Hall (NASA, 1982), 1–15, on 11. An earlier statement of this policy can be found at least a year earlier in NASA technical documents such as "Statement of Work for the Space Telescope Science Institute (STScI)," 15 April 1981, 8; copies of both documents are in the library at the STScI in Baltimore.

[101] NASA's original press release is available at http://hubblesite.org/newscenter/archive/releases /1996/01/text/ (accessed 1 July 2014), while technical details are at http://www.stsci.edu/ftp/science /hdf/hdf.html (accessed 1 July 2014) and http://www.stsci.edu/stsci/meetings/shst2/williamsr.html (accessed 1 July 2014). See also Elizabeth A. Kessler, *Picturing the Cosmos: Hubble Space Telescope Images and the Astronomical Sublime* (Minneapolis, 2012).

[102] The original Hubble Deep Field paper—Robert E. Williams, Brett Blacker, Mark Dickinson, W. Van Dyke Dixon, Henry C. Ferguson, Andrew S. Fruchter, Mauro Giavalisco, et al., "The Hubble Deep Field: Observations, Data Reduction, and Galaxy Photometry," *Astron. J.* 112 (1996): 1335–89—has been, according to Google Scholar (7 August 2016), cited more than 1,300 times. STScI director Williams even noted that some requests for HST time offered shorter proprietary data periods to get "an edge" vis-à-vis more stingy proposals; Toni Feder, "Space Scientists Split of Proprietary Data Rights," *Phys. Today* 51 (1998): 52–3.

[103] Henry Norris Russell to Ira S. Bowen, 3 November 1945, cited in David DeVorkin, "Electronics in Astronomy: Early Applications of the Photoelectric Cell and Photomultiplier for Studies of Point-Source Celestial Phenomena," *Proc. IEEE* 73 (1985): 1205–20, on 1220.

[104] Jesse L. Greenstein, oral history interview with Rachel Prud'homme, 16 March 1982, Center for History of Physics, American Institute of Physics, College Park, Md.

[105] Quote from https://asaip.psu.edu/Articles/astroinformatics-in-a-nutshell (accessed 15 June 2015). See also Kirk Borne, Alberto Accomazzi, Joshua Bloom, Robert Brunner, Douglas Burke, Nathaniel Butler, David F. Chernoff, et al., "Astroinformatics: A 21st Century Approach to Astronomy," *Astronomy and Astrophysics Decadal Survey, Position Papers*, no. 6 (2009), http://adsabs.harvard.edu/abs /2009astro2010P...6B (accessed 15 June 2015).

informatics, genomics, proteomics, and so forth.[106] Whether astroinformatics coalesces into a sustainable community is too early to tell. But one outcome is already clear: contemporary astronomers need no longer be producers of scientific data but can instead make careers as consumers of it.

Looking across astronomy's three overlapping data eras—analog, electronic, and digital—one can make a broader observation about data friction and the norms and practices about data sharing. In the traditional photographic era, data friction was high. It was difficult and expensive to trade, share, move, and reproduce raw data. Yet the "rules" of the community's moral economy were relatively simple—data almost always belonged to the individual who collected it; what circulated was mostly processed data and findings. As astronomers entered the born-digital era, new hardware and software substantially reduced data friction, making it easier to move and share information in the form of raw data. But, at this point, navigating astronomy's moral economy also became harder as issues around data sharing, ownership, and access became more complex.[107]

Astronomers' experiences navigating their community's norms and expectations about data sharing invite comparisons to those of scientists from other disciplines. A tempting topic for future research, a full exegesis is not possible here. However, a few observations can be made. Kohler's detailed exploration of the drosophilists' network of sharing and exchange, for example, highlights the importance of free exchange of fruit fly stocks, the role of "enlightened self-interest" in promoting this exchange, and the "unspoken rules of etiquette" that governed this circulation.[108] One factor that facilitated the acceptance of these rules, he suggests, was the relative abundance of *Drosophila* as a research material, a situation made possible by large-scale breeding and stock keeping.

In comparison, astronomers' perception that the amount of data available to them was expanding at a rate that seemed overwhelming did not, by itself, reshape their moral economy. Instead, technological interventions in the form of data standards and new data-handling tools were required. An abundance of research material alone—in this case, the rapidly rising flood of data—did not, *pace* Kohler, suddenly alter astronomers' long-held if often unspoken rules and expectations about sharing it.[109]

In this essay, I have discussed representative examples of astronomers' engagement with their data: the development of machines in the early 1970s to convert analog data to digital format, computer-savvy astronomers working at national observatories in the United States and Europe in the late 1970s to create a community-wide data standard to facilitate sharing, attempts in the United Kingdom circa 1980 to share software tools for data processing, and early twenty-first-century international efforts to build

---

[106] Previously, other fields have also been portrayed as extensions of information science; see Timothy Lenoir, "Shaping Biomedicine as an Information Science," in *Proceedings of the 1998 Conference on the History and Heritage of Science Information Systems*, ed. Mary Ellen Bowden, Trudi Bellardo Hahn, and Robert V. Williams (Medford, N.J., 1999), 27–45; David Baltimore, "How Biology Became an Information Science," in *The Invisible Future: The Seamless Integration of Technology into Everyday Life*, ed. Peter J. Denning (New York, 2001), 43–55.

[107] My thanks to David C. Brock for valuable comments on this point.

[108] Kohler, *Lords of the Fly* (cit. n. 15).

[109] My thanks to the perceptive reviewer who suggested this as a point of comparison. The effect of data's "bigness" is a point taken up more recently in Sabina Leonelli, "What Difference Does Quantity Make? On the Epistemology of Big Data in Biology," *Big Data & Soc.* 1 (2014): 1–11.

publicly accessible online data archives. From these vignettes, a more general typology emerges of astronomers' data-driven working world beyond the telescope.

Data conversion. Astronomers saw value in transforming data recorded on traditional photographic plates into digital data that they could then interact with. Subsequently, researchers constructed increasingly sophisticated systems to render analog data into a more pliable digital format.

Data standardization. In all technological systems, agreed-upon standards serve as political as well as technical tools—the means to discipline unruly arrays of measurements, signals, and so forth, as well as the communities associated with them.[110] Likewise, in astronomy, seemingly mundane standards served social and technical purposes by encouraging additional order and rationality.

Data processing. Once "raw" data had been collected, a key task for the astronomer was processing it. This general term encompasses a wide range of specific actions, but, in all cases, the goal was to interact with it so as to produce meaningful scientific information.

Data archiving and access. Traditionally, optical astronomers imagined their data as personal property. In some cases, observatories maintained physical libraries of data in the form of photographic plates to which staff contributed. However, as scientists gained access to increasingly expensive facilities, particularly those funded with public money, the idea of data as a public good became more powerful and widespread. Subsequently, some astronomers began to envision a more seamless system in which data collected by many different telescopes are managed and stored.

All of these activities are embedded in a larger framework of circulation and sharing and conditioned by astronomy's political and moral economies. To be sure, while new tools and new technologies can help reduce data friction, they cannot by themselves eliminate its accompanying social friction. To paraphrase Shakespeare, that particular fault lies not in the stars but in ourselves.

---

[110] Andrew L. Russell, "Standardization in History: A Review Essay with an Eye to the Future," in *The Standards Edge: Future Generations*, ed. Sherrie Bolin (Ann Arbor, Mich., 2005), 247–60.

The Field in the Museum:
Puzzling Out Babylon in Berlin

by Mirjam Brusius*

ABSTRACT

Can objects be data? Under what conditions do they become data? Looking at the Prussian-led excavations in Babylon around 1900, this essay will investigate processes of knowledge production in relation to archaeological objects between their excavation and their incorporation in museum collections. It will focus on the nature of the connection between data and the physical objects they represent. How much data does an object yield when it is "found"? What role do epistemic practices, such as the application of visual media in the field and the museum, play in turning objects into data? And finally, will the data remain the same once the objects have reached their final destination?

In April 1898, the architect and archaeologist Robert Koldewey (1855–1925) returned to Berlin from a preliminary expedition to Babylon in modern-day Iraq, where he surveyed potential sites for future excavations to be led by the Prussian state.[1] Pressure was high for Prussia to find an appropriate excavation site in Mesopotamia. The Prussians were leading authorities in philology, but France and especially Britain were at an advantage in that they possessed cuneiform tablets, which aided in the decipherment of cuneiform script and led to the development of Assyriology as a field of study. Although the Prussians had the scholarly skills and infrastructure to examine cuneiform script, they lacked the material required to further their research program. Around the turn of the century, the new "German Empire" overcame this limitation by making significant advances in historical and archaeological scholarship and exploration, especially in the Ottoman Empire.[2] Cultural expansion was a high

I would like to thank Christine von Oertzen, David Sepkoski, Elena Aronova, William Carruthers, and the anonymous reviewers for inspiring editorial remarks. Funding for this research was provided by the Max Planck Institute for the History of Science and by the Kunsthistorisches Institut in Florenz, Max-Planck-Institut.

* The Oxford Research Centre in the Humanities (TORCH), University of Oxford, Oxford, United Kingdom; mbrusius@cantab.net.

[1] On Koldewey, see Ralf-B. Wartke, ed., *Auf dem Weg nach Babylon: Robert Koldewey—Ein Archäologenleben* (Mainz, 2008).

[2] On Germany's orientalism, see Andrea Polaschegg, *Der andere Orientalismus: Regeln deutschmorgenländischer Imagination im 19. Jahrhundert* (Berlin, 2005). On Babylon, Assyriology, and scholarship in Germany, see esp. Suzanne Marchand, *Down from Olympus: Archaeology and Philhellenism in Germany, 1750–1970* (Princeton, N.J., 1996), 188–227; Marchand, *German Orientalism in the Age of Empire: Religion, Race, and Scholarship* (Cambridge, 2009), 36–51, 195–202. See also Can Bilsel, *Antiquity on Display: Regimes of the Authentic in Berlin's Pergamon Museum* (Oxford, 2012), 159–88.

priority in the hierarchy of colonial strategies, and Prussian research expeditions simultaneously served science, technology, economy, politics, society, and the national psyche. Just as the Middle East was a source of raw material like oil, the region was also a treasure trove of archaeological artifacts that could potentially bolster Prussia's own cultural identity.[3] Mesopotamia was not just any land: it was considered the *Urquelle*, that is, the cradle, of European culture, and even of civilization.[4]

From his preliminary expedition, Koldewey brought three small, brightly colored, enameled fragments to Berlin. However, in moving from the field to the museum, the three tiles set in motion a significant epistemic transition, which turned these simple fragments into objects of potential value. When Koldeway presented the bricks to Richard Schöne, the director of the Königliche Museen Berlin,[5] Schöne attributed such extraordinary distinctiveness to the finds that he ordered Prussia's excavation efforts to include Babylon. Schöne's decision resulted in the accumulation of thousands of fragments of what would become one of Berlin's most popular tourist attractions: Babylon's Ishtar Gate and Processional Way, currently on display in the city's Pergamonmuseum.

In the photograph in figure 1, tables covered with fragments stretch out in the arcades in front of the museum building. Workers in white lab coats, perhaps museum curators, are trying to make sense of them. Paintings on the walls supposedly serve as tools for the reconstruction. A neoclassical bust at the top right reminds the viewer of the canonical order that until then had dominated Berlin's museum landscape: classical antiquity. This antiquity was now contested by an alternative antiquity, embodied in the Babylonian fragments.

While this photograph is usually used to support a triumphalist narrative of the reconstruction of the Ishtar Gate, I will take the photograph as my starting point to describe the epistemological challenges and frictions preceding the gate's reassembly. I will focus on the materiality of the fragments and trace their journey from their discovery, to their display on the working tables we see depicted in the photograph, and beyond. This slow and convoluted process from excavation to exposition provides ample opportunity to reflect on the tenuous relationship between objects, data, and

---

[3] Thomas W. Gaehtgens, *Die Berliner Museumsinsel im Deutschen Kaiserreich: Zur Kulturpolitik der Museen in der wilhelminischen Epoche* (Munich, 1992), 94. Excavations were funded partly by the state and partly through the newly founded Deutsche Orient-Gesellschaft (DOG). The DOG's aims, according to its statutes, were to support the study of antiquities and to publish research in this area. Although it was a private organization, it was supported by Kaiser Wilhelm II, who expressed special interest in the study of antiquities because of their relation to the Old Testament. While the state, the DOG, and the museums collaborated closely on excavations, it was clear that all finds were to be turned over to the museums. See Archives of the Deutsche-Orientgesellschaft, Zentralarchiv der Staatlichen Museen zu Berlin (hereafter cited as "DOG"), file II, 1.2.1., 25 November 1897; Nicola Crüsemann, *Vom Zweistromland zum Kupfergraben* (Berlin, 2001), 212.

[4] The response from the Ottoman Empire was by no means passive or indifferent. The Ottoman state pursued its own imperial project of archaeology. See Bilsel, *Antiquity on Display* (cit. n. 2), 24–5.

[5] On the history of the Vorderasiatisches Museum, see Vorderasiatisches Museum and Nicola Crüsemann, *Vorderasiatisches Museum Berlin: Geschichte und Geschichten zum hundertjährigen Bestehen* (Berlin, 2000). On ancient Near Eastern Art on the Berlin Museum Island, see the detailed study by Crüsemann, *Zweistromland* (cit. n. 3). On Museum Island, see Gaehtgens, *Die Berliner Museumsinsel* (cit. n. 3). On the reconstruction of the Ishtar Gate, see, most recently, Nikolaus Bernau and Nadine Riedl, "Für Kaiser und Reich: Die Antikenabteilung im Pergamonmuseum," in *Museumsinszenierungen: Zur Geschichte der Institution des Kunstmuseums. Die Berliner Museumslandschaft 1830–1990*, ed. Sven Kuhrau, Alexis Joachimides, Viola Vahrson, and Nikolaus Bernau (Dresden, 1995), 171–89.

Figure 1. Blick in die 1928 provisorisch in der Säulenhalle am Neuen Museum eingerichtete Werkstatt, in der die Ziegelbrocken vom Ischtar-Tor sortiert und zugeordnet wurden [View of the workshop provisionally set up in the arcades near the Neues Museum, where the bricks of the Ishtar Gate were classified.]. *Staatliche Museen zu Berlin—Vorderasiatisches Museum, Fotoarchiv, Bab Ph 3700.*

the knowledge that can be drawn from them. The term "data" was not used by the historical actors when they described the material as it was excavated, measured, drawn, photographed, described, and copied on paper; sorted, packaged, and transported to the museum; and subjected to another set of procedures that awaited the rubble for reconstruction. However, framing this long and complex history as one of objects, data, and data processing proves useful heuristically for exploring the material culture of archaeology on the one hand and the relationship between objects and data on the other. If we were to define data in this context, it could be defined as a set of values, based on collected information (in this case of material nature), from which conclusions, so-called facts, can be drawn. But artifacts, in this case archaeological finds, are not always already facts, nor can they easily become facts. In the context of materiality, where does data start, and where does it end? Even experts needed additional information for a find to serve and be deployed as data. An archaeological find, for example, an artifact made of specific material, might represent potential data in the eyes of the specialist who collected it, but without context, it remains meaningless as a scholarly object to an average museum visitor. In turn, this very same object can take on other functions (e.g., of aesthetic nature), removed from its original surroundings and—in some cases—any epistemic meaning, in which case an object represents no data at all.

In tracing the path of archaeological finds and their inclusion in the life cycles of knowledge making by different parties, I will invoke Hans-Jörg Rheinberger's remarks (in reference to Jacques Derrida) on the meaning of "traces" in understanding the role of materiality in knowledge production. The trace of a potential epistemic object is initially of unstable nature and must be made "durable" in order to turn objects into data. "Epistemic" here refers to the impact of visual and textual records on knowledge production; chains of reference among these paper tools stabilize their status as data. Traces, for example, can appear in the form of preliminary notes and experimental sketches. They are forms of material manifestation that are "palpable" for the production of knowledge but have not yet been turned into more universally comprehensible (visual or written) forms of representation, for example, what might appear in printed communication for the scientific community.[6] Traces, Rheinberger argued, only acquire meaning as data if one can "relate and condense them into an epistemic thing, a suspected *Sachverhalt*, a possible 'fact' in the everyday language of the sciences. What was looked for was traces that had the potential to become data in the sense described, and that in turn could be used to create data patterns: curves, maps, schemes, [etc.] . . . Once stabilized, they would go for scientific facts. . . . Facts would then appear as stabilized arrangements of data."[7]

Archaeology was always situated at the interface between measurable science and social practice, a conflict that enhanced the multiple meanings of the objects discussed here through the lens of different parties of interest. The different discourses determined what kind of knowledge the finds should produce. In what sense the bricks and tiles themselves were data, or what kinds of data they carried, was always a subject of debate. For the excavators, the rubble itself unearthed in Mesopotamia could no doubt constitute historical knowledge about Babylon and ancient Mesopotamia at large. For archaeologists, the objects thus had the potential to become an "epistemic thing" that could yield historical data, whereas for others, their status as data was not of primary concern. In other words, different knowledge was imposed on the finds from the Prussian expedition. These objects were to be studied and showcased in the newly constructed Berlin Royal Museums, where they would celebrate, and also legitimize, a newly founded empire.[8] Three parties were involved in this undertaking—archaeologists, philologists, and museum curators—and these groups altered the indexical traces of the archaeological finds in very different ways. As a result, the case discussed here is particularly intricate.

As the fragments moved farther from the excavation site toward their final destination in Berlin, attitudes about their status as material data also underwent changes. In what follows, I argue that objects have an undefined and versatile potential as data but that this is precisely the reason why they face semantic difficulties when they move into different contexts.

The ways in which the status of archaeological objects changes before and when they arrive at a museum and become objects of study—something that arguably had an impact on the reconstruction practices of the Ishtar Gate—have rarely been discussed. By uncovering these shifts in the objects' status, I first intend to stimulate

[6] Hans-Jörg Rheinberger, "Infra-Experimentality: From Traces to Data, from Data to Patterning Facts," *Hist. Sci.* 49 (2011): 337–48, on 338.

[7] Ibid.

[8] Gaehtgens, *Die Berliner Museumsinsel* (cit. n. 3), 67; Gregor Schöllgen, "'Dann müssen wir uns aber Mesopotamien sichern!' Motive deutscher Türkeipolitik zur Zeit Wilhelms II in zeitgenössischen Darstellungen," *Saeculum* 32 (1981): 130–45. On the Pergamonmuseum, see Bilsel, *Antiquity on Display* (cit. n. 2); Alina Payne, "Portable Ruins: The Pergamon Altar, Heinrich Wölfflin and German Art History at the Fin de Siècle," *Res* 54/55 (2008): 168–89.

discussion of the material culture of archaeology with the awareness that research of-ten disembodies knowledge from the concrete objects through which it circulates. In the case of the excavations and the processing of the bricks and tiles that eventually led to the reconstruction of the Ishtar Gate in Berlin, it becomes clear that the mate-riality of the fragments never ceased to matter (to this day).

Against prevailing assumptions that the size of architectural elements prevents them from traveling, recent studies on the mobility of knowledge have claimed that "the more unlikely a scenario the more powerful its consequences will be."[9] In the case of the Ishtar Gate, nothing exemplifies this unlikelihood more than the original fragments spread out in the photographs in figure 1. The scenario might evoke stabil-ity, but it is a photograph about instability and the spaces of transit. Yet the space be-tween the excavation site and the arrival of the objects in the exhibition hall of a mu-seum—in short, the period when the objects were in transit—is seldom discussed in epistemological terms. In the nineteenth and early twentieth century, the mobility and portability of the objects, however, was often the foremost practical concern. It is pre-cisely here, when objects were in transit, that knowledge about them was negotiated and constituted. Though the objects the excavators held in their hands were "authen-tic," their authenticity was at stake the very moment they were removed from their original site: their surroundings—their original traces—were slowly detached from the objects as they were packed in boxes, shipped down the Euphrates, and transported to Europe. When the gate was finally reconstructed, many portions of it were re-constructed with replicas.[10] The Ishtar Gate's final reconstruction had very little to do with authenticity as it is defined today, yet the reconstruction became the "real" Babylon to visitors who would never glimpse the original. For the public, the arrival of objects and the reconstruction of Babylon within the museum became a mere spec-tacle, which occurred at the expense of material authenticity. What was once a poten-tial source for reliable information about the surrounding Babylonian city structure in the eyes of the excavators was partly replaced by fake material in order to ensure a "more authentic" image of the original gate. In what follows, I would also like to consider the conditions of data mining in material culture in the spaces in transit or, in other words, when and where objects can become data—and under what circumstances.

THINGS AND DATA PART I

The preliminary expedition from which the three auspicious bricks were brought back laid the groundwork not only for the excavation of Babylon, but also for a di-

[9] Alina Payne, "Introduction: The Republic of the Sea," in *Dalmatia and the Mediterranean: Por-table Archaeology and the Politics of Influence*, ed. Alina Payne (Leiden, 2014), 1–18. On the move-ment of objects, see Igor Kopytoff, "The Cultural Biography of Things: Commoditization as Process," in *The Social Life of Things: Commodities in Cultural Perspective*, ed. Arjun Appadurai (Cambridge, 1986), 64–91. See also Lorraine Daston, ed., *Biographies of Scientific Objects* (Chicago, 2000); James A. Secord, "Knowledge in Transit," *Isis* 95 (2004): 654–72.

[10] Authenticity is a slippery term in the long nineteenth century, as Kate Nichols showed recently in reference to reproductions of Antique monuments in the Victorian Crystal Palace. How could some-thing that is "not the real brick at all" appear "more authentic" than a genuine archaeological artifact? If "authentic" is defined as being closest to an object's original appearance, the restored and painted "all Greek" Parthenon frieze in the Crystal Palace might "well offer a more authentic view than the battered original in the British Museum." Nichols, *Greece and Rome at the Crystal Palace: Classical Sculpture and Modern Britain, 1854–1936* (Oxford, 2015), 90. See also Bilsel, *Antiquity on Display* (cit. n. 2).

vision that still dominates the cultural exploration of the ancient Near East in Europe: the museum, archaeology, and philology. Each field had strong and rancorously competitive advocates. While they all relied on one another, each one followed its own logic and had its own views on what would be worthwhile to collect, and which aspect of the finds should be valued the most as the rubble made its journey to Europe. What was partly a fight between the field and the museum also constituted a controversy about objects in relation to texts, as the text inscribed on these objects provided valuable data for the historical and philological study of Babylonia.

The history of science and the history of material culture have often seen "text" and "object" as two distinct categories. While text is usually deemed more important, it is often forgotten that text in the ancient Near East was materialized through archaeological objects. All parties involved in the excavations—including philologists—were thus looking for objects, but they had their own (rough) vision of which objects should be collected, which properties of an object were potentially of value, and where the objects should be transported to in order for their value to be recognized. For archaeologists, an object gained—and in fact only possessed—value once its relation to the space where it was excavated was documented. By way of contrast, philologists saw knowledge materialized in the form of inscriptions on clay tablets. These inscriptions did not always require a direct connection to the tablet's physical origin (in fact, leaving the original context almost seemed to be a condition for their further endurance as scientific objects).[11] Finally, museum curators concerned with expectations of the public relied on the "ultimate destination," Berlin's Museum Island, to turn the finds into objects of public interest.[12] To them, the three bricks were to become samples that set standards for what the excavators should be looking for to achieve the final goal: the reconstruction of one of the most significant palace sections from ancient Babylonia. What kinds of material contained or presented knowledge and information of potential value was thus a matter of perspective and subject to constant disputes. These disputes erupted right at the start of the expedition and continued over the next three decades, never leading to a satisfactory compromise for all parties.

During the expedition, Koldewey never ignored places of potential importance. Constantly measuring, drawing, and describing promising sites in notebooks—thus producing durable "traces"—the archaeologist was convinced that it was not enough simply to find "pretty ruins and return a bag full of cullets, only to discard them thereafter."[13] But Eduard Sachau, an orientalist in search of tablets with cuneiform inscriptions, opposed Koldewey's view from the start.[14] The two men leading the exploration hardly agreed on anything. In Koldewey's view, the elderly Sachau hindered his adventurous and younger colleague rather than helping him. Consequently, Koldewey

[11] See Joanna Radin, "'Digital Natives': How Medical and Indigenous Histories Matter for Big Data," in this volume.

[12] See Etienne Benson, "A Centrifuge of Calculation: Managing Data and Enthusiasm in Early Twentieth-Century Bird Banding," in this volume.

[13] Walter Andrae, *Babylon: Die versunkene Weltstadt und ihr Ausgräber Robert Koldewey* (Berlin, 1952), 85.

[14] Crüsemann, *Zweistromland* (cit. n. 3), 132–7. Koldewey's first description of the excavation site after the digging had officially started can be found in his letter to Otto Puchstein, 5 April 1899, in Robert Koldewey, *Heitere und ernste Briefe aus einem deutschen Archäologenleben* (Berlin, 1925), 136–7.

developed a strong personal antipathy toward Sachau, which began to carry weight when it came to judging the value of finds: the philologist valued cuneiform tablets with inscriptions, whereas the archaeologist was looking for aesthetically appealing bricks and sought to understand them in the context of the site. Tensions between philology and archaeology, scholarship and museum display, script and object, and isolation and context were not new: when leading excavations for Britain in the 1840s, Austen Henry Layard had considered Assyrian monuments as artistically inferior to classical art. For him the clay tablets were more relevant to study the past.[15]

Koldewey was certain that he would have discovered twice as much in Babylon without Sachau, whose scholarly endeavors and approaches to fieldwork the archaeologist deemed irrelevant.[16] While Koldewey's strategy was to examine as many sites as possible to determine their potential to yield objects of interest, he was cynically amused by the philologist's decision to study each find in detail in order to decide whether continuing the excavations would be worthwhile. This took time. Meanwhile, the excavation team had already moved on to a different site.[17] Philology thus dragged behind the actual fieldwork and had no impact on future digs. In spite of all their differences, however, Koldewey and the philologist had one view in common: while they knew roughly what they were interested in—portable objects—they did not really know what exactly they were looking for. The contingency involved in these early excavations is a little-studied aspect, but it enforced the instability of the finds from the moment they were unearthed. There was no clearly defined goal when the first fragments were excavated. In order to make the finds commensurable within the larger canonical order of Prussia's already existing collection, the excavators relied on unstable standards and taxonomies related to the fluctuating cultural ideals of the Wilhelminian Empire. Lacking definite objectives from the museum or the government about what should be collected and why, the battle over scholarly/epistemic objects and potential museological objects suitable for display became a key debate transcending the field, the museum, as well as the scholarly institutions such as universities where the objects could potentially be studied. This constellation provokes a deeper discussion about what constitutes valuable data in the form of material objects. In this dispute, what mattered was not only the objects' origin but also their final destination, that is, the context in which the objects might potentially yield useful data—or no data at all.

Unable to overcome his disagreement with Sachau, Koldewey continued the excavations with a younger companion, architect Walter Andrae (1875–1956), who had also been hired because of his astounding drawing skills. This choice was noteworthy because drawings became indispensable tools for recording the fragments' original context on paper, a process that in turn also determined what would be collected. Koldewey moved away from the study of tablets and from a biblical agenda that sought to prove the veracity of the Old Testament, both of which had dominated the French and British hunt for archaeological finds in the region. Instead, Koldewey

---

[15] Mirjam Brusius, "Misfit Objects: Excavations in Mesopotamia and Biblical Imagination in Mid-19th Century Britain," *J. Lit. Sci.* 5 (2012): 38–52.

[16] Koldewey to Puchstein, 7 May 1889, in Koldewey, *Heitere und ernste Briefe* (cit. n. 14), 123–4. See also Andrae, *Babylon* (cit. n. 13), 84. Koldewey underestimated Sachau, who pushed for Babylon in a letter to the museum. DOG, file II, 1.1.3., 10 March 1898.

[17] Koldewey to Puchstein, 27 November 1901, in Koldewey, *Heitere und ernste Briefe* (cit. n. 14, 142–4).

slowly started to make sense of the colored tiles that so intrigued him. With Andrae's assistance, he began indexing human and animal body parts that he believed were depicted on the fragments, keeping various lists and accounts at the same time.[18] These notes belonged, in Rheinberger's terms, "to the lab's knowledge regime"; comparable to "tentative interpretations of experimental results," which "lie *between* the materialities of experimental systems and the conceptual constructs that leave the immediate laboratory context behind in the guise of sanctioned research reports."[19] This process can be described as a decisive step in the definition of the objects as data, to which I will return below.

THINGS AND DATA PART II

A similar dispute about which objects were of use and should be collected—in particular in light of the object-text division—continued on another battleground in Berlin. In his attempt to strengthen the scholarly significance of the royal museum collections at the university and reciprocally involve academic experts in the curatorial tasks of the museums, Schöne appointed Friedrich Delitzsch (1850–1922), a leading Assyriologist and professor at Berlin University, as director of the Museum of Ancient Near Eastern Antiquities. At the time, this small collection had no finds from Prussia's own national excavations on display. When Delitzsch took office in 1899, several parties expressed concerns that an Assyriologist would not take good care of the "non-inscribed" objects, let alone their public display. In 1903, plans for an independent ancient Near Eastern museum in Berlin had emerged, though they were not carried out until much later. Between 1899 and 1911, the Mesopotamian objects were housed in an attic and then in the storerooms of the Kaiser-Friedrich-Museum. For experts like Delitzsch, museums were not only about exhibition and public access but also about storage; they were archives of potentially relevant material for future scholarship. Consequently, much of the collection was not made accessible to the public until Delitzsch's term ended in 1917. After 1918, some objects were accessible, but only to a limited number of individuals, in particular, leading scholars.[20]

For the eccentric, ingenious, and self-assured Koldewey, the Assyriologist's appointment was a tragedy. Additional news that his adversary Sachau had become chair of the Deutsche Orient-Gesellschaft (DOG) did not delight him either. In Koldeway's eyes, DOG members were a boring and uninspiring lot. Even wine, he reported, did not improve their tiresome meetings. Unimpressed with the DOG's decision to engage in excavations focused on the targeted removal of objects of interest such as tablets, and very different in scope from the large topographical digs he had in mind for Babylon, Koldewey signed a contract with the Berlin museums to move forward with a project of his own.

[18] "2. Auszug aus zwei Briefen Dr. Koldewey's," *Mitt. Deutsch. Orient-Gesell.* 13 (1902): 12–4, report, 4 April 1902. See also Kay Kohlmeyer, Eva Strommenger, and Hansjörg Schmid, eds., *Wiederentstehendes Babylon: Eine antike Weltstadt im Blick der Forschung* (Berlin, 1991), 50. See also Koldewey to Königliche Museen, 4 April 1899, DOG, file II, 1.1.5.1.
[19] Hans-Jörg Rheinberger, *An Epistemology of the Concrete: Twentieth-Century Histories of Life* (Durham, N.C., 2010), 233–43, on 245; emphasis in the original.
[20] On Delitzsch's museum directorship and related conflicts, see Crüsemann, *Zweistromland* (cit. n. 3), 137–42, 152–64; Frederick Bohrer, *Orientalism and Visual Culture: Imagining Mesopotamia in Nineteenth-Century Europe* (Cambridge, 2003), 275–304.

With his affection for the three small fragments from Babylon, Schöne represented a powerful advocate for Koldewey's plans. However, the director of Berlin's royal museums was far more committed to his own vision of displaying Prussian achievements than to Babylonian archaeology as such. To complicate things even further, Delitzsch, the newly appointed head of Berlin's Museum of Ancient Near Eastern Antiquities, personally visited the excavation sites of Babylon and Assur in 1902 and 1905 to "help out" and work on cuneiform inscriptions.[21] With Delitzsch's arrival in the field, the ongoing dispute between philologists and archaeologists deteriorated further. Reinforcing the Assyriologists's view in regarding objects without script as of little interest, Delitzsch pushed to release the inscribed cuneiform objects from Babylon immediately.[22] While everyone else deemed it safer to make sure that the objects could be exported according to the new antiquity laws of the Ottoman Empire, Delitzsch was even willing to risk the loss of tablets to the Ottoman Museum in Istanbul as long as they could be copied beforehand.

It is worth noting that for Delitzsch, a drawing or photograph of an inscription could serve his aims just as well as the actual tablet. In other words, assyriological scholarship could be carried out with a replica if the tablet could not be transported to Berlin. For the philologists, the actual object became increasingly irrelevant. What mattered was the text transcribed from the object. These copying practices were part of the process of turning the finds into durable facts (e.g., in the form of publications). This indifferent attitude toward material culture seems surprising for a museum director. For Delitzsch, however, valuable data was not necessarily embedded in the originality of the object but could be mediated. As for the colored tiles, he did not disapprove of their excavation, supporting the view of his Berlin colleagues that they might help the public image of the museum. However, as discrete objects, Delitzsch deemed them entirely insignificant because they contained no textual information. His lack of appreciation for their historical value was reinforced by his view that he wanted the reconstructed animals to be "beautiful and shiny"—simply the products of fine artisanship. With representatives of the museum in the field, the dispute between philology and archaeology escalated, leading to a decade of hostilities. Apart from the personal animosities within the field itself (Koldewey vs. Delitzsch), what was at stake here was the relationship between the museum and the field, as well as between different viewpoints about what kinds of knowledge were to be produced through the excavated finds, and where they were to be produced.

THE MUSEUM IN THE FIELD

The archaeologist Koldewey was not fond of either museums or the institutionalized scholarly world. Though witty, humorous, and charming, he lacked the diplomatic skills necessary for negotiations with the Ottoman Empire. However, his language skills in Arabic and his excellent connections in the region made him indispensable as head of the Prussian excavations in the region. In a sense, Koldewey was a local; during twenty years of fieldwork, he left only two times for a holiday and thus became more familiar with the local customs than he was with recent developments in his former home country.

---

[21] Archives of the Vorderasiatisches Museum (hereafter cited as "VAM"), file 51, fol. 6, Staatliche Museen zu Berlin Zentralarchiv.

[22] Delitzsch to Generalverwaltung, 12 May 1910, DOG, file II, 1.2.1.2.

As much as he seemed remote from several of the parties involved, Koldewey and his companion Andrae were united by three convictions. First, they both believed in the intertwined relationship between fieldworker and object. Second, each object represented potential data that had to be secured through documentation by the people who excavated it. And third, cuneiform tablets were not a priority in their collection efforts. Working side by side, the drawing artist/architect and the archaeologist agreed that extracting and processing knowledge gained from the rubble started on-site. They were deeply convinced that contextual knowledge was essential to understanding the excavated finds. In particular, Koldewey believed strongly that the more work he did on the site itself in terms of classifying and recording—in other words, "tracing"—the more knowledge that could be gained and later retrieved about the object. This tracing activity was thus a key aspect of the transformation of objects into epistemic things.[23] First, the fragments were washed in the Euphrates. Then they were dried, sorted, numbered, inventoried, and drawn. In the process, workers would determine whether they represented parts of a lion, bull, or dragon. Initially drawn on paper, several model animals were constructed on-site to facilitate the collecting and processing effort.[24] In particular, these immediate visual records were crucial to imposing meaning on the fragments.[25]

Living amidst the archaeological rubble, Koldewey experimented with excavation methods and developed new techniques that allowed him to examine and document each cultural surface systematically and meticulously.[26] Besides removing individual objects, he examined their material texture (e.g., in the case of the bricks, their enamel surface) to gain knowledge about their surroundings as well as about technologies and materials used at the time. He did so by taking careful notes and measurements, which in turn allowed him to gain exact knowledge about the structure and history of entire buildings and the layout and material legacies of the ancient city as a whole. In doing so, Koldewey's methods marked a turning point in the development of archaeology as a discipline: they forged a model for the exact archaeological field sciences in Mesopotamia.[27] Koldewey's stratigraphic excavation methods had little in common with his European predecessors' unsystematic "plundering" in the area—by way of contrast, he established a system within which archaeological objects could themselves potentially become data, as long as he traced the material within its physical context. In other words, for Koldewey, context and topography mattered more than the mere recovery of objects. He treated the sites themselves as scientific objects. Pushing the notion of archaeology as a spatial science, an object could no longer be "understood" without collecting physical information about the material culture that

[23] Rheinberger, "Infra-Experimentality" (cit. n. 6), 338.

[24] Typescript of a publication in "Broschüren zur Museologie," May 1946, "Nachlass Walter Andrae," 65, no. 7b, Staatsbibliothek zu Berlin.

[25] Walter Andrae, quoted in Beate Salje, "Robert Koldewey und das Vorderasiatische Museum Berlin," in Wartke, *Auf dem Weg nach Babylon* (cit. n. 1), 124–47, on 141. On "material chains" and "paper tools" in relation to data, see Christine von Oertzen, "Machineries of Data Power: Manual versus Mechanical Census Compilation in Nineteenth-Century Europe"; David Sepkoski, "The Database before the Computer?"; and Staffan Müller-Wille, "Names and Numbers: 'Data' in Classical Natural History, 1758–1859," all in this volume.

[26] Koldewey also showed affinities for medical self-experiments with tobacco and for dissecting animals in front of visitors; see Wartke, *Auf dem Weg nach Babylon* (cit. n. 1), 66, 104.

[27] Margarete von Ess, "Koldewey—Pionier systematischer Ausgrabungen im Orient," in ibid., 91–103.

surrounded it. This time-consuming procedure was problematic for Koldewey's museum counterparts. Even though they acknowledged Koldewey's expertise, they often wished he would simply cut out "the cake" of the archaeological site rather than painstakingly trying to understand every single detail.[28] In their view, knowledge making was not to take place on-site, but rather in the "safe space" of Berlin's Museum Island, amidst scholarly institutions and experts. For the museum, objects became meaningful—that is, objects of potential insight—once they had arrived at their final destination. There they could be used in spectacular displays or turned into objects of study by scholars who imposed meaning upon them.[29]

Rheinberger notes that laboratory notes and write-ups in their preliminary form were long regarded "as simple means for recording and storing data, as mere 'temporary storage bins' along the way to some 'final result.' In this purely retrospective vision, however, only the definitive data counted; and these appeared as results painted directly by nature's brush, transparent and vanishing intermediaries between the material under investigation and its conceptualization."[30] Highlighting the provisional and experimental spaces between the beginning (in this case, unearthing objects) and the final, official end (in this case, the arrival of objects in the exhibition hall or the scholarly study room) in the process of knowledge formation, he argues that "the 'containers' in which they are 'temporarily stored' by no means compose a neutral scaffolding in an otherwise purely intellectual process of knowledge formation."[31] These "'temporary storage bins' are and remain—literally—*between* the epistemic objects and the knowledge processes bound up with them."[32]

One such "temporary container" was the site itself. The historiography of museums has long reflected a centripetal view of the relationship between museum and field. Recent work, however, has started to highlight the fact that museums are themselves fields of knowledge negation and creation, in effect blurring the lines between field and museum science.[33] Along these lines, however, the field—as Koldewey's view conveyed—should also be seen as a site of museological knowledge production. Koldewey believed that the methods he used on-site could turn objects into meaningful data about the site's historical and material past at large. Political circumstances reinforced his view, as the new Ottoman antiquity laws hampered the export of archaeological objects to Berlin's museums. The new regulations forced Koldewey to establish an on-site "museum," a so-called premuseum [*Vormuseum*] in Babylon, which might have resembled more of a storage space. This museum-in-the-field housed a preliminary collection of excavated objects alongside inventories also produced there. The Prussian expedition team sent only some of these objects to Berlin;

---

[28] Paul Rohrbach Schöneberg to Koldewey, 25 May 1903, DOG, file II, 1.2.17.2.; letter written by Güterbock, 21 April 1906, DOG, file II, 1.7.37.

[29] On these practices, see James Delbourgo, "What's in the Box?," *Cabinet Magazine* 41 (2011): 47–50. See also Delbourgo's unpublished draft of "Return of the Sloane Ranger," based on a presentation at the conference "Historicizing Big Data," 31 October 2013, Max Planck Institute for the History of Science (with thanks to James Delbourgo).

[30] Rheinberger, *Epistemology* (cit. n. 19), 245.

[31] Ibid.

[32] Ibid.; emphasis in the original.

[33] See, e.g., Sarah Byrne, Anne Clarke, Rodney Harrison, and Robin Torrence, eds., *Unpacking the Collection: Networks of Material and Social Agency in the Museum* (New York, 2011); Tony Bennett, "Making and Mobilising Worlds: Assembling and Governing the Other," in *Material Powers: Essays beyond the Cultural Turn*, ed. Tony Bennett and Patrick Joyce (London, 2010), 190–208. See also Bruno Strasser, "Collecting Nature: Practices, Styles, Narratives," *Osiris* 27 (2012): 303–40, on 310.

the objects were selected on the basis of their condition, the climatic conditions, and available modes of transport.[34]

Although Koldewey's museum-in-the-field increasingly blurred the lines between archaeological fieldwork in Mesopotamia, on the one hand, and curation work in Berlin, on the other, both sites continued to contest the information they wished to see produced, the knowledge to be extracted, and the techniques and technologies to be used. Berlin's museum directors wanted the objects themselves and the knowledge drawn from them to be shared in publications and publicly displayed. Koldewey was convinced that only people like himself, who had seen excavation sites with their own eyes, were capable of publishing sophisticated reports on their finds. However, he had little interest in publishing all of his results immediately, for in his view, contextual information was required to make sense of the object. He published only with reluctance and seldom beyond the mandatory reports he was forced to submit, regardless of the pressure put on him by his Berlin superiors.

Rheinberger has described these kinds of transformations as transitions from "trace" to "data." However, for long periods, Koldewey's thorough personal excavation diaries, logbooks, maps, lists, and drawings—the traces—were not fully synthesized anywhere but in his own memory. In fact, Koldewey once stated that no logbook or diary could replace an excavator's memory, reflecting his attitude of superiority toward excavation and practices related to it.[35] In some cases, he kept what could have become official output, that is, published data in printed form, under tight wraps. This led to yet another conflict. His Prussian superiors sought timely evaluation and representative results in the form of prestigious volumes (so-called facts), a format that Koldewey had produced successfully in the past, not least to legitimize the funding for the excavations. As soon as a project was under way, though, writing and publishing were not his priority. He took many of his insights to his grave when he died in 1925, after his return to Berlin.

After Koldewey's death, Andrae was responsible for directing Babylon's resurrection in Prussia. The much younger architect had come to appreciate Koldewey's meticulous technique during his own excavation efforts in Assur. In a manifesto on fieldwork, Andrae declared that excavations were the best way to obtain ancient material (as opposed to acquiring objects through sales, trades, or donations). He argued that only objects from the field were "real" and disputed the possibility that terms such as "authenticity" [Echtheit] could adhere outside the context from which the objects derived.

Andrae's descriptions of fieldwork often took on odd forms, more reminiscent of love letters than handbooks on excavating. For example, he often described the knowledge that could potentially be extracted from objects as something akin to a treasure, accessible only to the most devoted. The genius loci of an object could still be traced within it, in particular by the person excavating it. This person, Andrae argued, "fell in love" with the object on-site and cared for it affectionately.[36] Excerpts from another manifesto are even more passionate: he wrote that objects could be a salvation for excavators, and that excavators could develop a passionate affection toward seemingly "dead" things, perceiving them as vivacious and radiant examples of human craft. Only

---

[34] On the distribution of finds [Fundteilung] between Prussia and the Ottoman Empire, see also DOG, file II, 1.2.3.6. 1909/10, 1.2.3.7. 1911/12, 1.2.3.8. 1913/14.
[35] VAM, file 51, fols. 44–45, Staatliche Museen zu Berlin Zentralarchiv.
[36] "Nachlass Walter Andrae" (cit. n. 24).

scholars who could carry out their excavations with such passion, Andrae argued, had the ability to find true authenticity in objects. The object's radiance would reflect on the excavator, who in turn possessed the ability to "read" the object.[37] The excavator touched the objects, smelled the soil in which they were covered, stood on the ground, felt the desert wind and the intensity of the sun, and oversaw the local workers. Excavators thought of themselves as "museum inaugurators," Andrae claimed, but he ascribed additional skills to them that went beyond the mere ability to convey the materiality [*das Stoffliche*] of what they excavated.

In short, for Andrae, the bricks became meaningful, much as they had for Koldewey, through systematic documentation on the one hand, but also through personal (eye)witnessing on the other. For both Koldewey and Andrae, the past was stored in objects themselves and could be retrieved by excavating and documenting the object's surrounding context. Above all, however, the object was processed by their personal experience, for them a sacred, exclusive, and in fact the only way of understanding the object. The finds, according to the excavators, had history inscribed into the material. The past that these bricks "stored" in the eyes of these two men, however, was not only the ancient past of their original making. As Andrae's references to contemporary life and multisensory experience reveal, the objects also stored traces of their own life's work, experience, and achievements. This makes it questionable whether their potential as data would have been accessible to anyone but them.

We see here the contingency involved in the process by which objects acquire meaning. Certain properties of objects define how they operate in social life, how they are valued and understood, read and interpreted. Yet value is also externalized and depends on epistemic, social, political, and personal context. Meaning is inscribed not only in the forms of objects but also in the trajectories of their use. In short, meaning cannot be found solely in the distant past. While at first Andrae and Koldewey shared the museum's affinity for colored bricks, their insistence on context and experience sharply contrasted with the museum's appreciation of the finds, which in Berlin were detached in real time and space from (literally) distant Babylon. These views became particularly manifest when the museum restricted its "orders" to decorative "animal figures." Koldewey had been frustrated that his thorough excavations were undermined by the museums' exclusive focus on figurative ensembles to the point that he begged his Berlin superiors to put a stop to the constant "cry for animals."[38] For the curators in Berlin, the bricks were meaningful if they were somehow "recognizable" as the original specimens when they had first arrived. For Koldewey, however, who had seen millions of bricks excavated in context, scratched the dirt off their surface, and held many of them in his own hands, objects embodied information of all kinds, regardless of whether they represented part of a lion's paw, a bull's tail, or no animal at all.

TRACES

Traces, according to Rheinberger, only acquired meaning as data if they could be condensed into an epistemic thing (a possible fact) "that in turn could be used to create data patterns: curves, maps, schemes, and the like—potentially the whole plethora of

---

[37] Ibid.
[38] Koldewey to Königliche Museen, 14 August 1910, DOG, file II, 1.2.5.16.

forms of visualization the sciences have been so inventive to generate."[39] But even on their way to becoming epistemic things, a wide range of media were (and still are) involved in the making of archaeological data, as is the case with other disciplines that rely on field methods. The archival documentation of the Babylon expedition contains plenty of sources indicating the epistemic potential of field records, which initiated independent knowledge production. Even though the Babylonian bricks were portable, they still traveled through surrogates, including text, drawing, and photography—in other words, "portable proxies."[40]

Drawings, maps, photographs, indices, and logbooks were at least as important for the excavators to gain insights as the bricks themselves. In fact, in some cases (e.g., in the event of loss), these visual and textual records mattered more than the actual objects. They, too, like the museum practices surrounding the previously discussed *Vormuseum* on the excavation site, fall into the category of temporary storage, which promised stability to turn objects into durable epistemic things.

In other words, archaeological data moving between Berlin and Babylon had many guises, which changed frequently during the transmission period. Furthermore, the individual steps of mediation sometimes failed to work. Different types of media had different roles and functions in this chain. Koldewey and Andrae were avid proponents of drawings, which were important tools for understanding the physical site before further steps were taken. Andrae believed that everything should be "drawn from nature," the method he thought was best in terms of avoiding errors that often appeared when sketches were started on-site and then completed away from the field. Visual documentation was thus considered the first and central step of fieldwork. Excavations should only begin after all relevant structures were incorporated into a grid system on paper, so that the dig could start at the spot considered best. Going hand in hand with observation and note-taking techniques (a combination that in Andrae's view would eventually lead to a perceptive gaze and an astute mind), drawings determined how buildings were "understood."[41] Drawings made on-site also ensured that, whenever possible, excavators had an "overview of the site." In the museum, such panoramic drawings aided reconstruction efforts. In fact, drawings were often the only seemingly stable and coherent element in an otherwise contingent process, providing orientation amidst the flood of fragments, which themselves became simply too malleable. The drawings were traces that contributed to turning objects into facts.

Photography was an additional tool, although Koldewey remained ambivalent about its merits. When he first left for Mesopotamia, he did not bring a camera.[42] In his view, photographs produced "indifferent" representations: because the camera recorded everything mechanically, photographs lacked the ability to help excavators understand their object of study the way drawings did, which could highlight individual features.[43] Photography, however, proved useful for other reasons related to objects' status in transition. One of its functions was to overcome time, for example,

[39] Rheinberger, "Infra-Experimentality" (cit. n. 6), 345–6.

[40] Payne, "Introduction" (cit. n. 9), 3. Payne differentiates between "mobile" and "portable."

[41] Walter Andrae, *Lebenserinnerungen eines Ausgräbers* (1961; repr., Stuttgart, 1988), 253–4. See also Ernst Walter Andrae and R. M. Boehmer, *Bilder eines Ausgräbers: die Orientbilder von Walter Andrae, 1898–1919* (Berlin, 1989), 8.

[42] Andrae, *Babylon* (cit. n. 13), 15.

[43] The DOG archives contain a variety of sources on the issue of photography; I cite only some of them in subsequent notes.

by transmitting records from the excavation site to the museum, often long before the objects completed their arduous journey to Europe. Photographs were also considered a reliable way to transfer information in the event that an object was lost, and they were also used for publication and classification.[44] In practice, using photography came with a price. As much as the photographs were meant to replace lost objects, the photographic equipment sent to Babylon was itself often lost on its way. Other problems concerned the complex nature of photographic image making. Koldewey tried to develop the photographs himself, but he failed on various levels. Sometimes the negatives were scratched or accidentally exposed to light. Sometimes it was simply too hot for the chemical developer to work, making either negatives or the prints illegible.[45] No less challenging was having the negatives developed in Baghdad; unhappy with the results, Koldewey sent his film to Berlin to be developed.[46] These photographic negatives were encoded traces from the field, the content of which was revealed only on a developed positive print. By the time they arrived in Berlin, these negatives were often corrupted or even destroyed.[47]

Even when photography worked, it proved not to be the most convenient of tools: not only were the developed prints to be returned to Koldewey in Babylon in order to be numbered and indexed by him personally, but the same photographs had to be sent back to Berlin along with the index. Without an index or a contextualizing map of the excavation site, Koldewey's colleagues in the museum did not have the slightest idea what the pictures represented. Several times, errors in the index made in Babylon forced Koldewey and his Berlin colleagues to send the prints back and forth more than once. Time became a key issue: when the photographs arrived in Berlin, the images often no longer matched the current state of the constantly changing excavation site.

Yet even if the prints were satisfactory, for Koldewey photographs remained inferior tools. Without his critical eye, they presented traces without meaning. When Koldewey judged submissions for an illustrated publication series in 1911, he pointed to the risk of accepting too much "rubbish," implying that photographs alone would not do the job. He deemed it more important that people saw and noticed what was novel and important in individual objects. Rather than reinforcing old tropes—"photographs, caravans, coffee-cooking turks"—new knowledge could be gained only in conjunction with other forms of documentation.[48] Drawings and traditional paper squeezes or molds were considered essential to turn photographs into valuable documentation.[49] Photographs continued to be seen as complementary; they merely gave a better impression of what the objects might have looked like.

In short, a complex chain involving different tools and media was thought to be the key to overcoming difficulties concerning time and space. Techniques such as mold-

---

[44] VAM, file 51, fols. 28–30 (8 January 1899), Staatliche Museen zu Berlin Zentralarchiv. On photography and archaeology, see Frederick Bohrer, *Photography and Archaeology* (London, 2011). On the loss of objects in transit and photography, see Mirjam Brusius, "Le Tigre, le Louvre et les échanges de connaissances archéologiques visuelles entre la France et la Grande-Bretagne aux alentours de 1850," *Cah. Ecole Louvre* 5 (2014): 34–46.

[45] Koldewey to Königliche Museen, 10 June 1912, DOG, file II, 1.2.5.18.

[46] Koldewey to Königliche Museen, 1 May 1899, DOG, file II, 1.1.5.1. See also the letters written on 29 May 1899 and 17 July 1899.

[47] Schöne to Koldewey, 21 August 1900, DOG, file II, 1.2.3.2.

[48] Koldewey to Königliche Museen, 12 August 1911, DOG, file II, 1.2.5.17.

[49] Weissbach to Koldewey, 9 September 1903, DOG, file II, 1.2.5.11.

ing also presented not only visible but also direct, physical, and thus tactile traces of the objects in an attempt to turn them into objects of knowledge. Their frequent use suggests that objects did not actually serve as inherent and unmediated data immediately after they were excavated on-site, but rather that paper technologies functioned as carriers stabilizing and mediating between the field and the museum—and even within the field itself—what in the excavators' eyes could be potential data. Once the actual objects started their journey to Europe, however, their meaning took on other forms. In fact, the personal experiences described by both Andrae and Koldewey showed that there was never only one meaning attached to the finds, even though their teamwork on-site might suggest that there was. The objects were ambiguous from the very outset, and thus their value varied according to the ideas and motivations of other interested parties.

THE FIELD IN THE MUSEUM

Not least because photographs proved to be unsatisfying proxies for those who had never been to Babylon, the original objects, in particular the tiles, somehow had to make it to the museum. In 1903, 399 boxes, each containing 250 tiles, were sent to Berlin from Babylon, first by *keleks* (rafts) via Basra, and then by ship via Hamburg. A memorandum written by Delitzsch and Koldewey in June 1902 stated that workers should keep the 20,000 numbered and 100,000 unnumbered tiles, some with notes, in order and unpack them only with supervision. The missive contained detailed instructions with the tiles on how to unpack, index, and put them together. They were divided into (*a*) reliefs, (*b*) ornaments, (*c*) colored enamels, and (*d*) wall pieces found in situ. In Berlin, a large room was to be reserved for the unpacking and demineralization.[50] This project was bound to face problems. On the archaeological site, every item was wrapped in paper with a number, most of which were ignored in Berlin when the bricks were unwrapped.[51] The undeniable gap between field and museum was once again apparent: Koldewey's meticulous work in Babylon did not carry much meaning for the curators who had not been involved in the field.

Aided by Andrae's drawings, the fragments were initially to be classified and then sorted according to type. Altogether, Koldewey had identified 540 brick types.[52] This process of "unpacking" the objects from the temporary paper wrap in which they were stored was literally and metaphorically meaningful: what was to be unpacked was not only the object according to its assigned taxonomical classification, but also its meaning in a wider context, made durable by Koldewey through the numbered paper wrap. Thus, this paper wrap may be considered the last trace that physically linked the object with the field, and thus the field with the museum.

Archaeologists routinely describe sites as composed of assemblages encountered in deposits. Curators encountered a similar situation when the finds arrived in Berlin. The museum turned into a field. However, this field now lacked context, in the form of traces, such as the numbered paper wraps as well other contextualizing documentation on paper. What was about to become data in the field entered another realm with its own rules and goals.

[50] Published in Crüsemann, *Zweistromland* (cit. n. 3), 184–5. Crüsemann also describes the transfer of the bricks in detail; see 184–7.
[51] Walter Andrae, quoted in Salje, "Robert Koldewey" (cit. n. 25), 141.
[52] Letter by Koldewey, 22 June 1903, quoted in Crüsemann, *Zweistromland* (cit. n. 3), 184–7, 242–54.

Although the bricks made sense to those who had unearthed them, the helpers in the museum lacked the knowledge and the imagination needed to achieve a taxonomical organization of the bricks according to the ideas of those who worked in the field. In order to master the extraordinary challenge of reunifying the decontextualized rubble, the helpers in the museum adapted different methods remote from the context-based approach the excavators had in mind. The museum started with one "sample lion."[53] Without the excavators to oversee the project, workers used color to "prettify" damaged areas of the lion. These colorful bricks thus became, for the public, symbolic of Babylon.[54] Trouble was inevitable: one party (the museum and supporters of Koldewey) pushed for "authenticity" (which meant not adding color), and the other (the DOG) argued for aesthetics and public spectacle, even if this involved drastic changes to the brick's material composition.[55]

The situation escalated when Andrae visited Berlin in 1908 and was utterly shocked by the fate of his Babylonian lions, bulls, and dragons: some fragments had been trimmed with a saw to fit. Spaces with missing fragments, which were meant to be left blank, had been embellished with oil paint. While Delitzsch and the general secretary of the museum, Wilhelm von Bode, were impressed and proud, Andrae was devastated, calling the display "barbarian." He himself had prepared a model lion without any modified bricks in order to make it look "natural" while at the same time revealing the damage to the object.[56] The brick was authentic material for Andrae and as such under no circumstances was to be equated with "fake" material. His attitude did not resonate with most of the Berlin officials involved in reconstructing Babylon. For them, the main goal was to produce something suitable for display. If the bricks did not fit, they had to find a solution.

The matter became so charged that, in 1908, a so-called beast conference under the patronage of Kaiser Wilhelm II was held so that the parties involved could find a consensus. The Kaiser supported Andrae's view and decreed that all tiles should remain in their original form. However, disagreements persisted.[57] The originality of the material might matter a great deal for Andrae, but the museum's curators valued the original much less, since missing fragments could easily be replaced with new material. In the process, the actual object lost significance and became detached from the process of reconstructing Babylon in the museum.

BEAST PUZZLE

On 20 January 1927, the last 536 boxes with finds from Babylon finally arrived in Berlin.[58] The enameled tiles accounted for around 400 boxes of the finds. As in 1903, all items were immediately sent to the chemical laboratory for demineralization and waxing (this took about eighteen months) to prepare the material for reconstruc-

---

[53] Stated by the directorate of the DOG, Heinrich Prinz zu Schönaich-Carolath, and General Verwaltung der Königlichen Museen, 11 December 1905, DOG, file II, 1.7.37.

[54] Wiegand to Koldewey, 15 September 1905, DOG, file II, 1.2.5.4. 1904/05.

[55] Crüsemann, *Zweistromland* (cit. n. 3), 245.

[56] Andrae, *Babylon* (cit. n. 13), 94; Robert Koldewey, "2. Ausgrabungsberichte aus Babylon," *Mitt. Deutsch. Orient-Gesell.* 32 (1906): 1–3. See also Andrae, *Lebenserinnerungen* (cit. n. 41), 196–7.

[57] Kohlmeyer, Strommenger, and Schmid, *Wiederentstehendes Babylon* (cit. n. 18), 53.

[58] W. Andrae, "2. Reise nach Babylon zur Teilung der Babylon-Funde," *Mitt. Deutsch. Orient-Gesell.* 65 (1927): 7–27; Bilsel, *Antiquity on Display* (cit. n. 2), 182–3. On the "Fundteilung," see also DOG, file II, 1.2.3.6. 1909/10, 1.2.3.7. 1911/12, and 1.2.3.8. 1913/14.

tion. Andrae mourned the fact that some tiles were destroyed by the time they arrived in Berlin. Others were in the process of deteriorating. Still others contained finger-prints of the workers in Nebuchadnezzar's workshop, many of whom must have been women and children.[59] Up to thirty workers from the museum's staff were involved in arranging the fragments and looking for animals and ornaments. The photograph of this event bears witness to the team's struggle in organizing the animal tiles: eye to eye, paw to paw.

The original plan of the Pergamonmuseum did not include the display of the new Babylon collection.[60] Several life-size paper models of the Ishtar Gate and Babylon's Processional Way were reconstructed (partly in the state opera workshop) to convince the museum architects that a larger structure was undoubtedly necessary in order to house the treasures captured from the Ottoman Empire (fig. 2). In the end, however, the reconstruction was customized to the museum's needs rather than the other way around. Consequently, the Processional Way became much narrower than Andrae had hoped, and the Ishtar Gate much smaller.

Further concessions were made to meet the expectations of the museum's audi-ence. Andrae assumed that missing details (e.g., in an animal figure) would disturb the viewer. This created a dilemma. The new context of the tiles in the museum forced him to rethink his relationship to the very material he had excavated. With tiles missing from the animal figures, the rubble became an incoherent and potentially less useful archive. Simply storing this archive was not an option. Realizing that it was impossible to rebuild the Ishtar Gate and Processional Way entirely from the original bricks, Andrae commissioned a workshop outside of Berlin to produce Babylonian enamel colors according to a recipe found in an inscription on the bricks. Andrae in-sisted that tiles should be replaced only if, after a thorough search, the original ones could not be found. Convinced of the importance of originality, Andrae also felt obliged to show the viewer what was authentic and what was not. He did everything in his power to limit the reconstruction to the original material but ultimately had to give up.[61] If originals were unavailable, they replaced them with "local species." Echo-ing Bruno Strasser's remarks on the transition of objects into scientific collections, the bricks were subjected to both an "ontological reduction and a formal standardiza-tion":[62] they were "formatted identically" but also became taxonomically distinct from one another.

The team juggled 30,000 bricks of a puzzle. Can Bilsel has noted that the classifi-cation of the fragments according to types on the museum's worktables turned the cu-rators into re-creators of their own vision of the archaeological monument. According to Bilsel, "each time the restorers picked a lion's eye from the desk, they had to choose from a large pool of fragments which could have belonged to any of about one hundred lions on one side of the Processional Way. The face of each brick was reconstituted from a combination of six to seven fragments, which further increased the choices that the restorers had to make each time they assembled a figure. Unlike a jigsaw puzzle, the Babylonian walls had not one but many solutions."[63] Bilsel concludes that rather

[59] Walter Andrae, "2. Von der Arbeit an den Altertümern aus Assur und Babylon," *Mitt. Deutsch. Orient-Gesell.* 66 (1928): 20–22.

[60] Carola Wedel, ed., *Das Pergamonmuseum: Menschen, Mythen, Meisterwerke* (Berlin 2003), 98.

[61] Rheinberger, *Epistemology* (cit. n. 19), 233–43.

[62] Strasser, "Collecting Nature" (cit. n. 33), 321.

[63] Bilsel, *Antiquity on Display* (cit. n. 2), 180.

Figure 2. *Ischtar-Tor von Babylon, Aufstellungsversuch in Pappe (Theater-Kulisse)* [Ishtar Gate of Babylon, Attempt at a possible setup with cardboard (theater scenery)], *1928/29. Staatliche Museen zu Berlin—Vorderasiatisches Museum, Fotoarchiv, Bab Ph 3702.*

than a copy true to its original nature, the reconstructions in the Pergamonmuseum should be taken as a fascinating re-constellation of antique fragments aligned with fin de siècle preoccupations of *Jugendstil* and Art Nouveau.[64]

Perhaps out of desperation, even Andrae stopped making strict distinctions between original, replica, and copy. Indeed, at this point, for the museum and the public, it hardly mattered whether the majority of the bricks were originals or copies. The

[64] Ibid.

original tiles were certainly authentic, and yet they were physically and epistemolog-ically remote from their original context. Although a substantial number of the bricks used for the reconstruction in Berlin were from Babylon, they still failed to convey the scope and nature of the actual site. Andrae's dream of conveying the quantity and monumentality involved in the site (and its excavation) by means of scholarly meth-ods, never came true.[65] "Authenticity" thus became something associated with imag-ination, rather than originality. Worst of all in their view, the data Koldewey and Andrae had collected over decades, including detailed measurements, was eventually ignored and became obsolete. Space restrictions made it impossible for the builders to replicate the originals in the museum. Indeed, those epistemic things had, in the words of Rheinberger, "the capacity to become detached not only from their initial referent, that to which they originally referred, but also from the one who writes, the one who produces the trace."[66]

The museum was not about Babylon, but about Berlin. In the Ishtar Gate and the adjunct reconstruction of the Processional Way, objects became alienated from their original context as they traveled from the field to the museum. The idea of Babylon as a site was reduced by its own alienation. Altered to match the local needs in what had now become interwar Germany, the erected reconstruction now seemed like a remote reverberation of ambitious imperial efforts that took place at the dawn of a new cen-tury.

CONCLUSION

In 1928, Andrae became director of Berlin's Museum of Ancient Near Eastern Art. Under his directorship, the collection—now incorporated into Berlin's new Perga-monmuseum—completed its shift in emphasis from philology and cuneiform to ar-chitecture. By 1930, Andrae and his team had reconstructed thirty lions, twenty-six bulls, and sixteen dragons from Babylon, in addition to other massive architectural structures from the site.[67] The opening of the Processional Way and the Ishtar Gate co-incided with the one-hundredth anniversary of the Prussian Royal Museums in 1930. Unable to display all of its animals in the exhibition space, the museum loaned or sold some of the Babylonian beasts to other museums.[68]

The gate may have come together as an impressive architectural whole, but it dif-fered substantially from the original, few of the actors involved in its reconstruction had ever seen. The completed reconstruction embodied a shift in expectations that occurred when the objects were transferred from the field to the museum. For Koldewey and Andrae, the traces extracted from the original site using various tools and methods (a process that Rheinberger described as turning something into possi-ble "facts") were meant to establish durable links between the object and its material context in order to draw wider historical conclusions. In the process, the excavators tried to make the objects meaningful in a nonarbitrary way. For them, the objects themselves represented data about Babylon and its surroundings while they were still on-site. But many of the maps and measurements, as well as classifying information

[65] Quoted in Andrae, *Lebenserinnerungen* (cit. n. 41), 274. Andrae wrote an article about the recon-struction of the gate; see Andrae, "2. Von der Arbeit" (cit. n. 59), 20.

[66] Rheinberger, *Epistemology* (cit. n. 19), 244.

[67] Tim Ingold, "Materials against Materiality," *Archaeol. Dialogues* 14 (2007): 1–16, on 15.

[68] Andrae, *Lebenserinnerungen* (cit. n. 41), 277; Bilsel, *Antiquity on Display* (cit. n. 2), 180.

attached to each brick—traces to help the data come alive—were ultimately ignored in the context of the gate's reconstruction, as were many of the original bricks. For the curators, whose main concern was putting the fragments on museum display, the bricks were just that: actual fragments of the gate with no additional epistemic meaning. What mattered more than their original structure was whether objects could be turned into abstract formations suiting the purpose of their museological or scholarly endeavors. This material detachment could also be observed in the context of philology, in spite of efforts to build an actual collection of cuneiform tablets. In some cases, however, Delitzsch promoted recorded information in the form of copied inscriptions at the expense of original objects, simply for the sake of the "textual data" he needed. Like for Koldewey and Andrae, the objects represented data, but as Delitzsch was extracting the required information—the inscriptions—the object's actual presence was no longer essential.

The bricks from Babylon show how the material context changed meaning as the pieces made their way to Berlin. What remained was a projection that eventually revealed more about the place where the finds arrived than about the place from whence they came. While the majority of the public in Berlin seemed entirely satisfied with having substitute bricks on display, a large number of the original archaeological fragments were and continue to be housed in recently renewed storage facilities.[69] Meanwhile, various attempts were made to reconstruct the royal palace in Babylon, Iraq, notably without those bricks that are now safely stored in Berlin. In other words, the value of the Berlin reconstruction may have been produced in contexts detached from the material site and no longer depending on physical materiality. What made the reconstruction authentic, however, was the material trace embodied through those original bricks, many of them safely archived within the museum store, away from public view.

Focusing on the ambiguity of evidence and debates over its mediation, this essay sought to unpack the contingency of data in relation to material objects. What makes the "puzzle image" discussed at the beginning of this essay so compelling is the contrast between the piles of fragmented material and the seeming architectural and museological coherence of the final reconstruction of Babylon in the museum. It highlights the competing aspects of archaeology: on the one hand, as a scientific field, and on the other, as a social and museological practice related to public presentation, where ample room for interpretation and imagination matters more than measurable context. The bricks were mutable and palpable—even replaceable—objects, deemed to serve the need to reconstruct the Babylonian palace structures.

Data evolves in its surroundings. Its relationship with material is never fixed but always elusive. These surroundings include users and the manifold ways in which the data are engaged in their context. The properties of materials are thus not fixed attributes without the potential to change their meaning. Rather, they present unstable histories. The transformation of "pure knowledge" as an intrinsic property of things seems to be an illusion. Instead, knowledge is the result of a complex selection and trans-

---

[69] For storage, see Mirjam Brusius, "The Ancient Near East in Storage: Assyrian Museum Objects as a Cultural Challenge in Victorian England," in *The Museum Is Open: Towards a Transnational History of Museums, 1750–1940*, ed. Andrea Meyer and Bénédicte Savoy (Berlin, 2013), 19–30; see also Mirjam Brusius and Kavita Singh, eds., *Museum Storage and Meaning: Tales from the Crypt* (London, 2017).

formation process in which people (e.g., archaeologists at excavation sites or scholars at a museum) retrospectively imprint meanings on objects. The objects had the potential to become data as they were unearthed. They became data when they began to be processed and contextualized. Yet their ultimate journey took them to a different destination, distinct from the context the excavators originally had in mind. If this destination is ill-conceived, the meaning of objects is in flux. In other words, notwithstanding the different expectations people had in relation to the objects as they made their way to Europe, there was also no stable space where the finds could be integrated. When the bricks arrived, no one knew where or how they should be displayed. There was no museum to house them, and their relationship to the local collection remained unclear for a long time. Although taken at the brick's final destination in Berlin, the photograph of the bricks from Babylon spread out on worktables presents the space "in between," the last trace linking the field with the museum. While the bricks had become data in their short life on-site, these data were about to be embedded in a new argument when they arrived on Berlin's Museum Island.

A Centrifuge of Calculation:
Managing Data and Enthusiasm in Early Twentieth-Century Bird Banding

by Etienne S. Benson*

ABSTRACT

Beginning in 1920, bird banding in the United States was coordinated by an office within the U.S. Biological Survey that recruited volunteers, issued permits, distributed bands and reporting forms, and collected and organized the data that resulted. In the 1920s and 1930s, data from thousands of volunteers banding millions of birds helped ornithologists map migratory flyways and census bird populations on a continental scale. This essay argues that the success of the bird-banding program depended on a fragile balance between the centripetal effects of national coordination and the centrifugal effects of volunteer enthusiasm. For various reasons, efforts to maintain this balance were largely abandoned by the Bird-Banding Office from the late 1930s onward. Nonetheless, the first two decades of the national bird-banding effort provide an example of how a "citizen-science" project that generates "Big Data" can produce significant scientific results without subordinating the enthusiasms of volunteers to the data-collecting needs of professional scientists.

Bird banding is the practice of attaching metal rings stamped with serial numbers to the legs of birds so that they can be identified at a later date. First adopted by European and North American ornithologists on a large scale in the early twentieth century, it can seem rather quaint in comparison to some of the techniques for studying birds that have since become available. It relies on decidedly low-tech instruments—bands, nets, pliers, notebooks—and on time spent in the field rather than seated before a computer. The amount of data produced by the technique was astounding and even sometimes overwhelming by the standards of the early twentieth century, but it can seem small next to the "Big Data" produced by some contemporary ornithological research methods. A satellite-linked electronic tag, for example, is capable of

* Department of History and Sociology of Science, University of Pennsylvania, Claudia Cohen Hall, Suite 303, 249 South 36th Street, Philadelphia, PA 19104-6304; ebenson@sas.upenn.edu.

I would like to thank Lisa Ruth Rand for assistance in obtaining archival materials as well as the participants in the Max Planck Institute for the History of Science reading groups and workshops on histories of data, particularly Elena Aronova, Christine von Oertzen, and David Sepkoski, for their many helpful comments and suggestions.

tracking the geographical position of a migrating bird on a daily or even hourly basis even in regions of the world, such as the high seas, where human observers are sparse. Such tags can produce hundreds or thousands of data points per tagged bird, in contrast to the one or two typically provided by banding. Meanwhile, citizen-science projects recruit thousands of bird-watchers with smartphones and Internet connections to share their observations of bird migrations and other phenomena. Such electronic instruments and networked citizen scientists produce much of the data upon which today's ornithological discoveries rest.[1]

Nonetheless, bird banding continues to play an important role in studies of bird behavior, and its history offers insights into the dynamics of cooperation and conflict between professional ornithologists and amateur volunteers that are relevant to many of today's citizen-science projects. As I describe below, the American bird-banding effort of the 1920s and 1930s relied heavily on volunteer banders, whose efforts were coordinated by professional ornithologists at the Bird Banding Office (BBO) of the U.S. Biological Survey in Washington, D.C. Indeed, bird-banding efforts helped to strengthen the professional-amateur distinction in ornithology, where such boundaries remained fluid much longer than in many other scientific disciplines.[2] By the 1920s, however, the lines were being increasingly starkly drawn, so that professionals such as Frederick C. Lincoln, who headed the BBO from 1920 to 1946, could lay claim to greater authority than most amateurs who lacked specialized training and institutional support. In this context, the bird-banding effort was a typical early twentieth-century elaboration on the long tradition of observation networks in the sciences—one that harnessed the power of the state, ideals of civic participation, and the advantages of specialization to create a centripetal knowledge infrastructure that funneled data from peripheral amateurs to centrally located professionals.[3]

Narratives of professionalization and centralization do not, however, account for the full complexity of the early twentieth-century bird-banding effort, which cannot be reduced to a system for exploiting amateur enthusiasms for the accumulation of data in a center of calculation.[4] The BBO's dependence on volunteers, whose main reason for participating was their "passion for birds,"[5] meant that professional ornithologists at the center of the network spent a great deal of their time, especially in the early years of the project, encouraging activities that sustained volunteer enthusiasm

[1] On the history of bird banding in the United States, see Jerome A. Jackson, William E. Davis Jr., and John Tautin, eds., *Bird Banding in North America: The First Hundred Years*, Memoirs of the Nuttall Ornithological Club, no. 15 (Cambridge, Mass., 2008); Mark V. Barrow, *A Passion for Birds: American Ornithology after Audubon* (Princeton, N.J., 1998), 169–71; Robert M. Wilson, *Seeking Refuge: Birds and Landscapes of the Pacific Flyway* (Seattle, 2010), 72–5.

[2] Barrow, *Passion for Birds* (cit. n. 1).

[3] The history of observation networks and knowledge infrastructures is large and growing. For a survey with a focus on nonprofessional observers, see Jeremy Vetter, "Introduction: Lay Participation in the History of Scientific Observation," *Sci. Context* 24 (2011): 127–41. On the relevance of nineteenth-century historical examples to present-day citizen-science projects, see Gowan Dawson, Chris Lintott, and Sally Shuttleworth, "Constructing Scientific Communities: Citizen Science in the Nineteenth and Twenty-First Centuries," *J. Victorian Cult.* 20 (2015): 246–54.

[4] On the BBO as a "centre of calculation" that allowed the Biological Survey to "centralize the production of bird-banding data," see Wilson, *Seeking Refuge* (cit. n. 1), 74. See also Bruno Latour, *Science in Action: How to Follow Scientists and Engineers through Society* (Cambridge, Mass., 1987), 219–37; Susan Leigh Star and James R. Griesemer, "Institutional Ecology, 'Translations,' and Boundary Objects: Amateurs and Professionals in Berkeley's Museum of Vertebrate Zoology, 1907–39," *Soc. Stud. Sci.* 19 (1989): 387–420.

[5] Barrow, *Passion for Birds* (cit. n. 1).

but also generated a variety of centrifugal forces that ultimately worked against the BBO's own aims. One of the most prominent volunteer "cooperators," Margaret Morse Nice, publicly criticized the forms of banding research that mostly directly supported Lincoln's aims, for example, while regional associations of volunteer bird-banders established local data repositories and "emergency" stores of bands in direct violation of the BBO's wishes. The success of bird banding in the 1920s and 1930s depended on an inescapable interplay of centripetal and centrifugal forces.

Inescapable, that is, under particular social, economic, technical, and scientific conditions. Over the course of the 1920s and 1930s, it gradually became both desirable and possible for the BBO to reduce its dependence on volunteers and its vulnerability to their centrifugal enthusiasms. It is here that the issue of data management becomes central. In the early years of the BBO, there were clear links between volunteer enthusiasm, data collection, and scientific results: the more the BBO fostered the enthusiasm of volunteers, the more data was generated, and the easier it became to map migration pathways and census bird populations on a continental scale. As the volume of data flooding into the BBO increased, however, it became impossible for its limited staff to stay afloat; at times they fell literally years behind. For this and other reasons, the BBO was, by the 1930s, more concerned with disciplining volunteers than with fostering their enthusiasms. By the end of the 1940s, a dramatic shift away from reliance on volunteers had taken place. The history of the U.S. bird-banding program's embrace of and subsequent retreat from a balance between centrifugal and centripetal forces suggests the viability as well as the limits of alternatives to the centripetally oriented projects that dominate "citizen science" today.[6]

THE EMERGENCE OF ENTHUSIASM FOR BIRD BANDING IN NORTH AMERICA

Enthusiasm for bird banding emerged in Europe and North America at the turn of the twentieth century primarily in the realm of civic ornithology—that is, the study of birds outside of the universities and government agencies where highly trained professionals pursued scientific research.[7] In Europe, the Danish schoolteacher Hans Christian Cornelius Mortensen's banding studies of the migration of starlings and other species inspired a wave of imitators in Germany, Great Britain, Hungary, Russia, and elsewhere.[8] The ornithologist Johannes Thienemann, for example, recruited a network of volunteers to support banding-based studies of migration from his base at the Rossitten bird observatory in East Prussia.[9] In North America, civic ornithologists similarly took the lead in developing bird-banding techniques and recruiting fellow enthusiasts into ad hoc regional networks. In 1902, for example, Paul Bartsch of the Smithsonian Institution began attaching aluminum bands inscribed with a serial num-

---

[6] On issues raised by the boom in citizen-science projects since the 1990s, see Janis L. Dickinson and Rick Bonney, eds., *Citizen Science: Public Participation in Environmental Research* (Ithaca, N.Y., 2012); Abraham Miller-Rushing, Richard Primack, and Rick Bonney, "The History of Public Participation in Ecological Research," *Frontiers Ecol. Environ.* 10 (2012): 285–90.

[7] Raf de Bont, *Stations in the Field: A History of Place-Based Animal Research* (Chicago, 2015), 147–74. See also Lynn Nyhart, "Civic and Economic Zoology in Nineteenth-Century Germany: The 'Living Communities' of Karl Möbius," *Isis* 98 (1998): 605–30.

[8] Many of these European projects were cut short by World War I; Frederick C. Lincoln, "The History and Purposes of Bird Banding," *Auk* 38 (1921): 217–28, on 219. See also Harold B. Wood, "The History of Bird Banding," *Auk* 62 (1945): 256–65.

[9] De Bont, *Stations in the Field* (cit. n. 7), 161.

ber, the year, and the instruction to "Return to Smithsonian Institution" onto herons at a breeding colony near Washington, D.C.[10] Soon after, P. A. Taverner, a young Canadian architect and amateur ornithologist living in Ann Arbor, Michigan, began distributing bands to a small network of fellow enthusiasts.[11] In 1909, Jack Miner began banding geese and ducks at his farm in Ontario with bands handstamped with a serial number, his address, and a Bible quotation.[12]

In these early efforts, the frequency of "returns"—observations of previously banded birds—was quite low, and most returns concerned birds recaptured at the site of initial banding. While this discouraged some ornithologists from adopting the technique, it motivated others to coordinate banding over increasingly wide geographical scales. One of the key figures in this scaling-up process was Leon J. Cole, who had begun advocating bird banding in 1901.[13] In 1908, Cole recruited fellow enthusiasts in the New Haven Bird Club to join him in a collective banding project. A year later, disappointed at the paltry rate of returns, he proposed establishing an association to carry forward the effort on a national scale at a meeting of the American Ornithological Union.[14] By the spring of 1910, the American Bird Banding Association (ABBA) had distributed approximately 5,000 bands, of which 911 had been reported as used and thirty-one had subsequently been recovered. A recovery rate of 3.4 percent was still low, but it was already a significant improvement over earlier efforts.[15] Over the following decade, approximately 22,500 birds were tagged with bands distributed by the ABBA.[16]

Sources of enthusiasm for bird banding during this early period can be divided into two kinds: an enthusiasm for amassing vast collections of data on bird populations, on one hand, and an enthusiasm for acquiring intimate acquaintance with individual birds, on the other. Cole was among those who were excited by bird banding as a technique for amassing vast quantities of data on many birds rather than as an opportunity to study a few birds in great detail. Looking back on the ABBA's work from the perspective of the early 1920s, Cole described it as a scheme to "gradually build up an accumulation of accurate data on bird movements which would be of the greatest scientific value" by drawing on "the interest and coöperation of the large number of amateur bird students in the country."[17] After Cole took a position at the University of Wisconsin in 1910 and shifted his focus to genetics, the management of the ABBA fell to Howard H. Cleaves, who arranged for a ledger and filing

---

[10] Paul Bartsch, "Notes on the Herons of the District of Columbia," *Smithsonian Misc. Collections* 45 (1904): 104–11; Bartsch, "A Note on the First Bird-Banding in America," *Bird-Banding* 23 (1952): 59–60.

[11] P. A. Taverner, "A Tagged Flicker," *Wilson Bull.* 18 (1906): 21–2; Taverner, "Tagging Migrants," *Auk* 23 (1906): 232; W. L. McAtee, "Percy Algernon Taverner, 1875–1947," *Auk* 65 (1948): 85–106.

[12] On Miner's banding work in relation to his conservation philosophy, see Tina Loo, *States of Nature: Conserving Canada's Wildlife in the Twentieth Century* (Vancouver, 2006), 63–92.

[13] Cole later claimed to have been inspired by the use of numbered tags for research on lobsters; Leon J. Cole, "The Early History of Bird Banding in America," *Wilson Bull.* 34, no. 2 (1922): 108–44. See also Wood, "History" (cit. n. 8), 260.

[14] Leon J. Cole, "The Tagging of Wild Birds: Report of Progress in 1909," *Auk* 27 (1910): 153–68; Letter, Leon J. Cole, C. J. Pennock, Louis B. Bishop, Glover M. Allen, and Thomas S. Roberts (Executive Committee of the American Bird Banding Association), New Haven, Conn., 10 February 1910, *Wilson Bull.* 22 (1910): 53–5.

[15] Cole, "Tagging" (cit. n. 14), 157–9.

[16] Frederick C. Lincoln, "Some Causes of Mortality among Birds," *Auk* 48 (1931): 538–46, on 539.

[17] Cole, "Early History" (cit. n. 13), 109.

cabinet for banding data to be installed at the Linnaean Society of New York.[18] Like Cole, Cleaves believed that "bird banding is not the work of a limited circle but the duty of many, and it is only by extensive banding that results of value can be obtained."[19] This was a clearly centripetal view of bird banding, in which a network of amateurs contribute data to a center of calculation where it will be used by a few highly trained professionals.

Through the 1910s, however, this centripetal vision remained more aspirational than real, with the ABBA's slowly growing collection of bird-banding data languishing mostly unused except by banders eager to discover the fate of the birds they had banded themselves. One reason was that the rate of returns continued to be so low and the distribution of banders across the continent so uneven as to make any general conclusions drawn from them highly speculative. This was a function of the small number of people banding birds and of their tendency to band nestlings. Birds too young to fly were easy to catch and band but also very likely to die before they could be recaptured. Moreover, most of those involved in bird banding at this time seem to have had little interest in using the data for panoptic purposes. For them, the ABBA served as a registry or clearinghouse through which they could discover the fate of those few birds that they had banded themselves. The ABBA's central files seem never to have been used as the basis for a published study, although the private data collections assembled by particular banders or research stations often were.[20]

For most banders, the dream of accumulating massive amounts of data in a repository for centralized study was less appealing than the prospect of receiving free bird bands and instruction in the proper techniques for using them, which would allow them to learn more about the fate of individual birds, as well as to take personal credit for the scientific findings that resulted. S. Prentiss Baldwin, a businessman and amateur ornithologist who played a leading role in the development of bird banding in the United States between the 1910s and 1930s, is a case in point. Baldwin had first begun to trap live birds in the process of exterminating nonnative sparrows on his estate near Cleveland in the early 1910s. In 1914, after learning about the methods promoted by the ABBA, he began banding on what was then considered a very large scale, resulting in about 1,600 birds banded between 1914 and 1918.[21] The publication of his methods and findings in 1919 was widely credited with inspiring a revival of interest in bird banding.[22] In particular, by developing trapping and banding techniques for adult birds, which were harder to catch but more likely than nestlings to be recaptured, Baldwin's research reopened the possibility of recovery rates that could sustain the enthusiasm of amateurs.[23]

---

[18] Howard H. Cleaves, "What the American Bird Banding Association Has Accomplished during 1912," *Auk* 30 (1913): 248–61, on 248–9.

[19] Ibid., 253.

[20] For an example of a study that used the association's bands but not the association's collection of records, see Edward A. McIlhenny, "An Early Experiment in the Homing Ability of Wildfowl," *Bird-Banding* 11 (1940): 58–60.

[21] S. Charles Kendeigh, "In Memoriam: Samuel Prentiss Baldwin," *Auk* 57 (1940): 1–12, on 4.

[22] S. Prentiss Baldwin, "Bird-Banding by Means of Systematic Trapping," *Abstr. Proc. Linn. Soc. N.Y.* 31 (1919): 23–56.

[23] Baldwin assisted with the preparation of the first instructional pamphlet issued by the BBO; Frederick C. Lincoln, *Instructions for Bird Banding*, U.S. Department of Agriculture, Department Circular 170, Bureau of Biological Survey (Washington, D.C., 1921).

Baldwin's interest in bird banding cannot be entirely separated from the dream of a massive, centralized, continent-wide data set, but neither can it be reduced to it. The research program he developed eventually led to the banding of tens of thousands of birds, and in the 1920s and 1930s he contributed significantly to the coordination of effort among bird banders across North America and to the growth of the federal bird-banding program. Still, the centripetal dreams that got Cole's and Cleaves's hearts pumping faster seem to have left Baldwin cold. As one of his field assistants recalled after his death, his goal was never to band "large numbers of individuals for the sake of a big record" but rather to "obtain all possible information from those that were handled."[24] Data was vitally important, in other words, but it was small data kept locally rather than Big Data gathered in Washington that mattered most. For such work, a centralized registry might be helpful, but a centralized data set was largely irrelevant. Like many other banders, Baldwin's enthusiasm for banding was sustained by the experience of handling individual birds, investigating their life histories, and associating with like-minded enthusiasts—not by the idea of handing data over to a distant center of calculation.

MANAGING DATA AND ENTHUSIASM AT THE BBO

In 1920, the ABBA was disbanded and its functions were taken over by the new Bird-Banding Office in the United States and by an equivalent office in Canada.[25] For its first two and a half decades, from 1920 to 1946, the U.S. office operated under the direction of Frederick C. Lincoln. In a memoriam written after Lincoln's death in 1960, one leading American wildlife conservationist described him as someone who had had "no hobbies, other than work in his chosen field."[26] This paradoxical formulation suggests the depth of Lincoln's commitment to ornithology and the blurring of lines between professional and amateur that it entailed. His passion for birds was already evident during his adolescence in Colorado, where he spent summers working at the Colorado Museum of Natural History and made the acquaintance of Alexander Wetmore, an ornithologist with the Biological Survey. In 1920, after becoming the museum's curator of ornithology and serving as a pigeon expert in the U.S. Signal Corps, Lincoln was recruited by the Biological Survey to head the bird-banding effort. The U.S. government's interest in bird banding was a result of the Convention for the Protection of Migratory Birds signed with Canada in 1916, which obligated both nations to manage bird populations with "due regard to the zones of temperature and to the distribution, abundance, economic value, breeding habits, and times of migratory flight" of the species concerned.[27] Bird banding promised to provide the data required for such management, but only if it could be successfully coordinated on a national scale.

---

[24] Kendeigh, "In Memoriam" (cit. n. 21), 5.

[25] E. W. Nelson, "Bird Banding Work Being Taken over by the Biological Survey," *Wilson Bull.* 32 (1920): 63–4.

[26] Ira N. Gabrielson, "Obituary," *Auk* 79 (1962): 495–9, on 495.

[27] U.S. Department of Agriculture, Bureau of Biological Survey, *Migratory Bird Treaty, Act, and Regulations* (Washington, D.C., 1920), 5. See Kurkpatrick Dorsey, *The Dawn of Conservation Diplomacy: U.S.-Canadian Wildlife Protection Treaties in the Progressive Era* (Seattle, 1998), 165–237; Mark Cioc, *The Game of Conservation: International Treaties to Protect the World's Migratory Animals* (Athens, Ohio, 2009), 58–103.

After arriving in Washington, Lincoln threw himself into the work of promoting bird banding as a revolutionary technique for ornithology. "Occupying a position with the shotgun, the field glass and the scalpel," he wrote in 1926, banding "bids fair to become the court of last appeal in the determination of many facts concerned with the migration and life histories of birds." (fig. 1)[28] Lincoln's proselytizing efforts were most intense during his first few years on the job, when he devoted himself to inspiring and recruiting a voluntary corps of bird banders while also educating hunters and bird lovers about the importance of reporting recovered bands. In addition to writing popular articles, he gave interviews to the press and delivered numerous presentations before Boy Scout troops, Audubon societies, nature clubs, and other organizations. Lincoln emphasized the utilitarian purposes of bird banding but also sought to position it as a way for ordinary citizens to contribute to science while enhancing their own enjoyment of birds. In an account of the origins of bird banding in the United States titled "The Romance of the Numbered Band," most likely written in the late 1930s, Lincoln noted that the "charm of intimate acquaintance with birds, brought about by the repeated handling of the same individuals, started a wave of interest and enthusiasm unparalleled in the history of American ornithology."[29]

Rather than discouraging volunteers' enthusiasm for "intimate acquaintance" with individual birds, Lincoln actively encouraged it, even while insisting that banders send their reports to Washington in formats that could be efficiently processed for the purposes of waterfowl management. This balance between encouraging volunteer enthusiasm and disciplining their data collection was reflected in an article on "The History and Purposes of Bird Banding" that Lincoln published in *The Auk* in the spring of 1921.[30] The article explained the importance of bird banding for large-scale migration studies, but it also sought to link bird banding research to popular enthusiasm for bird-watching. It noted the ease with which the nation's thousands of bird-feeding stations could be converted into banding stations, the variety of research questions that could be answered, and the harmlessness of the practice. Even the bird lover most concerned about the welfare of humanity's feathered friends, it suggested, could band birds in good conscience. By way of conclusion, the article predicted that with "the element of anticipation always present, expecting and watching for a bird marked at some other station or during a previous year, it seems that this new system of bird banding should find a host of enthusiastic participants."[31] This was a message that Lincoln repeated before a variety of audiences in the early years of the federal bird-banding program.[32]

Lincoln's advocacy efforts soon paid off. In fact, the practice had soon garnered so much interest that the BBO was forced to limit the number of banding permits it is-

[28] Frederick C. Lincoln, "Bird Banding—In Progress and Prospect," *Auk* 43 (1926): 153–61, on 153.

[29] The manuscript was written and submitted to *Collier's Magazine* but never published; Robert McCormick to Frederick C. Lincoln, 2 June 1941; Lincoln, "Bird Banding," n.d., 34 pp. [pages missing], on 13, Box 37, Folder: The Romance of the Numbered Band, Record Group 22, Office Files of Dr. Frederick C. Lincoln, 1917–1960, National Archives and Records Administration, College Park, Md. (hereafter cited as "Lincoln Files").

[30] Lincoln, "History" (cit. n. 8).

[31] Ibid., 227.

[32] See, e.g., his presentation to the Naturalist Field Club of Johns Hopkins University in 1924; Frederick C. Lincoln, "Report of Trip to Baltimore, Maryland, Apr. 30, 1924," n.d., manuscript, 3 pp., on 2, Box 36, Folder: Reports 1924, Lincoln Files.

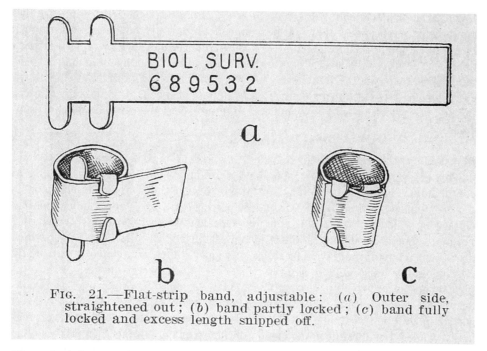

FIG. 21.—Flat-strip band, adjustable: (*a*) Outer side, straightened out; (*b*) band partly locked; (*c*) band fully locked and excess length snipped off.

Figure 1. A diagram showing one of the stamped, numbered bands distributed by the Bird-Banding Office to its cooperators in both open and locked (i.e., attached) states. Source: Frederick C. Lincoln, Instructions for Banding Birds *(Washington, D.C., 1924), 20, fig. 21.*

sued. This was not because the would-be banders were unqualified or because the data they planned to collect would not have been useful, but because the office was unable to process the flood of data with which it was already being inundated. By the summer of 1926 there were 1,133 active cooperators and around one hundred more planning to begin banding soon. Since the establishment of the BBO, volunteers had banded more than 203,000 birds, from which more than 9,300 returns had been received (a rate of about 4.6 percent).[33] The BBO's inability to manage this flow had direct consequences for Lincoln's efforts to generate enthusiasm for the new technique. In the late 1920s, the BBO began to "handle the banding work 'with brakes on,' that is, definite effort [was] made to refrain from undue publicity or campaigns that would have the effect of greatly increasing the number of cooperators."[34]

Facing practical limits to the amount of data that could be collected, processed, stored, and analyzed given available resources, Lincoln shifted his focus from inspiring interest in banding to recruiting precisely the right kind of cooperator and ensuring that the work of existing cooperators was maximally useful for the BBO. In light of the office's origins in the 1916 convention, Lincoln believed the ideal bander was one who operated a feeding station at which hundreds or thousands of waterfowl were banded yearly and who submitted accurate banding reports and returns in a timely manner, thereby contributing to the continental-scale surveys upon which water-

[33] Frederick C. Lincoln, Memorandum for Chief of Bureau, 19 June 1926, 1 p., Box 36, Folder: Reports 1926, Lincoln Files.
[34] Frederick C. Lincoln, Memorandum for Dr. Bell, 13 July 1928, 3 pp., on 2, Box 36, Folder: Reports 1928, Lincoln Files.

fowl management depended. From the mid-1920s onwards, the BBO sought to dis-
courage the applications of would-be cooperators who did not intend to establish
large-scale trapping stations and to revoke the permits of those whose work it deemed
to be of a "desultory character."[35] By the late 1930s, it was rejecting more than one permit application per day.[36] Nonetheless, because Lincoln found it difficult to turn down
candidates who promised to add coverage of important species or geographical areas,
the number of permit-holding volunteers and unprocessed reports continued to increase.

Over the course of the 1920s and 1930s, the focus of Lincoln's attention thus shifted
from publicity and recruitment toward clerical workflow and data management. Within
the first few years of the BBO's founding, for example, he recognized that writing to
contributors individually to acknowledge receipt of their data was consuming an inor-
dinate amount of his time. By 1923, about 59 percent of the 4,687 communications that
crossed his desk were in the form of preprinted postcards rather than letters or memos.[37]
For more complex communications, Lincoln established *Bird Banding Notes*, which
disseminated trapping and banding techniques and reported exemplary or particularly
exciting returns. *Bird Banding Notes* was also used to encourage cooperators to complete reporting cards with pen rather than pencil, to submit reports at regular time intervals, to avoid duplicating reports, and to ensure that all information sent was correct
and complete. Failing to comply with any of these requirements, cooperators were regularly warned, could result in revocation of their banding permits.[38] In this way, the
challenges of data management helped to reorient the relationship between the BBO
and its cooperators from individual, idiosyncratic, and encouraging to collective, standardized, and disciplinary.

The flood of data also inspired a series of efforts aimed at reforming data-
management practices within the BBO. At first, when the number of banders remained in the hundreds rather than thousands, the clerical work of transcribing and
filing banding reports seemed manageable (fig. 2). In 1922, Lincoln confidently told
his supervisor that the "system for the mechanical part of the work is now very nearly
automatic, and appears to admit of infinite expansion."[39] As the number of cooperators' reports grew, however, he was forced to implement further reforms because of
"the greatly increased volume of work and lack of additional assistance."[40] Some of
these reforms addressed minor matters of workflow; others involved significant re-
organizations of the data. In 1925, for example, the BBO began separating the "dead"
files of birds banded so long ago that they were unlikely to be observed again from
more recent "live" records.[41] In 1926, Lincoln reported that "a complete change in

[35] Frederick C. Lincoln, Memorandum for the Chief of Bureau [E. W. Nelson], 20 September 1924,
2 pp., on 2, Box 36, Folder: Reports 1925, Lincoln Files.
[36] Frederick C. Lincoln, Memorandum for Mr. Leichhardt, 18 September 1939, 2 pp., on 1, Box 39,
Folder: Statements of Policy, Lincoln Files.
[37] Frederick C. Lincoln, Memorandum for Major Goldman, 19 December 1923, Box 36, Folder:
Reports, Memos, etc., FY 1921 and 1922, Lincoln Files.
[38] "General Information to Cooperators," *Bird Banding Notes* 10 (1924): 1–4, on 1.
[39] Frederick C. Lincoln, Memorandum for Major Goldman, 22 July 1922, 2 pp., on 2, Box 36, Folder:
Reports, Memos, etc., FY 1921 and 1922, Lincoln Files.
[40] Frederick C. Lincoln, Memorandum for Chief of Bureau [E. W. Nelson], 2 July 1924, 3 pp., Box 36,
Folder: Reports 1924, Lincoln Files.
[41] Frederick C. Lincoln, Memorandum for Dr. Jackson, 18 November 1925, 2 pp., Box 36, Folder:
Reports 1926, Lincoln Files.

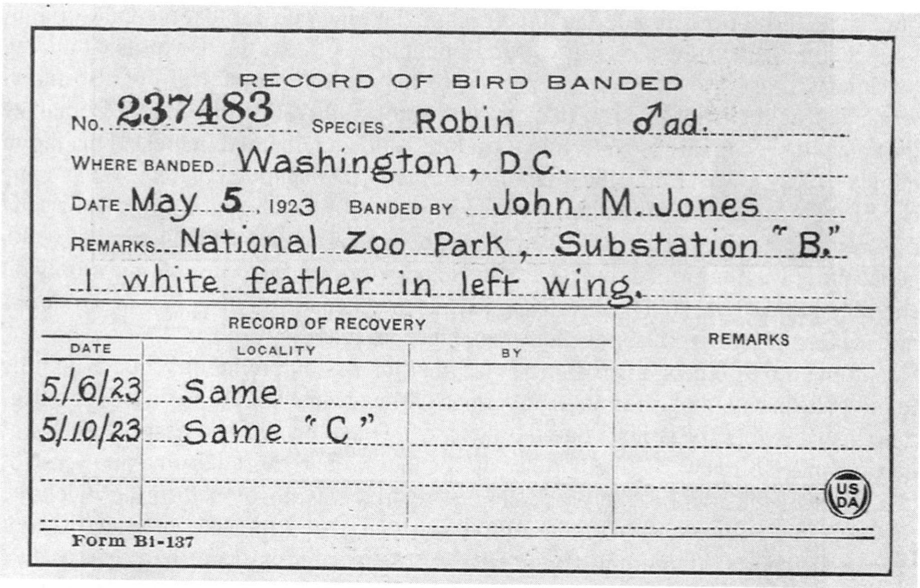

RECORD OF BIRD BANDED
No. **237483** SPECIES Robin ♂ ad.
WHERE BANDED Washington, D.C.
DATE May 5, 1923 BANDED BY John M. Jones
REMARKS National Zoo Park, Substation "B."
1 white feather in left wing.

RECORD OF RECOVERY

DATE	LOCALITY	BY	REMARKS
5/6/23	Same		
5/10/23	Same "C"		

US DA

Form B1-137

Figure 2. *A sample record card reporting the tagging of an adult male robin at the National Zoological Park in Washington, D.C., on 5 May 1923, and its recapture at the same site on 6 and 10 May. Source: Frederick C. Lincoln,* Instructions for Banding Birds *(Washington, D.C., 1924), 24, fig. 27.*

system" for the banding files was speeding the work of banders and clerical staff.[42] But none of these reforms remained satisfactory for long. In 1929, the office fell a full year behind schedule, and Lincoln admitted that "we are totally unable to adequately handle the volume of work with the personnel available."[43] In the midst of an unprecedented economic disaster, funds for additional personnel were not forthcoming. The BBO continued to fall behind, and the tone of Lincoln's memos became increasingly desperate.[44]

With no additional personnel forthcoming, mechanization offered a ray of hope. In 1929, the BBO acquired an electric card-sorting machine that promised to facilitate the censusing of bird populations.[45] The American Wild Fowlers, a private conservation association, donated funds to hire a card puncher who spent five weeks coding,

[42] Frederick C. Lincoln, Memorandum for Dr. Jackson, 24 June 1926, 3 pp., Box 36, Folder: Reports 1926, Lincoln Files.

[43] Frederick C. Lincoln, Memorandum for Dr. Bell, Bird Banding Report—Fiscal Year 1929, 12 July 1929, 4 pp., on 1, Box 36, Folder: Reports ~1929 (Memos), Lincoln Files.

[44] Frederick C. Lincoln, Memorandum for Dr. Bell, Bird Banding Report—Fiscal Year 1930, 9 July 1930, 4 pp., Box 36, Folder: Reports, Memos, etc., FY 1930, Lincoln Files.

[45] C. Stuart Houston, M. Kathleen Klimekiewicz, and Chandler S. Robbins, "History of 'Computerization' of Bird-Banding Records," *North Amer. Bird-Bander* 33 (2008): 53–65, on 55. As Christine von Oertzen argues, the success of mechanical punch card systems depended on particular configurations of human labor and on the more fundamental innovation of using "movable paper tools" to represent information about individuals; see von Oertzen, "Machineries of Data Power: Manual versus Mechanical Census Compilation in Nineteenth-Century Europe," in this volume. See also Markus Krajewski, "Tell Data from Meta: Tracing the Origins of Big Data, Bibliometrics, and the OPAC," in this volume; Lars Heide, *Punched Card Systems and the Early Information Explosion, 1880–1945* (Baltimore, 2009).

punching, and sorting about 30,000 cards, and the Machine Tabulating and Computing Section of the Bureau of Agricultural Economics offered the use of its tabulating machine.[46] Lincoln described the turn to mechanization in practical but also moral terms; it would allow the BBO to meet its "responsibility" to cooperators to analyze the data it had accumulated.[47] While most bird banders remained interested mainly in the fates of their own birds, the credibility of the BBO depended on its making some use of the data it was collecting. At the same time, mechanization provided a new opportunity to tighten the BBO's control over cooperators. At the beginning of 1930, cooperators were informed that new reports and returns were to be submitted only on simplified cards that could be directly coded and punched at the BBO.[48] As data processing became mechanized, cooperators' flexibility decreased.

Automation offered some real increases in efficiency and reliability, but it did little to solve the problem of information overload. A few years later, Lincoln was still begging his superiors for at least one more clerk.[49] Triage became necessary: reports of newly banded birds were given priority, since they were required for making sense of later returns; meanwhile, the processing of returns was postponed to the indefinite future. By the summer of 1938, the BBO had accumulated approximately 40,000 unprocessed reports, amounting to the entirety of the returns from the previous two years.[50] This could be seen as a product of poor management and insufficient resources, but it can also be understood as the almost inevitable result of seeking to accumulate as much data as possible. Given the frictions and unexpected challenges of data management, it was easy for "as much as possible" to become "a bit more than possible." Lincoln shared some responsibility for this state of things, inasmuch as his annual request for additional staff was usually accompanied by the admission that the number of cooperators and banded birds continued to increase.[51] Information overload and data hunger went hand in hand.

CENTRIFUGAL ENTHUSIASMS ON THE AMATEUR PERIPHERY

The volunteers who flooded the BBO with data were often highly educated, passionately committed to ornithology, and hungry for data in their own way. Among them was Margaret Morse Nice, who, like many other American women of her generation, was forced to end her formal scientific training before completing her doctorate. Nonetheless, she succeeded in becoming one of the most respected American orni-

---

[46] Lincoln, Memorandum for Dr. Bell, Bird Banding Report—Fiscal Year 1930 (cit. n. 44). See also J. E. Law, "Electric Sorting Machine for Banding Data," *News from the Bird Banders* 5 (1930): 15.

[47] Lincoln, Memorandum for Dr. Bell, Bird Banding Report—Fiscal Year 1929 (cit. n. 43). On moral and emotional economies of science, see Lorraine Daston, "The Moral Economy of Science," *Osiris* 10 (1995): 3–24; Lorraine Daston and Peter Galison, *Objectivity* (Cambridge, Mass., 2007); Paul White, "Introduction" to Focus Section on "The Emotional Economy of Science," *Isis* 100 (2009): 792–7.

[48] Lincoln, Memorandum for Dr. Bell, Bird Banding Report—Fiscal Year 1930 (cit. n. 44). See also Law, "Electric Sorting Machine" (cit. n. 46), 15.

[49] Frederick C. Lincoln, Memorandum for Doctor Bell, Re: Bird Banding Report; Fiscal Year, 1931, 2 July 1931, 6 pp., on 5, Box 36, Folder: Reports, Memos, FY 1931, Lincoln Files.

[50] Frederick C. Lincoln, Annual Report—Fiscal Year 1938, Distribution and Migration of Birds, Division of Wildlife Research, n.d., 13 pp., on 10, Box 38, Folder: Section Reports, Memos, etc. (Including memos from Div. Chief) FY 1938, Lincoln Files.

[51] I have found no evidence that volunteers were ever asked to assist with clerical tasks in the BBO.

thologists of the mid-twentieth century on the basis of rigorous empirical studies and wide reading of North American and European ornithological literature.[52] Nice achieved this status while serving as one of the BBO's cooperators and, for many years, as an associate editor of the journal *Bird-Banding*. Like many of her fellow banders, Nice both depended on and contributed to the BBO's centralizing efforts, which exploited the enthusiasm of volunteers like herself to produce data for professionals while also providing tools that enabled amateurs to pursue their own research interests. Nice played a leading role in fostering practices that pulled away from centralized data gathering and encouraged each bander to tackle important scientific problems that could be addressed using a small set of carefully observed birds. Though harder to track through the historical record, such centrifugal practices were as important as the centripetal practices that pulled banding data into Lincoln's office in Washington.

Implicit in many of her research articles, Nice's vision of banding as a centrifugal, peripheral, and individual practice was rendered explicit in an article titled "The Opportunity of Bird-Banding," which appeared in *Bird-Banding* in 1934. As exhortatory in tone as some of the newsletters published by the BBO around the same time, the article upbraided Nice's fellow banders for "apparently believing that banding is merely a means of studying migration" and thus being satisfied to "capture as many birds as they possibly can, simply attaching the numbered band and that is the last they see or hear of the vast majority of their subjects."[53] Although Nice mentioned neither Lincoln nor the BBO by name, it was precisely their approach to banding that she was attacking. Willing to sacrifice quantity for quality and breadth for depth, Nice argued that significant scientific discoveries would emerge if banders would only "study a few birds carefully."[54] What Nice meant by this was life-history research of the kind that had occupied Baldwin, whose most striking result had concerned the mate fidelity of house wrens—a topic for which the individual identification provided by banding was essential but the collection of data in a centralized repository was irrelevant.[55]

Nice's studies of song sparrows similarly depended on individual identification over multiple seasons, for which bands from the BBO and a federal banding permit were essential but the centralized data set that Lincoln was painstakingly assembling was not.[56] Using additional colored celluloid bands that allowed her to identify individual birds from a distance, Nice conducted a study of breeding behavior that was widely recognized upon its publication in 1937 as "the finest and most comprehen-

[52] Nice later received an honorary doctorate from Mount Holyoke College; David W. Johnston, "Margaret Morse Nice (1883–1974)," *Bird-Banding* 45 (1974): 360. On the position of women in American science during this period, see Margaret W. Rossiter, *Women Scientists in America: Struggles and Strategies to 1940*, vol. 1 (Baltimore, 1982), 129. On Nice's career, see Margaret Morse Nice, *Research Is a Passion with Me* (Toronto, 1979); Gregg Mitman and Richard W. Burkhardt Jr., "Struggling for Identity: The Study of Animal Behavior in America, 1930–1945," in *The Expansion of American Biology*, ed. Keith R. Benson, Jane Maienschein, and Ronald Rainger (New Brunswick, N.J., 1991), 164–94; Barrow, *Passion for Birds* (cit. n. 1), 195–8.

[53] Margaret Morse Nice, "The Opportunity of Bird-Banding," *Bird-Banding* 5 (1934): 64–9, on 64.

[54] Ibid.

[55] S. Prentiss Baldwin, "The Marriage Relations of the House Wren (Troglodytes a. aedon)," *Auk* 38 (1921): 237–44.

[56] Margaret Morse Nice, "Studies in the Life History of the Song Sparrow, Vol. I: A Population Study of the Song Sparrow," *Trans. Linn. Soc. N.Y.* 4 (1937): 1–247.

sive study ever made of any North American bird."[57] By disseminating her own note-taking and data-recording methods through *Bird-Banding*, Nice helped to promote such studies among her fellow volunteer banders.[58] She was not opposed to studies of migration, to cooperative work, or to submitting data to the BBO, but she was most passionate about the opportunities that banding opened up for local observation of phenomena such as territoriality. Sending reports to Washington was a bureaucratic obligation that had to be fulfilled before the real work of closely observing a few individual birds could begin.

Nice's example suggests that cooperators could participate in the federal banding program even while pursuing research objectives that were independent of it or even opposed to it. The centrifugal forces that created tension between the BBO and the volunteers upon whom it depended are also visible in the role played by regional bird-banding associations in the 1920s and 1930s. In addition to administering a registry of bird bands, the ABBA had served as a form of sociability for bird banders. After its dissolution in 1920, a social void was created that the BBO could not fill, even though it recognized that such associations could help it achieve its data-gathering goals. In the early 1920s, with Lincoln's encouragement, volunteers filled this void by establishing the Western Bird Banding Association, the Northeastern Bird Banding Association, the Eastern Bird Banding Association, and the Inland Bird Banding Association. These associations collected dues, distributed bulletins or newsletters, organized meetings of banders, publicized banding in the press, and defended bird banding against its critics.[59]

Lincoln encouraged these organizations as a way to swell the ranks of banders, coordinate "chains of stations" that could trap migrating birds repeatedly, and encourage the public to report found bands.[60] In small ways, the BBO even offered material support for the regional associations. In 1923, for example, to help the Inland Bird Banding Association recruit members who were not themselves active banders, Lincoln offered to include them on the mailing list for *Bird Banding Notes*.[61] Nonetheless, tensions between the regional associations and the BBO sometimes posed a real threat to the harmonious functioning of the federal bird-banding program. The Western Bird Banding Association, for example, whose members tended to see their needs as distinct from those in other regions of the country, established an "emergency" reserve of bird bands that it distributed to its members during periods of heavy migration. The BBO only grudgingly tolerated such emergency stores, since they put the re-

---

[57] Lawrence E. Hicks, Review of "Studies in the Life History of the Song Sparrow I," *Bird-Banding* 8 (1937): 137–8, on 137.

[58] See, e.g., Margaret Morse Nice, "The Technique of Studying Nesting Song-Sparrows," *Bird-Banding* 1 (1930): 177–81.

[59] Sara R. Morris, Brenda Dale, and Mary Gustafson, "Roles and Contributions of Banding Organizations to the North American Banding Program," in Jackson, Davis, and Tautin, *Bird Banding* (cit. n. 1), 31–64. For a defense of bird banding from the director of the Biological Survey against criticisms that it harmed birds, see E. W. Nelson, "Danger in Bird Traps," *Auk* 42 (1925): 304–7.

[60] On the BBO's support of regional associations, see "Bird Banding Associations," *Bird Banding Notes* 4 (1923): 5. On "chains of stations," see "Directory of Bird Band Cooperators," *Bird Banding Notes* 11 (1924): 3.

[61] Frederick C. Lincoln, "Report of the Annual Meeting of the Inland Bird Banding Association, held at Indianapolis, Ind., Nov. 2 and 3, 1923," n.d., unpublished manuscript, 5 pp., on 2, Box 36, Folder: Reports 1924, Lincoln Files.

gional association in the position of knowing who was using a particular set of bands before the office itself did.[62]

In addition to tensions over emergency bands, the establishment of regional data repositories also raised hackles at the BBO, particularly when cooperators began sending their original data first to the regional repository and only subsequently to Washington, if at all. "There is no objection to cooperators' sending copies of their records to their local organizations if they so desire," the BBO stressed, "but the original records should be sent direct to the Bureau."[63] The regional associations also sometimes resisted the BBO's attempts to limit the number of cooperators. In 1930, for example, as the BBO was turning down applications from would-be banders on a daily basis, one of the leaders of the Northeastern Bird-Banding Association argued in the pages of *Bird-Banding* that "more collaborators are needed" to fill the gaps in the existing coverage of trapping stations.[64] Volunteers who responded to this call must have been disappointed when they received a form letter from the BBO informing them that it could not accept new cooperators. Volunteers' desire to study the life histories of individual birds rather than the migrations of populations, to maintain local autonomy and the ability to respond quickly to changing conditions, and to expand coverage in their own regions regardless of the number of cooperators nationally thus sometimes came into direct conflict with the BBO's aims.

THE TURN AWAY FROM VOLUNTEERS

Throughout the 1920s and 1930s, the BBO sought to balance its desire for data suited to the management of continental waterfowl populations with the necessity of sustaining the enthusiasm of its volunteer cooperators, whose own priorities sometimes pulled in contradictory directions. Certain frictions were generated by the very activities that helped to lubricate the flow of data.[65] The result of this conflictual cooperation was several decades of remarkable scientific productivity. Under Lincoln's leadership, the BBO used banding data collected by volunteers to make two major contributions to the challenges of continental waterfowl management. The first was the mapping of waterfowl migration along four North American "flyways," which made it possible to develop management plans tailored to each population. The management of volunteer banders on a continental scale thus made possible the management of waterfowl on a continental scale.[66] The second major contribution was a method for calculating the size of each of these distinct populations on the basis of the ratio between banded and unbanded birds in a particular sample. The fact that

[62] On the Western Bird Banding Association's "emergency" band store, see Harlan H. Edwards, "Emergency Supply of Bands," *News from the Bird Banders* 2 (1927): 4. The BBO urged cooperators to instead send requests for new bands marked "Emergency" to Washington, to which it promised to give "prompt attention"; "Bands," *Bird Banding Notes* 6 (1923): 11–12.

[63] "General Information to Cooperators," *Bird Banding Notes* 9 (1924): 2. See also "General Information to Cooperators," *Bird Banding Notes* 10 (1924): 1.

[64] Charles B. Floyd, "A List of the Active Banding Stations in the Territory of the Northeastern Bird-Banding Association," *Bird-Banding* 1 (1930): 127–36.

[65] Cf. the varying uses of the concept of "friction" in Anna Lowenhaupt Tsing, *Friction: An Ethnography of Global Connection* (Princeton, N.J., 2005); Paul N. Edwards, *A Vast Machine: Computer Models, Climate Data, and the Politics of Global Warming* (Cambridge, Mass., 2010), 83–110.

[66] For a popular summary of the flyway findings, see Frederick C. Lincoln, *The Migration of American Birds* (New York, 1939). On the origins of flyway-based management, see Wilson, *Seeking Refuge* (cit. n. 1), 75.

this statistic is still known among wildlife biologists as the "Lincoln index" is largely a consequence of his ability to marshal and discipline the enthusiasm of the volunteers who produced the data that made it useful.[67] At the same time, volunteer banders such as Nice made major contributions to the study of life histories and behavior that dovetailed with the emerging field of ethology in Europe.[68]

The dynamic tension between centripetal and centrifugal forces that characterized the bird-banding program during its first two decades was neither inevitable nor permanent. It was the result of particular conditions and choices that began to change in the late 1930s. In 1938, the BBO decided to concentrate the banding of ducks, geese, and other waterfowl in federal and state wildlife refuges, effectively limiting volunteers to the banding of nongame birds. This was a response to a prohibition on the use of decoys and bait by hunters that had been implemented following a catastrophic drop in continental waterfowl populations during the drought years of the early 1930s.[69] Without decoys or bait, many banding stations could no longer attract waterfowl. A decade earlier, the loss of these volunteer-operated banding stations would have severely damaged the bird-banding program, but by the late 1930s an expanding federal and state bureaucracy for wildlife management was available to take up the slack. Ironically, this wildlife bureaucracy was one of the results of the success of the volunteer-driven bird-banding program, which had provided data supporting the establishment of new regulations and refuges.[70]

In the early 1940s, the shift from volunteers to professional banders continued with the reorganization of the federal government's conservation agencies and the onset of World War II. In 1940, the Biological Survey was moved from the Department of Agriculture to the Department of the Interior, where it became part of the new Fish and Wildlife Service.[71] After the move, survey employees faced increased pressure to focus on management-relevant research. Within the BBO, this had the effect of accelerating the shift to large-scale, professional waterfowl banding. The onset of war further de-emphasized volunteers and research on nongame birds, while also hampering the bird-banding effort as a whole. Following the U.S. declaration of war in December 1941, Lincoln attempted to justify bird banding by characterizing waterfowl as a meat source that would help to "fill the national larder," but bird banding was not a high priority for the U.S. government in the early 1940s.[72] Even before 1941, many

---

[67] For an early, unpublished discussion of both the flyway idea and the Lincoln index, see Frederick C. Lincoln, Memorandum for Mr. Redington, 9 November 1927, 2 pp., Box 36, Folder: Reports 1928, Lincoln Files. For an example of a late twentieth-century reference to the Lincoln index, see John R. Skalski and Douglas S. Robson, *Techniques for Wildlife Investigations: Design and Analysis of Capture Data* (San Diego, 1992), 62.

[68] On Nice's mediating role between American animal behaviorists and European ethologists, see Richard W. Burkhardt Jr., *Patterns of Behavior: Konrad Lorenz, Niko Tinbergen, and the Founding of Ethology* (Chicago, 2005), 58.

[69] Frederick C. Lincoln, Annual Report, Section of Distribution and Migration of Birds, Division of Wildlife Research, Fiscal Year 1939, n.d., 37 pp., on 26, in Box 38, Folder: Section Reports, Memos, etc. (Including memos from Div. Chief), FY 1939, Lincoln Files.

[70] On the expansion of the wildlife refuge system, see Wilson, *Seeking Refuge* (cit. n. 1), 66–71; Charles G. Curtin, "The Evolution of the U.S. National Wildlife Refuge System and the Doctrine of Compatibility," *Conserv. Biol.* 7 (1993): 29–38.

[71] On the early years of the Fish and Wildlife Service, see Ira N. Gabrielson, "The Fish and Wildlife Service," *Sci. Monthly* 65 (1947): 181–98.

[72] Frederick C. Lincoln, Memorandum for Dr. Bell, Re: Changes in emphasis on research due to the defense program as per memo of the Director dated 31 December 1941, 2 pp., on 1, Box 38, Folder: Section Reports, Memos, etc. (Including memos from Div. Chief), FY 1942, Lincoln Files.

volunteer banders had already been drafted into the military or were engaged in other defense-related activities that prevented them from contributing.[73] During the war, aluminum for bird bands became scarce, and it became virtually impossible for BBO staff to travel.[74] Reduced to a skeleton crew, the BBO suspended its research work and devoted itself entirely to data management in the hope that the records would be useful when research resumed after the war.[75]

Under these conditions, the BBO relied even more heavily than before on banding work at federal wildlife refuges, and it contemplated establishing new banding stations of its own to "replace volunteer stations that were formerly in operation," although these plans were also hindered by the war.[76] After wartime restrictions were lifted, the number of birds banded annually quickly recovered, but the shift from volunteers to professionals that had begun in the late 1930s continued. In 1949, the BBO reported that the total number of cooperators had doubled over the previous year, bringing it nearly back to prewar levels. One thing had changed, however: "Whereas before the war the cooperators were almost all volunteers, more and more professional wildlife workers are now utilizing banding."[77] More than half of the banding was now being done by employees of the Fish and Wildlife Service, state game agencies, or university-based research units.[78]

The professionalization of bird banding that began in the late 1930s and was accelerated by wartime conditions had the effect of changing the BBO's cost-benefit calculation such that the cost of discouraging volunteers was now outweighed by the benefits of standardizing data. This changing calculation was reflected in the tone of the BBO's communications with cooperators. The first postwar issue of *Bird Banding Notes* in 1946 included several pages of cautions, warnings, and requests, among them the reminder that "bird-banding cooperators should remember that this is a scientific study and that their own pleasure in the work is incidental."[79] Admonishments of this kind were not entirely unprecedented, but they were soon accompanied by strictly enforced new reporting requirements that took even some veteran banders aback. In the June 1948 issue of *Bird Banding Notes*, the BBO introduced a number of new rules effective by the end of the month and threatened to withhold new bands from cooperators who failed to immediately comply. At the same time, it

[73] Lincoln, Annual Report, F.Y. 1941, n.d., 11 pp., on 9, Box 38, Folder: Section Reports, Memos, etc. (Including memos from Div. Chief), FY 1941, Lincoln Files.

[74] On aluminum, see Frederick C. Lincoln and John W. Aldrich, Annual Report, Section of Distribution and Migration of Birds, Division of Wildlife Research, Fish and Wildlife Service, [Fiscal Year 1945], n.d. [August 1945], 15 pp., on 5; on travel funds, see Letter, Charles B. Floyd [Secretary, Northeastern Bird-Banding Association] to Director, Fish and Wildlife Service, 22 November 1944, both in Box 38, Folder: Section Reports, Memos, etc., FY 1945, Lincoln Files.

[75] This was a variety of "planned hindsight" under conditions of labor shortage and limited resources; see Joanna Radin, "Planned Hindsight: The Vital Valuations of Frozen Tissue at the Zoo and the Natural History Museum," *J. Cult. Econ.* 8 (2015): 361–78.

[76] Lincoln, Annual Report F.Y. 1941, n.d., 11 pp., on 10, in Box 38, Folder: Section Reports, Memos, etc. (Including memos from Div. Chief), FY 1941, Lincoln Files.

[77] "Where Do We Stand?," *Bird Banding Notes* 4 (1949): 2–4, on 3.

[78] Between May 1949 and April 1950 (omitting schedules submitted late), 318,221 birds were banded by fewer than 700 individual and institutional cooperators. Of these, more than 55 percent were banded by state game departments, national wildlife refuges, or "special units"—a mixed category covering banding stations associated with universities, private conservation organizations, and the federal government; Seth H. Low, "Cooperator Participation for the 1950 Bird Banding Year," *Bird-Banding* 22 (1951): 64–71, numbers drawn from table on 71.

[79] "General Information to Cooperators," *Bird Banding Notes* 3 (1946): 71.

pruned from its mailing list hundreds of cooperators who had been "delinquent" in submitting data.[80] Volunteers who did not hastily compile and submit their records in the new format found themselves suddenly cut off from the free bands, banding permits, forms of sociality, and scientific legitimation offered by the bird-banding program.

The new tone of the BBO's communications did not go unnoticed by its volunteer cooperators. In the wake of the June 1948 announcement, the editor of the Western Bird Banding Association's newsletter complained that the office's high-handed approach "tends to make the volunteer give thought to other fields of endeavor where the demands are less binding." After all, he added, volunteers were contributing their time, expertise, and often significant financial investments in equipment and supplies to the national bird-banding program; it hardly seemed fair to ask them to reformat their data on a moment's notice.[81] Lincoln's replacement as head of the BBO, Seth H. Low, defended the need for strict standards given the limited staff available to handle reports, asking volunteers to "place the good of the program as a whole ahead of individual inconveniences."[82] In subsequent years, when Low canceled plans to attend their annual meeting or denied their requests for "emergency" stores of bands, members of the Western Bird Banding Association detected clear signs that the value of their contributions had declined in the eyes of the BBO.[83]

In the postwar decades, many American bird lovers continued to find banding rewarding, but they no longer played the central role in the BBO's research efforts that they had in the interwar years. Volunteers continued to make unique contributions by focusing on nongame birds and providing coverage in regions where professional and institutionally based banders were thin on the ground. As its dependence on volunteers decreased, however, the BBO became less interested in facilitating centrifugal activities that had been a source of both tension and enthusiasm in the prewar decades. The postwar expansion in federal support for wildlife research flooded the BBO with reports from professional banders and exacerbated the sense among volunteers that their work was of secondary importance.[84] Meanwhile, employees of state fish and game agencies began to take a serious interest in banding when they realized it could help them challenge restrictive federal hunting regulations.[85] In the 1950s, the Bird Banding Laboratory (as it had been renamed after moving into new facilities at the Patuxent Wildlife Research Center in Maryland) devoted an enor-

[80] "Schedules," *Bird Banding Notes* 4 (1948): 2–5; "Mailing List," ibid., 8.

[81] Paul H. Steele, "Memorandum from Banding Headquarters," *News from the Bird Banders* 23 (1948): 27.

[82] Seth H. Low, "A Report to the Cooperators from the BBO," *News from the Bird Banders* 24 (1949): 31–2, on 32.

[83] On Low's canceled visit, see Mrs. E. C. Baltzar, "Notice of the Annual Meeting," *Notes from the Bird Banders* 26 (1951): 1; Evelyn Baltzar, "Items from the Los Angeles Chapter Meetings," *News from the Bird Banders* 26 (1951): 29–31, on 29. On emergency bands, see "Minutes of the Annual Meeting," *News from the Bird Banders* 25 (July 1950): 30–2.

[84] See, e.g., the following letter from the president of the Northeastern Bird-Banding Association to the chairman of the Senate Committee on Expenditures in the Executive Departments: Charles H. Blake to John L. McLellan, 15 February 1952, in *Federal Wildlife Conservation Activities, 1951: Report of the Committee on Government Operations*, 82d Congress, 2d Session, Report No. 1457 (Washington, D.C., 1952), 282–3.

[85] Remarks of Albert M. Day, Director, at Branch of Game Management Session, Fish and Wildlife Service Conference, Washington, D.C., 22 January 1952, 8 pp., on 2–3, Box 36, Folder: Reorganization through 1955, Lincoln Files.

mous amount of effort to recoding and punching records of waterfowl and other game birds gathered since 1921, but it decided not to do so for nongame birds.[86] This data-management decision reflected the main organizational challenge of the Bird Banding Laboratory in the postwar decades, which was to coordinate the work of government agencies and university research units and to resolve tensions between state and federal wildlife management—not to balance the needs of the federal data-collection effort with the enthusiasms of amateur volunteers.[87]

The change in the BBO's approach to its volunteers resonated with broader postwar shifts in the role of experts in government and their relation to nonexpert citizens. As the cultivation of a corps of volunteer cooperators faded in importance for the Fish and Wildlife Service, and as new technologies of electronic tracking were developed that made it possible to study bird migration without needing to rely on a network of human observers, wildlife biologists and ornithologists increasingly directed their efforts toward "selling" conservation to an undifferentiated public that was increasingly emerging as the target of "public relations."[88] The Fish and Wildlife Service's postwar leadership decided that fostering a network of thousands of committed, enthusiastic, and highly trained—if also sometimes fractious and unreliable—volunteers was a lower priority than it had been during the interwar years. Instead it focused on calculating ever more accurate estimates of waterfowl populations and maps of migratory pathways that could be used to overcome the resistance of state governments to federal regulation, as well as to sell conservation to a sometimes skeptical public.

DATA, DEMOCRACY, AND CITIZEN SCIENCE

Large-scale bird-banding programs have now been active in North America for nearly a century, and there are no indications that the practice will soon be abandoned. In recent years, more than a million birds have been banded annually in the United States.[89] In absolute terms, the number of banders is significantly larger than it was in Lincoln's day; one estimate published in the early 2000s put the total number of North American bird banders at more than 6,100 individuals.[90] Even as other techniques have grown in importance, the practice continues to be seen as essential to

[86] In 1959, many of these and other punch cards that had been painstakingly accumulated since the end of the 1920s were damaged by a warehouse fire. In subsequent years, records were increasingly recorded in digital form for easier processing; Daniel L. Leedy, "Some Federal Contributions to Bird Conservation during the Period 1885 to 1960," *Auk* 78 (1961): 167–75, on 170; C. Stuart Houston, M. Kathleen Klimekiewicz, and Chandler S. Robbins, "History of 'Computerization' of Bird-Banding Records," *North American Bird-Bander* 33 (2008): 53–65.

[87] E.g., in 1954 the Bird Banding Laboratory organized a cooperative project with wildlife biologists from twelve states; "Waterfowl Banding Operations Expanded," Press Release, 2 June 1954, Department of the Interior Information Service, Fish and Wildlife Service, 1 p., Box 41, Folder: Waterfowl—Banding, Lincoln Files.

[88] For an example of the language of "selling conservation to the public," see, e.g., Clarence Cottam, "Progress in Wildlife Research and Training," Presented at National Wildlife Conference Montreal, Canada, 14 March 1955, 25 pp., on 14, Box 43, Folder: Wildlife Training, Lincoln Files.

[89] As of 2011, the Bird Banding Laboratory estimated that it received reports of approximately 1.2 million banded birds each year, along with 87,000 "encounter" reports of previously banded birds; "How Many Birds Are Banded?," USGS Bird Banding Laboratory, https://www.pwrc.usgs.gov/bbl/homepage/howmany.cfm (accessed 30 June 2015).

[90] John Tautin, "One Hundred Years of Bird Banding in North America," USDA Forest Service Gen. Tech. Rep. PSW-GTR-191, 2005, 815–6, http://www.fs.fed.us/psw/publications/documents/psw_gtr191/psw_gtr191_0815-0816_tatuin.pdf (accessed 6 July 2017).

the management and conservation of waterfowl and other bird populations.[91] More-over, it continues to attract volunteers with deep expertise who are passionate about both birds and data. Nonetheless, much has changed since the interwar years. Not only is most banding done by professionals, as it has been since the late 1940s, but even volunteer banding has become more hierarchical and expert led, in part because of efforts to minimize the risk of harm to banded birds. Applicants for "master bander" permits are now required to go through extensive training and submit a "complete research proposal" to the Bird Banding Laboratory before they are issued a permit.[92] Tellingly, the estimate of 6,100 banders cited above includes "federal and state conservation agencies; university associates; amateur ornithologists; bird observatories; environmental centers; non-governmental organizations; environmental consulting firms, and other private sector businesses."[93] Volunteer banders of the kind who dominated the practice in the interwar years—particularly those who are unaffiliated with a research institution or nongovernmental conservation organization—make up a small proportion of this total.

Given these changes and the development of alternate ways of studying birds, the ongoing relevance of the early decades of large-scale bird banding in the United States might not be immediately apparent, especially given the extent to which the social conditions for volunteer participation in scientific research have changed. In the first half of the twentieth century, figures such as Baldwin, the retired business-man devoted to scientific research, and Nice, the highly trained ornithologist unable to advance professionally because of her gender, had the financial and intellectual resources necessary to transform their passion for birds into world-class research. To-day, retired businessmen are more likely to donate money to scientific institutions than they are to practice science themselves, and it is less likely that a woman not only as passionate about research but also as highly talented and trained for it as Nice was would be relegated to the position of volunteer or amateur. Moreover, today's advocates of citizen science are eager to engage a much broader range of people than were targeted by bird-banding efforts in the early twentieth century—for example, those with casual interest, minimal scientific training, or social and economic disadvantages. The emergence of the concept of "citizen science" in the 1990s reflects the democratization of participatory hopes beyond the elitism of much early twentieth-century amateur science.[94]

What, then, is to be learned from a way of structuring the relationship between volunteer amateurs and professional experts that depended, in part, on social conditions and aspirations that no longer exist? I have tried to capture one lesson with the awk-ward metaphors of centrifugal and centripetal force—awkward not least because the former is usually understood to be a fictitious or illusory force. The metaphor is none-

---

[91] P. A. Buckley, Charles M. Francis, Peter Blancher, David F. DeSante, Chandler S. Robbins, Gra-ham Smith, and Peter Cannell, "The North American Bird Banding Program: Into the 21st Century," *J. Field Ornithol.* 69 (1998): 511–29.

[92] "Banding Permit General Information," Bird Banding Laboratory, Patuxent Wildlife Research Center, U.S. Geological Survey, Laurel, Md., https://www.pwrc.usgs.gov/bbL/homepage/gen_info .cfm (accessed 29 October 2015).

[93] Tautin, "One Hundred Years" (cit. n. 90), 815.

[94] The *Oxford English Dictionary* cites a first use of "citizen science" in an ornithological context in the late 1980s, but it only came into wide use a decade later; see also Dickinson and Bonney, *Citizen Science* (cit. n. 6).

theless apposite inasmuch as it expresses the twofold productivity and inherent con-
tradictions of the nationally coordinated bird-banding program that was launched under the aegis of the BBO in 1920. The BBO operated as a classic center of calculation
in Bruno Latour's sense of the term, accumulating data and power through the circulation and accumulation of "immutable mobiles" at an institution of centralized power.[95]
At the same time, however—and in some cases despite itself—it also inspired and facilitated a burst in technical creativity and scientific sociality at the peripheries of professional ornithology, one that resulted in scientific insights that were no less valid and
influential than those achieved by Lincoln in Washington. Arising within a single system of relationships and from some of the very same causes, these forces often pushed
in opposite directions. Bird banding in the interwar period was simultaneously a means
of collecting data and of empowering citizens, of making professionals and of making
amateurs.

Today's citizen-science projects are motivated both by a desire to collect data by
taking advantage of amateur enthusiasms and by a desire to educate and engage citizens in the production of scientific knowledge. Much of the scholarship on citizen
science over the past decade or so seeks to address the tensions between these two
goals. Projects that are equally successful at meeting professional scientists' data needs
and at educating and engaging the public have proven difficult to fashion, though some
do better than others. The professional scientists who lead some of the most successful
centripetally oriented projects, such as the Cornell Laboratory of Ornithology's mapping of bird distributions on the basis of volunteer observations, tend to characterize
the citizen scientist as a data collector rather than as a data interpreter, let alone independent initiator of scientific investigations. Citizen science becomes the process of
"enlist[ing] the public in collecting large quantities of data across an array of habitats
and locations over long spans of time."[96] In projects that take this approach, citizens can
easily be reduced to sensors and the relationship between scientist and volunteer to
quality control.[97] Meanwhile, centrifugal projects such as do-it-yourself pollution monitoring seek to provide tools to nonprofessionals in response to community needs. But
the research they enable is similarly limited, in most cases, to the production of data that
must be interpreted and endorsed by experts to become authoritative.[98]

The boom in citizen-science projects since the 1990s, which has often been framed
as a democratizing move, has thus had some ironically centralizing and hierarchizing
consequences. Even as more and more citizens have been inspired to participate in
scientific research, the scope of their participation has been increasingly narrowed.
This narrowing is linked to the idea that the greatest opportunities and challenges

[95] Latour, *Science in Action* (cit. n. 4), 215–57.
[96] Rick Bonney, Caren B. Cooper, Janis Dickinson, Steve Kelling, Tina Phillips, Kenneth V. Rosenberg, and Jennifer Shirk, "Citizen Science: A Developing Tool for Expanding Science Knowledge and
Scientific Literacy," *BioScience* 59 (2009): 977–84, on 977. See also Florian Charvolin, "Le Programme Feederwatch et la Politique des Grands Nombres," *Développement Durable et Territoires*
(2004), https://developpementdurable.revues.org/687 (accessed 6 July 2017).
[97] For an early example of positive use of "citizen sensor" language, albeit one in which "sensor"
and "observer" are used more or less interchangeably, see Michael F. Goodchild, "Citizens as Voluntary Sensors: Spatial Data Infrastructure in the World of Web 2.0," *Int. J. Spatial Data Infrastructures
Res.* 2 (2007): 24–32. See also Jennifer Gabrys, *Program Earth: Environmental Sensing Technology
and the Making of a Computational Planet* (Minneapolis, 2016).
[98] See Gwen Ottinger, *Refining Expertise: How Responsible Engineers Subvert Environmental Justice Challenges* (New York, 2013).

for scientific research today arise from the possibility of accumulating vast amounts of data. In this data-centric vision of science, where quantity is privileged over quality and pattern matching is privileged over interpretation, citizens are seen as capable of contributing to science either as collectors of otherwise inaccessible data or as cognitive resources to be harnessed for purposes of classification and description. These activities are, of course, only a small subset of the range of activities that have conventionally been thought to constitute "science." The interwar bird-banding program described here provides some perspective on this narrowing of scope for nonprofessional participation in science. It offers an example of an approach that explicitly, if sometimes begrudgingly, sought to satisfy both the centripetal data hunger of professional scientists and the centrifugal investigations and enthusiasms of amateurs. In practice, it seems likely that many of today's centripetal citizen-science projects are, intentionally or not, already being subverted for purposes other than those intended by their designers. Embracing that possibility might be a way for such projects to avoid reinforcing existing hierarchies of expertise and contributing to the impoverishment of our collective notion of citizenship.

Geophysical Datascapes of the Cold War:
Politics and Practices of the World Data Centers in the 1950s and 1960s

by Elena Aronova*

ABSTRACT

The International Geophysical Year or IGY (1957–8), conceived against a background of nuclear secrecy intensified by Cold War political tensions, enabled the distinct data regime that took hold in Soviet and American World Data Centers in the 1950s and 1960s—a regime that turned data into a form of currency, traded by the political players of the Cold War. This essay examines this data regime in detail, considering, in turn, the issues of secrecy and access, sharing and exchange, accumulation and archiving, and, finally, handling and use of the IGY data. Features of the IGY's data centers, such as the notion of centralized storage of open data, freely accessible to users from around the world, played an important role in establishing the practices of data governance that continue today. These practices, however, were outcomes of the politics, visions, and accompanying technologies that were embedded in a supportive political culture of the Cold War. By revisiting the drawbacks and challenges that accompanied the Big Data moment in the early Cold War, this essay explores multiple meanings of data, and the ways in which data participated in a subtle Cold War political economy, beyond their use (or the lack thereof) in the production of knowledge.

INTRODUCTION

Remember the IGY was a big data collecting binge.
—Alan Shapley, 1960[1]

The International Geophysical Year (IGY; 1957–8) can easily appear as preelectronic "Big Data." Extending the postwar scientific largesse to the earth sciences, the IGY involved thousands of scientists in sixty-seven countries in collaborative observational

* Department of History, University of California, Santa Barbara, CA 93106-9410; earonova@history.ucsb.edu.

I would like to thank Michael Gordin, Lorraine Daston, Sally Gregory Kohlstedt, James Fleming, James Secord, and John Krige for many useful comments on early versions of this essay, and Azarii G. Gamburtsev and Nina G. Gamburtseva for helpful conversations and for sharing materials from the Gamburtsev family archive.

[1] Cited in Fae L. Korsmo, "The Origins and Principles of the World Data Center System," *Data Sci. J.* 8 (2010): 55–65.

programs throughout the world.[2] Large volumes of data on various aspects of the physical environment were collected and stored in World Data Centers (WDCs), created, for
the first time, to assemble data in all branches of geophysics. Data were not only collected
but shared, disseminated, and made widely accessible through the WDC system. As the
executive director of the U.S. national committee for the IGY and the coordinator of
American data centers, Hugh Odishaw, underscored, "Data interchange was the immediate and specific *end* of [this] vast scientific program."[3] Free and open data exchange
between geopolitical allies and adversaries alike was the central feature and the raison
d'être of the program. The IGY's announced aim was the creation of a data archive that
could be mined endlessly, for future uses not yet known.[4]

The IGY, in other words, was "a big data collecting binge," as one of its main architects, Alan Shapley, has put it. It was also, as some scholars argue today, an important
antecedent of today's Big Data practices: the notion of centralized storage of open data,
freely accessible to users from around the world, which the IGY institutionalized, played
an important role in establishing the practices of data governance that continue today.[5]
Embedded in the supportive political culture of the Cold War, the IGY contributed to giving Big Data its momentum in the 1950s and 1960s. At the same time, the IGY "big data
collecting binge" had an uneven effect: the very technological choices that made the
program possible resulted in large-scale data loss and the limited use of the IGY data.

In this essay I examine the continuities and ruptures that bind the preelectronic past
and postelectronic present together by looking at the specific data regime that the IGY
worked within and created. This data regime was marked by (1) the politicization of
data, which turned data into a form of currency, traded and bartered by two main holders of planetary geophysical data, the United States and the USSR; (2) an emphasis
on the accumulation of data rather than on their immediate use, which resulted in the
stockpiling of large volumes of data recorded in an analog format and stored on
microfilm; (3) the investment in and promotion of microfilm-based data technologies,
perceived as a viable and valuable alternative to electronic digital computers.

In what follows I examine the IGY data regime in detail, considering, in turn, secrecy
and access; sharing and exchange; accumulation and archiving; and, finally, handling
and use of the IGY data, while illustrating each thematic strand with snapshots from the
history of Soviet and American WDCs. These topics are of special interest today, when

---

[2] On the history of the IGY, see Roger D. Launius, James Rodger Fleming, and David H. Devorkin,
eds., *Globalizing Polar Science: Reconsidering the International Polar and Geophysical Years* (New
York, 2010); Dian Olson Belanger, *Deep Freeze: The United States, the International Geophysical
Year, and the Origins of Antarctica's Age of Science* (Boulder, Colo., 2006); Allan A. Needell, *Sci-
ence, Cold War, and the American State: Lloyd V. Berkner and the Balance of Professional Ideals*
(Amsterdam, 2000); Jacob Darwin Hamblin, *Oceanographers and the Cold War: Disciples of Marine
Science* (Seattle, 2005). See also Elena Aronova, Karen S. Baker, and Naomi Oreskes, "Big Science
and Big Data in Biology: From the International Geophysical Year through the International Biological Program to the Long Term Ecological Research (LTER) Network, 1957–Present," *Hist. Stud. Nat.
Sci.* 40 (2010): 183–224.
[3] Hugh Odishaw, "What Shall We Save in the Geophysical Sciences?," *Isis* 53 (1962): 80–6, on 81;
emphasis added.
[4] Today the WDCs have a massive online presence; see http://www.icsu-wds.org/ (accessed 15 August 2016).
[5] For example, Christine Borgman argued that the WDC system, which formalized "exchanges of
data that previously were bartered," can be seen as a historical antecedent of modern data governance
models, as well as a reminder of some of the "older roots" of open-access publishing and open-data
movements; see Borgman, *Big Data, Little Data, No Data: Scholarship in the Networked World*
(Cambridge, Mass., 2015), 7, 72.

open access, data sharing, archiving, and reuse are presented as positive, novel, and ineluctable features of today's Big Data practices enabled by digital computers. The history of the WDC suggests a more complicated story. Rather than a result of a technological revolution, the IGY Big Data binge was an outcome of contested visions and oftentimes dead-ended technological solutions.[6] The history of the IGY's data centers highlights the multiple meanings of data in Cold War science and the ways in which geophysical data participated in a subtle Cold War political economy, beyond their use (or lack thereof) in the production of knowledge.

CLOSING THE DATA GAP: NEGOTIATING THE STATUS OF DATA IN COLD WAR GEOPHYSICS

A distinctive feature of the IGY was its dual—scientific-cum-military—agenda.[7] All of IGY's thirteen disciplines had direct military implications.[8] While geophysics has a long history of serving military needs, this relation became especially tight with the advent of nuclear weapons and the Cold War.[9] The symbiotic relationship with secret military science had the greatest immediate impact on access to geophysical data. In the aftermath of World War II, geophysics expanded, but open geophysical data were in increasingly short supply. By 1950, most data in the disciplines identified as IGY themes were classified in both the United States and the USSR, and data exchange between the Cold War adversaries was reduced to a limited number of standard meteorological observations. Closing the data gap was one of the incentives for the geophysicists' concerted drive for the IGY, and for the military and quasi-military agencies to back it up, on both sides of the "Iron Curtain."

---

[6] For the purposes of this essay, I focus on forgotten visions and abandoned technologies rather than on the IGY's better-known highlights and achievements, which included extensive exploration of the oceans and polar regions, Arctic and Antarctic, and the launch of the first artificial satellite, *Sputnik*, on 4 October 1957, which marked the beginning of the space age. The scientific highlights of the IGY included the crucial evidence for plate tectonics theory and the discovery of Van Allen radiation belts. For a discussion of the scientific achievements of the IGY, see Jorge Berguño and Aant Elzinga, "The Achievements of the IGY," in *History of the International Polar Years (IPYs): From Pole to Pole*, ed. Susan Barr and Cornelia Lüdecke (Berlin, 2010), 259–78; Erik M. Conway, "The International Geophysical Year and Planetary Science," in Launius, Fleming, and Devorkin, *Globalizing Polar Science* (cit. n. 2), 331–42; Dasan M. Thamattoor, "Stratospheric Ozone Depletion and Greenhouse Gases since the International Geophysical Year: F. Sherwood Rowland and the Evolution of Earth Science," in ibid., 355–72.

[7] See Needell, *Science* (cit. n. 2). See also Jacob Darwin Hamblin, *Arming Mother Nature: The Birth of Catastrophic Environmentalism* (Oxford, 2013); Paul Edwards, *A Vast Machine: Computer Models, Climate Data, and the Politics of Global Warming* (Cambridge, Mass., 2010).

[8] The IGY disciplines included aurora and airglow, cosmic rays, geomagnetism, glaciology, gravity, ionospheric physics, longitudes and latitudes, meteorology, nuclear radiation, oceanography, seismology, solar activity, and upper atmospheric studies using rockets and satellites.

[9] The literature on the vital importance of the geophysical sciences for the development of modern warfare is vast but focused mostly on the United States and, to a lesser extent, Western Europe. For a general discussion, see Ronald E. Doel, "Constituting the Postwar Earth Sciences: The Military's Influence on the Environmental Sciences in the USA after 1945," *Soc. Stud. Sci.* 33 (2003): 635–66. For a more recent account, see Simone Turchetti and Peder Roberts, eds., *The Surveillance Imperative: Geosciences during the Cold War and Beyond* (Basingstoke, 2014). The historiography of the military agenda of Soviet geophysics is scarce, but the existent studies suggest a similarly significant military involvement in the development of geophysical disciplines in the Soviet Union during the Cold War; see Irina V. Bystrova, *Voenno-promyshlennyi kompleks SSSR v gody kholodnoi voiny* (Moscow, 2006).

The example of seismology illustrates how the two imperatives—the strategic importance of geophysical data with the advent of nuclear weapons, and the excessively secretive regime established in geophysics as a result—coalesced to make geophysical data a site of both political struggle and international cooperation. In the Soviet Union, seismologists were involved in the atomic bomb project from its inception, initially as experts in seismic methods of geophysical prospecting. In 1946, Grigorii Gamburtsev, the country's leading expert in seismic methods of oil and gas prospecting, joined the top-secret atomic team, charged with organizing geophysical prospecting of uranium in the territory of the Soviet Union.[10] Starting in 1951, Gamburtsev lobbied for the organization of a special seismological service that would work on the development of coherent nuclear test detection methods. The beginning of underground nuclear testing in the United States gave the proposal momentum, and a classified seismological branch of the Geophysical Institute, headed by Gamburtsev and his former student Ivan Pasechnik, was inaugurated to work on the detection of nuclear tests at a distance. Writing in 1954, Gamburtsev noted: "At present, the seismological method of detection and localization of the nuclear tests is adopted by Soviet Army services."[11] After Gamburtsev's death in 1955, the scientific director of the Semipalatinsk nuclear test ground, seismologist Mikhail Sadovskii, became the head of the Geophysical Institute, institutionalizing the crucial link between the military and academic seismology.[12]

The dual, scientific-cum-military, agenda gave a major boost to both civilian and military branches of Soviet seismology.[13] One of the consequences of these developments was the introduction of a stringent "nuclear secrecy" regime in Soviet seismology.[14] In April 1953, referring to the main regulatory document that defined classified information in the Soviet Union, the "Glavlit list," Gamburtsev complained about the strange situation that strangled seismologists' work:

> According to . . . Glavlit list No. 2, "all material, acts and information on earthquakes" are considered state secrets and are banned from publication in the open press. . . . Consequently, all scientific results produced at the Geophysical Institute that mentioned

---

[10] A. P. Vasiliev, "Ob osnovopolagaiushchem vklade akademika G. A. Gamburtseva v sozdanii sistemy dal'nego obnaruzheniia iadernykh vzryvov," in *Aktual'nost' idei G.A. Gamburtseva v geofizike XXI veka*, ed. A. O. Gliko (Moscow, 2003), 64–80.

[11] Grigorii Gamburtsev, "Seismicheskie metody obnaruzheniia vzryvov iadernogo oruzhiia" (ca. 1954–5), unfinished draft, "Zakluchenie" (Gamburtsev Family Archive, Moscow).

[12] On Sadovskii, see Kai-Henrik Barth, "Detecting the Cold War: Seismology and Nuclear Weapons Testing, 1945–1970" (PhD diss., Univ. of Minnesota, 2000); A. V. Nikolaev, ed., *Mikhail Aleksandrovich Sadovskii: Ocherki, vospominaniia, materialy* (Moscow, 2004).

[13] On seismic monitoring in the United States, see Michael Gordin, *Red Cloud at Dawn: Truman, Stalin, and the End of the Atomic Monopoly* (New York, 2009), esp. chap. 5. In the 1950s, American experts, unlike their Soviet counterparts, ruled out seismic methods of detection as unreliable and operational failures. It was not until the 1960s that seismology became important as a monitoring technology throughout the world, enabling verification of compliance with the Atmospheric Test Ban Treaty of 1963. On the possibility to discriminate between earthquakes and explosions as a new research area, see Kai-Henrik Barth, "The Politics of Seismology: Nuclear Testing, Arms Control, and the Transformation of a Discipline," *Soc. Stud. Sci.* 33 (2003): 743–81; Axel Volmar, "Listening to the Cold War: The Nuclear Test Ban Negotiations, Seismology, and Psychoacoustics, 1958–1963," *Osiris* 28 (2013): 80–102. On the contestation of the reliability of seismic detection methods, see Anna Amramina, "Political Seismology or Seismological Politics: Natural Resources Defense Council–USSR Experiments in Underground Nuclear Test Verification," *Seismol. Res. Lett.* 86 (2015): 451–57.

[14] On "nuclear secrecy," see Peter Galison, "Removing Knowledge," *Crit. Inq.* 31 (2004): 229–43; Alex Wellerstein, "Knowledge and the Bomb: Nuclear Secrecy in the United States, 1939–2008" (PhD diss., Harvard Univ., 2010).

earthquakes have been banned from publication. The publication of the seismological bulletin has been stopped, too.[15]

It is instructive to compare different editions of the Glavlit list (*perechen' Glavlita* in Russian), as these were mentioned in Gamburtsev's correspondence, to see how closely the tightening of the secrecy regime in seismology followed the developments in nuclear weaponry.[16] First compiled in 1948, the Glavlit list did not mention earthquakes. The list was updated in the aftermath of the first American underground test in 1951. The 1952 edition included a separate section on seismology, which classified as state secrets:

> All materials, acts and information on earthquakes;
> Seismic regional maps;
> Data on the effects of major earthquakes in the past 3 years;
> Any information on damage caused by earthquakes;
> Any information on earthquake-resistant constructions.[17]

The formulations used by the Glavlit were totalizing and general. As Gamburtsev complained, just mentioning the word "earthquake" could result in a publication ban.[18] Gamburtsev's letter, written shortly after Stalin's death, typified a larger campaign by Soviet scientists, particularly nuclear physicists, calling for reforms of Soviet science policy that would open up Soviet science for international cooperation and data exchange.[19] In the case of geophysics, just a few months after Stalin's death, the Soviet Academy of Sciences signaled its participation in the IGY, and by 1954 it had set up the Soviet IGY national committee, chaired, not surprisingly, by Gamburtsev.

Access to geophysical information was one of the main arguments for Soviet participation in the IGY. In a secret report submitted to the Soviet leading authority, the Central Committee of the Communist Party, scientists emphasized that the IGY created vital opportunities for strengthening national defense and the development of modern weapons: oceanographic data were critical for the Soviet naval fleet and for the "solution of the practical questions concerning the disposal of nuclear waste in the deep sea";[20] in meteorology, IGY data opened possibilities for long-term global weather forecasts, which "would become critical in case a country refused to provide

[15] G. A. Gamburtsev to A. N. Nesmeyanov, 18 April 1953, Rossiiskii gosudarstvennyi arkhiv noveishei istorii, Moscow (Russian State Archive of Contemporary History; hereafter cited as "RGANI"), fond 5, opis 17, delo 417.

[16] On Glavlit, see V. R. Firsov, ed., *Tsenzura v Rossii: istoriia i sovremennost'* (Moscow, 2013); A. V. Blum, *Sovetskaya tsenzura v epokhu total'nogo terrora, 1929–1953* (St. Petersburg, 2000); Michael S. Fox, "Glavlit, Censorship and the Problem of Party Policy in Cultural Affairs, 1922–28," *Soviet Stud.* 44 (1992): 1045–68.

[17] G. A. Gamburtsev to A. N. Nesmeyanov, 18 April 1953, RGANI, fond 5, opis 17, delo 417.

[18] Ibid.

[19] In the aftermath of Stalin's death, leading Soviet scientists, particularly nuclear physicists, capitalized on their power to push for major reforms in Soviet science policy, enabling "the shift from autarchy to restricted internationalism in Soviet science"; see Konstantin Ivanov, "Science after Stalin: Forging a New Image of Soviet Science," *Sci. Context* 15 (2002): 317–38. On the similar campaign in mathematics, see Slava Gerovitch, *From Newspeak to Cyberspeak: A History of Soviet Cybernetics* (Cambridge, Mass., 2002).

[20] For a discussion of radioactive waste dumping in Western countries during the Cold War, see Jacob D. Hamblin, *Poison in the Well: Radioactive Waste in the Oceans at the Dawn of the Nuclear Age* (New Brunswick, N.J., 2008).

meteorological forecasts to our aircraft in its airspace," while other data were necessary for current needs of military aviation and future needs of ballistic missiles and
spacecraft. [21] In a memo to party authorities, the chairman of the Soviet IGY national
committee argued that "participation in the program allows our scientists to get access
to important and up-to-date scientific information from abroad."[22]

Access to data was the powerful imperative behind geophysicists' campaign for the
IGY on both sides of the "Iron Curtain."[23] While emphasizing that sharing data would
benefit their respective intelligence and military services, geophysicists in both the
U.S. and the USSR used the IGY to renegotiate the boundary between open and classified data in geophysics. American IGY planners emphasized the distinction between
"basic data"—"the elementary building blocks of scientific progress in the earth sciences," in the words of one IGY leader—and "end products."[24] Basic data, the argument went, should be open and shared between allies and adversaries alike, while data
products that might be used for military purposes should be classified. The distinction
between basic data and data products echoed the distinction between "basic" and "applied" science, which gained a renewed prominence in the context of the Cold War
among scientists turned public intellectuals such as Michael Polanyi who insisted that
"pure" or "basic" science should be defended as an essential liberal value.[25] Similar to
the distinction between basic and applied science, which was used as an ideological
resource to achieve different ends, the distinction between basic data and data products was emphasized by scientists to strengthen their case for the IGY.

The IGY was conceived against the background of the already existing strategy of
Cold War internationalism in nuclear physics and driven by many of the same factors
and motivations: aspirations to ease international tensions through science diplomacy,
the promotion of national scientific leadership, and intelligence gathering under the
guise of scientific cooperation agreements, all at once.[26] In an effort that coincided
with the planning for the IGY in the early 1950s, U.S. physicists renegotiated the nuclear secrecy regime in Cold War America.[27] The excessive nuclear secrecy embodied
in the 1946 Atomic Energy Act was relaxed in a new, less restrictive Atomic Energy
Act, signed into law in August 1954. The new regime, which regulated the boundary

---

[21] A. N. Nesmeyanov and I. P. Bardin, October 1958, Memo for the Central Committee of the Communist Party, RGANI, fond 5, opis 35, delo 74.
[22] I. Bardin to E. Furtseva, December 1957, Memo for the Central Committee of the Communist
Party prepared by the Soviet IGY National Committee, RGANI, fond 5, opis 35, delo 74.
[23] As Jacob Hamblin argued, the idea of an international program of global geophysical observations plotted by a few American geophysicists in 1950 gained momentum because the U.S. military
needed data on their Cold War adversaries' environment; see Hamblin, *Arming Mother Nature* (cit.
n. 7).
[24] Korsmo, "Origins and Principles" (cit. n. 1).
[25] Mary Jo Nye, *Michael Polanyi and His Generation: Origins of the Social Construction of Science* (Chicago, 2011). On the origin of the distinction between pure and applied science, see Robert
Bud, "'Applied Science': A Phrase in Search of a Meaning," *Isis* 103 (2012): 537–45; Graeme Gooday,
"'Vague and Artificial': The Historically Elusive Distinction between Pure and Applied Science," *Isis* 103
(2012): 546–54; and Sabine Clarke, "Pure Science with a Practical Aim: The Meanings of Fundamental
Research in Britain, circa 1916–1950," *Isis* 101 (2010): 285–311. For the ideological reversal of the distinction in the Soviet context, see Ivanov, "Science after Stalin" (cit. n. 19).
[26] On Cold War–type internationalism, see John Krige, "Atoms for Peace, Scientific Internationalism, and Scientific Intelligence," *Osiris* 21 (2006): 161–81.
[27] See Wellerstein, "Knowledge" (cit. n. 14); David Holloway, *Stalin and the Bomb: The Soviet Union
and Atomic Energy, 1939–1956* (New Haven, Conn., 1994); Jessica Wang, *American Science in an Age
of Anxiety: Scientists, Anticommunism, and the Cold War* (Charlotte, N.C., 1999).

between "open" and "classified" information by distinguishing the civilian "peaceful atom" from its "military uses" in nuclear weapons, permitted the declassification and release of a significant amount of nuclear information, for instance, at the 1955 international conference in Geneva.[28]

While the nuclear physics example set the stage for IGY data exchange, geophysics differed in at least one important respect. In contrast to the case of nuclear physics, IGY data exchange was institutionalized through the system of World Data Centers (WDCs). The WDC consisted of two centers for all IGY disciplines WDC-A in the United States, set up as a set of twelve primary data centers corresponding to different IGY disciplines, and WDC-B in the USSR, made up of two subcenters, the main center in Moscow and another one in Novosibirsk, along with the third center, distributed between different institutions in Western Europe, Australia, and Japan (WDC-C).[29] All three WDCs were expected to collect a complete set of IGY data so that by the end of the program three identical data sets would be compiled. IGY planners emphasized that the triplicate organization of the WDC was justified by the need for data preservation ("to insure against catastrophic destruction of a single center") and data use (to assure the geographic accessibility of data to users in different parts of the world).[30] Political analysts, however, were quick to observe that the structure of the WDCs mirrored Cold War geopolitical rivalries and alliances.[31] These tensions and diplomacies were reflected in the practices of data exchange.

The fact that the data were "opened," with the procedures of their exchange formalized in the IGY *Guide to Data Exchange*, did not necessarily mean that all "open data" were automatically shared between the Cold War adversaries. To be sure, in the case of nuclear physics, too, data sharing and exchange were a matter of tensions and negotiations since scientists were in a position to influence particular decisions regarding when to share which data, and with whom. In the case of the IGY, however, the institutionalization of data exchange made these subtle politics and practices of data exchange more tangible than in other similar cases. It is these politics and practices of IGY data centers, determined by the Cold War political economy rather than by formal regulations, that I will be concerned with in the rest of the essay.

THE COLD WAR POLITICAL ECONOMY OF DATA EXCHANGE

The implicit assumption in the IGY *Guide to Data Exchange* was that, once "opened," data would automatically "flow" to the data centers. Geophysicists often referred to

---

[28] Krige, "Atoms for Peace" (cit. n. 26).

[29] For an internal account of the history of the WDC system, see S. Ruttenberg and H. Rishbeth, "World Data Centers—Past, Present, and Future," *J. Atmos. Terr. Phys.* 56 (1994): 865–70. On the origins of the WDC-A in the United States, see Korsmo, "Origins and Principles" (cit. n. 1). On the history of WDC-B, see Aleksandr Povzner, "Istoriia podgotovki i osushchestvleniia nauchnykh issledovanii po programme Mezhdunarodnogo Geofizicheskogo Goda" (PhD diss., State Univ. Moscow, 1966); A. S. Kudashin, "Sozdanie v SSSR mirovogo tsentra dannykh po planetarnoi, 1957–1960," *Voprosy Istorii Estestvoznaniia i Tekhniki* 36 (2015): 368–76.

[30] Hugh Odishaw, "International Geophysical Year: A Report on the United States Program," *Science* 127 (1958): 115–28; Odishaw, "What Shall We Save?" (cit. n. 3); Walter Sullivan, *Assault on the Unknown* (New York, 1961), 35.

[31] Diana Crane, "Transnational Networks in Basic Science," *Int. Org.* 25 (1971): 585–601. After the IGY, the data exchange continued, but the data policy regarding the dissemination of three sets of identical data to different data centers was not extended to newer types of data. See *Guide to International Data Exchange through the World Data Center for the period 1960 onward* (London, 1963).

"torrents" of data flowing in and out of the WDCs.[32] In practice, the IGY data did not flow effortlessly and smoothly like a fluid. Making data flow rather than trickle (or, indeed, come to a halt) required constant negotiation and oversight. The assessment and evaluation of "data flow" was one of the important, albeit not widely publicized, functions of the data centers. The data flow was much smoother between political allies, but even in these cases scientists often tried to send data directly to their IGY collaborators in another country, bypassing the WDC. In most cases the stations sent the data to their national IGY committees, which were instructed to copy and distribute them to the other two WDCs. However, only a few national committees outside Eastern Europe dutifully sent their data to the Soviet center. Despite the agreements and the rules of exchange formalized in the IGY *Guide*, the Soviets had to press the other WDCs in order to make data flow to Moscow.

At the end of 1959, the WDC in Moscow was still missing lots of data. The Soviet managers of the center sent requests for data to the top organizational bodies of the IGY, to the annoyance of their American counterparts. Jack Lumby, the director of the WDC-A for oceanography, had to explain the situation in his report to the Office of Naval Research, where one of the Soviet requests ultimately landed:

> It seems to me that what is happening is that the Russian WDC is shopping around to try and induce people to send in their IGY data—more especially the odd bits and pieces which did not accompany the main body of data but of the existence of which they learnt, I suppose, from programs, reports, and so on.[33]

"We have done much the same thing ourselves," Lumby admitted.[34] Indeed, to make sure that all "Russian data" (or "Soviet data" as these were interchangeably called in WDC-A reports) were duly transmitted to the American data center was a matter of special concern in Washington. The coordination office of the WDC-A assessed the "data flow" from the Soviet bloc countries separately from other countries, keeping a detailed record of the flow of "Russian data." All in all, the data flow looked good: it was characterized as "copious," "reasonably prompt," with data "of very good quality and reliable" for most IGY disciplines.[35]

Data on rockets and satellites presented a more contentious issue. The difficulty in obtaining Soviet data from the satellites program caused open tensions between the American and Soviet IGY national committees. The Soviet Union launched its first two Sputniks as part of the Soviet contribution to the IGY. Thousands of professional and amateur astronomers and radio operators, as well as scientific stations from all

---

[32] The phrase "torrents of raw data" is from Lloyd Berkner's proposal on the need for a national geophysical institute (J. Merton England, *A Patron for Pure Science: The National Science Foundation's Formative Years, 1945–57* [Washington, D.C., 1983], 307). "Flow of data" was one of the rubrics used in the internal reports of WDC-A; see IGY Papers, National Academy of Sciences Archives, Washington, D.C. (hereafter cited as "IGY Papers"), Series 8.

[33] J. R. Lumby to Gordon Lill, 29 October 1959, World Data Centers and Data Processing: Oceanography: USSR correspondence, 1957–9, IGY Papers.

[34] Ibid. A few months earlier, Lumby, in his turn, had been fishing for Russian data. In a letter to his counterpart, the director of WDC-B for oceanography in Moscow, he inquired about a "few items of Russian data . . . which we have not yet received"; J. R. Lumby to P. Evseev, 4 May 1959, World Data Centers and Data Processing: Oceanography: USSR correspondence, 1957–9, IGY Papers.

[35] Richard T. Hansen to Hugh Odishaw, 20 February 1959, World Data Centers and Data Processing: Solar Activity: USSR correspondence, 1957–9; IGY Papers.

over the globe, were tracking *Sputnik*'s trajectory, and the WDC-A duly sent their tracking data to the Soviet IGY committee.[36] By contrast, the flow of Soviet satellite data to the United States was "very slow," "ineffective," and "somewhat incomplete," according to the assessment of the WDC-A for rockets and satellites.[37]

How was the data flow measured to ensure fair exchange? There was no standard-ized method of measuring the volume of data across the IGY disciplines. Data were counted in "packages," in "data-pieces" or "data-sheets," and kilometers/miles of mi-crofilm.[38] Although somewhat archaic, this system worked for the purposes of mon-itoring the data exchange. The Soviet reports on data flow demonstrate that the frus-trations of American rocket scientists were substantiated: a tally of data received from WDC-A by the fall of 1958, for example, shows that while the Soviets received data on the satellite launches from the United States, the UK, Japan, Argentina, South Africa, Belgium, France, and Pakistan, they had not sent anything in return.[39] The reason was simple: the WDC-B did not have these data. Data on Soviet rockets and satellites, most proximately tied to the top secret Soviet military missile program, were handled sep-arately from other IGY programs. The Soviet WDC, however, was able to reciprocate its American counterpart with ample data supply in other IGY disciplines, making the data exchange work.

The asymmetry of exchange was the pump that made data flow. The geophysicists in charge of the Soviet WDC never missed an opportunity to point this out to party officials, emphasizing that their center received more data from other centers than it sent in exchange. The head of the Soviet IGY committee highlighted this disparity in a report:

> The amount of geophysical data that we receive from other countries is much larger than the amount of data that we give away. . . . According to our data center WDC-B, the amount of data on the ionosphere that we have received from Western [data centers] so far is 7 times larger than the corresponding Soviet data we have sent to the WDC-A; the cosmic rays data are 6 times larger, and meteorological data are almost 10 times larger.[40]

The IGY created the possibility of receiving data without sharing the same kind of data in return. Thus, the Soviet Union did not participate in the IGY program on nuclear radiation, yet the Soviet WDC was receiving data collected by the IGY nuclear radi-ation programs via American and British subcenters.[41] In return, in some IGY disci-plines the Russian WDC provided its American counterpart with "extra data," initiat-ing the exchange of data beyond the scope of the IGY programs. As the head of the

---

[36] On the involvement of amateur astronomers in tracking *Sputnik*'s course, see W. Patrick McCray, *Keep Watching the Skies! The Story of Operation Moonwatch and the Dawn of the Space Age* (Prince-ton, N.J., 2008).

[37] "World Data Centers and Data Processing: Data Center: 1956–1957," IGY Papers. See also Rip Bulkeley, "The Sputniks and the IGY," in *Reconsidering Sputnik: Forty Years since the Soviet Satel-lite*, ed. Roger D. Launius, John Logsdon, and Robert W. Smith (Amsterdam, 2000), 125–60, on 148.

[38] Paul J. Kellog to Pembroke Hart, 23 May 1958, World Data Centers and Data Processing: Cos-mic Rays: USSR correspondence, 1957–9, IGY Papers.

[39] A. Nikolaev, "Report on the work of the WDC-B1, as of 21 Oct 1958," Archive of the Russian Academy of Sciences, Moscow (hereafter cited as "ARAN"), fond 683, opis 1, delo 9.

[40] I. Bardin to E. Furtseva, December 1957, RGANI, fond 5, opis 35, delo 74.

[41] "We are receiving interesting data in the section 'nuclear radiation' from the USA and the UK. The data sets are not complete, but unfortunately we cannot request them" (since the Soviet Union did not participate in the program on nuclear radiation); Transcript of discussions of data flow, Protokoly MGG, 1960, 3 March 1960, ARAN, fond 683, opis 1, delo 19.

WDC-A for geomagnetism noted, "the flow of geomagnetic data from the USSR was practically non-existent prior to IGY; now, in addition to the IGY data, mean station values have been obtained for several years prior to IGY."[42] In a metaphorical sense, the IGY exchange pump made data flow against the concentration gradient, and along previously improbable transport routes.

With these practices in place, data came to be regarded as an "exchange currency," as Soviet oceanographer Grigorii Udintsev put it at a meeting discussing the reluctance of some oceanographers to share data on projects conducted outside the scope of IGY programs: "Our data are our exchange currency! It is necessary to make sure that all data [from Soviet oceanographic surveys] collected during the IGY would be transmitted to the WDC-B."[43] As another Soviet IGY scientist underscored, "Thanks to the system of exchange . . . the USSR has become [one of the biggest] holders of . . . planetary geophysical data, which helps to exert our influence over other countries."[44]

While data exchange was recognized as a stunning success of the IGY, the politicization of data on which it was built had unanticipated consequences for the data centers and their practices. Data having become a currency, the WDCs' main function had become the accumulation of data, for the most part recorded in an analog format. Consequently, in the 1960s the WDCs were facing an analog data deluge.

DATA ACCUMULATION: THE ANALOG DATA DELUGE

IGY planners anticipated that computer methods of data processing would soon be widely accepted. Yet the idea of using a machine-readable format for IGY data—meaning punch cards at the time—was rejected, since this would greatly impede data exchange, the central feature of the program.[45] Microphotography—microcards and microfilm—was chosen as a medium that allowed data centers to not only store but also efficiently disseminate large volumes of data.[46] The World Meteorological Organization, which served as one of the WDC-C subcenters for the meteorological program of the IGY, estimated that the complete set of IGY meteorological data would require more than 50 million punch cards, hardly manageable for circulation around the world. The same volume of data could fit on only 18,500 microcards, standard-size library catalog cards made from a special high-resolution dye-coated paper. A microcard, on which up to ninety-six pages of text could be reproduced in a greatly reduced size, would fit in a standard-size envelope.[47] Other IGY disciplines used microfilm, which was less expensive than microcards.[48] In the United States, the data centers ar-

[42] Pembroke J. Hart, "Report on IGY WDC-A: Geomagnetism, Gravity and Seismology," 8 February 1960, WDC-A: General Correspondence, January–March 1960, IGY Papers.

[43] "Protokol, 15 December 1960, "Protokoly MGG, 1960," ARAN, fond 683, opis 1, delo 19.

[44] V. V. Belousov to M. D. Millionshchikov, 27 June 1963, "Perepiska ob organizatcii mezhdunarodnogo tcentra geofizicheskikh dannykh," ARAN, fond 683, opis 1, delo 43.

[45] Hugh Odishaw to Robert C. Ridings, 26 March 1959, Data Center Coordination Office: Chron File, January–June 1959, IGY Papers.

[46] On the role of microfilm in postwar American social sciences, see Rebecca Lemov, *World as Laboratory: Experiments with Mice, Mazes, and Men* (New York, 2006).

[47] "Microcards of IGY Meteorological Data," Report no. 7, August 1957, World Meteorological Organisation, Meteorological Data Center, Geneva (World Data Centers and Data Processing: World Meteorological Organisation: Documents on Data, 1956–1957; IGY Papers). See discussion in Edwards, *A Vast Machine* (cit. n. 7), 202–7.

[48] "Report by Schilling on a Visit to the National Weather Records Center at Asheville, North Carolina, on June 27 and 28, 1956," Vestine and Schilling Visits, 1956, IGY Papers.

ranged with the Library of Congress for microfilm copying services.[49] In the USSR, all microfilming was done at the data center itself, which by 1958 moved into a spacious new facility near Moscow State University equipped with copying presses, microfilm cameras, viewers, and printers.[50] In either case, miniaturization—the reduction of the physical size of the "data pieces"—was critical for the WDCs, which combined the functions of an archive, a copy center, and a postal service.

It was hardly a coincidence that the data dissemination system was adopted by some of the recent users of "V-mail," a postal system used overseas by the military during World War II in the United States and in Britain. V-mail letters were written on standard-size letterforms, then microfilmed, airmailed (150,000 letters in one mailbag), and delivered as printout air-graph letters.[51] This process reduced the bulk and weight of mail in difficult wartime conditions and was transferred to international data exchange. Microfilming was originally adopted as the intermediary step in the transmission of data between centers during the eighteen-month-long "year" of massive data gathering. It turned out, however, that microfilm as a medium for geophysical data had a longer lifetime.

In the 1960s, with computer-centered digital data acquisition systems increasingly replacing traditional data techniques, microfilm continued to be used as data storage medium for geophysical data. A 1964 report of the National Bureau of Standards underscored that "the use of microfilm as a storage medium facilitates access to data" and recommended that it be used as a medium for satellite data collected during the International Year of the Quiet Sun (IYQS, 1964–5), the program most directly modeled after the IGY and designed to take advantage of the years of minimal solar activity.[52] During the 1960s, the IGY became a model of data collection and exchange across geophysical disciplines and beyond.[53] Many of these programs sent their data to the WDCs, to be microfilmed and stored in the data vaults for future researchers.

Microfilming reduced the physical size of "data pieces" without having to convert the data to numeric values, which seemed especially appropriate for WDCs given the variety of data formats used in different IGY disciplines. Most data collected during the program were nonnumerical and highly heterogeneous in their types and forms. For example, the data sent to Aurora subcenters included all-sky camera photographs, copies of original spectrograms, maps, monthly logs of aurora observations and monthly bulletins of observations from the stations, visoplots (standardized forms to record observational data on auroras designed specifically for the IGY), tables and charts, films of auroras, technical reports, preprints, and reprints of scientific publications.[54] A wide range of data formats were used in other IGY disciplines as well.[55] Underscoring the diversity of types and physical forms of data, Odishaw, the executive director of WDC-A's

[49] Pembroke J. Hart, "Report on IGY WDC-A: Geomagnetism, Gravity and Seismology," 8 February 1960, WDC-A, General Correspondence, January–March 1960, IGY Papers.

[50] Kudashin, "Sozdanie v SSSR mirovogo tsentra" (cit. n. 29).

[51] Jonathan Auerbach and Lisa Gitelman, "Microfilm, Containment, and the Cold War," *Amer. Lit. Hist.* 19 (2007): 745–68.

[52] *Technical Highlights of the US National Bureau of Standards: Annual Report* (Washington, D.C., 1964), 188.

[53] Aronova, Baker, and Orestes, "Big Science" (cit. n. 2).

[54] "Forms of Data," 6 March 1957, Aurora II: Cornell U: 1956–9, IGY Papers.

[55] E. H. Vestine and G. F. Schilling, "Plans for the U.S. World Data Center," 9 October 1956, Status Report on US Data Center: 1956–8, IGY Papers.

coordination office in Washington, called the WDC an "archive of numerical data, curves, and maps."[56]

How big was the IGY analog data universe? In 1960, just one subcenter of WDC-A for the ionosphere produced more than 200 miles of microfilm in one year, which was but a small fraction of all films and ionograms collected by this one subcenter from IGY programs.[57] The WDC-B in Moscow reported that it received 275,000 data sheets and 300,000 photo stills from IYQS programs, from which rockets and satellites alone yielded 190,000 data sheets.[58] Over the course of three years, from 1964 to 1967, WDC-B gathered data from ninety countries, accumulating more than 12 million "data sheets" and producing thousands of kilometers of microfilm in addition to the bound booklets with printed data that WDC-B distributed on demand.[59]

In the 1960s, the WDC was the largest repository of data ever seen in geophysics. With concern heightening over environmental issues, IGY data centers started to be seen as a treasure trove of data on various aspects of the environment with the potential to answer questions beyond the immediate concerns of geophysics. Yet, because of the massive volume of data and its diversity, extracting useful information from the accumulated data turned out to be an extremely daunting task. Only a small part of the data holdings in the WDCs was ever used or even processed. With the exchange pump not slowing down and the data holdings piling up, the data flow felt more like a flood. In the aftermath of the IGY, the U.S. national committee appointed a special board on geophysical data chaired by one of the IGY's leaders, Alan Shapley, to study what began to be referred to in internal correspondence as "the problem of geophysical data."[60] While everyone on the committee agreed that the IGY had launched an "unprecedented international exchange" of "great value to science and to mankind," they felt that the sheer "scope and size of the monumental body of data" posed the new problem of extracting that value, not in a distant future but at that moment.[61]

By the early 1960s, with the nuclear arms race furthered and the Soviet Union apparently on the winning side in the space race, the "data race" seemed to be next on the agenda. A number of American scientists felt that the "problem of geophysical data" required urgent attention. One of the first actions of the Federal Council for Science and Technology (FCST), formed in 1959 under the chairmanship of President Eisenhower's Science Advisor, was the appointment of a standing committee to consider the problem of "Environmental Data Acquisition, Handling, and Dissemination."[62] The U.S. Air Force weather agency's meteorologist Marshall Jamison, a liaison between academic and military geophysicists during the IGY, insisted that the committee should "urge" the National Academy of Sciences to call on all major federal scientific agencies in the United States—the NSF, NASA, the Office of the Secretary of Defense (OSD), the

---

[56] Hugh Odishaw to Nisson A. Finkelstein, 4 January 1960, WDC-A: General Correspondence, January–March 1960, IGY Papers.

[57] Hugh Odishaw to Robert C. Ridings, 26 March 1959, Data Center Coordination Office: Chron File, January–June 1959, IGY Papers.

[58] "Otchet o rabote MTsD-B1 za period MGSS (I.1964–II. 1967)," Otchet MTcD, 1964–7, ARAN, fond 683, opis 1, delo 89.

[59] "Rabota MTcD-B1," Otchet MTcD, 1964–7, ARAN, fond 683, opis 1, delo 89.

[60] A. H. Shapley to Joseph Kaplan, 25 October 1960, Commission on Geophysical Data: Report 1960, IGY Papers.

[61] Commission on Geophysical Data, "Draft of the Report," 25 October 1960, Commission on Geophysical Data: Report 1960, IGY Papers.

[62] Draft memo, Standing Committee of the Federal Council for Science and Technology, 13 October 1960, WDC-A: General Correspondence, April–December 1960, IGY Papers.

U.S. Air Force, the U.S. Navy, the Department of Commerce housing the U.S. Weather Bureau, and the Library of Congress—to "consider this problem jointly in a vigorous frontal attack." According to Jamison, "The Air Force feels that . . . the treatment of environmental data . . . deserve[s] attention at the national level."[63] Addressing the agencies on behalf of the committee, Jamison stated:

> The recent IGY accumulations of data and the enormous data gathering capability of satellites, constant level balloons, and fixed and moving observing locations "of opportunity" like Naval and merchant vessels and commercial and military aircraft highlight [the fact] . . . that now is the time to insure that present and future data do not overwhelm us, but rather are acquired, screened, processed, and distributed through structure(s) designed for maximum economy and usefulness from a national viewpoint.[64]

The "problem of the IGY data," Jamison argued, had become a national problem: data needed a "national policy" in order to be mobilized "rapidly for national use." As Jamison aptly put it,

> The IGY . . . produced literally tons of synoptic data on all facets of our environment. These data, readily available to the scientific community, represent a great national asset. Stored in a relatively irretrievable form, they represent a collection of waste paper.[65]

This sentiment was shared on both sides of the Cold War. In 1958, the Soviet national committee pushed for more efficient processing of IGY data. Addressing the party officials, the chair of the Soviet IGY national committee wrote, "The first country that will process and analyze these important data will profit the most from the . . . IGY."[66] The fact that the Soviet WDC was centralized, and not dispersed in many subcenters like its American counterpart, gave the Soviet Union an advantage, he argued. It was now necessary to use that advantage since "the time gain from processing and analyzing the IGY data would allow the Soviet Union to get the upper hand" in harvesting the IGY data before the United States.[67]

In both the United States and the Soviet Union, the question of how the IGY data could be put to use and turned into an "asset" was on the minds of geophysicists and science policy makers. The technological solutions each side came up with, however, differed.

COMPUTER VERSUS MICROFILM: THE TECHNOLOGIES
OF GEOPHYSICAL DATA ARCHIVES

By the end of the 1960s, its holdings still growing, the WDCs continued to accumulate data in an analog format, just as computers started to take center stage across the geo-

---

[63] "The Treatment of Environmental Data—A National Problem," Standing Committee of the Federal Council for Science and Technology, 13 October 1960, WDC-A: General Correspondence, April–December 1960, IGY Papers.
[64] "Environmental Data Acquisition, Handling, and Dissemination," Standing Committee of the Federal Council for Science and Technology, 13 October 1960, WDC-A: General Correspondence, April–December 1960, IGY Papers.
[65] "The Treatment of Environmental Data—A National Problem," appendix, 13 October 1960, WDC-A: General Correspondence, April–December 1960, IGY Papers.
[66] I. P. Bardin to E. A. Furtseva, 20 March 1958, RGANI, fond 5, opis 35, delo 74.
[67] Ibid.

physical disciplines.[68] To be sure, American IGY planners had considered the issues of mechanization, automatization, and even computerization of IGY data processing since the inception of the program.[69] To suppliers of data-processing systems in the United States, the WDC-A had been a desired customer ever since the program was widely publicized in the mid-1950s. For example, a California-based company, Data Services, repeatedly approached Odishaw, emphasizing that the company would be "most anxious" to offer electronic equipment, expertise, and experience "with data reduction services for the government, electronics industry and the oil industry," as well as a "supply of part-time personnel consisting of engineers, students, and housewives," to help with IGY data handling.[70] Odishaw politely declined this and other similar offers, explaining that data reduction and analysis was primarily the responsibility of the institutions conducting the IGY programs rather than of the data center.[71]

The organization of the WDC subcenter for Aurora and Airglow (visual) at Cornell University illustrates how American data centers dealt with data processing through the division of labor between the data centers and data users—academic scientists at universities. In 1956, when the plans for data centers were announced, Carl W. Gartlein, a physics professor at Cornell who had been working in aurora research since the 1930s, volunteered to take on the job of organizing the WDC-A for auroral observational data.[72] Gartlein developed an elaborate punch card system for processing the data on visoplots (a special form he developed for the volunteer aurora reporters) using computers. The initial processing was done in Gartlein's lab in the Physics Department. A researcher—typically a young PhD or a graduate student—punched the cards marking the features of reported auroras such as type, position, and time. The cards were then delivered to the basement of the same building, Rockefeller Hall, the home of the Physics Department, where two rooms were designated as the WDC-A for Aurora and Airglow. Gartlein did not need more space for the data center: his WDC staff consisted of only a couple of technicians who copied the punch cards onto IBM cards and sent them to other Cornell departments, where they were processed on an IBM computer.[73] The processed data—synoptic maps of auroras and other statistical compilations—were then exchanged through the WDC system.

Visiting WDC-A subcenters in 1957, Edward Hulburt, a physicist at the Naval Research Laboratory who chaired the U.S. national committee for the IGY, found Cornell's arrangements quite satisfactory. Data processing, he wrote in his report, was in the hands of a scientist who was "very enthusiastic about the opportunity" to have firsthand access to IGY data, and who had "access to a complete battery of various types of IBM machines" at Cornell.[74] From Gartlein's perspective, as he emphasized

---

[68] Patrick McCray shows that in the 1960s, even the most ingrained and traditional data practices in the optical astronomy community were shifting from analog to digital, especially in the United States; see McCray, "The Biggest Data of All: Making and Sharing a Digital Universe," in this volume.

[69] See, e.g., Joshua Stern, Reynold Greenstone, and J. Howard Wright, "Data Processing Devices and Systems," Report, September 1955, WDC-A: Data Handling, IGY Papers.

[70] James T. O'Dea to Hugh Odishaw, 28 February 1956, Data Center Coordination Office: Misc Correspondence, 1955–9, IGY Papers.

[71] Hugh Odishaw to James T. O'Dea, 7 March 1956, Data Center Coordination Office: Misc Correspondence, 1955–9, IGY Papers.

[72] Frederick P. Siegal, "Gartlein Heads IGY Program," *Cornell Daily Sun*, 20 January 1959; Arnold H. Herman, "Gartlein Views IGY, Cornell Contribution," *Cornell Daily Sun*, 16 January 1959.

[73] E. O. Hulburt, "Visit to Dr. C.W. Gartlein," 5 February 1957, Data Center A, 1956–7, IGY Papers.

[74] Ibid.

in his initial proposal for the data center, the IGY data handling required not only "the technique" and enthusiasm, but also many hands to do the keypunching.[75] To this end, Gartlein recruited doctoral students who were processing IGY data to gather material for their dissertations. As Gartlein estimated, "the mass of material obtained during the IGY will supply thesis material for a large number of doctoral investigations."[76]

Thus, the massive influx of IGY auroral observational data was met, on the receiving end, by the manpower of physics PhD students, whose numbers skyrocketed during the early Cold War.[77] The symbiosis between data and students is a strategy widely used in the sciences.[78] In the case of IGY data, however, the ability to turn the daunting task of massive data processing into PhD dissertation projects, as well as views about the best use of IGY data, varied across disciplines. In meteorology, data reduction to numerical values on punch cards and further processing on IBM computers was widely used in the World Meteorological Organization, which served as a WDC for the IGY meteorological programs. In ionospheric physics, the National Bureau of Standards served as the WDC-A subcenter for ionospheric IGY programs. The ionograms received from the stations were reduced to numerical values on punch cards and further processed on IBM computers of the Central Radio Propagation Lab in Boulder, Colorado. Unlike the auroral data center, however, in ionosphere programs, the exchanged data included not only compilations of processed or "semiprocessed" data but also copies of unprocessed data such as 35-mm film ionograms.[79]

In contrast to these disciplines, glaciologists felt that routine data processing and reduction was meaningless in their field. While in meteorology the routine reduction of meteorological data had become a way to establish explanations for global atmospheric phenomena that would be applicable anywhere in the world, glaciology was concerned with the explanations of the local particularities of ice and glacier formations. As glaciologists participating in the IGY argued, data reduction did not make much sense for them because "the nature of glaciological data calls for the qualitative interpretation of the data by the investigators themselves."[80] For the most part, glaciologists exchanged preprints and reprints, arguing that in their field "raw data is of little value away from its collector."[81]

Either way, IGY data processing was the responsibility of data producers and data users—the geophysicists with their own research interests in these data. Their experiences with IGY data led some of them to become "data scientists" dealing largely with problems of data reduction and processing on IBM computers. These experiences and trajectories were distinct from data practices at the data centers, however. WDC's data managers were dealing with more traditional problems of archives—those related to the storage, organization, and retrieval of information in their holdings. In the 1960s,

[75] "Proposal to Operate Primary World Data Center for Aurora and to Conduct Analyses of Auroral and Related Data," n.d., Data Center A: Aurora: U of Alaska, 1956–60, IGY Papers.

[76] Ibid.

[77] David Kaiser, "Cold War Requisitions, Scientific Manpower, and the Production of American Physicists after World War II," *Hist. Stud. Phys. Biol. Sci.* 33 (2002): 131–59.

[78] See Robert E. Kohler, *Lords of the Fly*: Drosophila *Genetics and the Experimental Life* (Chicago, 1994).

[79] R. W. Porter to Paul Smith, 16 November 1960, WDC-A: General Correspondence: April–December 1960, IGY Papers.

[80] Ibid.

[81] Wallace W. Atwood to D. C. Hartin, 19 October 1956, Correspondence with National Committees (Except Communist): 1956–9, IGY Papers.

the storage medium of IGY data—microfilm—turned these traditional archival problems into a whole new and exciting area of "data science" that sometimes complemented computer-based data science and sometimes competed with it.

Microfilming, or "microphotography," had existed since 1839, when it was first done in England. In the 1950s and 1960s, microfilm captured the public imagination. Auerbach and Gitelman have argued that microfilm was "transformed by Cold War optics."[82] Like computers, microfilm technologies became vehicles for discourses and imageries that bore the distinct imprint of the Cold War. The cultural imagery of computers, Paul Edwards has argued, was embroiled in Cold War semiotic spaces dominated by the "closed-world discourse" of global surveillance and control, the deterministic logic and optimistic vision of the centralized military command.[83] As Auerbach and Gitelman have pointed out, "the case of microfilm suggests an additional and vernacular Cold War discourse that was optical in nature," enmeshed in the imageries of microfilm as a thrilling spying technology, and as a means of preserving the "total archive" of a nation in the case of nuclear holocaust.[84]

Vannevar Bush's Memex—a hypothetical machine that organized all knowledge on microfilm—was the technodream of the time, which reflected a sense of optimism and enthusiasm around microfilm.[85] Speaking of a real device designed by Bush, the Rapid Selector, the biologist Julian Huxley wrote:

> The most Wellsian thing, which gave me more ideas of the mechanical devices possible in the future . . . was a device, which was being perfected at the Massachusetts Institute of Technology, for increasing the efficiency with which a microfilm selected items of a bibliography. The device is very complicated, but the results [are] very simple, namely that with the system set up, if you want to have all the information in the world on, say, the incidence of tuberculosis among polygamous females in Polynesia, you could run all the microfilms in the machine and have all the information in one minute.[86]

Occupying similar semiotic spaces during the 1950s and 1960s, microfilm and computers were seen as information technologies complementing each other. Some of the early computer journals, such as *Datamation*, covered both computing and microfilming technologies. The microfilm industry and its users created specialized journals,

---

[82] Auerbach and Gitelman, "Microfilm" (cit. n. 51).
[83] Paul Edwards, *The Closed World: Computers and the Politics of Discourse in Cold War America* (Cambridge, Mass., 1996).
[84] Auerbach and Gitelman, "Microfilm" (cit. n. 51).
[85] On the technological history of Memex, see Colin Burks, *Information and Secrecy: Vannevar Bush, Ultra, and the Other Memex* (Metuchen, N.J., 1994). On Memex-inspired imaginaries and microfilm-based aspirations of American anthropologists in the 1950s, see Rebecca Lemov, "Towards a Data Base of Dreams: Assembling an Archive of Elusive Materials, c. 1947–61," *Hist. Workshop J.* 67 (2009): 44–68. On the appeal of analog computers and other alternatives to electronic digital computers during the Cold War, see Charles Care, *Technology for Modelling: Electrical Analogies, Engineering Practice, and the Development of Analogue Computing* (London, 2010); David A. Mindell, *Between Human and Machine: Feedback, Control, and Computing before Cybernetics* (Baltimore, 2002); Manfred Sapper, *Kooperation trotz Konfrontation: Wissenschaft und Technik im Kalten Krieg* (Berlin, 2009); James S. Small, *The Analogue Alternative: The Electronic Analogue Computer in Britain and the USA, 1930–1975* (London, 2001). For an example of an "information revolution" that was driven by the resistance to contemporary mechanization of information processing, see Christine von Oertzen, "Machineries of Data Power: Manual versus Mechanical Census Compilation in Nineteenth-Century Europe," in this volume.
[86] "Verbatim report of talk by Dr. Huxley at the Sorbonne University, Paris, 26 Feb 1948," Julian S. Huxley Papers, box 66, folder 7, Rice University, Houston.

such as *Microdoc*, *Data Systems News*, *National Micro-News*, *Microfiche Found Newsletter*, and *Reprographics* (published in three languages: English, German, and Japanese). These periodicals, and a whole host of other journals, disseminated up-to-date information on microfilm, microfiche and microcard technologies, and microfilm-based automatic and semiautomatic search systems, as well as advertisements by companies manufacturing these devices.[87]

In the 1960s, microfilm was a sophisticated technology that allowed not only storage but also search, organization, and analysis of analog data without turning them into digital form. The MIRACODE system produced by the Eastman Kodak Company in the early 1960s is an example of the application of a microfilm-based device for processing large volumes of information stored on microfilm, developed for small companies. MIRACODE, an acronym for "Microfilm Information Retrieval Access CODE," was a comparatively simple device that consisted of two main parts—a microfilm camera and a microfilm reader. The input material, for example, a text, was first marked over with three-digit code numbers corresponding to different topics in the texts. At the microfilming stage, the MIRACODE camera transformed the code numbers into machine-readable binary codes, recorded on film next to the appropriate page image. The film then could be searched for logical combinations of code numbers at the retrieval station, with retrieved images projected on a viewing screen and printed as needed.[88]

For the data aspirations of many users in the 1950s and 1960s, microfilm-based systems such as MIRACODE, which combined information-retrieval functions with information-logical systems, offered better options for data management than a computer—a large, cumbersome calculating machine that often did not work well. At a conference of the Council of Social Science Data Archives, a University of California political scientist emphasized that microfilm-based information technologies had many "advantages over certain drawbacks in contemporary computer technology":

> [MIRACODE] allows direct man-machine interaction with browsing capabilities that are impossible with computer systems operating in a batch-processing mode. Furthermore, its relatively low purchase price . . . permits its consideration as a system for tasks, which could not justify the expense of renting and operating a suitable computer. Finally, it has some of the same powerful searching capabilities of a computer, employing Boolean logic on machine-readable optical codes.[89]

To some observers, geophysical data, which were already on microfilm, looked like a perfect area in which the promising technology could be used. In the mid-1950s, following the announcement of the plans for the IGY, the U.S. national committee for the IGY was showered by proposals from innumerable enthusiasts suggesting how best to organize the IGY data to make them more usable. One of the proposals came from a then unknown "documentation consultant," Eugene Garfield, who later gained renown as the founder of the Institute of Scientific Information (started as Garfield's private firm, Documentation Inc., in 1954) and the creator of the Science Citation

---

[87] *Mikrofil'mirovanie tekhnicheskoi dokumentatcii i mikrofil'mirovanie materialov. Annotirovannyi bibliograficheskii ukazatel' literatury za 1966–1968 gg.* (Moscow, 1969).

[88] Earl S. Daniel, "Solving Information Storage and Retrieval Problems with Miracode," *J. Chem. Document.* 6 (1966): 147–8.

[89] Kenneth Janda, "Political Research with MIRACODE: A 16 mm Microfilm Information Retrieval System," *Soc. Sci. Inform.* 6 (1967): 169–81.

Index (first published in 1964). In 1956, Garfield wrote to the U.S. national committee for the IGY offering "assistance in handling the scientific information and records which will accrue from the [IGY] programs."[90] Garfield enclosed his recent conference paper in which he called for nothing less than the realization of Vannevar Bush's Memex. Bush himself, Garfield argued, proposed Memex "perhaps as a joke, . . . something like H. G. Wells' World Brain in the form of an electromechanic desk."[91] As a result, "the merits of his proposal have been lost in the plethora of idealistic comment which it provoked." However, Garfield continued, "an actual realization of Memex" should be pursued, by tackling the "mechanical problem"—the mechanization of all information work and the "elimination of manpower"—and by "coordination of documentation on a national level." A practical realization of the Memex technodream, Garfield argued, would be the establishment of one national center to ensure the centralized management of scientific data, as had been done in the Soviet Union:

> In the Soviet Union, documentation retrieval, like other sciences, is receiving support of the State apparatus. . . . In a few years we shall also be hard pressed to duplicate the documentation facilities available to the Russian scientist. I do not look to such centralized, governmental control as an answer. We must, however, fully appreciate the dangers of being left behind. . . . The time has come for the creation of . . . a scientific body whose scope is the totality of information pertinent to science and research.[92]

In the Soviet Union, the possibilities of electro-mechanical microfilm-based technologies were, indeed, thoroughly explored. The Soviet Union had far fewer all-purpose electronic computers than the United States, and it struggled to catch up.[93] In both the United States and the USSR, computers were vital components of the weapons systems and were developed as products of military-sponsored research. In contrast to the United States, where computers were available for nonmilitary projects, in the Soviet Union access to computers was severely restricted. While computers were tightly linked to military research, microfilm technologies were developed as library technologies and were accessible to Soviet users of all kinds. Moreover, although the United States was the world leader in computer manufacturing, microfilm-based technologies were produced in many countries, including East Germany, which provided a steady supply of microfilm-based technologies within the Soviet bloc. Western literature on both microfilm and computers was closely followed, abstracted, and translated.[94] Specialized periodicals, such as the East German journal *Zeitschrift für Datenverarbeitung*, published translations of articles from both Western and Soviet periodicals on developments in data processing, computing, and microfilm technologies.

The Soviet Union started to manufacture its own microfilm-based devices in the 1950s, and by the mid-1960s had a range of Soviet brands available for purchase. Popular brochures explained the general principle of microfilm systems and their advan-

---

[90] R. C. Peavey to Eugene Garfield, 4 August 1956, Correspondence and Reports: 1955–7, IGY Papers.
[91] Eugene Garfield and Robert Hayne, "Needed—A National Science Intelligence and Documentation Center," December 1955, Correspondence and Reports: 1955–7, IGY Papers.
[92] Ibid.
[93] On "catching up" as an imperative for developing Soviet computing, see Gerovitch, *From Newspeak to Cyberspeak* (cit. n. 19).
[94] On abstracting and Soviet information science efforts, see Michael D. Gordin, *Scientific Babel: How Science Was Done Before and After Global English* (Chicago, 2015), 248–51.

tages, presenting microfilm as a technology of the not-so-distant future, when all information would be organized in microfilm archives, "microbooks" would replace books, and pocket microfilm readers, such as the Soviet-made *Lootch* (Russian: Луч, ray, beam), no larger than a cigarette pack, would be as common as reading glasses.[95] Professional manuals intended for specialists, on the other hand, reviewed the devices produced in the Soviet Union and abroad, encouraging Soviet entrepreneurial engineers and inventors to creatively adopt them for various needs.[96]

The All-Union Institute of Scientific and Technical Information in Moscow (VINITI), which Eugene Garfield referred to as a model in his proposal for the IGY, and which inspired him to change the name of his own firm to the Institute of Scientific Information, was a clearing house of information on microfilm available for scientists and engineers across the Soviet Union. The envy of science planners in the United States and the UK, VINITI was regarded as "the largest scientific information centre in the world," in the words of the British delegation that visited its facilities in 1963.[97] Best known for its abstracting of scientific literature, VINITI also experimented with various automatic and semiautomatic information processing systems, both digital and analog, as well as manufacturing and testing microfilm-based devices in its facilities.[98] In the words of the president of Soviet Academy of Sciences, Aleksandr Nesmeyanov, who was much involved in setting up of VINITI, "the future path of scientific information is in the machine methods."[99] Accordingly, VINITI's conferences, focused on "information retrieval systems and automatic processing of scientific and technical information," provided a forum for discussing the newest developments in both computing and microfilming technologies.[100]

Geophysical data kept at the WDCs were one of the areas of creative applications and adaptations of microfilm-based information processing technology. Thus, at a conference organized by VINITI in 1967, a representative of WDC-B described their experiences with collecting, organizing, and processing geophysical data. The speaker discussed technical characteristics of a device constructed specifically for the needs of WDC at the construction bureau of the Institute of Earth Magnetism, Ionosphere, and Radiowave Propagation, which curated the WDC-B in Moscow. The device was based on a principle similar to Kodak's MIRACODE, combining microfilm with information-logical retrieval systems.[101] For the WDC, with its "millions of cadres of microfilmed data," the microfilm-based technology seemed to meet the data aspirations better than a computer. In addition to saving time and expenses over keypunch-

[95] Ia. I. Leikina and A. M. Kristalinskii, *Kopirovanie i operativnoe razmnozhenie proektno-tekhnicheskoi dokumentatsii: Posobie* (Moscow, 1968).
[96] For example, S. N. Neiman, *Elektrofotograficheskaia kopiroval'naia apparatura nepreryvnogo deistviia otechestvennogo proizvodstva* (Leningrad, 1966); *Elektrofotograficheskoe kopirovanie proektno-konstruktorskoi dokumentatsii* (Moscow, 1963).
[97] Gordin, *Scientific Babel* (cit. n. 94). On VINITI and its history, see A. I. Chernyi, *Vserossiiskii institut nauchnoi i tekhnicheskoi informatsii: 50 let sluzheniia nauke* (Moscow, 2005).
[98] *Scientific and Technical Information in the Soviet Union, Report of the D.S.I.R.-Asib Delegation to Moscow and Leningrad, 7–24 June, 1963* (London, 1964), 39.
[99] Ibid.
[100] See, e.g., *Trudy III vsesoiuznoi konferentcii po informatsionno-poiskovym sistemam i avtomatizirovannoi obrabotke nauchno-tekhnicheskoi informatcii*, vol. 4, Tekhnicheskie ustroistva informatsionnogo obsluzhivaniia i operativno-mnozhitel'naia tekhnika (Moscow, 1967).
[101] O. K. Skikevich, "O registratsii i reprodutsirovanii pervichnykh geofizicheskikh dannykh," in ibid., 285–99.

ing the original data or abstracting the information and then keypunching, microfilm allowed the retrieval and linking of entire two-dimensional data spaces (or "data-scapes"—the functions of two or more variables) revealing associative clusters and patterns of relations.[102] With microfilm, the entire two-dimensional "datascape" could be stored, processed, and analyzed using optical scanning devices that read code numbers on the film as they flashed past the optical scanner.

The Soviet solutions and experiences challenged American attitudes toward the organization of data. To American administrators and policy makers, the Soviet model of organization of science, with its centralization and top-down planning, combined with the Memex idea of organizing all knowledge on microfilm, just as Garfield suggested, in one national center, disturbingly contrasted with what appeared to be the wastefulness of the American laissez-faire data marketplace. In the words of Marshall Jamison in 1960, when he advised the FCST-appointed study group "to plan imaginatively and boldly" when exploring the problem of "how to handle the costly, numerous [environmental] data of the future,"

> a grand, electromechanical central warehouse containing all types of environmental data is probably not a solution [to the problem of geophysical data]. The team is free, however, to consider [the Soviet model] . . . [since] it is equally improbable that a laissez-faire, topsy-growing expansion of present activities will turn out to be the best solution.[103]

A "grand electromechanical central warehouse" of environmental data on microfilm was, like Memex itself, a vision never realized. Even the most sophisticated microfilm-based devices promised more than they delivered. The imagined futures of geophysical datascapes on microfilm turned out to be extraordinarily short-lived. For a short while, however, the "data race" with the Soviet Union fueled the visions of microfilm-based environmental archives that made data transfer from an analog medium to a digital one look redundant.

CONCLUSION

The case discussed in this essay suggests that the phrase "political economy of data exchange" is not simply a quasi-economic metaphor but can be taken quite literally, to denote an extremely productive mechanism of value generation in a scientific data marketplace enabled by the supportive culture of the Cold War. Bruno Latour and Steve Woolgar argued in their *Laboratory Life* that contemporary scientific practices epitomize a symbolic capitalist marketplace, in which the conversion of credits and rewards in a cyclical exchange becomes the driver of today's competitive scientific enterprise.[104] Later historians reinterpreted the circulation of scientific materials, technologies, tools, and data as systems of primarily noneconomic exchange. Robert Kohler, examining the customs of data exchange in the community of *Drosophila* geneticists, adopted the notion of moral economy developed by historian E. P. Thompson to describe how relations in the moral community functioned as an alternative to

---

[102] Ibid.
[103] "Environmental Data Handling Study," October 1960, WDC-A: General Correspondence, April–December 1960, IGY Papers.
[104] Bruno Latour and Steve Woolgar, *Laboratory Life: The Construction of Scientific Facts* (Beverly Hills, Calif., 1979).

the capitalist market economy. Kohler underscored the fact that that free exchange and sharing of data in the community of *Drosophila* geneticists was based on shared moralities, values, and norms.[105] The account of data exchange presented in this essay brings the political and moral economies in closer proximity, by showing that the processes of the IGY data exchange involved the circulation of data as a distinct form of currency, as sources of scientific or symbolic value, as well as of moral or ethical value.

The Cold War politicized data exchange, turning it into a mechanism of soft power that reproduced the geopolitical rivalries on a smaller stage of global data rivalry. In this regime of exchange, data became a currency, with the value defined by the Cold War political economy. As I have shown in this essay, the politicization of data had unanticipated consequences for the data centers. In the data regime that prioritized accumulation and exchange over the use of the data, microfilm-based devices became the epitome of information technology that were expected to revolutionize data practices in the not-so-distant future. In the world of the WDCs and especially in the Soviet Union, the tiny, sophisticated microfilm and the microfilm-based devices, for a short while, overshadowed the computer. While the role of computers in opening up new possibilities in large-scale information processing has been the subject of a number of recent historical studies,[106] the history of data practices in the WDCs sheds light on a less familiar story—the parallel rise, in the aftermath of World War II, of microfilm-based information technologies. In the case of the WDCs, with their holdings of geophysical data already on microfilm, data transfer from an analog medium to a digital one appeared redundant.

The strategic significance of geophysical data did not translate directly and deterministically into making "data products." While giving a major boost to ambitious data-centered science, the IGY's World Data Center system did not become a harbinger of new data practices. The data regime that the IGY shaped, and was shaped by, sheds light on the roads that were abandoned in the end. Exploring these abandoned roads, however, helps make our histories of data more inclusive and helps us see what was at stake at the crossroads.

---

[105] Kohler, *Lords of the Fly* (cit. n. 78), 137, 141. Historians of science have since extensively expanded this notion; for discussion and the relevant literature, see McCray, "The Biggest Data" (cit. n. 68).

[106] See, e.g., Hallam Stevens, "A Feeling for the Algorithm: Working Knowledge and Big Data in Biology," in this volume.

Big Data Is the Answer . . . But What Is the Question?

by Bruno J. Strasser* and Paul N. Edwards§

ABSTRACT

Rethinking histories of data requires not only better answers to existing questions, but also better questions. We suggest eight such questions here. What counts as data? How are objects related to data? What are digital data? What makes data measurable, and what does quantification do to data? What counts as an "information age"? Why do we keep data, and how do we decide which data to lose or forget? Who owns data, and who uses them? Finally, how does "Big Data" transform the geography of science? Each question is a provocation to reconsider the meanings and uses of "data" not only in the past but in the present as well.

Unquestionably the most vocal exponent of epistemic hierarchy, the American writer, performer, and all-around wild man Frank Zappa frequently declaimed on stage that "information is not knowledge, knowledge is not wisdom . . . and music is the best."[1] Zappa was just one among many to adopt the "DIKW" (data, information, knowledge, wisdom) hierarchy. Long before rock 'n' roll, T. S. Eliot may have been DIKW's first exponent: "Where is the wisdom we have lost in knowledge? Where is the knowledge we have lost in information?" he asked, in *The Rock* (1934).[2] But whereas early authors focused on the relationships between information, knowledge, and wisdom (and music), since the late 1980s "data" has taken center stage, especially after philosopher of science and systems researcher Russell L. Ackoff published his short article "From Data to Wisdom."[3]

The contributions in this volume offer a unique opportunity not only to rethink the meanings of "data," today and in the past, but also to consider why data has become

* Section of Biology, University of Geneva, CH-1211 Geneva 4, Switzerland, and Section of the History of Medicine, Yale University School of Medicine, New Haven, CT 06510; bruno.strasser @unige.ch.

§ Center for International Security and Cooperation, Stanford University, C-226 Encina Hall, Stanford, CA 94309, and School of Information, University of Michigan, 105 South State Street, Ann Arbor, MI 48109; pedwards@stanford.edu.

We would like to thank all the participants at the two workshops on "Historicizing Big Data" at the Max Planck Institute for the History of Science, especially the organizers, Elena Aronova, Christine von Oertzen, and David Sepkoski. We also thank Jérôme Baudry, Dana Mahr, and two anonymous reviewers for helpful comments and suggestions.

[1] Frank Zappa, "Packard Goose," *Joe's Garage: Act III*, 1978.
[2] Jennifer Rowley, "The Wisdom Hierarchy: Representations of the DIKW Hierarchy," *J. Inform. Sci.* 33 (2007): 163–80.
[3] Russell L. Ackoff, "From Data to Wisdom," *J. Appl. Syst. Analysis* 16 (1989): 3–9.

so central to contemporary discussions about the production of knowledge. A genuine rethinking implies not only better answers to existing questions, but also better questions (including those proposed in this volume's introduction). Here we suggest eight further ways to reframe the discussion about data, especially "Big Data," in its historical contexts.

1. WHAT COUNTS AS DATA?

What are data? This obviously fundamental question recurs in virtually all discussions that attempt to historicize Big Data (including many of those in this volume).[4] One answer—also recurrent—is that the category of data describes something basic, foundational, elemental, or *Ur-* (the convenient German prefix). Labeling something "data" usually serves the rhetorical purpose of qualifying some trace or sample as objective, precognitive, unanalyzed, a sure foundation for knowledge claims. The traction of the expression "raw data" comes precisely from this claim that data are unaltered, uncooked, unprocessed by human subjectivity, and thus beyond doubt.[5] Labeling something "data" carries epistemic weight. Dropping the phrase "I have data" in a conversation often seems sufficient to end debate.

More interesting than searching for an ontological conception of data, as if a datum were an atom of knowledge (now that even atoms are no longer atoms), is to ask how *what counts as data* has changed over time. In crystallography, for example, x-rays diffracted by a crystal produce an image containing dark spots, whose intensities are used to calculate "structure factors," which in turn are used to determine the coordinates of each atom composing the crystal. But where are the data? Crystallographers were first content to publish atomic coordinates as the "data" supporting a proposed structure, before they were asked to provide more foundational data, the "structure factors," and eventually yet more foundational data, the original diffraction images.[6] To take another example, since the 1980s climate scientists have referred not only to recorded instrument readings, but also to simulation model outputs as "data," full stop. Today the volume of data from simulation models vastly exceeds that of the entire historical instrument record. Furthermore, data from simulation models may be created to represent either the real past, including the distant past of paleoclimates, or an experimental past (an Earth with different continental positions, different atmospheric composition, etc.), or possible planetary futures.[7]

Thus "data" is not a natural kind, but a property attributed at a given moment in history to a set of signs, traces, indices ("inscriptions," if one insists on Latourian language). As philosopher Sabina Leonelli writes, data are "fungible objects defined by their portability and prospective usefulness as evidence."[8] To attach the label "data" to something is to place that thing specifically in the long chain of transformations that

[4] Rob Kitchin, *The Data Revolution* (London, 2014).

[5] On the critique of "raw data," see Lisa Gitelman, ed., *Raw Data Is an Oxymoron* (Cambridge, Mass., 2013). On the rhetorical power of "data," see Daniel Rosenberg's chapter in that volume, "Data before the Fact," 15–40.

[6] Bruno J. Strasser, "The 'Data Deluge': Turning Private Data into Public Archives," in *Science in the Archives: Pasts, Presents, Futures*, ed. Lorraine Daston (Chicago, 2017), 185–202.

[7] Paul N. Edwards, *A Vast Machine: Computer Models, Climate Data, and the Politics of Global Warming* (Cambridge, Mass., 2010).

[8] Sabina Leonelli, "What Counts as Scientific Data? A Relational Framework," *Phil. Sci.* 82 (2015): 810–21.

moves from nature to knowledge;[9] this act of categorization marks a particular moment in time when someone thought some inscription or object could serve to ground a knowledge claim. At any such moment, "data" is the closest thing to nature that is no longer "natural"—or, to put it differently, data is the first transformation of nature in the production chain that culminates in knowledge. Thus seeing "data" almost as an adjective, as a *relational* property (like being the youngest child in a family), makes apparent why what counts as data changes over time: as the frontier between nature and knowledge evolves, so do the data that inhabit this moving frontier.

2. HOW ARE OBJECTS RELATED TO DATA?

Less often discussed are questions about the materiality of data. A revealing opening move can be to ask, *What are data made of?* From texts and numbers carved in stone to digital bits stored as electromagnetic charges on silicon chips, all data without exception have a material aspect. The rapidly rising energy consumption of "cloud" servers around the world (currently equal to the output of more than fifty average-sized nuclear power plants; i.e., twelve percent of the world's nuclear electricity) serves as a useful reminder that data can exist only within a material infrastructure.[10]

Thinking about data as objects can focus attention on "data friction": the forms of resistance data offer to circulation and to transformation, and the means developed to smooth their path.[11] A key goal of the 1957–58 International Geophysical Year (IGY) was to share the billions of data records, such as meteorological observations and seismographs, collected around the world. In principle, the IGY's World Data Centers were supposed to provide IGY data on request—but how? By the time the IGY took place, many scientists already saw computers as the future of data processing, but computer technology was young, and few standards for computer-readable records had been established. Rather than decree such a standard themselves, IGY planners chose to distribute them on printed "microcards" or microfilm, instead of on punch cards or magnetic tapes, the two major computer media of the day, as Elena Aronova (this volume) mentions in her discussion of IGY data regimes.[12] This fateful choice forced anyone wishing not just to examine, but also to use IGY data to first transcribe the microcard or microfilm records onto computer-readable media. This created a substantial barrier to data reuse. Computer-readable data are, of course, hardly immune to data friction. Everyday examples abound, while large-scale friction in computerized data can be glimpsed in such cases as the periodic "migration" (to new tapes) of the National Oceanic and Atmospheric Administration's vast libraries of data tapes. Without the migration, data would be lost to the rapid technological obsolescence of tape readers and the physical decay of the underlying media.[13] In these examples, we see

---

[9] Bruno Latour, *Pandora's Hope: Essays on the Reality of Science Studies* (Boston, 1999), chap. 2.

[10] Jonathan G. Koomey estimated the worldwide server electricity consumption at 271.8 terawatt hours (TWh) in 2010. World nuclear power plant electricity production was 2,364 TWh in 2014. Koomey, *Growth in Data Center Electricity Use 2005 to 2010* (El Dorado Hills, Calif., 2011).

[11] The concept of data friction is developed at length in Edwards, *A Vast Machine* (cit. n. 7).

[12] Elena Aronova, "Geophysical Datascapes of the Cold War: Politics and Practices of the World Data Centers in the 1950s and 1960s," in this volume.

[13] M. Halem, F. Shaffer, N. Palm, E. Salmon, S. Raghavan, and L. Kempster, "Technology Assessment of High Capacity Data Storage Systems: Can We Avoid a Data Survivability Crisis?," in *Government Information Technology Issues 1999: A View to the Future* (Washington, D.C., 1999).

that data-as-objects—even electronic ones—have mass and momentum. Their material characteristics strongly shape the cost and physical possibilities of storage, retrieval, and use, as most of the papers in this volume demonstrate in one way or another. Even today, in the age of cloud storage, when scientists need to move the largest datasets to another physical location, they put them on disk drives and ship them off by mail. The data are then delivered to their destination the old-fashioned way—by a mail carrier on foot.

What happens if, instead of asking after data-as-objects, we shift to the relational view discussed above, asking, *How are objects related to data?* We say, without thinking, that we "collect" or "assemble" data, as if they were shells on the beach or a drawer full of random Lego pieces. Such locutions almost certainly descend from practices of collecting objects for scientific analysis. Historically, collections of objects in a single place created unique opportunities for classification, comparison, and analysis. Indeed, it would be difficult to overstate the centrality to science of "collecting" and "collections." They lie at the core of modern knowledge institutions, from the all-important library to museums to laboratories.[14]

Collected objects such as tissue samples, plant specimens, rocks, and molecules *become* data by being brought into a collection, that is, into relationships with other objects and with a knowledge institution that considers them to represent nature, to be a "second nature." Their material characteristics matter (so to speak) to their lives as data. Mirjam Brusius's chapter in this volume details the complexity of transforming fragments found at an archeological site into reconstructed artifacts, a process that involved extensive detours into mapping, drawing, and cataloging, thereby arousing tensions over the status of these derivative forms of data.[15] To take a related but different example, specimens have carefully defined locations in a natural history museum; the museum itself exists to solve the problem of locating, comparing, storing, and preserving them—not a trivial issue given that (for example) many biological species are known only from a single specimen.[16] In such museums, specimens are physically moved around in order to compare them visually or examine them with special instruments. Transporting specimens between physically distant museums may sometimes be necessary, but in many places and times it has also been perilous and expensive. As a result, practices such as scientific and medical illustration, taking plaster casts of specimens, scale model making, and taxidermy developed to obviate this problem: a simulacrum, model, or representation of a specimen could be moved without risking the destruction or loss of the original. An ever-increasing effort to transform objects-as-data into more portable data-as-objects, including physical simulacra, images, texts, numbers, and digital bits, has been fundamental to the history of science.[17] Crucially, these data-as-objects need not necessarily dwell in just one place; reproducible representations (including electronic ones) can be here *and* be there, allowing the simultaneous and coordinated production of knowledge in many places at once.[18]

---

[14] Bruno J. Strasser, "Collecting Nature: Practices, Styles, and Narratives," *Osiris* 27 (2012): 303–40.

[15] Mirjam Brusius, "The Field in the Museum: Puzzling Out Babylon in Berlin," in this volume.

[16] Geoffrey C. Bowker, "Biodiversity Datadiversity," *Soc. Stud. Sci.* 30 (2000): 643–83; Lorraine Daston, "Type Specimens and Scientific Memory," *Crit. Inq.* 31 (2004): 153–82.

[17] See, e.g., Martin Rudwick, "George Cuvier's Paper Museum of Fossil Bones," *Arch. Natur. Hist.* 27 (2000): 51–68.

[18] David Weinberger, *Too Big to Know: Rethinking Knowledge Now That the Facts Aren't the Facts, Experts Are Everywhere, and the Smartest Person in the Room Is the Room* (New York, 2011).

Data-as-objects can also, of course, be collected, arranged, compared, and analyzed. Historically, this has been the role of librarians and archivists, for whom the materiality, weight, and volume of information have always been foreground concerns.[19] The ease and low cost of copying and distributing digital information has far-reaching consequences for assembling and accessing collections—or, we might say, meta-collections—of data, and thus to participation in the production of knowledge. So the question of data's material infrastructures can be reformulated, at least in part, as *What do collections do to (and for) data?* How do they allow data to circulate, or limit their circulation? What kinds of processing do they require? How do they create relationships among individual data items? Keeping their Aristotelian duality in mind—data are both form and matter—can reveal both similarities and differences among collections and data practices across time. Where data (and in-*form*-ation) are conceptualized primarily as abstract forms (texts, numbers, sequences, etc.), historians should ask after their material basis, what we have called "data-as-objects." Where data are conceptualized primarily as material objects (specimens, bones, rocks, etc.), historians can query the structures and practices that transform these objects into data, what we have called "objects-as-data."

3. WHAT ARE DIGITAL DATA?

Today we refer to almost anything computers can process as "digital," but this is a relatively recent convention. Through the 1990s, the terms "electronic" or "computerized" were more common, and even in the early 2000s, "electronic" was a viable alternative to "digital" (e.g., "e-science"). Today's designation of computable electronic formats as "digital" proves problematic for historians of data, because it implies a temporal divide between pre- and postcomputer data. That divide is very real for some kinds of data but not for others: hence our question, *What are "digital" data?*

The root meaning of "digital" refers not to electronic but to numerical (or more loosely, symbolic) representations based on a finite set of discrete elements. In this sense, all scientific data originally recorded as numbers or text, whether in lab notebooks, printed books and articles, or other media, were already digital. And indeed, preelectronic digital data storage and processing technologies, especially punch cards, were robust and widely used. Many of the first computerized databases were simply electronic versions of data collections originally stored on cards or microfilm.[20] Long before World War II, libraries at weather centers around the world already held tens of millions of punch cards containing weather data. By 1960, the data library at the U.S. National Weather Records Center contained over 400 million cards, so many that administrators feared the building might collapse under their weight. Transcribed onto magnetic tapes, these punch card libraries became the first electronic databases for global climatology. Similarly, the National Library of Medicine created MEDLARDS in 1964 as the largest publicly accessible electronic database. MEDLARDS was essentially a digitized version of the Index Medicus, a printed bibliographic index to medical literature whose advent dated to 1879 (MEDLARDS eventually became today's enormous PubMed). Finally, the *Atlas of Protein Sequences and Structures,*

---

[19] Michael K. Buckland, "Information as Thing," *J. Amer. Soc. Inform. Sci.* 42 (1991): 351–60.
[20] Markus Krajewski, *Paper Machines: About Cards and Catalogs, 1548–1929* (Cambridge, Mass., 2011); Lisa Gitelman, *Paper Knowledge* (Durham, N.C., 2014).

launched in 1965 as the first electronic database of protein sequence data, was a computerized version of its author's collection of printed sequences. Thus, although databases seem synonymous with electronic, computerized collections, many of them were in fact born long before the computer age (as David Sepkoski reminds us in this volume for the case of paleontology).[21] In other words, many early electronic databases began as straightforward transcriptions of older collections. Any data that already existed in the form of numbers, or indeed of discrete symbols of any kind (such as texts), had to be changed into an electronic form for the computer to use it. In these cases, the transition from paper, microfilm, and so on, to magnetic tape, disk, and so on, was merely a change of substrate: they became electronic, but they were already born digital in the most direct sense of the term. Of course, inscriptions often have many properties other than the symbols they contain, and these sometimes matter. Paul Duguid recounts the story of a researcher he once encountered in an archive, not just reading old documents but also sniffing them. By detecting the smell of vinegar, used in the nineteenth century as a disinfectant against cholera, he could date and sometimes place an epidemic.[22]

The question, *What are digital data?* can also help us reflect on the more significant digital data revolution, namely, analog-to-digital conversion. This data transition was far more difficult and took much longer than the mere transcription of numbers and text, which were (as we have been saying) in an important sense already digital. Scientists, programmers, and engineers had to figure out how to represent continuous signals such as waveforms, images, video, and sound in discrete (digital) forms, as well as how to analyze and manipulate such analog data with a digital computer. This required a revolution in numerical methods and discrete mathematics, which began in the 1940s and was substantially complete by the late 1960s, though refinements continue to this day.[23] The transition from analog to digital data and computing brought innovations in analog-digital conversion. In the 1960s and 1970s, for example, the isolines of hand-drawn synoptic maps used in weather forecasting were digitized for computer processing, requiring a human operator to manually follow the map's grid and enter (via a keyboard) the value of the nearest isoline, using such devices as the Bendix Datagrid Digitizer (figs. 1, 2). A similar system was used to digitize early Landsat photographs.[24]

In the same period, digital sensors designed to produce numerical outputs directly began to replace analog instruments that required human interpretation, for example, reading off numerical values from the scales marked on mercury-column thermometers, or estimating the intensities of x-ray diffraction spots from a crystallographic image. Recorded on magnetic tapes, these newly digitized data could be more readily stored, reproduced, and transported, and of course—the ultimate payoff—much more

[21] David Sepkoski, "The Database before the Computer?," in this volume.

[22] John Seely Brown and Paul Duguid, *The Social Life of Information* (Boston, 2000).

[23] William Aspray, *John Von Neumann and the Origins of Modern Computing* (Cambridge, Mass., 1990); Paul N. Edwards, *The Closed World: Computers and the Politics of Discourse in Cold War America* (Cambridge, Mass., 1996), chaps. 2, 3; Michael R. Williams, *A History of Computing Technology* (Englewood Cliffs, N.J., 1985); Joseph A November, *Biomedical Computing: Digitizing Life in the United States* (Baltimore, 2012), chap. 2.

[24] R. H. Rogers, C. L. Wilson, L. E. Reed, N. J. Shah, R. Akeley, T. G. Mara, and V. Elliott Smith, "Environmental Monitoring from Spacecraft Data," Laboratory for Applications of Remote Sensing Symposia, Purdue University, Paper 76, 1975, http://docs.lib.purdue.edu/lars_symp/76 (accessed 20 April 2017).

Figure 1. A Bendix Datagrid Digitizer in use at the U.S. National Center for Atmospheric Research (NCAR) in the 1970s. The operator moved a crosshair-style cursor across analog maps placed on the table's gridded surface. Centering the cursor on a gridpoint, she then used a keyboard (see fig. 2) to enter the value of the isoline nearest to the gridpoint. Image courtesy of NCAR.

readily manipulated by computers. This "digital convergence" has, of course, continued into the Internet era, when virtually all media use digital formats. Data are increasingly "born digital" (and electronic), a fact well known to all photographers who mourn the loss of analog film. So while the growth of digital data started long before the advent of computers and the Internet, its most dramatic expansion began with the full-scale conversion of previously analog techniques and instruments to "digital" (numerical, discrete) ones starting in the 1940s.

Digital convergence seems to render all data equivalent, and this is certainly the rhetoric of our times. Yet numbers and symbols, no matter what their medium of storage and treatment, remain interestingly distinct from born-analog information. Here we offer just one example. About one megabyte of computer storage can contain the entire text of a 500-page book, where "text" means characters, numbers, and other discrete symbols only. Yet just one page-size (analog) photograph, scanned at reasonably high resolution (1,700 × 1,000 pixels) as a JPEG digital file, is also one megabyte (or more) in size—even though the JPEG format's "lossy" compression algorithms delete irrelevant and redundant information to conserve storage space. Thus, all the text contained in the U.S. Library of Congress's 32 million books and other printed items could be represented in about 20 terabytes (at this writing, four matchbox-sized disk drives). Yet when printed pages are treated as *analog* data—for example, scanned as grayscale images to capture illustrations, marginalia, and other potentially important features— the resulting files are 100–1,000 times larger. Digitizing the approximately 110 million nonprint (analog) items in the Library of Congress's physical collections, such as pho-

Figure 2. A Bendix Datagrid keyboard and cursor with crosshairs. Image courtesy of NCAR.

tographs, sound recordings, and films, to an archival standard of quality would require hundreds of petabytes, perhaps even an exabyte or two—tens to hundreds of thousands of times more storage than all the books in the Library of Congress.[25] Analog data proves remarkably resistant to its digital reduction.

The digital revolution has affected most of the arenas that previously produced and used analog data: photographs, sounds, drawings, and countless scientific recordings, such as electrocardiograms, bubble chamber traces, or electrophoresis gels. In some cases, such as photography, digital sensors have almost completely replaced their analog counterparts, but others are still digitized after the fact from analog media. Once digitized, these data enjoy a greatly increased potential for circulation, calculation, and comparison. At the same time, they have lost the trace of their physical and intimate contact with nature, upon which rested belief in their authenticity.[26] The potential for fraud, in science and elsewhere, is increased by the lack of a unique analog original that could vouch for the reality of the phenomenon being represented.[27] Some new

[25] One exabyte = 1,000 petabytes = 1,000,000 terabytes = 1,000,000,000 gigabytes. The calculation in this paragraph is highly approximate but probably correct to an order of magnitude. For its basis, see the following blog posts: Matt Raymond, "How Big Is the Library of Congress?," *Library of Congress Blog*, 11 February 2009, http://blogs.loc.gov/loc/2009/02/how-big-is-the-library-of-congress/ (accessed 20 April 2017); Michael Lesk, "How Much Information Is There in the World?," 1997, http://www.lesk.com/mlesk/ksg97/ksg.html (accessed 20 April 2017); and Nicholas Taylor, "Transferring 'Libraries of Congress' of Data," *The Signal* (blog), 11 July 2011, http://blogs.loc.gov/digitalpreservation/2011/07/transferring-libraries-of-congress-of-data/ (accessed 20 April 2017).

[26] Roland Barthes, *La chambre claire: Note sur la photographie* (Paris, 1980).

[27] Thus returning to the essential problem of trust in human testimony, inescapable before photography and the age of "mechanical reproduction"; see Lorraine Daston and Peter Galison, *Objectivity* (New York, 2007); and Steven Shapin, *A Social History of Truth: Civility and Science in Seventeenth-Century England* (Chicago, 1995). On fraud and trust in images, see Nick Hopwood, *Haeckel's Embryos: Images, Evolution, and Fraud* (Chicago, 2015).

digital data have also lost the remarkable "depth" that allowed the photographer in Michelangelo Antonioni's 1966 movie, *Blow-Up*, to enlarge his negative repeatedly and thereby to discover a dead body in the background of a romantic park scene. In the digital age of big but "thin" data, his enlargement might only have revealed meaningless gray pixels.

4. WHAT MAKES DATA MEASURABLE? WHAT DOES QUANTIFICATION DO TO DATA?

Our own foregoing discussion is an instance of a widespread practice: the quantification of data, treating all data as having a measurable and comparable size. Today these quantities are usually expressed in bytes, or units of digital storage capable of encoding one character, whether a numeral, letter, or some other symbol. A byte is made up of bits (short for binary digits; i.e., the famous zeros and ones digital computers work with). In an Ur-document of information theory, "A Mathematical Theory of Communication" (1948), Claude Shannon introduced bits as a universal measure of information content.[28] Like money, bits rapidly became a kind of currency used to compare computer storage and processor capability.

The byte superseded the bit as a standard measure when the American Standard Code for Information Interchange (ASCII) text encoding system gained widespread currency in the 1960s. Although 6- and 7-bit bytes were once used, today standard measures such as kilobytes (Kb) and megabytes (Mb) always reference 8-bit bytes. Each 8-bit byte can represent a maximum of 256 characters. This is enough to handle the alphanumeric systems used by many Western languages, but the 8-bit byte cannot accommodate the thousands of characters in logographic languages such as Chinese, or even the many diacritical marks used in such Western languages as Swedish and Czech. Well into the 1970s, this problem stymied the introduction of native-language computing in Asia and later created a technical bottleneck in the early spread of the Internet.[29] Further, as a universal unit for quantifying data, the byte responds poorly to human experience. As discussed above, a 500-page book and a single scanned photograph require the same number of bytes of computer memory, yet from a human point of view, the book usually contains far more information. The conundrums presented by this seeming paradox were evident even at the dawn of information theory. Warren Weaver's exposition in *Scientific American* (1949) noted that Shannon's clarity came at the expense of separating the relatively easy "technical problem" (how information travels from transmitter to receiver; how transmission errors can be prevented or corrected) from two counterparts, the "semantic problem" (how "the information" conveys meaning) and the "effectiveness problem" (how a message affects the recipient's behavior).[30]

The quantification of data almost invariably serves to support a narrative about an "explosion" of data, and it rarely extends much beyond an exclamation of wonder. Yet

[28] Claude Shannon, "A Mathematical Theory of Communication," *Bell Syst. Tech. J.* 27 (1948): 379–423, 623–656.
[29] Jeffrey Shapard, "Islands in the (Data) Stream: Language, Character Codes, and Electronic Isolation in Japan," in *Global Networks: Computers and International Communication*, ed. Linda Harasim (Cambridge, Mass., 1993). For computing in Asia, see James W. Cortada, *The Digital Flood: The Diffusion of Information Technology across the US, Europe, and Asia* (New York, 2012); and Basile Zimmermann, *Waves and Forms: Electronic Music Devices and Computer Encodings in China* (Boston, 2015), chap. 12.
[30] Warren Weaver, "The Mathematics of Communication," *Sci. Amer.* 181 (1949): 11–5.

to a historian or an anthropologist, the idea of a single unit of measure for everything that has ever counted as data—whether object, inscription, specimen, or sample; analog or electronic; and so on—seems fraught with puzzles, if not patently absurd.[31] The fact that many kinds of scientific data, but also so many aspects of our informational lives—from family pictures to favorite music, to epistolary relations—have come to be quantified, and quantified using the *same* metric, constitutes a historically significant turning point deserving of scholarly attention. So when thinking about data, instead of marveling at its size, one might ask such questions as, *How did data become a measurable quantity? What does quantification do to data? More broadly, how does data measurement fit into the history of the relationship between quantification, trust, and objectivity?*[32] Elena Aronova offers one answer in the context of the IGY, where national contributions to data collections were quantified in service of the Cold War competition among the scientific powers (including the invention of a "data gap" echoing the "missile gap").[33] To take another example, in the history of taxonomy, counting species has seemed like an obvious thing to do, but as Staffan Müller-Wille has convincingly shown, many ways of doing taxonomy simply described species without quantifying their numbers.[34] The very impulse to measure data—as well as the highly abstract, and misleading, currency of bits and bytes—can thus be questioned, and their history traced.

5. WHAT KIND OF INFORMATION AGE IS THIS, ANYWAY?

Discourses about the unprecedented quantity of data—Terabytes! Petabytes! Exabytes!—serve to justify the diffuse feeling (most likely shared by anyone with an email account) that we are suffering from an "information overload" resulting from a "data deluge," a unique feature of our "information society." A more useful perspective might stem from cultural historian Robert Darnton's remark that "every age was an age of information, each in its own way."[35] Claims of a period-specific information overload have a long history. In his remarkable *Avatars of the Word*, classicist James O'Donnell showed that contemporaries of every major transition to a new technology of writing—from papyrus scrolls to the codex, to the printing press, to the computer age—expressed exhaustion and despair when confronting the increasing quantity of written information.[36] In the Renaissance, botanists complained about the vast number of new species to be described.[37] Darnton and other scholars of the Enlightenment similarly point to the challenges of collecting, storing, and making sense of an increasing

---

[31] Bruno J. Strasser, "Data-Driven Sciences: From Wonder Cabinets to Electronic Databases," *Stud. Hist. Phil. Biol. Biomed. Sci.* 43 (2011): 85–7.

[32] Theodore M. Porter, *Trust in Numbers: The Pursuit of Objectivity in Science and Public Life* (Princeton, N.J., 1995)

[33] Aronova, "Geophysical Datascapes" (cit. n. 12).

[34] Staffan Müller-Wille, "Names and Numbers: 'Data' in Classical Natural History, 1758–1859," in this volume; see also Staffan Müller-Wille and Isabelle Charmantier, "Natural History and Information Overload: the Case of Linnaeus," *Stud. Hist. Phil. Biol. Biomed. Sci.* 43 (2011): 4–15.

[35] Robert Darnton, "An Early Information Society: News and the Media in Eighteenth-Century Paris," *Amer. Hist. Rev.* 105 (2000): 1–35, on 1.

[36] James J. O'Donnell, *Avatars of the Word: From Papyrus to Cyberspace* (Cambridge, Mass., 1998).

[37] Brian W. Ogilvie, *The Science of Describing: Natural History in Renaissance Europe* (Chicago, 2006); Daniel Rosenberg, "Early Modern Information Overload," *J. Hist. Ideas* 64 (2003): 1–9.

quantity of knowledge—a principal justification of the encyclopedia movements of the eighteenth and nineteenth centuries. A question for today's Big Data could thus be, *What kind of information age is this?*

If we focus solely on the twentieth century, complaints of an information overload or a data deluge abound. In 1934, for example, the Belgian documentalist Paul Otlet highlighted the rapidly increasing number of written documents: "Their enormous mass, accumulated in the past, grows daily, hourly, by new additions in disconcerting, sometimes frightening numbers. . . . Of them, as of rain falling from the sky, one can say that they could set off a flood and a deluge, or trickle away as beneficial irrigation" (fig. 3).[38] The abundance of information called for new ways of organizing knowledge, the topic of Otlet's illuminating essay, as discussed by Markus Krajewski in this volume.[39] Much like Otlet's heroic Mundaneum, which undertook to store and catalog all the world's knowledge as well as to provide it on demand to researchers, both Vannevar Bush's imaginary Memex (1945), a microfilm-based, proto-hypertext retrieval system, and computer science pioneer J. C. R. Licklider's vision for *Libraries of the Future* (1965) promised to tame the rising tide through finer-grained classification, division of knowledge into atomic units, and associative linking, search, and retrieval.[40] Social commentator Alvin Toffler's influential *Future Shock*, published in 1970, popularized the notion of information overload and explored its psychological consequences.[41] Since then, such claims have become too numerous to count.

Understandably, a number of commentators have attempted to explain today's perception of an information overload by pointing to our time's unique features, such as the ability to produce, store, and share digital data with devices embedded in networked infrastructures. By asking *How is Big Data produced?* they have traced the history of the technologies responsible for the data deluge, from high-energy particle detectors to massive relational databases, and produced stories that fit into standard narratives of technological progress.[42] This often technologically deterministic perspective helps with the "how," yet it does not explain *why* data were deemed sufficiently valuable to store.

6. WHY DO WE KEEP DATA?

Even if one should remain critical about narratives of an unprecedented Big Data era, one must concede that data volumes *are* growing rapidly, while means of producing digital data are proliferating (from high-throughput DNA sequencers to high-definition smartphone cameras). Yet one should not assume that this increase in the amount of data, by itself, leads to Big Data or an information overload. To historicize twenty-first-century Big Data and its specific kind of information overload, it may be more meaningful to examine the relationships among the amount of data considered signif-

---

[38] Paul Otlet, *Traité de Documentation: Le Livre sur le Livre* (Brussels, 1934), 3; translation by Paul N. Edwards.

[39] Markus Krajewski, "Tell Data from Meta: Tracing the Origins of Big Data, Bibliometrics, and the OPAC," in this volume.

[40] Vannevar Bush, "As We May Think," *Atlantic Monthly* 176 (1945): 101–8; J. C. R. Licklider, *Libraries of the Future* (Cambridge, Mass., 1965).

[41] Alvin Toffler, *Future Shock* (New York, 1970).

[42] For example, Viktor Mayer-Schönberger and Keneth Cukier, *Big Data: A Revolution That Will Transform How We Live, Work, and Think* (Boston, 2013).

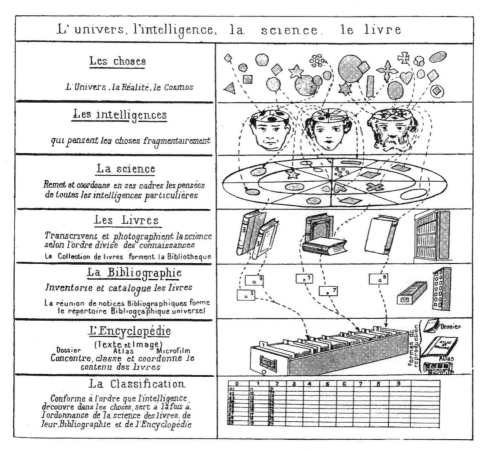

Figure 3. *Paul Otlet's view of the chain of transformations from the world to the classified card indexes, in his* Traité de Documentation *(cit. n. 38), 34.*

icant, the technologies available to handle it, and the perceived benefits of mastering that information. Ann Blair's eloquent *Too Much to Know: Managing Scholarly Information before the Modern Age* epitomizes this approach.[43] From this perspective, key questions would be not only *How big is Big Data?* or *How much data are we producing?* but also *Why do we keep data?* In other words, what has made every shred of scientific information seem so valuable that it should be stored electronically, backed up multiple times, and often subjected to a DMP ("data management plan") for making it available to everyone, thus putting all scientists in the business of data librarianship? Does this value stem from the belief that mining data can produce scientifically or economically useful insights ("data is the new oil")?[44] Or perhaps from an anxiety about the fragility of digital media, which might lead to the instantaneous erasure of a decade's worth of family photographs or laboratory data? Or simply from the pack-rat impulse to keep everything, and thus avoid the challenging (and annoying) task of

---

[43] Ann Blair, *Too Much to Know: Managing Scholarly Information before the Modern Age* (New Haven, Conn., 2010).

[44] Perry Rotella, "Is Data the New Oil?," *Forbes.com*, 2 April 2012, http://www.forbes.com/sites/perryrotella/2012/04/02/is-data-the-new-oil/ (accessed 22 July 2015).

choosing what to discard? These questions redirect attention from the production of data to the uses (and non-uses) of data.

In the life sciences, it is the persistent belief, already common in the eighteenth century, that comparative perspectives can bring unique insights into the relationship between biological structures and functions that has made data about every single species worthy of being stored (think of Georges Cuvier's comparative anatomy). This perceived opportunity to derive knowledge (or value) from data goes further in explaining why we store so much of it than the simple fact that we can produce data in large amounts. After all, the vast majority of data produced is not preserved. Utopias of "ubiquitous computing" or a "quantified self," relying on clunky Google glasses, health-monitoring wristbands, wearable computers, and other devices, will never succeed in capturing more than a very small fraction of the data constantly streaming in through our senses—and will likely one day seem as naive as earlier attempts to create the databases of dreams described by Rebecca Lemov (this volume) or the cloud atlases of the nineteenth and early twentieth centuries.[45] Thus, questions of selection and valuation are perhaps more important than the simple fact of production, and these operations depend on imagined future uses. Thus, one important reason for the emergence of Big Data is the belief that data we store now can (and will?) be (re)used for the production of knowledge or value later on.

In their ongoing quest to reduce the cost and improve the quality of science, funders such as the U.S. National Science Foundation and the National Institutes of Health have come to the not unreasonable view that broad data sharing might strengthen science and save money. With wide access to other researchers' data, some scientists might avoid having to repeat costly observations or experiments, while others might discover new, unanticipated ways to analyze it. Yet the vast majority of data preserved is still used only once (if ever), and so far—with important exceptions—even most published data are never reused by anyone other than the original producer.[46]

Revelations about fraud and error in scientific publishing have generated another, essentially moral rationale for sharing data: replicability and accountability. If data and models were habitually published alongside the scientific articles that make use of them, other researchers could check the claimed results directly, or so the argument goes. This rationale plays into, and perhaps stems from, a larger ideology of "transparency" in public affairs, often pictured as an obvious, quasi-obligatory way to buttress trust and the moral value of openness in democratic societies. Yet publishing data is rarely as simple or as costless as proponents of data publication assume, while evidence for the putative benefits of reuse and replication remains minimal. To put it differently, in science (and in our everyday life more generally), data hoarding seems to reflect more a growing distrust and insecurity about people, institutions, and memory than some transcendental necessity to preserve human knowledge.

---

[45] Rebecca Lemov, "Anthropology's Most Documented Man, Ca. 1947: A Prefiguration of Big Data from the Big Social Science Era," in this volume; Lorraine Daston, "On Scientific Observation," *Isis* 99 (2008): 97–110.

[46] Christine L. Borgman, *Big Data, Little Data, No Data: Scholarship in the Networked World* (Cambridge, Mass., 2015); Jillian C. Wallis, Elizabeth Rolando, and Christine L. Borgman, "If We Share Data, Will Anyone Use Them? Data Sharing and Reuse in the Long Tail of Science and Technology," *PLOS ONE* 8 (2013): e67332, https://doi.org/10.1371/journal.pone.0067332 (accessed 20 April 2017).

Similarly, the cultural history of twentieth-century "time capsules," those buried boxes and coffee cans in which people consigned artifacts of their times to a future more or less distant, reveals much about what people of one time imagined would be most informative to their counterparts in the future.[47] Today, it seems obvious that many of these boxes in fact reveal very little about their times, while many things we would like to know about the past were never preserved because they were deemed uninteresting at the time the boxes were buried in the ground. The trash heaps of past societies are often more valuable to today's historians, as *Sacred Trash,* the history of the Cairo Genizah storing centuries of miscellaneous notes from Jewish life, so beautifully illustrates.[48] In 2008, Apple Computer marketed its backup drive as the "Time Capsule," which works together with the company's "Time Machine" software to preserve all the data contained in a computer. To future historians, the content of a given Apple Time Capsule—should any survive, along with the software required to read them—might be more revealing than that of its older eponym, but it will still capture only a thin slice from the fabric of our daily lives. Even Big Data will always be too small to contain everything we would like to know, especially about ourselves.

The question of the rise of Big Data can thus be reframed as one about the *reasons* we increasingly stockpile data. As the cultural historian Krzysztof Pomian pointed out, the growth of artifact collections in the sixteenth and seventeenth centuries reflected the tastes, passions, and curiosities of their times.[49] Today is no different. We keep data because we imagine future epistemic uses, or imagine that future people will find such uses, and because of the moral imperatives of sharing, openness, and accountability—creating a data deluge along the way. These issues suggest paying more attention to the actual uses and users of data and to the inevitable question of data ownership.

7. WHO OWNS DATA? WHO USES DATA?

Although the notion of data is often presented as impersonal and anonymous, the issue of *who owns data* matters deeply, either because data are about people, as Joanna Radin and Dan Bouk discuss in this volume, or because data were produced by people.[50] The "data exhaust" left behind by hundreds of millions in the course of daily online activity has become a supremely valuable resource for those—mostly large information technology companies—who know how to use it to target advertising, or even to predict what an individual will need before that individual herself is aware of the need.[51] Certain data about individuals is continually controversial, especially when it is collected without their consent or knowledge; among the numerous exam-

---

[47] William Jarvis, *Time Capsules: A Cultural History* (Jefferson, N.C., 2002).
[48] Adina Hoffman and Peter Cole, *Sacred Trash: The Lost and Found World of the Cairo Geniza* (New York, 2011).
[49] Krzysztof Pomian, *Collectors and Curiosities: Paris and Venice, 1500–1800* (Cambridge, Mass., 1990).
[50] Joanna Radin, " 'Digital Natives': How Medical and Indigenous Histories Matter for Big Data"; Dan Bouk, "The History and Political Economy of Personal Data over the Last Two Centuries in Three Acts," both in this volume.
[51] In one famous case, a man complained to the Target department store chain about baby-related advertising material sent to his teenage daughter. Unbeknownst to him, she was in fact pregnant. Her online behavior had triggered the store's algorithms to send the advertising. Charles Duhigg, "How Companies Learn Your Secrets," *New York Times Magazine*, 16 February 2012, http://www.nytimes.com/2012/02/19/magazine/shopping-habits.html (accessed 20 April 2017).

ples one could cite is the National Security Agency's surveillance of Americans' telephone conversations as revealed by Edward Snowden.

But the issue of data ownership goes far beyond issues of privacy related to data
about human beings.[52] Data are often the product of work processes, such as laboratory experiments or field studies, carried out by human beings employed by institutions. As a result, historically, scientists and even institutions have claimed at least
metaphorical, and often literal, ownership of "their" data, whether DNA sequences
or astronomical data, as W. Patrick McCray shows in this volume, typically through
authorship or patents.[53] In some scientific fields, data sharing and rights to reuse have
existed all along, however. Early Modern genealogists producing data did not seem to
clash with historians who made use of it, as Markus Friedrich points out.[54] Since the
nineteenth century, many governments have collected and shared certain kinds of data
very widely: censuses such as those discussed by Christine von Oertzen, economic
and labor statistics, geographical survey maps, weather data.[55] Geneticists working
on model organisms were famously open about sharing their data.[56] Yet in other fields,
data have been hoarded or jealously guarded, whether for profit or simply to retain
control. The consequences can sometimes be dramatic, even tragic, as when data
about clinical trials gone awry are concealed or surreptitiously altered, or damaging
for the scientific enterprise, as when scientific claims cannot be properly evaluated because data were withheld.

The question of data ownership has become more acute as what we might call "data
supply chains" become more visible and more differentiated.[57] Most data were once
produced by their eventual users and thus did not need to be shared; in that receding
world, observation and analysis were part of the same process, with storage and reuse
by others an afterthought at best. Across the last century or so, data suppliers, data
managers, and data users have become far more differentiated and specialized. As
the collection, organization, and curation of data become increasingly professional-
ized, a divide has appeared between the scientists who produce data, those who manage it, and those who analyze it.[58] Because data managers do not themselves seek to
exploit the data they handle, they usually do not conflict with data producers. Instead,
it is the professionalization of data analysis that has led to moral tensions with data
producers. This professionalization results in increasingly distinct communities, such
as "bioinformatics" or "lexomics," as discussed by Judith Kaplan in this volume, each

[52] For an overview of these issues, see Kitchin, *The Data Revolution* (cit. n. 4), chap. 10.

[53] W. Patrick McCray, "The Biggest Data of All: Making and Sharing a Digital Universe," in this
volume; Bruno J. Strasser, "The Experimenter's Museum: GenBank, Natural History, and the Moral
Economies of Biomedicine," *Isis* 102 (2011): 60–96.

[54] Markus Friedrich, "Genealogy as Archive-Driven Research Enterprise in Early Modern Europe,"
in this volume.

[55] Christine von Oertzen, "Machineries of Data Power: Manual versus Mechanical Census Compilation in Nineteenth-Century Europe," in this volume.

[56] Robert E. Kohler, *Lords of the Fly:* Drosophila *Genetics and the Experimental Life* (Chicago,
1994).

[57] Richard B. Rood and Paul N. Edwards, "Climate Informatics: Human Experts and the End-to-
End System," *Earthzine*, 22 May 2014.

[58] Information schools and other academic units have recently begun to offer degrees in "data science," a broad, unsettled category that includes data management as well as analysis, while job advertisements for "data scientists" are proliferating rapidly in industry. These programs and positions
are intriguingly agnostic with respect to the actual content of data, reflecting the commodification of
data mentioned earlier.

with its own rules and norms, including about authorship and credit.[59] Bioinformaticians routinely analyze other researchers' data, publish articles about the results of such analysis, and get tenure for their work (although they do not win Nobel prizes—yet), creating tensions or even "long strings of clashes"[60] with those who produce the data in the first place.

Although the social rewards attached to scientific activities have always been highly contingent, expert observations of nature have long held a central place in the very idea of scientific discovery.[61] As nature is being replaced by a "second nature" made of data, and as borrowed data become the primary object of investigation for a growing number of researchers, one wonders whether data analysis will increasingly be valued as *the* key intellectual activity in science—perhaps even eclipsing experimentation and theory.[62] Beyond the naive empiricism of "data-driven science" lies an entire world where the practices of correlation, comparison, and classification of data, as discussed by Hallam Stevens in this volume, are deeply embedded with experimentation and theory, and all play their role in the production of scientific knowledge.[63]

In such a world, published data take on characteristics of commodities, or commons, and the questions *Who owns data?* and *Who uses data?* coalesce, since data increasingly belong to those who use them. Today, there is an increasing expectation that scientific data should be free to use and transform, like software, where "free" should be conceptualized "as in 'free speech,' not as in 'free beer,'" as computer scientist Richard Stallman famously put it.[64] Thus data may now be used by people who have no connection to, or any practical knowledge of, the original conditions of data production, resulting in new epistemic risks.

8. HOW DOES BIG DATA TRANSFORM THE GEOGRAPHY OF SCIENCE?

The increased circulation of knowledge and the ability of computerized data to be in many places at the same time—to be virtually everywhere at the same time—should not obscure the fact that data still have a geography. The point is not simply that databases are located somewhere (the staff in one place, the servers often in another), or that users are located almost exclusively in places with Internet access (which now includes almost the entire world), but that the geography of science reflects the specific epistemic practices of Big Data, past and present,[65] requiring specific kinds of local

---

[59] Judith Kaplan, "From Lexicostatistics to Lexomics: Basic Vocabulary and the Study of Language Prehistory," in this volume; Strasser, "Collecting Nature" (cit. n. 14).

[60] Eliot Marshall, "Data Sharing—DNA Sequencer Protests Being Scooped with His Own Data," *Science* 295 (2002): 1206.

[61] Lorraine Daston and Elizabeth Lunbeck, *Histories of Scientific Observation* (Chicago, 2011).

[62] Tony Hey, Stewart Tansley, and Kristin Tolle, eds., *The Fourth Paradigm: Data-Intensive Scientific Discovery* (Redmond, Wash., 2009); Chris Anderson, "The End of Theory," *Wired* 16 (2008), https://www.wired.com/2008/06/pb-theory/ (accessed 20 April 2017). On "second nature," see Strasser, "Collecting Nature" (cit. n. 14).

[63] Hallam Stevens, "A Feeling for the Algorithm: Working Knowledge and Big Data in Biology," in this volume; Kitchin, *The Data Revolution* (cit. n. 4), chap. 8; Sabina Leonelli, "Integrating Data to Acquire New Knowledge: Three Modes of Integration in Plant Science," *Stud. Hist. Phil. Biol. Biomed. Sci.* 44 (2013): 503–14.

[64] Richard Stallman, "Free Software Philosophy," https://www.gnu.org/philosophy/free-sw.html (accessed 13 August 2015)

[65] David N. Livingstone, *Putting Science in Its Place: Geographies of Scientific Knowledge* (Chicago, 2003).

infrastructures. The American Museum of Natural History, then the largest natural history museum in the United States, opened in 1877 on New York City's Upper West Side, while the Rockefeller Institute for Medical Research, a temple of the new experimentalism made iconic by Sinclair Lewis's 1925 novel *Arrowsmith*, was founded in 1901 just a few blocks away, on the Upper East Side. The massive collections of specimens and the specialized laboratory instrumentation—that era's Big Data infrastructures—required both the physical spaces of large buildings and the social spaces of a modern urban environment, making natural history as much a field science as one performed in an urban museum.

The geographical logic of centralized depositories held so long as both data-as-objects and objects-as-data remained physically large, heavy, and difficult or impossible to reproduce. While these characteristics still hold for many collections, such as dinosaur fossils or the ice cores drilled from glaciers around the world, the rise of (electronic, digital) Big Data has brought the value of centralized research infrastructures into question. Yet they are unlikely to disappear anytime soon. Today's most powerful computers still require large, expensive, specialized facilities, and the biggest data of all—such as climate simulations, particle accelerator experiments, or satellite data streams—remain too cumbersome and expensive to move easily. Further, the social advantages of large research institutions, where face-to-face collaboration is easier and serendipitous interaction more likely, will probably never be entirely effaced by distant interactions or anonymous crowdsourcing.[66]

Still, data analysis is now also carried out in a distributed fashion, through crowdsourcing. Individual researchers, located outside the main research institutions, can access and use Big Data. As a result, the social landscape of science is changing. Lay citizens are beginning to contribute to the production of scientific knowledge by analyzing scientific data—often from home. In the citizen-science project Galaxy Zoo, citizens study telescope photographs to classify galaxies; in Eyewire, they map data about neuron networks; in Old Weather, they decipher weather data contained in maritime logbooks. These projects have already resulted in numerous publications, often in high-profile journals, and following every standard of a scientific publication—except that they include collective and distributed authors, such as "the Eyewirer" in a recent publication in *Nature*.[67] Big Data, because it is often too big to be fully exploited by those who have produced it, is increasingly made publicly available, opening the possibility of a new kind of citizen science in which lay people act not merely as observers or sensors, but also as analysts. Citizens have long contributed to the making of Big Data, as observers of ephemeral comets and migrating birds, and have even engaged in their own research projects, as Etienne Benson shows in this volume.[68] Today, citizens continue to generate large quantities of observational data, about their natural environment and their own bodies, contributing to the "data deluge" and to overcoming it at the same time through citizen projects controlled by scientific institutions. More important, perhaps, as Big Data escape the control of academic research

[66] Gary M. Olson and Judith S. Olson, "Distance Matters," *Human-Computer Interaction* 15 (2000): 139–78.

[67] Jinseop S. Kim et al., "Space-Time Wiring Specificity Supports Direction Selectivity in the Retina," *Nature* 509 (2014): 331–36, on 331.

[68] Etienne S. Benson, "A Centrifuge of Calculation: Managing Data and Enthusiasm in Early Twentieth-Century Bird Banding," in this volume.

institutions, corporations, and the state, they are produced, appropriated, and mobilized by other actors, such as Public Lab (best known for its participatory mapping of the Gulf Coast following the Deepwater Horizon oil spill) to weigh in on environmental and other issues. As a result, the geography of knowledge—and power—may begin to shift.

CONCLUSION

What makes the Big Data era seem revolutionary is not only a series of incremental technological changes, but also an erasure of Big Data's past. The contributions to this volume show that concerns regarding the collection, storage, and uses of vast amounts of data are not new, and that computers and the Internet are not necessary ingredients. In doing so, they help us distinguish what *is* new with today's Big Data. More important, these studies show that the current concerns with Big Data can be better understood by looking at past situations where, in widely different contexts, people were confronted with large amounts of data and devised solutions to deal with it. The rich historical literature on natural history collections, for example, offers insights into the moral and political economies of today's Big Data and helps us understand current tensions around property, privacy, and credit.[69] Conversely, current issues—the professionalization of "data science," the "placelessness" of data, and the measurement of data—help us revisit earlier historical studies with new eyes. Finally, questioning the current fascination with Big Data can contribute to understanding more broadly our present condition, with its anxieties about lost memories and its hopes for predictable futures. Although the power of data may never match that of music (or so Frank Zappa might proclaim), it is having tremendous consequences for how we produce knowledge and how we live our lives.

[69] Strasser, "The Experimenter's Museum" (cit. n. 53).

Notes on Contributors

Elena Aronova is Assistant Professor in the Department of History at the University of California, Santa Barbara. She has published articles on the history and politics of environmental data collection, history of evolutionary biology, and historiography of science during the Cold War. Most recently, she has coedited a collection of essays, *Science Studies during the Cold War and Beyond: Paradigms Defected* (Basingstoke, 2016). She is currently writing a book that compares the studies of science in the Soviet Union and the United States at the onset of the Cold War.

Etienne S. Benson is Assistant Professor in the Department of History and Sociology of Science at the University of Pennsylvania. He is the author of *Wired Wilderness: Technologies of Tracking and the Making of Modern Wildlife* (Baltimore, 2010) and of articles on the use of information technology in the earth and environmental sciences, the history of wildlife management and biodiversity conservation, and human-animal relations in urban and industrial spaces. He is currently writing a history of environmental techniques.

Dan Bouk is Associate Professor in the Department of History at Colgate University. He is the author of *How Our Days Became Numbered: Risk and the Rise of the Statistical Individual* (Chicago, 2015). He was a postdoctoral fellow and visiting scholar at the Max Planck Institute for the History of Science.

Mirjam Brusius is a historian of science and art, specializing in the history of collecting and visual culture across modern Europe and the Middle East. Her first books revisited the early history of photography, in particular the scholarly archive and network of the photographic pioneer W. H. F. Talbot. Her current project examines the value of Middle Eastern archaeological objects during the transition period on their way to Europe, when the finds seemed to have "no status." She recently coedited *Museum Storage and Meaning: Tales from the Crypt* (New York, 2017).

Paul N. Edwards is William J. Perry Fellow in International Security at Stanford University and Professor of Information at the University of Michigan. He is the author of *The Closed World: Computers and the Politics of Discourse in Cold War America* (Cambridge, Mass., 1996) and *A Vast Machine: Computer Models, Climate Data, and the Politics of Global Warming* (Cambridge, Mass., 2010), and coeditor of *Changing the Atmosphere:*

Expert Knowledge and Environmental Governance (Cambridge, Mass., 2001), as well as numerous articles.

Markus Friedrich is currently Professor of Early Modern European History at the University of Hamburg. His research interests include the history of archives, the history of knowledge and erudition, and the early modern history of Christianity. He is the author of several books and articles, including *Die Geburt des Archivs: Eine Wissensgeschichte* (Munich, 2013) and *Die Jesuiten: Aufstieg, Niedergang, Neubeginn* (Munich, 2016). English translations of both books are forthcoming.

Judith Kaplan studied history of science at the University of Wisconsin–Madison. She has held postdoctoral fellowships at the Max Planck Institute for the History of Science in Berlin and at the Penn Humanities Forum. She is currently a teaching fellow in Integrated Studies with the Benjamin Franklin Scholars Program at the University of Pennsylvania. Her work on the history of linguistics has appeared in the *Journal of the History of the Behavioral Sciences*, *History of the Human Sciences*, and *Studies in the History and Philosophy of Science*, among other outlets. Her research focuses on the development of comparative-historical linguistics over roughly the last two hundred years.

Markus Krajewski is Professor of History and Theory of Media in the Department of Arts, Media, Philosophy at the University of Basel, Switzerland. He is the author of *Der Diener: Mediengeschichte einer Figur zwischen König und Klient* (Frankfurt am Main, 2010; an English translation, *The Server: Media History of a Figure between King and Client*, is forthcoming from Yale University Press), *Paper Machines: About Cards and Catalogs, 1548–1929* (Cambridge, Mass., 2011), *World Projects: Global Information before WWI* (Minneapolis, 2014), and *Bauformen des Gewissens: Über Fassaden deutscher Nachkriegsarchitektur* (Stuttgart, 2015). Current research projects include planned obsolescence, media and architecture, epistemology of the peripheral, and the history of exactitude in scholarly and scientific contexts. He is the developer of the bibliography software *synapsen—a hypertextual card index* (www.synapsen.ch).

Rebecca Lemov is Professor in the Department of the History of Science at Harvard University. She is the author of *Database of Dreams* (New Haven, Conn., 2015) and *World as Laboratory*

(New York, 2006), and coauthor of *How Reason Almost Lost Its Mind* (Chicago, 2011).

W. Patrick McCray is Professor in the Department of History at the University of California, Santa Barbara, where he teaches and writes about modern science. He is currently writing a new book about collaborations between artists and engineers in the 1960s.

Staffan Müller-Wille is Associate Professor in the History and Philosophy of the Life Sciences at the University of Exeter (England). His research covers the history of the life sciences from the early modern period to the early twentieth century, with a focus on the history of natural history, anthropology, and genetics. Among more recent publications is a book coauthored with Hans-Jörg Rheinberger on *A Cultural History of Heredity* (Chicago, 2012) and two coedited collections on *Human Heredity in the Twentieth Century* (Abingdon, 2013) and *Heredity Explored: Between Public Domain and Experimental Science, 1850–1930* (Cambridge, Mass., 2016).

Christine von Oertzen is a senior research scholar at the Max Planck Institute for the History of Science in Berlin. She recently published a monograph entitled *Science, Gender, Internationalism: Women's Academic Networks, 1917–1955* (New York, 2014). Her current research revolves around the material culture of collecting and compiling data, big and small: from manual and machine processing of census data to at-home observation of infants. She is also coeditor (with Elaine Leong and Carla Bittel) of a volume entitled *Working with Paper: Gendered Practices in the History of Knowledge*, which will be published in 2018 by the University of Pittsburgh Press.

Joanna Radin is Assistant Professor of the History of Medicine at Yale University, where she is also affiliated with the departments of History, Anthropology, and American Studies, and the programs in History of Science and Medicine and Ethnicity, Race and Migration. She teaches feminist and indigenous STS and the history of biomedicine, anthropology, and information. Her recent research has been especially concerned with the politics of reuse, and the intersections of life, time, and temperature. She is the author of *Life on Ice: A History of New Uses for Cold Blood* (Chicago, 2017) and coeditor, with Emma Kowal, of *Cryopolitics: Frozen Life in a Melting World* (Cambridge, Mass., 2017).

David Sepkoski is Senior Research Scholar at the Max Planck Institute for the History of Science in Berlin. His most recent book is *Rereading the Fossil Record: The Growth of Paleobiology as an Evolutionary Discipline* (Chicago, 2012). His research focuses on the history of biological and environmental science in the nineteenth and twentieth centuries, and he is currently writing a book about extinction and biodiversity.

Hallam Stevens is Associate Professor of History and Biology and the Head of History at the School of Humanities at Nanyang Technological University (Singapore). He is the author of *Life Out of Sequence: A Data-Driven History of Bioinformatics* (Chicago, 2013), *Biotechnology and Society: An Introduction* (Chicago, 2016), and the coeditor of *Postgenomics: Perspectives on Biology after the Genome* (Durham, N.C., 2015).

Bruno J. Strasser is a historian of contemporary science and technology, full professor at the University of Geneva, and adjunct professor at Yale University. He is finalizing his second book, *Collecting and Experimenting: Making Big Data Biology*, and working on a new project, "The Rise of the Citizen Sciences: Rethinking Science and Public Participation." He has published on the history of international scientific cooperation during the Cold War, the development of molecular biology, and, more generally, the relationships between science and society.

Index

SUGGESTIONS FOR CONTRIBUTORS TO OSIRIS

OSIRIS is devoted to thematic issues, conceived and compiled by guest editors who submit volume proposals for review by the OSIRIS Editorial Board in advance of the annual meeting of the History of Science Society in November. For information on proposal submission, please write to the Editors at pmccray@history.ucsb.edu and ss536@cornell.edu.

1. Manuscripts should be submitted electronically in Rich Text Format using Times New Roman font, 12 point, and double-spaced throughout, including quotations and notes. Notes should be in the form of footnotes, also in 12 point and double-spaced. The manuscript style should follow *The Chicago Manual of Style*, 16th ed.

2. Bibliographic information should be given in the footnotes (not parenthetically in the text), numbered using Arabic numerals. The footnote number should appear as superscript. "Pp." and "p." are not used for page references.

   a. References to books should include the author's full name; complete title of book in *italics*; place of publication; date of publication, including the original date when a reprint is being cited; and, if required, number of the particular page cited (if a direct quote is used, the word "on" should precede the page number). *Example*:

   [1] Mary Lindemann, *Medicine and Society in Early Modern Europe* (Cambridge, 1999), 119.

   b. References to articles in periodicals or edited volumes should include the author's name; title of article in quotes; title of periodical or volume in *italics*; volume number in Arabic numerals; year in parentheses; page numbers of article; and, if required, number of the particular page cited. Journal titles are abbreviated according to the journal abbreviations listed in *Isis Current Bibliography*. *Example*:

   [2] Lynn K. Nyhart, "Civic and Economic Zoology in Nineteenth-Century Germany: The 'Living Communities' of Karl Möbius," *Isis* 89 (1999): 605–30, on 611.

   c. All citations are given in full in the first reference. For succeeding citations, use an abbreviated version of the title with the author's last name. *Example*:

   [3] Nyhart, "Civic and Economic Zoology" (cit. n. 2), 612.

3. Special characters and mathematical and scientific symbols should be entered electronically.

4. A small number of illustrations, including graphs and tables, may be used in each volume. Hard copies should accompany electronic images. Images must meet the specifications of The University of Chicago Press "Artwork General Guidelines" available from the Editor.

5. Manuscripts are submitted to OSIRIS with the understanding that upon publication copyright will be transferred to the History of Science Society. That understanding precludes consideration of material that has been previously published or submitted or accepted for publication elsewhere, in whole or in part. OSIRIS is a journal of first publication.

OSIRIS (ISSN 0369-7827) is published once a year.

Single copies are $35.00.

Address subscriptions, single issue orders, claims for missing issues, and advertising inquiries to *Osiris*, The University of Chicago Press, Journals Division, PO Box 37005, Chicago, IL 60637.

Postmaster: Send address changes to *Osiris*, The University of Chicago Press Subscription Fulfillment, 1427 E. 60th Street, Chicago, IL 60637-2902.

OSIRIS is indexed in major scientific and historical indexing services, including *Biological Abstracts, Current Contexts, Historical Abstracts*, and *America: History and Life*.

Paperback edition, ISBN 978-0-226-53877-8

Osiris

A RESEARCH JOURNAL DEVOTED TO THE HISTORY OF SCIENCE AND ITS CULTURAL INFLUENCES

A PUBLICATION OF THE HISTORY OF SCIENCE SOCIETY

CO-EDITORS
W. PATRICK MCCRAY
University of California Santa Barbara

SUMAN SETH
Cornell University

COPY EDITOR
BARBARA CONDON
Wellesley, MA

PAST EDITOR
ANDREA RUSNOCK
University of Rhode Island

PROOFREADER
JENNIFER PAXTON
The Catholic University of America

W. Patrick McCray
Co-Editor, Osiris
Department of History
University of California, Santa Barbara
Santa Barbara, CA 93106-9410 USA
pmccray@history.ucsb.edu

Suman Seth
Co-Editor, Osiris
Department of Science & Technology Studies
306 Rockefeller Hall
Cornell University
Ithaca, NY 14853 USA
ss536@cornell.edu